Advances in Oil and Gas Exploration & Production

The book series "Advances in Oil and Gas Exploration & Production" focuses on publishing scientific monographs covering a wide spectrum of topics related to geophysical and geological research within both conventional and unconventional oil and gas systems. These topics are approached from both exploration and production perspectives. The series aims to establish a diverse library of reference works that describe the current state of research on selected themes, such as specific techniques used in the petroleum geoscience industry or regional aspects. Notably, all books in this series are authored and edited by leading experts actively engaged in their respective fields.

The "Advances in Oil and Gas Exploration & Production" series encompasses single and multi-authored books, edited volumes, as well as Conference Proceedings. You can obtain a Book Proposal Form from our website or request one directly from the Publishing Editor, Qiao Shu, at qiao.shu@springernature.com.

Shell International B.V.
Development Research Centre of the State Council (DRC) of People's Republic of China
Editors

Embracing the Future, Powering Growth: An Energy System Renewed for China

in co-operation with
DRC 国务院发展研究中心
Development Research Centre of the State Council
Springer

Editors
Shell International B.V.
The Hague, The Netherlands

Development Research Centre of the State Council (DRC) of People's Republic of China
Beijing, China

ISSN 2509-372X ISSN 2509-3738 (electronic)
Advances in Oil and Gas Exploration & Production
ISBN 978-3-031-73077-1 ISBN 978-3-031-73075-7 (eBook)
https://doi.org/10.1007/978-3-031-73075-7

This Springer imprint is published by the registered company Springer Nature Switzerland AG
The registered company address is: Gewerbestrasse 11, 6330 Cham, Switzerland

Foreword by Jiantang Ma

According to the strategic plans of the Chinese government, high-quality and green development is the key to building China into a modern socialist country in all respects and to achieving the goals of peak carbon emissions and carbon neutrality. To accomplish high-quality and green development, it is imperative that China progress its energy revolution and transition, reshape the current energy system that is dominated by fossil fuels, and gradually shift to a new energy system.

In 2011, Development Research Centre of the State Council (DRC) of People's Republic of China and Shell started a series of strategic co-operation research projects on energy, as part of an important collaboration programme between China and the United Kingdom. To date, the co-operation has published three books: China's Medium- and Long-term Energy Development Strategy (published in Chinese only in 2013), China's Gas Development Strategies (2017) and China's Energy Revolution in the Context of the Global Energy Transition (2019). I consider this fourth report the greatest achievement so far in our collaborative research. It is a comprehensive study and assessment of the blueprint for China's energy development and it puts forward the vision and policy recommendations for building a new energy system by 2060.

To begin with, an era dominated by new energy is coming. In this respect, China needs to focus on the following four aspects.

First, energy efficiency is the most important aspect. To establish a new energy system in which non-fossil energy fully replaces fossil fuels in less than 40 years requires enormous challenges to be addressed, of which saving energy and improving energy efficiency are the most important. Second, continuously increase the electrification rate of end users. New energy is mostly used in the form of electricity. Given this, the shift of end-user energy uses towards electricity and the complete replacement of fossil fuels with electricity will lay a solid foundation for the large-scale development and use of new energy. The electrification rate in China is expected to increase to around 40% by 2035 and to more than 60% by 2060. Third, build wind and solar power facilities. China's power supply structure will gradually shift from one dominated by thermal generation to one composed largely of wind and solar power, with the installed capacity of wind and solar power forecast to increase exponentially. At the same time, hydropower, pumped storage and bioenergy will advance rapidly, with the development of nuclear power progressing gradually. Thermal power, which plays a fundamental role in ensuring energy supply security and balancing power supply and demand, will be added according to need. Fourth, develop hydrogen and biofuels. The pathway of "renewable energy + electrification" will not meet all energy demand. Deep decarbonisation of hard-to-abate sectors such as steel, chemical feedstocks, heavy-duty trucks, aviation and shipping is required. Non-electricity energy demand still needs to be supported by hydrogen and biofuels. By 2060, China's hydrogen demand is estimated to reach up to 120 million tonnes.

Moreover, we believe that the building of a new energy system should not compromise security of supply and that innovation is essential to build a better future. Ensuring energy security and promoting innovation should always be the priorities in the shift to the new energy system, and the principles of "building the new before discarding the old" and "pursuing progress while ensuring stability" should always be followed. First, controllable power supply cannot be ignored in the drive to meet basic energy supply security. While building a new power system, controllable power sources are needed to stabilise the system and provide supply security, and the role of different power sources needs to be understood. Judging by the operating conditions of the current power system, about 420 gigawatts of installed coal power capacity will still be needed across the country in 2060. Second, power system flexibility is essential. With more new energy in the power system whose output is unstable, controllable power sources, grid-to-grid interconnections, demand–response and flexible energy storage are needed to ensure system stability. Demand–response, energy storage and other balancing resources will play an increasingly important role in the power system. Third, the construction of new energy infrastructure needs to increase. The location of China's energy resources in one part of the country and demand centres in another will remain for a long time. New energy systems should be constructed in line with the idea of "large power generation clusters + strong power networks + distributed energy systems + local power generation facilities that ensure energy supply security". Meanwhile, a flexible interconnected grid framework of "large power grids + small and medium-sized regional power grids + smart distribution networks and microgrids" should be built at a faster pace. Fourth, progress in carbon capture, utilisation and storage should be made to transform fossil fuels into near-zero carbon energy forms so that some thermal power facilities can be retained. Fifth, new and safe nuclear power technologies should be developed. Efforts should be made to achieve breakthroughs in—and commercialise—controlled nuclear fusion.

We believe that a renewed energy system will trigger industrial revolution in a broader sense. The previous energy revolutions in human history have created new industries and new economies. A new energy revolution, characterised by zero carbon, will provide a new engine for economic growth in China and even for the world. According to estimates, China's total investment demand for energy infrastructure will exceed RMB 80 trillion by 2060. Construction of the new power system and the manufacture of the equipment needed can contribute more than 5% of GDP per year and create more jobs. In addition, the energy system revolution will generate demand for minerals, which will further develop the critical minerals industry.

As clearly stated in the report to the 20th National Congress of the Communist Party of China in 2022, China will speed up the planning and development of a system for new energy sources. The results of this fourth collaboration between the Development Research Centre and Shell represent a preliminary exploration of this crucial topic.

Jiantang Ma
Former Party Secretary and Vice President
Development Research Centre of the
State Council (DRC) of People's
Republic of China

Foreword by Ben van Beurden

It is, once more, an honour for Shell to have partnered with Development Research Centre of the State Council (DRC) of People's Republic of China. With this, the fourth piece of work we have completed together, the team has overcome challenges it did not expect. COVID-19 meant that the team could not meet in person, and this could have had a serious impact. But after working together so closely in a collaboration that stretches back more than a decade, they were able to work seamlessly, with deep mutual respect and understanding. Together they have produced a book of superb quality.

As ever, the team was able to draw on the DRC's comprehensive understanding of China's energy system and the development challenges which need to be addressed. It was also able to rely on Shell's international experience, knowledge of energy systems and our deepening understanding of energy transition pathways—not only for Shell, but also for specific countries and on a global basis.

In the first piece of work together, the DRC and Shell took a broad look at China's energy system. In the second, which was the first to be published, the work focused on the role of natural gas in diversifying China's energy mix. The third piece explored China's energy revolution in the context of a changing world energy system. Now, this latest book takes its inspiration from the historic announcement made by President Xi in September 2020: that China would reach peak emissions by 2030 and achieve carbon neutrality by 2060.

When President Xi set this course, many people outside China, who do not understand China, were surprised. They should not have been: the signs have been there for all to see. In 2017, at the 19th Party Congress, President Xi launched a "New Era" aimed at ensuring a better quality of life for the people of China. It was clear that this meant change for China's energy system: cleaner energy, with better air quality and lower greenhouse gas emissions. At that time, China had already made huge investments in wind and solar power. Since the dawn of the New Era, progress has accelerated at an impressive pace.

Today China is, by some distance, the leading global investor in solar, onshore and offshore wind, hydropower and nuclear generation technologies. China accounts for 75% of the world's production capacity for battery cells. More than 80% of all solar photovoltaic modules in the world are made in

China. One in four new cars bought in China in 2022 was an electric vehicle. These are extraordinary achievements that not only benefit China but, by helping bring down the cost of low-carbon technologies, should benefit the world as well.

These are all excellent foundations for the future success of China as it works towards net-zero emissions by 2060. It is by building on those foundations that China can succeed, and this book is an exploration of how to build upwards from today. Of course, it is important to acknowledge that there is no substitute for rapid action right now—both in terms of achieving domestic energy security and long-term decarbonisation goals. There is every reason to expect that, on the basis of China's track record on green energy in recent years, the nation will rise to this challenge.

As China faces the task of achieving carbon neutrality, this book is intended to act as a blueprint for action. It sets out a vision for the way forward in a clear-sighted manner. The book addresses both the risks and the opportunities involved for each topic so that judgements can be made on the basis of a full understanding of what is involved. Overall, the book systematically looks at the required transition in energy demand and energy supply, as well as the new energy infrastructure necessary to support these changes. In addition, the book explores the impact of the transition on regional economic development and on the wider economy.

In the pages that follow, all the critical principles for a successful transition in China are set out in detail. The importance of energy efficiency and the circular economy is central. An energy-efficient approach will make all other targets more manageable.

The book emphasises the crucial importance of electrifying energy use, combined with the decarbonisation of electricity supply. Of course, China has abundant renewable resources and the country has already made an excellent start in developing them. The book also looks at one of the main challenges facing China as it expands its renewable energy generation capacity: connecting the parts of the nation best-suited to generating zero carbon electricity to the high-demand population centres in the east.

This is already a lot to achieve, but it is clear from the pages of this book that there is much more to do. For sectors of the economy unable to electrify there needs to be a huge expansion of hydrogen production and use, as well as massive growth in the use of biomass, both to generate power and as a source of liquid fuels. Even with all these actions, some emissions will remain in the energy system. It is for this reason that the book goes on to recognise the important role that carbon capture and storage technologies will also have in meeting China's emissions target.

There is a lot to do: China is engaged with nothing less than building a new energy system. Energy is at the core of all economic activities so, in many ways, China is creating an entirely new economy, with action across both demand and supply alongside the infrastructure to support both. The opportunity in this task is as great as the task itself: the potential economic benefits to the country, as outlined in these pages, are substantial.

If China can go on and achieve its aims, if it can build on the foundations it has laid and if it can take the opportunities within its grasp, then it will stand as a powerful example to the world. Shell has been privileged to be part of the journey so far with the DRC. The greatest achievements are still to come. I look forward to seeing China succeed in its energy transition, as part of a world that also succeeds.

Ben van Beurden
Former Chief Executive Officer, Shell plc

Acknowledgements

Project Chairs

Jiantang Ma	Former Party Secretary and Vice President, Development Research Centre of the State Council (DRC) of People's Republic of China
Ben van Beurden	Former Chief Executive Officer, Shell plc
Wael Sawan	Chief Executive Officer, Shell plc

Project Executives

Guoqiang Long	Vice President and Research Fellow, Development Research Centre of the State Council (DRC) of People's Republic of China
Huibert Vigeveno	Downstream and Renewables Director, Shell plc
Edward Daniels	Former Strategy, Sustainability and Corporate Relations Director, Shell plc
Jason Wong	Former Executive Chairman of Shell Companies in China, EVP Global Lubricants
Lin Chen	Executive Chairman of Shell Companies in China

Project Team Leads

Jinzhao Wang	Former Director General and Research Fellow, Research Department of Industrial Economy, DRC, China
Laszlo Varro	Vice President, Strategy Insights and Scenarios, Shell International B.V.
Mallika Ishwaran	Chief Economist, Shell International Limited
Jeremy Bentham	Former Special Advisor, Shell International B.V.

Project Core Advisors

Jian Zheng	Director General of Infrastructure Development Department, National Development and Reform Commission (NDRC)
Wenbin Lv	Director General, Energy Research Institute, Academy of China Macroeconomic Research (formerly NDRC ERI)
Wen Song	Director General, Legal and Institutional Reform Department, China National Energy Administration (NEA)
Zhipeng Liang	Deputy Director General, Legal and Institutional Reform Department, NEA
Yi Wang	Member of the Standing Committee of the 13th National People's Congress and Vice President and Research Fellow of the Institutes of Science and Development, China Academy of Sciences (CAS)
Dan Shi	Former Director General and Research Fellow of the Institute of Industrial Economics of the Chinese Academy of Social Sciences (CASS)
Chengchuan Tian	Director General, Foreign Environmental Cooperation Center, Ministry of Ecology and Environment (MEE)
Huaqing Xu	President, National Center for Climate Change Strategy and International Cooperation (NCSC)
Dadi Zhou	Former President, Research Fellow, NDRC Energy Research Institute (NDRC ERI)
Yande Dai	Former President, Research Fellow, NDRC Energy Research Institute (NDRC ERI)
Yanbing Kan	Deputy Director General and Research Fellow, National Energy Conservation Center (NECC)
Zhixuan Wang	Former Vice President, China Electricity Council (CEC)
Liping Jiang	Former Vice President of State Grid Energy Research Institute (SGERI)
Baoshan Li	Vice President, China Renewable Energy Society (CRES)
Dayong Zhang	Deputy Secretary General, China Association for the Promotion of Industrial Development (CAPID)

Project Co-ordinators

Xu Zhaoyuan	Deputy Director General and Research Fellow, Research Department of Industrial Economy, DRC, China
Ling Wang	Government Relations Manager, Shell (China) Limited
Bin Lv	Associate Research Fellow, Research Department of Industrial Economy, DRC, China

DRC Project Team Members

Shiji Gao	Director General, Research Institute of Resource and Environment Policies, DRC, China
Jianlong Yang	Former Deputy Director General, Research Fellow, Research Department of Industrial Economy, DRC, China
Zifeng Song	Deputy Director General and Research Fellow, Research Department of Social Development, DRC, China
Jiaofeng Guo	Former Director General and Research Fellow, Research Institute of Resource and Environment Policies, DRC, China
Yi Zhou	Vice Director and Associate Research Fellow, Research Department of Industrial Economy, DRC, China
Qian Lu	Assistant Research Fellow, Research Department of Industrial Economy, DRC, China
Tao Hong	Director, Senior Economist, Research Institute of Resource and Environment Policies, DRC, China
Weiming Li	Director and Research Fellow, Research Institute of Resource and Environment Policies, DRC, China
Jifeng Li	Vice Director and Associate Research Fellow, Research Institute of Resource and Environment Policies, DRC, China
Han, Xue	Associate Research Fellow, Research Institute of Resources and Environment Policies, DRC, China
Xiurong Hu	Lecturer at the College of Economics and Management, Nanjing University of Aeronautics and Astronautics
Ling Cheng	Deputy Director and Senior Engineer, Research Department of Electricity Consumption and Energy Efficiency, China Electricity Power Research Institute (CEPRI)
Xinhua Zhang	Division Chief, Senior Engineer, Marketing and Sales Department, State Grid Corporation of China
Han Wu	Senior Engineer, State Power Rixin Technology Co., Ltd.
Yongwei Zhang	Vice President, 21st Century Research Institute of Tsinghua University, Vice President and Secretary General of China EV100
Bian Fu	Assistant Research Fellow, Energy Research Institute of China Macro-economic Research Institute
Tianzi Wang	Assistant Research Fellow, Energy Research Institute of China Macro-economic Research Institute
Caifu Zhong	Associate Research Fellow, Energy Research Institute of China Macro-economic Research Institute
Lei Tian	Vice Director, Energy Economics and Development Strategy Research Center, Energy Research Institute of China Macro-economic Research Institute
Huawen Xiong	Director, Research Fellow, Energy Environment and Climate Change Research Center, Energy Research Institute of China Macro-economic Research Institute

Dengfeng Liu	General Manager, Energy Internet Product Line, Xinjiang Goldwind Science & Technology Co., Ltd.
Jian Yu	Associate Professor, Central University of Finance and Economics
Kejun Dou	Deputy Secretary-General, Biomass Energy Industry Promotion Association, China
Yongliang Li	Vice Director and Senior Engineer, Industry Development Department, China Petroleum and Petrochemical Industry Federation (CPCIF)
Miao Li	Deputy Division Chief, Engineer, Energy Conservation and Low Carbon Development Division, Industry Development Department (CPCIF)
Ye Ning	Research Fellow, Senior Economist, Global Energy Interconnection Development and Cooperation Organization
Qian Zhou	Associate Professor, School of Economics and Management, North China Electric Power University
Xian Zhang	Division Director and Research Fellow, The Administrative Center for China's Agenda 21
Jingli Fan	Deputy Dean, Professor, School of Energy and Mining Engineering, China University of Mining and Technology
Jiabin Chen	Director and Research Fellow, Institute of Mineral Resource Economics, Chinese Academy of Natural Resources Economics
Chao Liu	Associate Research Fellow, Institute of Mineral Resource Economics, Chinese Academy of Natural Resources Economics
Binhan Nie	Assistant Research Fellow, Institute of Mineral Resource Economics, Chinese Academy of Natural Resources Economics
Dandan Feng	Associate Research Fellow, Institute of Mineral Resource Economics, Chinese Academy of Natural Resources Economics
Juan Guo	Associate Research Fellow, China Ministry of Natural Resources
Zhang Jinliang	Professor, School of Economics and Management, North China Electric Power University
Ke Wang	Professor, School of Management and Economics, Beijing Institute of Technology

Shell Project Team Members

William Wang	Head of Corporate Relations China, Shell (China) Limited
Georgios Bonias	Senior Energy Analyst, Shell International Limited
Joep Huijsmans	Senior Advisor, Shell Global Solutions International B.V.
Marcelo Espinoza	Manager, Power Fundamentals, Shell International B.V.
Burkard Schlange	Portfolio Manager, Shell Deutschland GmbH
Peter Webb	Former Manager, Finance and Taxation, Shell International B.V.
Qun Deng	Manager, Academic Partnerships & Grants Manager, Shell (Shanghai) Technology Limited
Tobias Chen	Shell Hydrogen Mobility Commercial Manager APAC
Juan Han	Business Development Manager, GTL, Shell (China) Limited
Yuying Jia	Shell Hydrogen Operations Team Lead Asia and China
Xiao Fu	Senior Lead Researcher, Energy System Modelling, Shell Global Solutions International B.V.

Contents

Abbreviations

BECCS	Bioenergy with carbon capture and storage
BEV	Batter-electric vehicle
CBAM	Carbon border adjustment mechanism
CCS	Carbon capture and storage
CCUS	Carbon capture, utilisation and storage
CfD	Contract for difference
CGE	Computable general equilibrium
CHP	Combined heat and power
CO_2/MJ	Carbon dioxide per megajoule
CPC	Communist Party of China
DACCS	Direct air capture with carbon storage
DRC	Development Research Centre of the State Council of China
ETS	Emissions trading system (or scheme)
FCEV	Fuel cell electric vehicle
FFV	Flexible fuel vehicles
GDP	Gross domestic product
GEC	Green electricity certificate
GNI	Gross national income
GW	Gigawatts
HEFA-SPK	Hydro-processed ester and fatty acid synthetic paraffin kerosene
IEA	International Energy Agency
IPCC	Intergovernmental Panel on Climate Change
IRENA	International Renewable Energy Agency
kcal/kg	Kilo calories per kilogram
kgce/m^2	Kilograms of coal equivalent per square metre
kWh	Kilowatt-hours
kW/L	Kilowatts per litre
LCOE	Levelised cost of electricity
LNG	Liquefied natural gas
MJ/m^2	Millijoules per square metre
MPa	Megapascals
Mt CO_2e/year	Million tonnes of carbon dioxide equivalent per year
MW	Megawatts
MWh	Megawatt-hours
NDRC	National Development and Reform Commission

NEA	National Energy Administration
NO_x	Nitrogen oxide
OECD	Organisation for Economic Co-operation and Development
PM	Particulate matter
R&D	Research and development
RED, RED II	EU Renewable Energy Directive
RMB	Renminbi, the currency of China
SO_2	Sulphur dioxide
t/a	Tonnes per annum
tce	Tonnes of coal equivalent
TFP	Total factor productivity
TWh	Terawatt-hours
WEC	Wave energy converter
W/m^2	Watts per square metre
Wh/kg	Watt-hours per kilogram

List of Figures

Chapter 4: Transport: The Transition to Electric Vehicles

List of Tables

Executive Summary

Jinzhao Wang, Zhaoyuan Xu, Bin Lv
and Mallika Ishwaran

1 Introduction

In September 2020, Chinese President Xi Jinping announced at the General Debate of the 75th Session of the United Nations General Assembly that China would strive to achieve peak CO_2 emissions before 2030 and carbon neutrality before 2060.[1] This means that China will reduce its carbon emissions from peak to near zero in about 30 years. There are both challenges and opportunities along this journey. China's efforts to achieve the goals of a carbon emissions peak and carbon neutrality (known as dual carbon) are expected to drive the development of and innovation in new energy, and increase economic growth.

This book uses various models—including the computable general equilibrium model, the long-range energy alternatives planning system, the integrated resource strategic planning model and the whole electricity system investment model—to explore how to build a new energy system and promote new energy development. This study also outlines a plan for China to move towards a carbon-neutral future and describes the new types of energy, new economy and new policies needed to reach that future.

[1] Xi Jinping's Speech at the General Debate of the 75th Session of the United Nations General Assembly, People's Daily, September 23, 2020.

J. Zh. Wang (✉) · Zh. Y. Xu · B. Lv
Department of Industry Economic Research, Development Research Centre of the State Council, Beijing, China

M. Ishwaran
Shell International Limited, The Hague, The Netherlands

2 China's Journey Towards Energy Transition and Carbon Neutrality

To achieve carbon neutrality, the priority is to develop zero-carbon energy and accelerate the energy transition. Based on currently foreseeable technology options, the basic strategies that will help China meet its goals of building a great modern socialist country and achieving carbon neutrality are to reduce energy demand and adjust the energy mix.

2.1 There Are Five Phases in the Transition to Carbon Neutrality

To set a good balance between energy security, economic growth, technological advance and other factors, China should move towards carbon neutrality in five phases. Between 2022 and 2025, the start-up phase, China should develop plans and design policies to reach its dual-carbon goals and unify thinking and understanding. From 2026 to 2030, the peak carbon emissions phase, China

Shell International B.V. and Development Research Centre of the State Council (DRC) of People's Republic of China, *Embracing the Future, Powering Growth: An Energy System Renewed for China*, Advances in Oil and Gas Exploration & Production,
https://doi.org/10.1007/978-3-031-73075-7_1

Table 1 The five phases of carbon neutrality

Phases	Phase 1	Phase 2	Phase 3	Phase 4	Phase 5
Period	2022–25	2026–30	2031–35	2036–55	2056–60
Description	Start-up	Peak carbon emissions	Technological breakthroughs	Expansion	Carbon neutrality
Priorities	Design and planning	Development of renewable energy and promotion of the energy transition	Large-scale rollout of innovative abatement technologies and other new energy technologies	Expansion of the new technology system	Negative emission technologies and carbon sinks

Source Research results of this report

should advance the integration and continuous development of existing and new energy carriers, allow non-fossil energy to become the absolute driver of growth, and ensure that total CO_2 emissions from energy peak before 2030. In the period 2031–2035, the technological breakthroughs phase, an array of abatement technologies—such as hydrogen, biomass and carbon capture, utilisation and storage (CCUS)—should achieve breakthroughs and large-scale rollout. In the years 2036 to 2055, the expansion phase, China should move into a new stage of development, and thanks to new technologies and total emissions control, CO_2 emissions should drop rapidly. And from 2056 to 2060, the carbon neutrality phase, China's CO_2 emissions from energy should decline to about 1.24 billion tonnes, and a new way of living take shape in society, characterised by carbon neutrality (see Table 1).

2.2 China's Economy Has Entered a New Stage of High-quality Development in Which Energy Efficiency Will Increase Significantly

In 2021, China's gross domestic product (GDP) was RMB 114.4 trillion ($17.7 trillion at a mid-2021 exchange rate: USD 1 = RMB 6.4515), equivalent to 77% of the GDP of the USA. Per capita GDP and per capita gross national income (GNI) were $12,600 and $12,400 respectively, which was close to the threshold of high-income countries defined by the World Bank ($12,700 in 2020). At the same time, China's position in the world economy rose. In 2010, the added value of China's manufacturing industry outperformed that of the USA for the first time, making China the world's largest manufacturer. In 2020, the value of China's manufacturing industry reached $3.85 trillion, 69.6% higher than the $2.27 trillion of the USA, achieving the strategic goal of industrialisation as defined by the 18th National Congress of the Communist Party of China. China's economy has entered a new stage of high-quality development.

Looking ahead to 2060, China's economy will continue on its path of sustained and stable growth. Average economic growth is expected to reach about 5.6% in the 14th Five-year Plan period (2021–25), before gradually declining to about 4.5% in the 15th Five-year Plan period (2026–30) and around 3% in 2030–60. The share of service industries in GDP will increase, from 54.5% in 2020 to about 68% in 2060. The structure of the manufacturing industry will be continuously optimised, with service-oriented manufacturing becoming an important channel for greater added value. The proportion of high-tech manufacturing will continue to rise and the conventional energy-intensive sectors, such as steel and cement, will slowly decline over time. Thanks to optimised industrial processes and high-quality development, energy efficiency will improve. China's energy consumption per unit of GDP is expected to decrease by 14.3% during the 14th Five-year Plan period, and drop by 60% and 70% respectively by 2050 and by 2060, compared with 2020.

2.3 To Achieve Carbon Neutrality, Total Primary Energy Demand Should Move in an Inverse U-shape and Non-fossil Energy Become the Main Driver of Growth

China's total primary energy demand will stabilise and peak at about 6.1 billion tonnes of coal equivalent (tce) in 2030–35 and will fall below 5 billion tce in 2060. The trend of the country's energy consumption growth is still difficult to reverse in the short term. With changes in urbanisation, coupled with industry's shift towards services, primary energy demand will continue to decline after its peak, dipping to 4.92 billion tce in 2060. In addition to the decline of coal, oil is expected to peak in 2025–30; natural gas, a low-carbon energy, will play an important role as a transition fuel and peak around 2035; and non-fossil energy will continue its rapid rise and reach about 85% of primary energy demand by 2060 (see Fig. 1).

2.4 In Terms of End-user Energy Demand, a Continuous Shift to Electrification Occurs After the Use of Fossil Fuels Peaks

End-use energy demand will adjust at a faster pace after 2030, as industry's share of energy demand declines and that of electricity increases significantly. China's total final energy use is expected to peak at about 4.3 billion tce around 2030, and then decline to around 2.8 billion tce by 2060. By sector, industry's share of total final energy use will continue to decrease from 70.3% in 2020 to

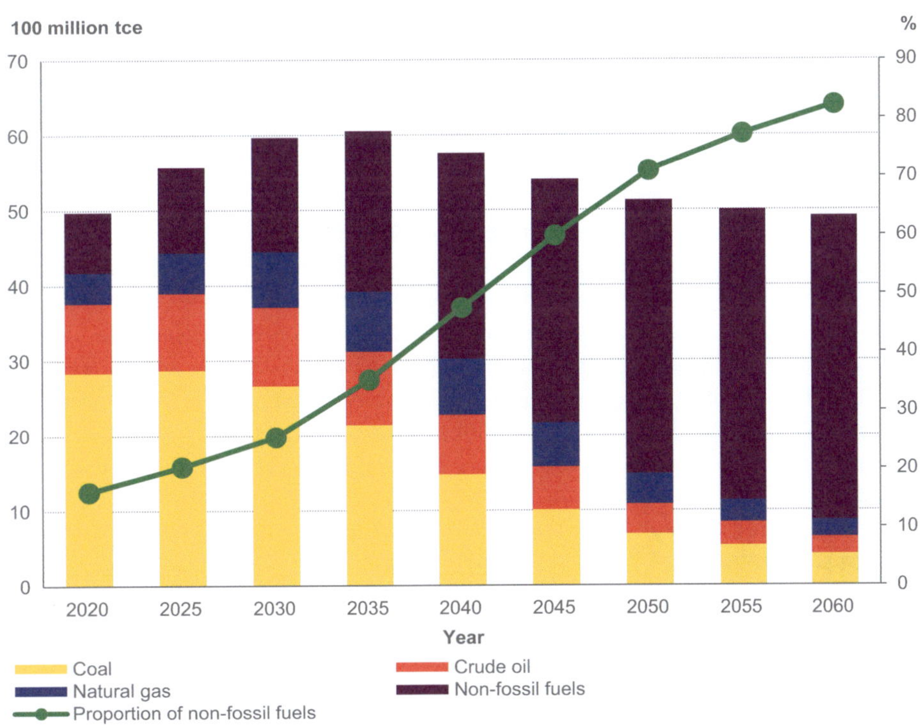

Fig. 1 Primary energy demand, 2020–60s. *Source* Calculations based on the models used by the project team

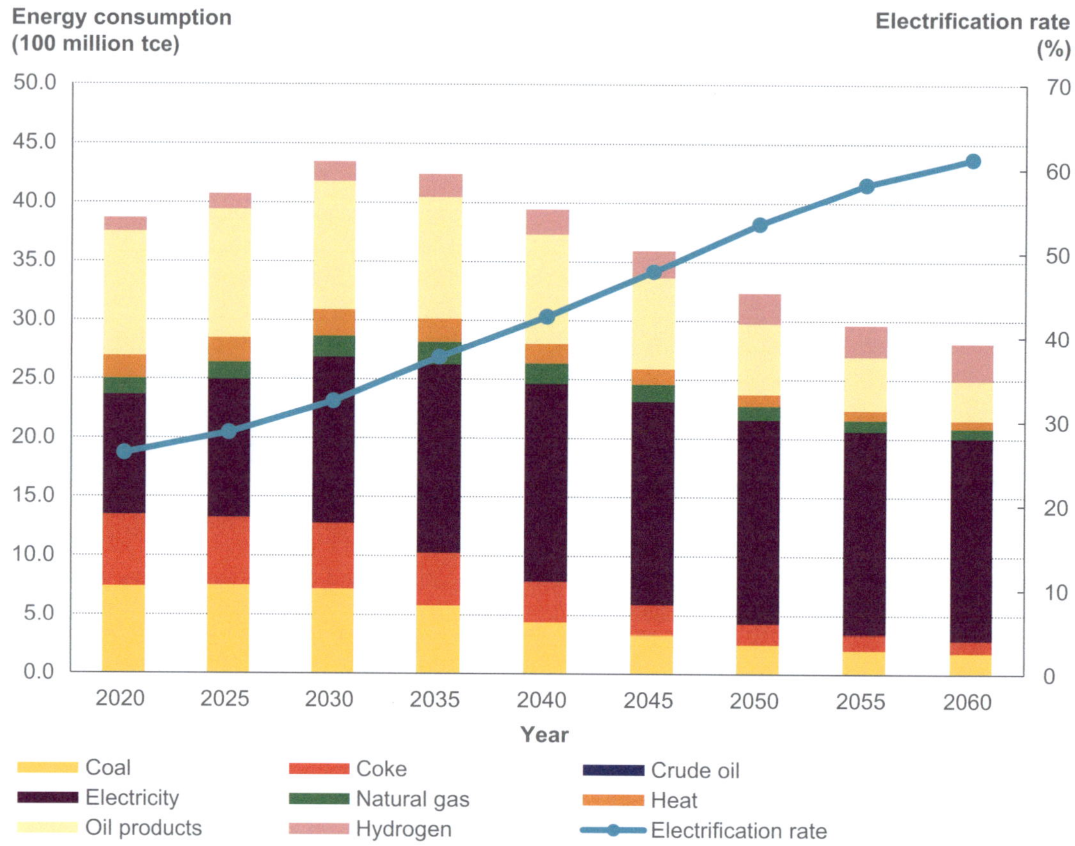

Fig. 2 End-user energy demand, 2020–60. *Source* Calculations based on the models used by the project team

52.1% in 2060; buildings' share will grow from 20.3% in 2020 to 30% in 2060; and transport's will reach 17.9% by 2060, which is 8.4 percentage points higher than in 2020. Electricity's share of end-user energy demand will rise rapidly to about 60% by 2060, with hydrogen playing an increasingly prominent role (see Fig. 2).

3 The Shift to Electricity from Fossil Fuels

In 2020, electricity accounted for 26.5% of China's total final energy consumption. Of industry's energy demand, 26.2% was electrified, with the electrification rate in the four energy-intensive industrial sectors (iron and steel, other metals, minerals and chemicals) amounting to 17.8%. Buildings had electrified rapidly at 44.1%. Transport had an electrification rate of 3.7%, implying huge growth potential. To achieve the goal of carbon neutrality, the electrification rate needs to increase significantly to about 40% by 2035 and around 60% by 2060. If electricity consumption from electrolysis-based hydrogen production is included, the share of electricity in total final energy use will be more than 78% by 2060.

3.1 The Electrification of Industry Is Expected to Reach 56% by 2060

China is entering the later stage of industrialisation. As a result, end-user energy demand in industry will plateau between 2028 and 2035. Energy consumption in industry is mainly driven

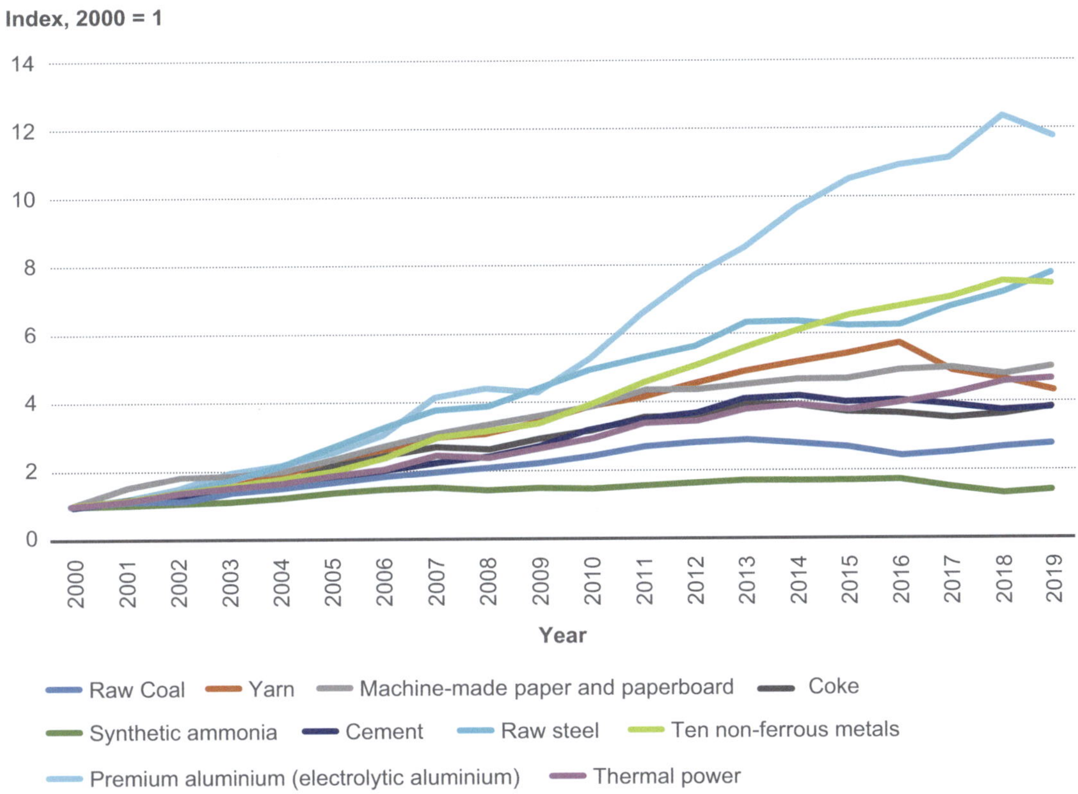

Fig. 3 Industrial output by product, 2000–19. *Source* National Bureau of Statistics of China after data normalisation (2000 = 1)

by energy-intensive sectors, including ferrous metal smelting and pressing, chemical feedstocks and chemical products, non-metallic mineral products and non-ferrous metal smelting and pressing. The output of energy-intensive industries is peaking. For example, cement and steel production have already peaked (see Fig. 3). The production of cement and raw steel is estimated to decrease to about 1 billion tonnes and 0.5 billion tonnes respectively by 2060. Final energy use in industry is forecast to plateau between 2028 and 2035 and plunge to 1.46 billion tonnes of coal equivalent by 2060.

To achieve a higher electrification rate in industry depends on technological innovation and electrifying more industrial processes. Because the substitution of electricity for fossil fuels at scale in steel, cement, petrochemicals, glass and textiles faces many challenges, the electrification rate in industry will increase at a slower pace than other sectors, reaching about 40% by 2040. With the increasing use of hydrogen, electric boilers and other technologies, coupled with the sharp decline in demand for steel and cement, the electrification rate in industry will climb at a faster pace after 2040 and is expected to reach 56% by 2060.

To improve the electrification rate in industry, efforts should be made in the following four areas. First, electrification should be technically feasible and economically viable. Electricity has been used to power a wide range of industrial processes and equipment, including motors, heat pumps and electric boilers for medium- and low-temperature heating, electric furnaces for steelmaking and a small number of electrolysers for hydrogen production. Second, plans for hard-to-abate sectors should be developed. China should channel more effort into research on new materials, new equipment and other base technologies related to the substitution of electricity for fossil fuels in industry, intensify efforts to make breakthroughs in critical technologies and core equipment related to electricity in industry, and focus on deploying demonstration projects to speed up larger-scale rollout. Third, demand should be reduced at source. China should set a trend of low-carbon consumption, encourage the use of low-carbon products and the construction of low-carbon buildings and reduce demand for energy-intensive products. China should also embrace the circular economy and improve resource recycling. Fourth, business models and pricing mechanisms should be optimised. China should emphasise the market's leading role, actively explore new business models based on market dynamics to electrify industry, and adopt pricing mechanisms like time-of-use and peak-valley tariffs to accelerate industry's shift to electricity.

3.2 The Electrification Rate of Buildings Could Exceed 70% by 2060

China's total building floor area will continue to grow. The floor area of urban residential buildings, rural residential buildings and public buildings is expected to increase to 47.8 billion square metres (m^2), 9.7 billion m^2 and 21.3 billion m^2 respectively. Of the total floor area of 78.8 billion m^2, about 24.2 billion m^2 has heating. Thanks to improvements in energy efficiency, the total energy consumption of buildings in China will peak around 2030 at 1.32 billion tonnes of tce, and continue to decline to about 840 million tce by 2060.

The electrification of buildings will increase rapidly, reaching about 73.3% by 2060. Electricity's share of urban heating in North China, urban residential buildings (excluding heating), public buildings (excluding heating) and rural residential buildings will increase to 65%, 78%, 80% and 70% respectively. The total electricity consumption of buildings will amount to 3.5 trillion kilowatt-hours (kWh) in 2030 and 4.6 trillion kWh in 2060, 2.1 times and 2.7 times respectively that of 2018.

To achieve a higher electrification rate in buildings, efforts need to focus on the following five technology areas. First, China should continue to encourage clean heating through the shift from coal to electricity in North China. The principle of "adapting the use of electricity, natural gas, coal and geothermal energy for heating to local conditions" should be followed, and total operating costs should be taken into consideration to gradually increase the amount of electric heating. Second, heat pumps should be adapted to local conditions and widely deployed in South China, which has hot summers and cold winters. Third, China should widely encourage the use of commercial and household electric cooking appliances. Fourth, the country should accelerate the deployment of renewable energy technologies in rural areas, including rooftop solar photovoltaic systems and solar power in agriculture. Fifth, China should roll out integrated, digitalised energy services for large public buildings to improve the buildings' energy efficiency.

3.3 The Shift to Electricity in Transport Will Speed up, but There Will Be Difficulties in Some Applications

Demand in the transport sector will continue to grow. As industrialisation and urbanisation advance, goods transport and intercity passenger movement will gradually shift from rapid to steady growth. Driven by globalisation, the volume of goods transported is expected to grow from about 20 trillion tonne–kilometres in 2019 to around 30 trillion tonne–kilometres in 2060. Domestic bulk cargo transport will peak before 2030, while small-volume, time-sensitive goods transport and international transport will grow rapidly. Passenger transport is expected to peak at 6 trillion passenger–kilometres around 2040, dropping to 5 trillion passenger–kilometres in 2060. Car ownership will grow continuously and is expected to reach 500 million vehicles by 2060. Total energy consumption in the transport sector will peak at about 700 million tce in 2030–35, decreasing to 500 million tce in 2060.

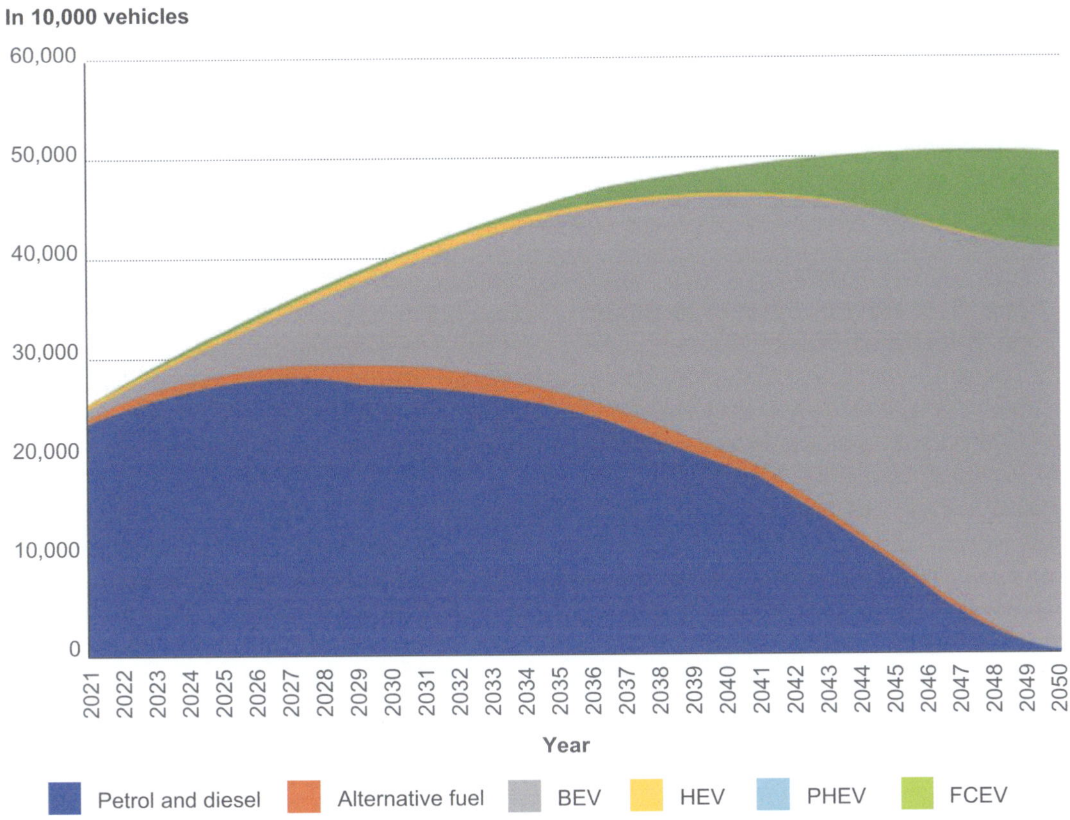

Fig. 4 Types of passenger vehicle in China, 2021–50. *Note* BEV = battery-electric vehicles; HEV = hybrid electric vehicles; PHEV = plug-in hybrid electric vehicles; FCEV = fuel cell electric vehicles. *Source* Data from the National Bureau of Statistics of China and research results of this report

There is much scope for transport electrification to increase. More than 60% of end-use energy demand in the sector is expected to be electrified by 2060. Final energy use in transport will shift from predominantly oil products to hybrids, and then to a low-carbon mix in which electricity accounts for a predominant share. Oil products' share will decrease from around 90% today to 56% in 2035 and 5% in 2060, whereas electricity will increase from around 3% today to 31% in 2035 and 61% in 2060. By 2060, hydrogen, biofuels and alternative (mainly hydrogen-based) fuels will be used in heavy-duty road freight, long-distance shipping, aviation and other types of transport. Of these, hydrogen will make up about 20%. Rail and road transport are expected to achieve the highest electrification rate at 100% and 90% respectively. The shift to electric passenger vehicles will occur at a faster pace, and internal combustion engine vehicles will exit the market by 2050 (see Fig. 4).

The passenger vehicle and rail segments are expected to lead electrification in transport. Steady advances in electrification should be made in short-haul and long-haul freight transport, and plans for electrification in other transport segments should be accelerated. Focus should be directed on developing electric vehicles. Electric vehicle uptake is expected to reach more than 25% by 2030; all passenger vehicles will be electric by 2050, with battery-electric vehicles and fuel cell electric vehicles accounting for 80% and 20% respectively. Rail electrification will continue to advance. At the end of 2019, China had 39,000 kilometres of unelectrified railway, which it aims to electrify by 2060.

Battery-electric trucks will dominate in medium- and short-distance goods transport, whereas hydrogen-fuelled trucks will be used mainly for heavy-duty transport. More efforts will be channelled into achieving breakthroughs in fuel cell technology and in alternative fuels for aviation and shipping.

4 The New Power System Is Based on Wind and Solar

By 2060, electricity supply is projected to reach 17.8 trillion kWh, of which about 3 trillion kWh will be used for hydrogen production. Be it installed capacity or power output, wind and solar power will predominate. However, an excessively high proportion of variable power generation will pose supply security and reliability challenges to the power system, so multiple measures should be taken and an overall plan developed to ensure reliability.

4.1 Net-zero Emissions in the Power System Can Be Achieved by 2045–50

China aims to achieve net-zero carbon emissions in the electricity sector ahead of schedule. With the slowdown in the use of coal and the rapid development of new energy, carbon dioxide emissions from the electricity sector are expected to peak and plateau between 2025 and 2030. By using a combination of renewables, thermal power with carbon capture, utilisation and storage, and bioenergy with carbon capture and storage, the power system is expected to reach net zero by 2045–50 and negative emissions after 2050.

4.2 The Installed Capacity of Wind and Solar Power Is Expected to Reach 6,700 GW by 2060

Power generation will gradually switch from thermal power to wind and solar power. The installed capacity of wind and solar will increase exponentially, reaching 3,200 GW and 3,500 GW respectively by 2060, which is 11.3 times and 13.8 times that of 2020. Of this, centralised wind farm clusters will contribute 68% of the total installed capacity of wind power. Centralised and distributed solar power systems will account for 43% and 57% respectively of the total installed capacity of solar power. The installed capacity of coal power will decrease from 1,330 GW in 2022 to about 420 GW in 2060. The potential of hydropower, pumped storage, nuclear power, biomass-fired power plants and other zero-carbon power generation technologies will be fully tapped.

4.3 Intermittent Power Generation Sources Will Predominate, While Controllable Thermal Generation Will Drop Below 15% by 2060

Thermal power plants will be essential to provide electricity supply security. According to demand–supply assessments, there will be challenges in addressing intermittency and peak demand in 2060, even if full use is made of pumped storage, grid-to-grid power transfers, demand–response, electric vehicles, electrochemical energy storage and other balancing options. To prevent system risks and counteract uncertain weather conditions, the electricity demand–supply balance calculations for all regions indicate that it is advisable to retain about 600 GW of installed thermal power capacity by 2060, including 420 GW of coal power and 180 GW of gas power (see Table 2). Use of these coal- and gas-fired plants will, however, only be about 1,500 hours and 4,000 hours per year respectively.

4.4 Power System Flexibility Is Key and the Need for Energy Storage Will Rise Significantly

Demand–response, energy storage and other flexibility measures will play an important role

Table 2 Pathways to power system transition

Year	2020	2030	2060
Installed power capacity by type in gigawatts			
Hydropower	340	420	500
Pumped storage	31	90	420
Coal	1,080	1,250	420
Gas	100	170	180
Nuclear	52	100	200
Wind	282	750	3,200
Solar	253	900	3,500
Biomass	30	55	66
Total installed power capacity	2,200	3,700	8,500
Power system flexibility demand			
Energy storage (including EVs)/GW	0.3	83	2,260
Demand response/GW	3.4	47	260
Total electricity/Trillion kWh	7.42	11	17.8

Source Data from the National Bureau of Statistics of China and research results of this report

in the power system. With increasing amounts of wind and solar power generated, the need for grid flexibility technologies rises. Even if the installed generating capacity of coal remains at 1,250 GW, the need for balancing capacity is expected to reach 270 GW in 2025 and exceed 700 GW in 2030. That is where grid-to-grid transfer, demand–response, electric vehicles and energy storage come in. To meet flexibility requirements and ensure supply security of a power system with a high proportion of renewables, energy storage capacity of up to 2,260 GW (including electric vehicles), pumped storage capacity of 420 GW and other resources will be required by 2060. Looking ahead, electric vehicles must become an important part of power system flexibility, with each vehicle potentially providing about 7 kilowatts of system support. If there are 200 million electric vehicles supporting the power system in 2060, they can provide as much as 1,400 GW of electricity. There will, however, still be a large gap between demand and supply in energy storage capacity.

4.5 New Energy Development and National Land Use Planning Need to Be Co-ordinated

Large-scale development of wind and solar power leads to rising demand for land, which makes planning imperative. The construction of wind and solar power plants requires a large amount of space. For example, large-scale solar photovoltaic power plants require 0.18–0.32 square kilometres for every 10 MW of installed solar power capacity. If the additional installed capacity required annually is 50–100 GW, a land area of about 1,250–2,500 square kilometres per year is needed. If 3,500 GW of solar power capacity is needed, around 87,500 square kilometres is required. With greater importance attached to ecological protection in China, coupled with increasingly less land available for development in some regions, land use is gradually becoming a barrier to the growth of wind and solar power. The land available for large-scale photovoltaic and wind power projects and supporting power transmission infrastructure in West China is heavily constrained by ecological red lines and in East China by land control. In the long term, it will be necessary to co-ordinate new energy development with national land planning. Combining wind and solar power generation with other land uses—such as solar photovoltaic and agriculture, solar photovoltaic and buildings, and wind power with pasture—will reduce the need for land.

5 Regional Energy Development Should Follow a Strategic Plan

The distribution of new energy resources like wind and solar is a challenge in China, as they are located far from demand centres. Given the potential supply security risks of these sources, an overall plan should be developed. New energy systems should be built according to the principle of "large power generation clusters + strong networks + distributed energy systems + local power generation facilities that ensure energy supply security". In addition, the distribution of industries should also be optimised.

5.1 The Uneven Distribution of Renewable Energy Resources Is Likely to Increase the Imbalance Between Regional Power Supply and Demand

China has abundant wind resources, which are mainly concentrated in the north-east, north-west and northern parts of the country. Based on current commercial technologies, exploitable wind power reserves are about 4,260 GW (calculated at a height of 100 metres). These include exploitable wind power resources of 3,900 GW onshore and 360 GW offshore (within 50 km of the coast). However, the distribution of wind resources is unbalanced, with onshore wind resources mainly concentrated in the three provinces of north-east China, as well as Hebei, Inner Mongolia, Gansu, Qinghai, Tibet and Xinjiang. Taking return on investment into account, the total onshore wind resources that are cost effective to exploit amount to 1,750 GW. The top 10 wind-rich provinces are: Inner Mongolia with 272 GW, Heilongjiang with 207 GW, Jilin with 121 GW, Xinjiang with 112 GW, Gansu with 105 GW, Liaoning with 87 GW, Qinghai with 72 GW, Guangxi with 66 GW, Shandong with 66 GW and Guangdong with 58 GW. For offshore wind power resources at a height of 100 metres and within 25 km and 25–50 km of shore, the exploitable resources are 190 GW and 170 GW respectively.

Solar energy resources are found mainly in West China. The installed capacity of exploitable solar energy in China is estimated to be about 15,600 GW, including 13,795 GW in West China, which is equivalent to 88.4%. The top five resource-rich provinces are: Xinjiang, with about 4,200 GW (26.92% of the country's total); Qinghai, with 3,400 GW (21.79% of China's total); western Inner Mongolia, with 2,600 GW (16.66% of the country's total); Gansu, with 2,130 GW (13.65% of the country's total); and Tibet with 700 GW (4.49% of China's total). According to the classification criteria of the Wind Energy and Solar Energy Resources Evaluation Center of the China Meteorological Administration, the following areas have poor solar energy resources: Hunan, Hubei, Guangxi, Jiangxi, Zhejiang, northern Fujian, northern Guangdong, southern Shaanxi, northern Jiangsu and southern Anhui, as well as Heilongjiang, Sichuan, Guizhou and northeastern Taiwan.

Hydropower resources are mainly concentrated in south-west and central–south China, while potential biomass resources are mainly found in South China, north-east China and parts of East China. South-west China accounts for 67.8% of China's total exploitable hydropower resources. In terms of biomass resources, South China is suitable for planting bioenergy crops, while north-east China has a high concentration of corn, rice and wheat, which are the main source of biomass resources like agricultural residues.

5.2 The Regional Energy System Should Be Planned in Line with the Principle of "Large Power Generation Clusters + Strong Networks + Distributed Energy Systems + Local Power Generation Facilities that Ensure Energy Supply Security" and Are Adapted to Local Industrial Demand

With a unique endowment of new energy resources, West China has great potential to develop important large-scale concentrations of power generation. China will make full use of its abundant wind, solar and hydropower resources in West China and continue to develop large wind and solar power clusters in Inner Mongolia, Gansu, Qinghai and Xinjiang. By 2060, the total installed capacity of wind and solar power in north-west China is expected to exceed 2,000 GW. China also aims to develop large hydropower clusters and pumped storage stations in south-west China. In addition, plans to increase local use of renewable energy should be made that include, for example, encouraging energy-intensive industries to move to those regions with large new energy clusters.

North-east China's future new energy investments will focus on power transmission capacity and hydrogen development. The north-east is relatively abundant in new energy resources. For example, Jilin province is located in a wind-rich belt across northern China, and is one of the country's nine largest wind power clusters. Moreover, Heilongjiang and Jilin also have abundant agricultural residues and other biomass resources. Because of limited local demand for electricity, north-east China will also become an important base of power transmission to other provinces. In view of the region's geographical advantages and industrial base, a hydrogen economy should also be developed to make good use of the diverse energy resources available.

Geographically, Central China connects North China and South China. Economically, it connects East China and West China. In terms of energy supply and demand, Central China serves as an important bridge between energy resources in West China and load centres in East China. The region needs to increase investment in energy infrastructure and play a co-ordinating role between the west and east. Bioenergy and low-speed wind power are likely to become two priorities in the new energy development of Central China.

East China should develop distributed power generation and microgrids to provide local supplies of electricity. For a long time to come, the eastern region will remain the centre of China's economic development and population distribution, as well as the energy load centre. East China is not rich in new energy resources, so it makes more sense to build distributed wind and solar power systems and offshore wind power facilities within the region. However, in the medium and long terms, distributed new energy systems cannot meet the power requirements of East China. In addition to the local supply of conventional power, west-to-east power transmission is still the most effective means to narrow the power supply gap in the east. Importing or transporting hydrogen and hydrogen-based fuels from West China is also an option.

China should focus on developing inter-regional power transmission networks and on improving its electricity trading mechanisms to promote the use of new energy. Looking ahead, the development of large new energy clusters in the central and western regions will require additional long-distance power transmission. Cross-regional and cross-provincial power flows are estimated to reach about 460 GW, 810 GW and 830 GW by 2030, 2050 and 2060 respectively. On the one hand, China should establish more secure and reliable power grids to ensure supply security of west-to-east power transmission. On the other hand, the country should continuously refine its electricity trading mechanisms and remove inter-provincial barriers that hinder the large-scale inter-regional transmission of new energy.

Table 3 Installed capacity of main power generation technologies by region (in GW)

Region	2030				2060			
	Conventional hydropower	Pumped storage	Nuclear power	Coal power	Conventional hydropower	Pumped storage	Nuclear power	Coal power
North China	4	20	8	304	4	35	19	100
East China	28	40	47	292	28	40	65	100
Central China	196	15	0	202	247	125	50	70
North-east China	7	16	9	111	9	60	9	40
North-west China	46	0	0	179	60	95	0	60
South China	149	29	41	162	149	66	60	50
Total	430	120	100	1,250	496	421	200	420

Source Research results of this report.

5.3 Although Large-scale Clusters of Wind and Solar Power Account for an Increasingly Large Share of the Power Mix, Coal Will Still Be Needed in 2060

Around 68% of wind power systems are concentrated in large clusters, while 57% of solar power plants are distributed. North-west and North China are key regions for wind and solar power facilities. Of China's total installed wind power capacity of 3,200 GW, onshore clusters predominate, supplemented by offshore wind farms. North-west China accounts for the largest proportion of total installed wind power capacity, followed by North China. These two regions together contribute about 47% of the country's total. Clustered and distributed solar power systems are almost equally important. North-west China contains the largest share of distributed solar at almost 35%, followed by North China at 22%.

Coal power capacity that stabilises the power system should be retained to ensure supply security and people's well-being. If there are no other feasible alternatives that provide 100% supply security, the stabilising role of coal is essential to counteract the supply risks of variable new energy. When all factors are taken into consideration, it would be prudent to retain at least 420 GW of installed coal power capacity across the country by 2060. By region, installed coal power capacity should be retained in East China (100 GW), North China (100 GW), Central China (70 GW), north-west China (60 GW), South China (50 GW) and north-east China (40 GW) (see Table 3).

5.4 Large Power Grids and Microgrids Play a Primary and Secondary Role Respectively

Efforts should be made to build a large national grid structure of west-to-east power transmission and north-to-south distributed power supply. The role of large power grids is to transport energy and provide supply security, while also integrating with microgrids, distributed power sources, energy storage facilities and electric vehicles to enhance capacity for regional grid-to-grid transfer. By 2030, 30 ultrahigh-voltage direct current (UHVDC) projects with a total transmission capacity of 240 GW will be in operation. A power grid structure comprising nine horizontal and five vertical transmission links in East China and three horizontal and two vertical transmission corridors in West China will be developed. By 2050, 61 UHVDC projects with a total transmission capacity of 490 GW will be completed. By 2060, this will increase to 64 UHVDC projects and 510 GW (see Fig. 5).

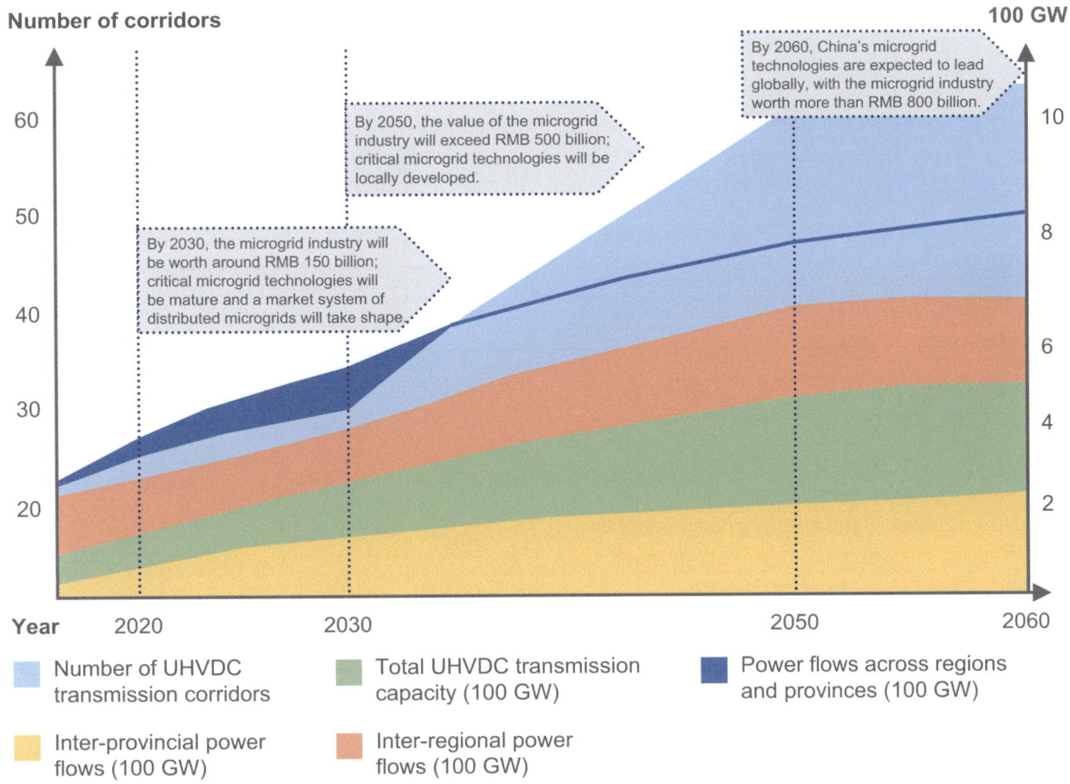

Fig. 5 Microgrids and power grids in 2020–60. *Source* Global Energy Interconnection Development and Cooperation Organization, The Road to China Carbon Neutrality, China Electric Power Press, 2021

A flexible interconnected structure of "large power grids + small and medium-sized regional power grids + smart distribution networks and microgrids", aided by digital control technology, will make grids more flexible and controllable (see Fig. 6). The flexible balancing capacity of microgrids enables distributed energy to be used locally, whereas the combination of microgrid and internet technologies makes the development of a regional "energy internet" possible. By 2030, investment in the microgrid industry is expected to exceed RMB 150 billion, and most microgrid projects will adopt the "solar power + energy storage" format. By 2050, critical microgrid technologies will be developed in China and the market size will reach RMB 500 billion. By 2060, Chinese microgrid technologies, especially smart technologies, will lead internationally and the market size of the microgrid industry will be worth more than RMB 800 billion.

6 Hydrogen Will Be an Important Energy Carrier in a Carbon-neutral World

The approach of "renewable energy + electrification" will not solve all carbon emission issues. Hydrogen is a clean, efficient, low-carbon and flexible energy carrier that can make up for the weak links in electrification and drive deep decarbonisation in hard-to-abate sectors such as steel, chemical feedstocks, heavy-duty trucks and high-temperature heat. Hydrogen can also become a key part of the future energy system.

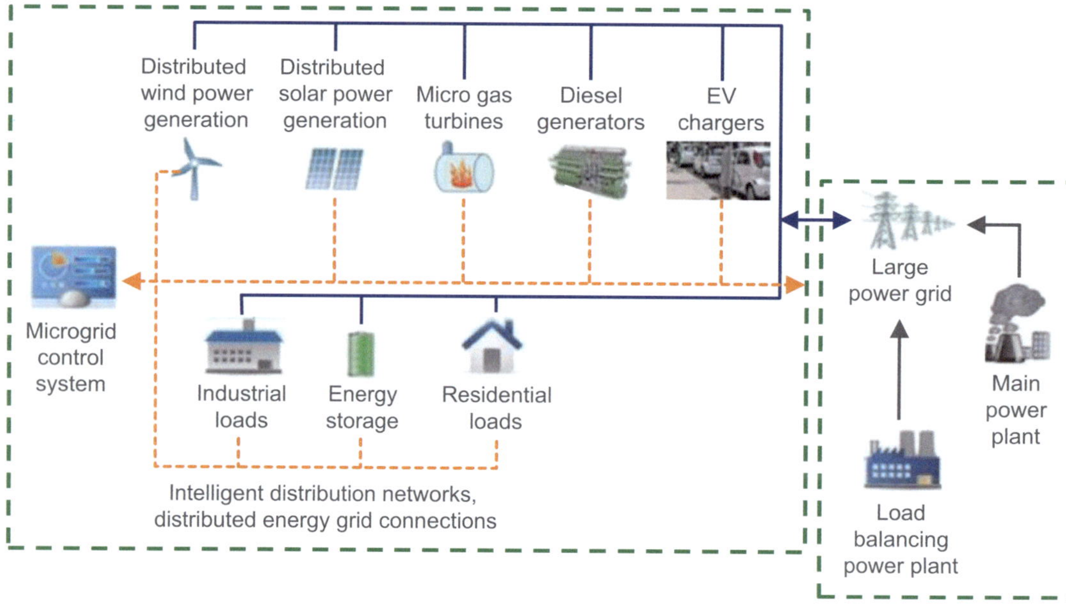

Fig. 6 Microgrids and large power grids. *Source* Research results of this report

6.1 The Main Role of Hydrogen Will Shift from that of a Raw Material to an Important Energy Carrier in a Carbon-neutral World

The International Energy Agency, the International Renewable Energy Agency, the Energy Transition Council and other organisations and companies have made optimistic forecasts about the future of hydrogen. According to the Energy Transition Council, hydrogen will become the most important non-electric energy carrier, with global hydrogen demand expected to be between five and seven times the current level by 2050. A report by McKinsey & Company forecasts that the world will use 660 million tonnes of hydrogen by 2050, accounting for 22% of global final energy use. The main application of hydrogen will gradually shift from that of an industrial raw material to a carbon-neutral energy carrier. In 2007, 99% of hydrogen globally was used as an industrial raw material and reducing agent. Of this, 66% was used in chemicals, 26% in oil refining and about 7% in metals and glass. Only about 1% of hydrogen was used as fuel in the transport and buildings sectors. To build a carbon-neutral world, new hydrogen-based technologies for industry, transport, buildings, energy and other applications are needed.

6.2 China's Hydrogen Demand Is Expected to Reach up to 120 Million Tonnes by 2060

Thanks to rapid technological advances and cost reductions, China's total hydrogen demand is estimated to rise to 120 million tonnes by 2060 in an ideal scenario. Industry will make up 73%, transport and buildings 17% and energy 10% of total hydrogen demand.

By 2060, total hydrogen demand in industry will be about 88 million tonnes. Hydrogen will be used as an industrial raw material and reducing agent, as an energy conversion medium (used to produce pure ammonia, methanol and other hydrogen-rich fuels) and as a high-grade fuel, mainly in steel, petrochemicals and chemicals. In our Dual-carbon Scenario, the production of raw steel is estimated to be 550 million tonnes. Half of China's raw steel output will be in electric furnaces and the remaining half will be produced using hydrogen, either by the direct reduction of iron in hydrogen-based shaft furnaces or through a combination of hydrogen-based blast furnaces with carbon capture, utilisation and storage. The use of hydrogen in hydrogen-based steelmaking will reach 20 million tonnes by 2060.

In petrochemicals, hydrogen is used mainly to make hydrogen-based chemicals. Almost 30% of the olefins and aromatics produced in China in 2060 will come from hydrogen-based chemicals, with hydrogen demand reaching 20 million tonnes. In the chemical industry, hydrogen is used mainly to produce synthetic ammonia, methanol and other products. In 2060, the yield of synthetic ammonia is expected to be 50 million tonnes, including 16 million tonnes of pure ammonia used as fuel. Methanol production is expected to be 140 million tonnes, including 40 million tonnes of methanol used as fuel. Hydrogen demand in the chemical industry is expected to be about 25 million tonnes in 2060. In addition, hydrogen used in the food, electronics, machinery and other sectors will be about 8 million tonnes in 2060, and demand for pure hydrogen as a fuel to provide high-temperature heat (steam) for industry, especially in energy-intensive industries, could reach 15 million tonnes.

By 2060, total hydrogen demand in the transport sector is expected to reach 20.5 million tonnes. Hydrogen can be used as fuel for heavy-duty trucks and as an alternative fuel for ships and aircraft. Hydrogen demand in transport will remain low in the near and medium terms. Hydrogen production, transport and storage technologies are expected to have progressed significantly by around 2035, with hydrogen increasingly used in hard-to-electrify transport segments. Alternative fuel solutions for heavy-duty trucks will likely be limited to two technologies: battery swapping and hydrogen fuel cells. In areas where distances are long or the battery swapping infrastructure is weak, hydrogen fuel cell heavy-duty trucks have clear advantages. By 2060, there will be about three million hydrogen fuel cell vehicles on the roads, and hydrogen demand will be about 8 million tonnes. In aviation, hydrogen can only be used to support the low-carbon development of short-haul regional and commuter aircraft, which have a small market share. Based on market demand forecasts, hydrogen demand for civil aviation will be about 5 million tonnes in 2060. Looking ahead, the clean energy carriers that are being developed and used by international aviation and shipping are mainly biofuels, methanol and ammonia. All things considered, ammonia fuel cells will be the preferred ship power system in the long run. In 2060, hydrogen demand from water-based transport and other transport segments is estimated to be about 6 million tonnes and 1.5 million tonnes respectively. By then, hydrogen could make up around 20% of total transport energy demand and play an important role as an alternative fuel.

Total final hydrogen demand in the buildings and energy sectors will be about 13 million tonnes, with energy storage becoming an increasingly important medium for hydrogen. By 2060, hydrogen demand in the buildings sector will be about 5 million tonnes, mainly from distributed combined heat and power systems. In the energy sector, demand will be about 8 million tonnes, mainly from its use in fuel cells as a standby power source. When intermittent renewable energy accounts for a high proportion of power generation, short-term energy storage is not enough to ensure power supply security. As a large-scale energy storage medium, hydrogen will have a lower marginal cost than electrochemical energy storage. Hydrogen could therefore meet the large-capacity and long-term (by day, month or season) energy storage requirements of a power system with a high proportion of renewable energy. The green hydrogen produced from renewable energy can be directly used in industry, transport, buildings and other end-use sectors, and can also be used for long-term storage in the form of hydrogen-based liquid fuels and cross-seasonal power generation. By 2060, about 3 trillion kWh of electricity are expected to be used to produce hydrogen from renewable energy.

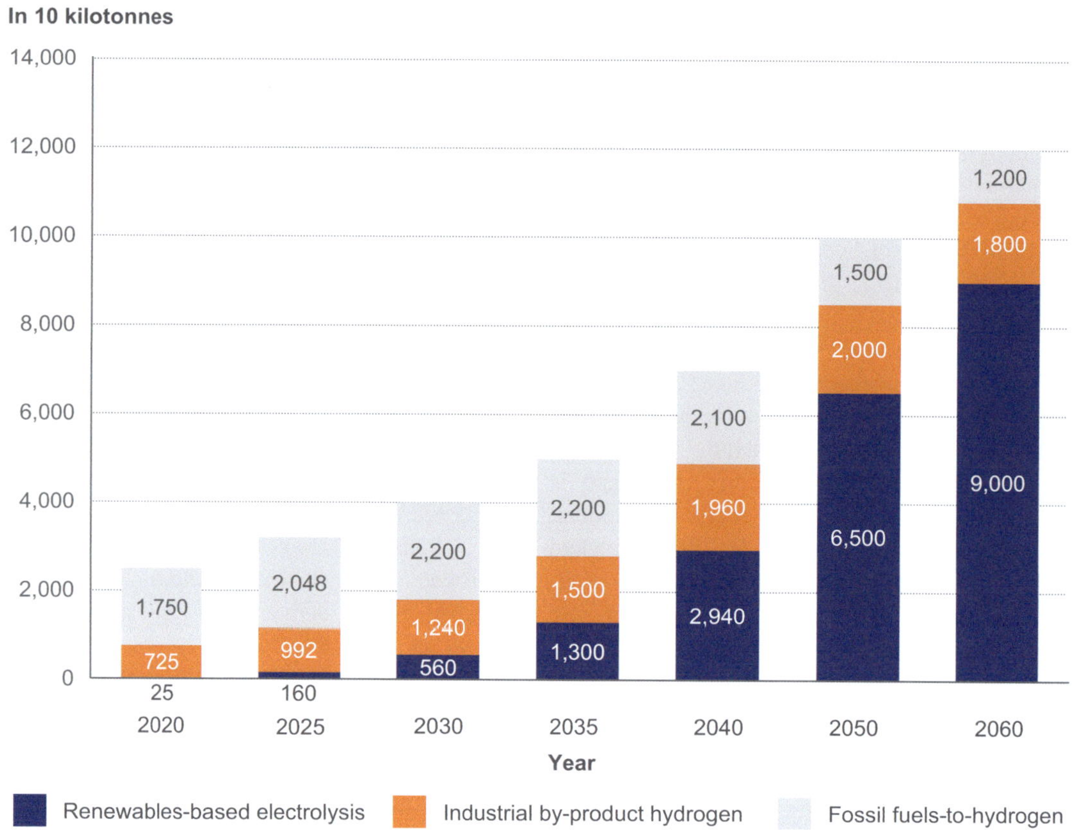

Fig. 7 Hydrogen supply structure, 2020–60. *Source* Research results of this report

6.3 Hydrogen Supply Will Gradually Shift from Fossil-fuel-based Production to Renewables-based Electrolysis

In the hydrogen supply structure, grey hydrogen made from natural gas or methane dominates with a share of about 70%. By 2060, green hydrogen made from renewable energy will account for more than 75% of demand. In the short term, green hydrogen is not sufficiently cost effective for large-scale production. Currently, coal-to-hydrogen and by-product hydrogen have a cost advantage. In the transition period to 2035, industrial by-product hydrogen will play an important role and account for around 30% of total hydrogen supply. However, in the longer term, the cost of production, storage, transport and refuelling technologies will decline, and green hydrogen will account for a continuously rising share of supply and gradually replace much of grey hydrogen (see Fig. 7).

The composite end-user selling price of hydrogen will drop below RMB 20 per kg by 2060. Costs in the hydrogen value chain—from production, storage, transport and refuelling to end-user application—will continue to decrease before reaching or falling below the cost boundary acceptable to each application. By 2060, renewables-based electrolysis will be the lowest-cost option, with a system supply cost of less than RMB 10 per kg (see Fig. 8).

7 Bioenergy Plays an Essential Role in the Future Energy System

As a zero-carbon energy carrier, biomass is globally recognised as an essential carbon abatement option. Especially when combined

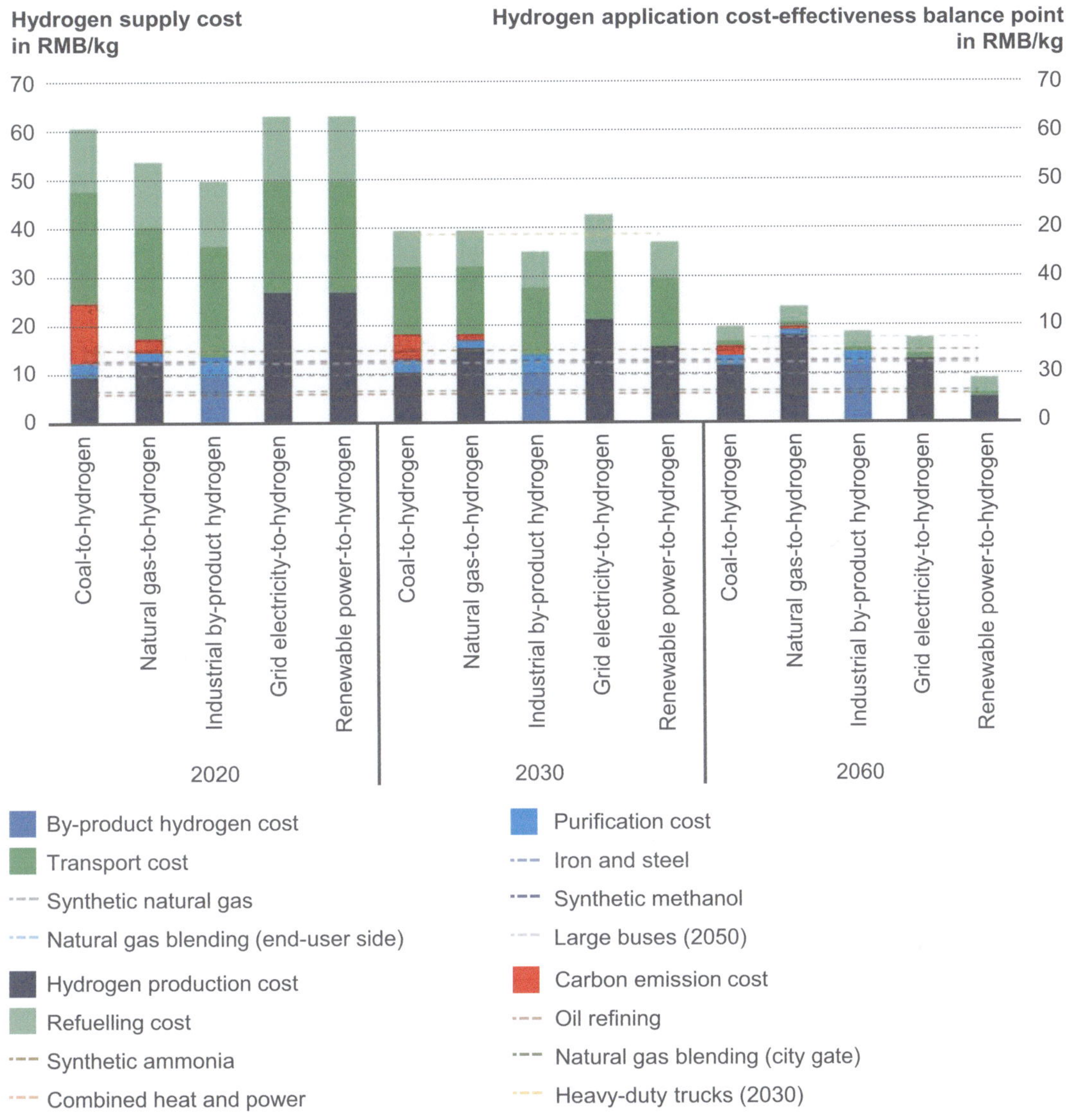

Fig. 8 Hydrogen supply costs, 2020–60. *Source* Calculations by the project team

with carbon capture, utilisation and storage, biomass can achieve negative carbon emissions. On the path towards carbon neutrality, bioenergy plays a significant role in supporting deep decarbonisation in aviation, shipping and other hard-to-abate sectors. Looking ahead, China should continue to promote biomass power generation, unlock the potential of biofuels and plan for a bioenergy future.

7.1 Bioenergy Provides Significant Carbon Abatement

Bioenergy is zero carbon or even negative carbon. Bioenergy is the chemical energy in biomass that is converted from the carbon dioxide stored by solar energy through photosynthesis. Biomass is generally regarded as a zero-carbon energy carrier. If a carbon capture and storage or carbon capture, utilisation and storage facility is installed at the end of a biomass use process to collect the CO_2 generated, negative carbon emissions can be achieved. This is an indispensable way to reduce greenhouse gases in the environment and achieve carbon neutrality.

Biomass can be used not only for zero-carbon power generation, but also for non-electricity alternative fuels. Direct combustion of agricultural and forestry residues, municipal waste incineration and biogas are used to generate power. In addition, ethanol can be produced from starch-rich grains, cassava and other crops, as well as cellulose. Biodiesel and biokerosene, which can be made from animal and vegetable oils, are important alternative fuels for aviation, shipping and other hard-to-electrify sectors.

Bioenergy is widely valued across the world. In North America, Europe and Latin America, bioenergy already accounts for 5%, 11% and 16% of total energy use respectively. The USA has long prioritised the development of bioenergy. In 2020, the installed capacity of biomass-fired power generation in the USA was about 16 GW, and the annual power output was 64,000 gigawatt-hours (GWh). The USA is also the world's largest producer and consumer of ethanol, accounting for half of global ethanol production and 14% of global biodiesel production.

7.2 China Has Relatively Large Biomass Resources

As of 2020, the annual production of large biomass resources in China was 3.494 billion tonnes, and its potential as energy was about 460 million tonnes of coal equivalent. China has a wide variety of biomass resources, including residues from crops and agriculture, forestry and the wood industry, livestock and poultry breeding, municipal waste and domestic sewage, industrial organic waste and high-concentration organic waste water.

Among the biomass resources that can be used for energy, agricultural residues, livestock and poultry manure and municipal waste account for the largest proportion. In 2020, agricultural residues were estimated to be around 820 million tonnes, of which about 694 million tonnes were recoverable. These were mainly distributed in 13 major grain-producing provinces, including the North China Plain, the Middle–Lower Yangtze Plain and the North-east China Plain. About 400 million tonnes of agricultural residues were used for fertiliser, feed, papermaking and other purposes, and about 420 million tonnes could be converted into energy. Livestock and poultry manure amounted to 1.868 billion tonnes (excluding waste water) and 211 million tonnes of manure were used to produce biogas. Forestry residues amounted to 350 million tonnes, of which 9.604 million tonnes were used for power generation. Of municipal waste, 310 million tonnes were cleaned up and transported, of which 143 million tons were incinerated.

The production of biomass will increase alongside economic growth and the development of new uses. China's total biomass resources will reach almost 5.4 billion tonnes by 2060. Agricultural residues will amount to about 1.234 billion tonnes, of which about 1 billion tonnes will be recoverable. Livestock and poultry manure will amount to 2.373 billion tonnes, and forestry residues 773 million tonnes. Municipal waste will plateau around 2045, reaching 1.005 billion tonnes in 2060. The annual production of waste oils and fats will be 13.947 million tonnes.

7.3 Biomass Will Play an Important Role in Clean Power Generation and Alternative Fuels

By 2060, the installed capacity of bioenergy will grow to 66 GW. By the end of 2020, China's total installed capacity of bioenergy was 29.52 GW and output was 132,600 GWh. Power generated from agricultural and forestry residues and municipal waste will continue to increase rapidly, from about 55 GW in 2030 to a peak of 66 GW around 2040 (see Fig. 9).

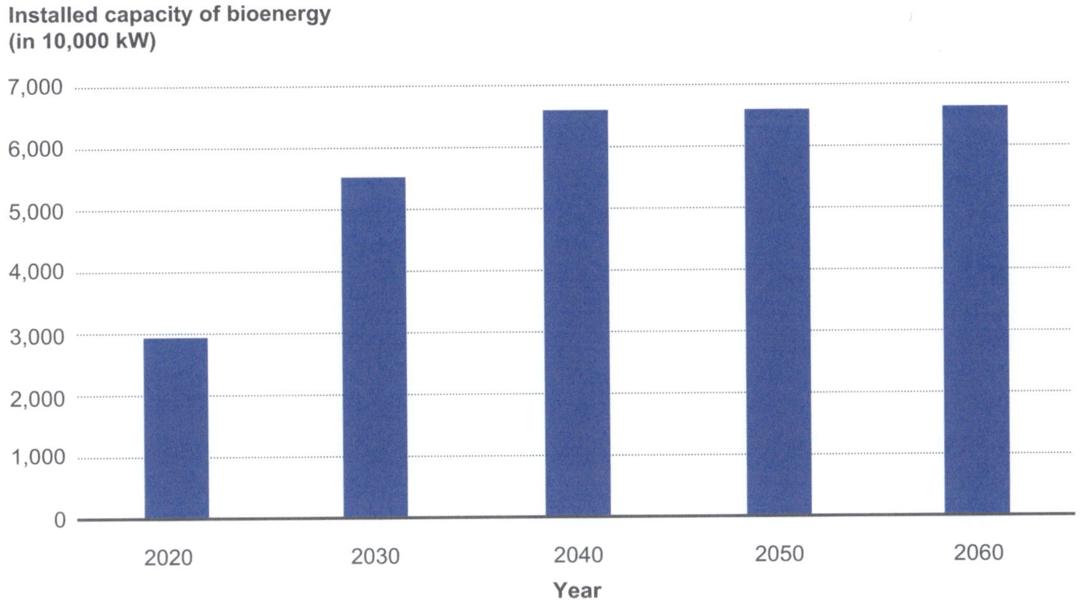

Fig. 9 Installed bioenergy capacity in China. *Source* Research results of this report

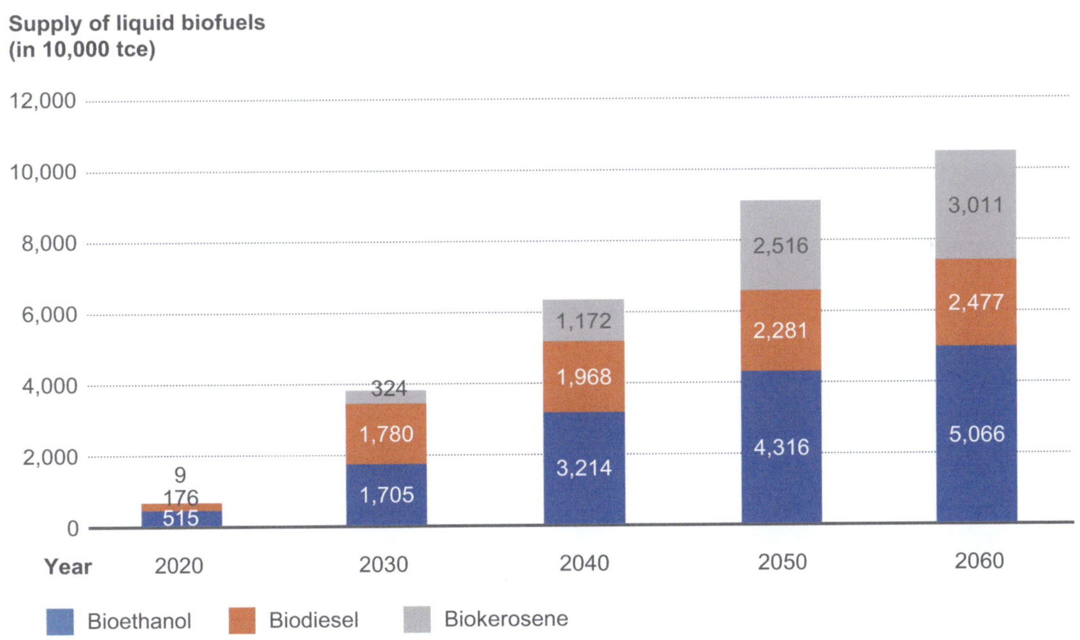

Fig. 10 Liquid biofuels in China, 2020–60. *Source* Research results of this report

Liquid biofuels will continue to increase to more than 100 million tonnes of coal equivalent (tce) by 2060. The current supply capacity of liquid biofuels is about 7 million tce, mainly bioethanol and biodiesel. Supplies of bioethanol, biodiesel and biokerosene will rise significantly from their current level, reaching 38.09 million tce by 2030. By 2060, 100 million tce of liquid biofuels will be used in non-electrified transport and aviation (see Fig. 10).

8 Investment in New Energy Infrastructure Increases

New infrastructure is needed when building an energy supply system that is dominated by new types of energy. It is also essential for advancing the energy revolution and digital transformation of the energy industry. New energy infrastructure will accelerate the development of a new economy and make economic growth and progress towards carbon neutrality possible.

8.1 Investment in New Energy Infrastructure Will Foster a Green and Low-carbon Economy

At present, China's annual investment in new energy infrastructure is about RMB 1.3 trillion. Estimates show that the total investment needed in low-carbon power systems, hydrogen, bioenergy, carbon capture, utilisation and storage and supporting infrastructure in industry will be at least RMB 80.6 trillion in the period 2020–60.

In general, the investment trend in new energy infrastructure will initially rise then decline. By 2030, investment will increase to about RMB 2.2 trillion per year; in 2031–40, investment will peak at about RMB 2.34 trillion per year; in 2041–50, it will drop to around RMB 1.79 trillion per year; and in 2051–60, it will continue to decrease to about RMB 1.63 trillion per year.

8.2 New Energy Infrastructure Investments Will Be Mainly in Four Areas

Between 2020 and 2060, total investment in new energy-based electricity and supporting flexibility technologies will reach RMB 54.78 trillion. This figure includes RMB 23.28 trillion in wind and solar power infrastructure, about RMB 3.15 trillion in new nuclear power plants, RMB 19.86 trillion in upgrading ultrahigh voltage and other long-distance transmission and distribution infrastructure, and about RMB 3.42 trillion in enhancing the flexibility of the power system with battery storage technologies.[2] The priorities of investment in new energy-based power and supporting flexibility capacity at different stages are shown in Fig. 11.

In 2020–60, total investment in non-electricity new energy infrastructure will be RMB 10.21 trillion. According to estimates, investment in infrastructure for hydrogen supply will be RMB 2.96 trillion; for hydrogen transport, storage and distribution RMB 2.11 trillion; and for biofuel infrastructure, it will be RMB 5.14 trillion. The non-electricity new energy infrastructure investments at different stages are shown in Fig. 12.

In 2020–60, total investment in carbon capture, utilisation and storage (CCUS) infrastructure will be RMB 16.4 trillion, including RMB 8.9 trillion in CCUS for the electricity sector, RMB 3.08 trillion in CCUS for industry and RMB 4.39 trillion in CCUS for transport and storage infrastructure. The CCUS investment scale and key priorities at different stages are shown in Fig. 13.

8.3 Multiple Measures Are Needed to Drive Investment in Infrastructure

Investment in new energy infrastructure is essential over the long term. The large size of the market will help boost economic growth and improve industrial competitiveness. China will unlock investment potential by combining the "plans-guided, government-led and market-oriented" approaches. For instance, China should research and formulate plans for investment in power generation, power grids, storage and other new energy infrastructure, and determine the investment policy. It should award government funds and assign the leading role to state-owned enterprises and launch a number of major new energy infrastructure projects during the 14th Five-year Plan period (2021–25). China

[2] Only major investments are included here.

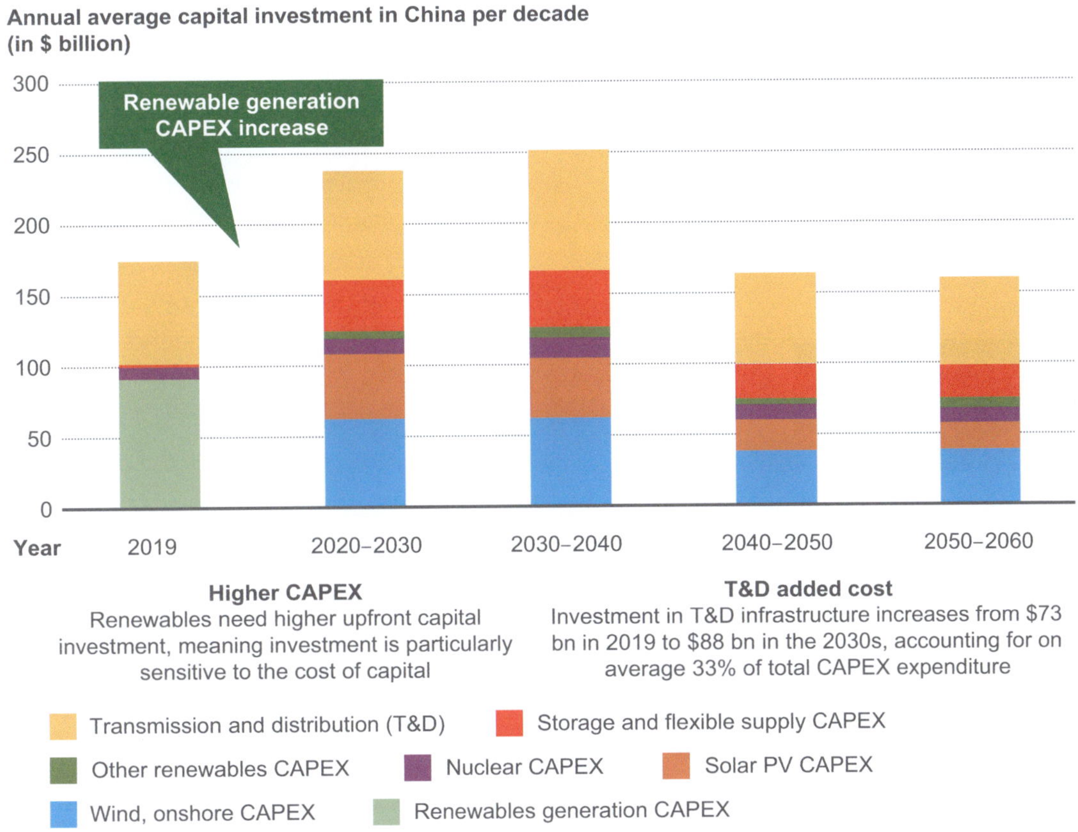

Fig. 11 Investment in new energy-based electricity and flexibility technologies, 2019–60. *Source* Research results of this report

should also use its electricity pricing and carbon market reforms and encourage private capital to take part in new infrastructure investment.

9 New Energy Can Co-ordinate and Drive Economic, Social and Environmental Development

A new industry can drive a new economy, and the development of new energy will unleash huge potential in boosting economic growth, creating jobs and co-ordinating environmental pollution control. However, it should also be noted that the development of the new energy industry requires the use of 26 mineral resources, most of which are imported. Resource security risks should, therefore, not be overlooked.

9.1 Carbon Neutrality Will Enable Sustained Growth of the New Economy

New energy will become a key driver of the new economy. First, new energy will enable more industries to shift to electrification and green innovation, revitalising many traditional sectors. For example, many traditional sectors will move away from combustion power to electric motor power, which will renew these industries. Second, thanks to the low-carbon transition, several emerging industries have boomed in recent years, such as new energy, intelligent connected vehicles, smart manufacturing and the digital economy. This will further drive the transformation of China's economy. Third, the development of new energy, such as wind and solar power, will

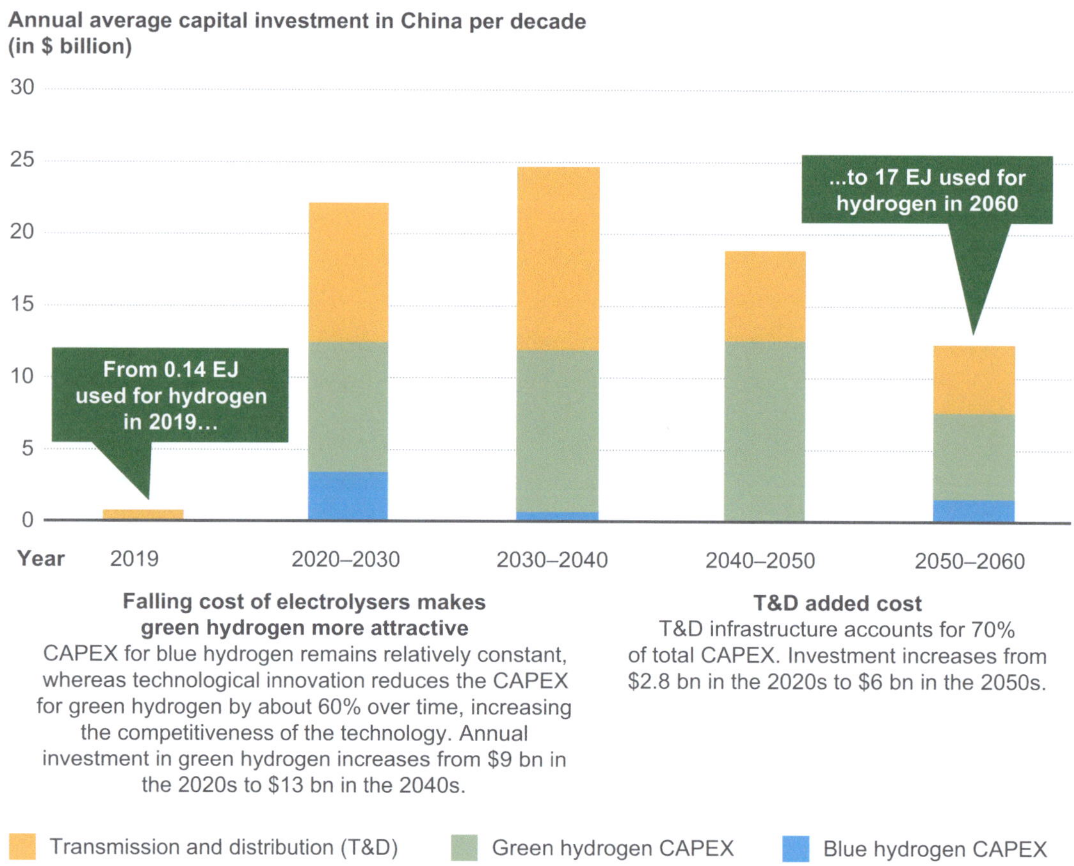

Fig. 12 Investment in non-electricity new energy infrastructure, 2019–60. *Source* Research results of this report

enable new optimisation opportunities. For example, north-west China, where traditional economic growth drivers are weak, may accommodate more energy-intensive industries.

Carbon emissions and economic growth will gradually be delinked and the carbon neutrality-based new economy will deliver significant benefits. According to our modelling, the new economy will drive sustained GDP growth. Measured in 2020 constant prices, China's GDP will nearly double by 2035 and almost quadruple by 2050, rising to more than RMB 400 trillion by 2060. Investment in new energy-based electricity infrastructure and the electrical equipment manufacturing industry will contribute more than 5% to GDP growth. The amount of added value created by the development of non-fossil energy in GDP will gradually increase from 1.3% in 2020 to 2.2% in 2030 and 5% in 2060 (see Fig. 14). If investment in electric vehicles, industrial carbon reduction and energy-efficient buildings are included, as well as the improved competitiveness delivered by technological change, the effects of the low-carbon transition on economic growth will double the above-mentioned figures. The shift to low-carbon energy will thus become an important driver of economic growth.

9.2 New Jobs Will Be Created

The number of jobs created in the wind and solar power value chains will increase significantly, according to the results of life-cycle jobs modelling. The number of jobs in the wind power sector will peak at about 3 million in 2045, while that in the solar power sector will peak at 7

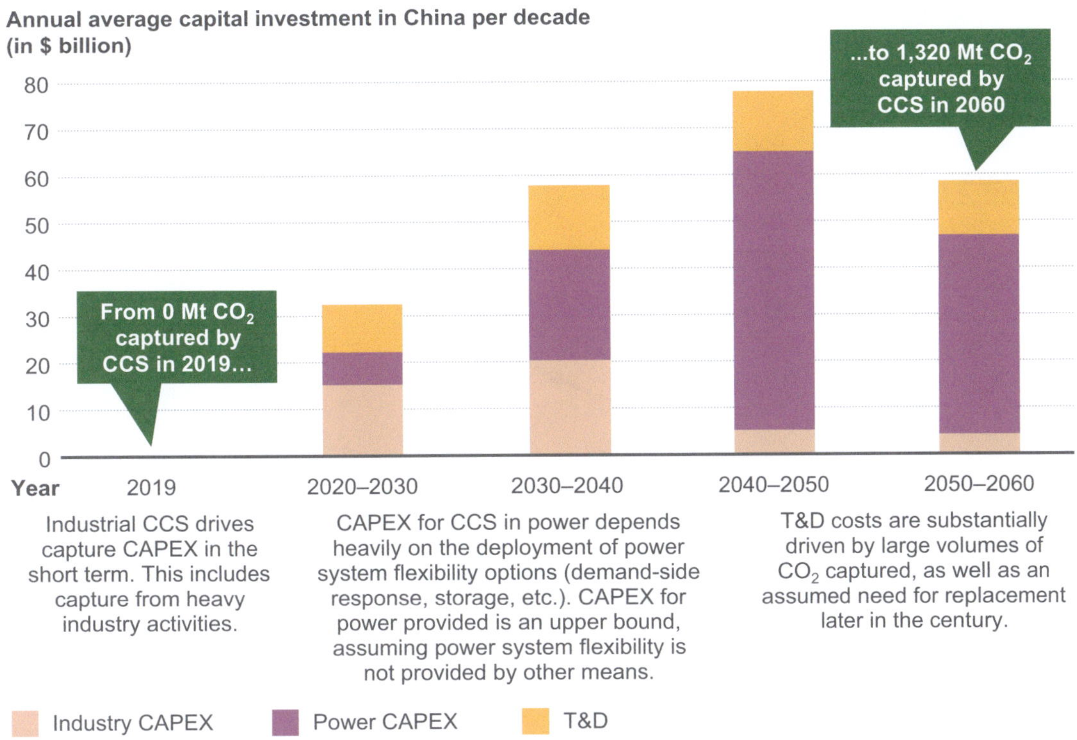

Fig. 13 Investment in CCS infrastructure, 2019–60. *Source* Research results of this report

million in 2050. Although the quantity of jobs will decline due to a decrease in new installed capacity after 2050, there will still be more than 2 million jobs in the wind power sector and more than 5 million jobs in the solar power sector by 2060 (see Fig. 15).

The total number of direct jobs associated with the energy transition is uncertain, but the structural adjustments are clear. During the transition to low-carbon power, the large-scale expansion of renewable energy will create new jobs, but the number of jobs in coal and other conventional power sectors will shrink. By 2060, the number of direct jobs is expected to reach up to 9.658 million, compared with 8.600 million jobs in 2020. However, due to technological advances and the economies of scale, the employment coefficient of various power generation technologies is estimated to continue to decline. Therefore, the total number of jobs in the power industry is highly uncertain and is likely to be lower than the upper limit mentioned above. In terms of the employment mix, the rapid shift of jobs from coal power to wind and solar power will be significant and certain. The total number of jobs in the fossil fuel power generation sector will decline from 4.677 million in 2020 to 333,000 by 2060, virtually close to zero. Total jobs in the wind and solar power sectors will rapidly increase, peaking at around 3 million in 2045 in the wind power sector and at about 7 million in 2050 in the solar power sector, before decreasing as new installed capacity declines. By 2060, there will still be about 2.3 million and about 5.2 million jobs in the wind and solar power sectors respectively (see Fig. 16).

9.3 Significant Environmental Benefits Will Result

The substitution of new energy for fossil fuels will help control air pollution and save water. Emissions of sulphur dioxide, nitrogen oxides,

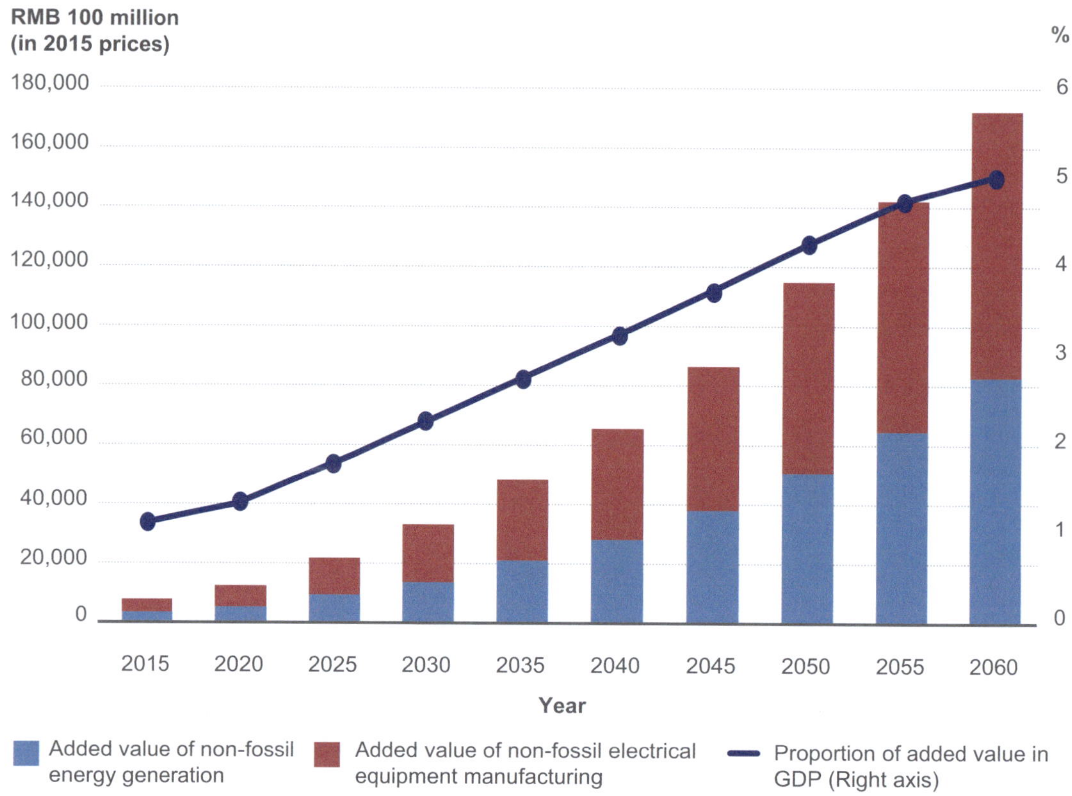

Fig. 14 The direct effects of new energy development on the economy, 2015–60. *Source* Research results of this report

particulate matter and other air pollutants will be effectively controlled. By 2060, emissions of sulphur dioxide, nitrogen oxides and particulate matter from coal power will decline significantly and stabilise at 184,000, 380,000 and 62,000 tonnes respectively.

The various power technologies have different impacts on water resources. In wind and solar power plants, water is used only when the wind turbine blades and solar photovoltaic panels are cleaned, which equates to 5 litres and 20 litres of water respectively per kilowatt-hour of electricity generated. The pressure on water resources can therefore be effectively mitigated. By 2060, the use of water in the power system will be about 5.97 billion cubic metres, less than the 2020 level. This indicates that the energy transition can help reduce water consumption.

10 Technological Innovation Enables Carbon Neutrality

There are many uncertainties and challenges in the technological pathways to carbon neutrality. China should always keep track of leading energy technologies and focus on achieving a smooth transition to the new energy system through technological innovation.

Taking into account the importance of technology and China's current stage of development, we make the following forecasts. By 2030, China should focus on the research, development and application of zero-carbon power technologies. By around 2035, hydrogen technology, biofuels and carbon capture and storage should be commercially viable. Around 2050, controlled

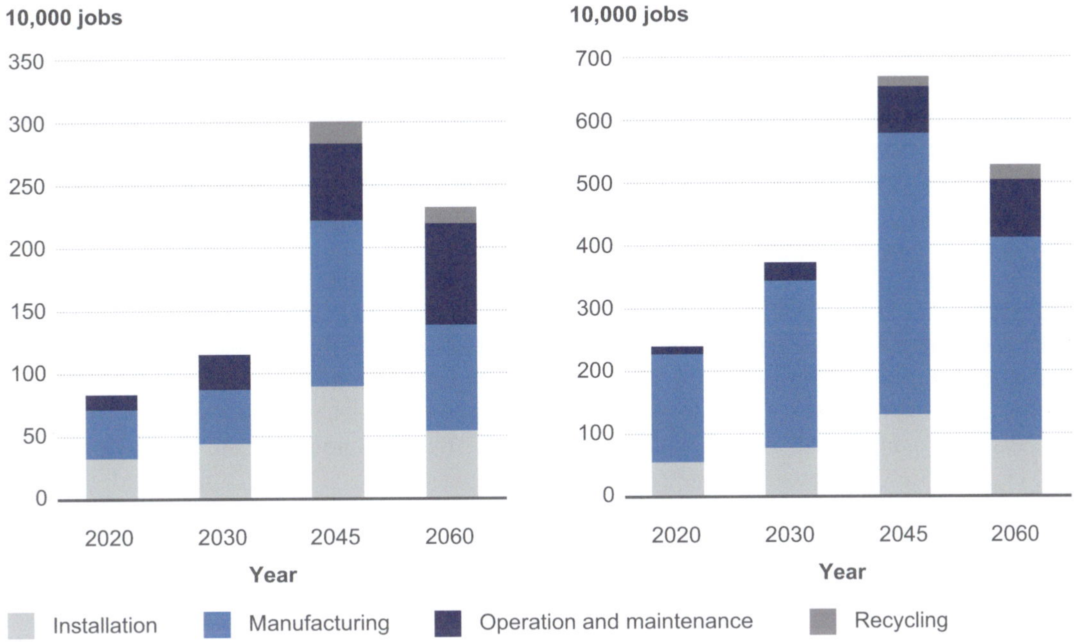

Fig. 15 Employment in the wind (left) and solar (right) power sectors by job type, 2020–60. *Source* Research results of this report

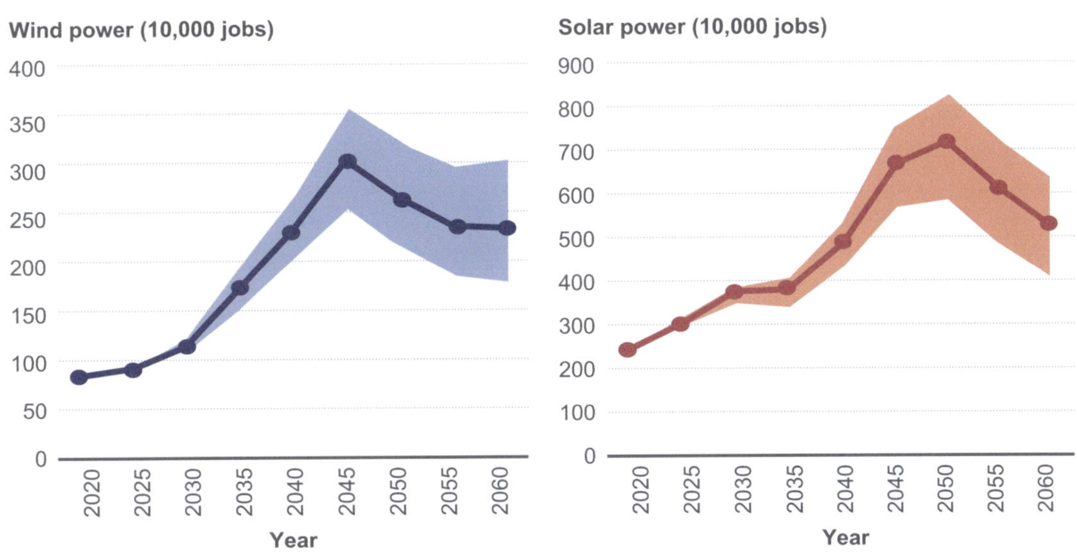

Fig. 16 Employment in the wind (left) and solar (right) power sectors, 2020–60. *Source* Research results of this report

nuclear fusion and other new nuclear power technologies should play a role in the energy system. And by 2060, carbon capture and utilisation recycling technology should be widely deployed.

10.1 Technologies for a Zero-carbon Power System

Hydropower, geothermal energy and conventional nuclear power are relatively mature. Wind power, solar photovoltaic power and large-scale heat pumps are in commercial operation. Concentrated solar power, ocean energy, coal power with carbon capture and storage (CCS), gas power with CCS, and bioenergy with CCS are still in the pilot stage. Ultrahigh-voltage direct current transmission technology is commercially viable, while flexible high-voltage alternating current transmission technology is in the verification stage. Fast frequency response and the use of virtual inertia to address the low rotational inertia caused by wind and solar power generation are in the large-scale prototype testing stage. Fast charging technology is undergoing commercial testing, while intelligent charging needs to integrate information and operational technologies to co-ordinate and manage electric vehicle charging. The in-process wireless charging that depends on the combination of induction coils and special road materials is a technology for the long term. Demand–response and battery storage technologies are in the commercial testing stage, and chemical energy storage technology is almost mature.

Of the technologies used in the electrification of transport, industry and buildings, electric rail technologies are relatively well proven. Electric passenger vehicle technologies have been commercialised, while ship electrification is at the verification stage. The aviation sector is developing and testing electric aircraft prototypes, but considering the constraints on battery energy density, the use of electric aircraft may be limited to short-haul flights. Conventional industrial processes require a lot of heat for low- and high-temperature heating, mineral electrolysis and other process requirements. Efforts to replace conventional fuels with electricity in heat supply are under development. At present, relatively mature electrification technology is used mainly to produce aluminium. The electrification of buildings requires a large number of advanced heat pumps for heating, cooling and air circulation. In addition, evaporated cooling is in commercial operation, while solid-state cooling is still in the small-scale prototype testing stage. Energy-efficient technologies for industry, buildings and transport, as well as materials, processes, design and manufacturing, must also be continuously improved.

10.2 Hydrogen

The use of hydrogen, especially low-carbon hydrogen, has not yet been commercialised. At present, many technologies essential to hydrogen production, storage, transport and use are at different stages of development and face different challenges.

Currently, grey hydrogen made from natural gas or methane is the mainstream hydrogen technology. For renewables-based electrolysis, challenges such as high costs and low efficiency remain. Faster technology innovation is imperative. Hydrogen transport and storage need to be integrated with infrastructure-related technologies. Relatively mature technologies include pipeline transport and tank storage. Hydrogen storage in ammonia tanks and salt caverns is at the early commercial pilot stage, while hydrogen blended with natural gas for transport in pipelines is under verification. In addition, liquid hydrogen storage and transport technologies are still undergoing large-scale testing.

Hydrogen is mainly used in industry, transport, buildings and power generation. In terms of the energy transition, synthetic methane technologies are in the testing and verification stage, while liquid hydrocarbon synthesis technologies have reached large-scale testing. In industry, the use of hydrogen in electrolytic methanol and ammonia production is undergoing commercial testing and verification, while the technologies for making steel using hydrogen are at the large-scale testing stage. In transport, light-duty passenger vehicles using hydrogen fuel cells have entered commercialisation, while the use of hydrogen in large vehicles, ships and railways is still being verified. Engine technology using hydrogen and ammonia-powered ships are still undergoing large-scale testing. In buildings, hydrogen-fired boilers and hydrogen fuel cells are in commercial operation, and hydrogen-driven heat pumps are also being tested and verified. In power generation, high-temperature fuel cells are in commercial use, and gas turbines using hydrogen have started commercial verification. Coal power technology that uses ammonia is still at the large-scale testing stage.

10.3 Bioenergy

Bioenergy mainly involves supply-side and application-side technologies. The supply-side technologies mainly concern the purchase of biomass, biofuel production and bioenergy power generation. On the application side, bioenergy has many potential uses in industry, transport and buildings.

An important part of growing biomass is double cropping, which is planting energy crops between periods of growing food crops. This approach is still at the small-scale pilot stage. Currently, the biofuels that are in commercial production include bioethanol, biomethane, biodiesel and biogas. Bioethanol and biomethane with carbon capture, utilisation and storage (CCUS) are undergoing commercial testing and verification. Biodiesel with CCUS, as well as algae-based biodiesel and biogas, are still at the small-scale pilot stage. Bioenergy power generation is relatively mature. Internal combustion engines that use biogas or bioliquids are in commercial use. Power generated by solid biomass with CCUS is still undergoing testing and verification.

Bioenergy has a wide range of applications, and some bioenergy technologies are relatively mature. In industry, bioenergy used as industrial fuel is well proven, and using biomass to produce methanol and other chemical products has entered commercial testing. Biomass-based hydrogen and ammonia production is still at the large-scale testing stage. In transport, passenger vehicles and light-duty trucks powered by biodiesel or bioethanol are in commercial use, while ships fuelled by biodiesel and aircraft powered by biofuel are still at the pilot and verification stage. In buildings, biofuel-based heating and biomass-fired stoves are in commercial use.

10.4 Carbon Capture and Storage

Carbon capture and storage (CCS) is the only technology to achieve net-zero emissions from fossil fuels. Even in a modern energy system based largely on renewable energy, CCS will still be needed to reduce the emissions from fossil fuels. Negative emission technologies, such as bioenergy with carbon capture and storage (BECCS) and direct air capture with carbon storage (DACCS), are also needed to achieve carbon neutrality.

Carbon capture technology is advancing rapidly. In the chemical industry, chemical carbon capture technology in the production of ammonia is relatively mature, while physical and chemical carbon capture in methanol production are undergoing commercialisation. In the iron and steel sector, chemical carbon capture in the direct reduction of iron is in commercial use, while physical carbon capture in oxygen-rich smelting reduction is still at the verification stage. Chemical carbon capture in hydrogen-rich blast furnace processes is undergoing large-scale testing. In the cement sector, chemical carbon capture and calcium looping have entered the commercial pilot stage, whereas other technologies are still undergoing large-scale testing. Carbon capture in power generation applies mainly to coal, gas and bioenergy. Carbon capture in coal-fired power generation has reached commercial testing, whereas the others are still undergoing verification. Direct air capture technologies under large-scale testing include solid state and liquid state alternatives.

China's geological carbon storage potential is about 1.21–4.13 trillion tonnes, second only to the USA. Pipeline transmission is a mature carbon transport mode, whereas sea transport needs to be verified. There are two main storage technologies: onshore carbon sequestration and offshore carbon storage. All technology types are at different stages of development. For example, enhanced oil recovery and leaching are the most mature and commercialised. CO_2-enhanced deep saline water recovery has evolved to the industrial demonstration stage. CO_2-enhanced coalbed methane recovery is at the early industrial demonstration stage.

Constrained by high costs and low technological maturity, CCS is not expected to mature and be deployed at scale until around 2035; it will, however, be important for deep decarbonisation by 2060. The main task before 2035 is to move CCS from the industrial demonstration to the commercialisation stage. It should be possible to control the cost of thermal power with CCS at RMB 0.15 per kWh after 2040. By 2060, CCUS technology will be widely deployed and negative emission technologies, such as BECCS and DACCS, will be rolled out at scale. According to preliminary estimates, the total emissions reduction of CCS will exceed 1.0 billion tonnes of CO_2 by 2060.

10.5 Controlled Nuclear Fusion

Nuclear fusion is the ideal technology for a green energy future. It involves the fusion of deuterium and tritium, which are isotopes of hydrogen, at high temperatures to form helium, which releases huge amounts of energy in the form of heat. Through nuclear fusion, a small amount of fuel can be extracted from inexpensive materials across the globe, and nearly unlimited clean electricity could be produced for a long time. According to existing research, more than 40 trillion tonnes of deuterium are stored in sea water, 1 litre of which can ionise 0.03 grams of deuterium. The energy released from the reaction of 0.03 grams of deuterium is equivalent to the combustion of 300 litres of petrol. If all natural deuterium and tritium resources are used for nuclear fusion, the energy released is sufficient to meet world energy demand for 10 billion years. More importantly, nuclear fusion is safer than nuclear fission, because it does not cause runaway chain reactions.

Countries pin great hope on controlled nuclear fusion and expect to continuously make breakthroughs in research and testing. According to data released by the International Atomic Energy Agency, there were 96 nuclear fusion plants in operation across the world at the end of 2021. That same year, a European research team broke the record for producing controlled fusion energy. In the Joint European Torus, the world's largest fusion reactor, a gas of hydrogen isotopes (deuterium and tritium) was heated to 150 million degrees Celsius and held steady for 5 seconds while nuclei fused in the reaction, releasing 59 megajoules of energy. The USA has also reported breakthroughs in another fusion approach, where 192 lasers were used to create a tiny hot spot with the diameter of a human hair that generated more than 10 trillion watts of energy in a billionth of a second. Since 2006, the EU, China, India, Japan, South Korea, Russia and the USA have been working on the International Thermonuclear Experimental Reactor project, also known as the world's biggest artificial sun. Phase 1 of operational testing is scheduled for 2025, with full operation targeted for 2035.

China's controlled nuclear fusion research is almost abreast of that of the international community, and some of its technologies even lead the way. In Chengdu in 2020, China's next-generation "artificial sun", known as HL-2M, achieved its first plasma discharge. This shows that China has independently mastered the design, manufacture and operation of large-scale advanced tokamak reactors. In 2021, China's experimental advanced superconducting tokamak achieved continuous high-temperature plasma operation for 1,056 seconds, a world record. According to the latest assessment, China expects to achieve a nuclear fusion input that is less than its output as early as 2035, paving the way for its commercialisation.

10.6 Carbon Capture and Utilisation

Carbon dioxide can be used as a raw material in the production of urea, concrete, methanol, synthetic methane and synthetic liquid hydrocarbons. The technology for methane production using CO_2 is relatively mature, while that for making concrete is already in commercial use. Using CO_2 to produce methanol and synthetic methane is still at the verification stage, while using CO_2 to make synthetic liquid hydrocarbons is still undergoing large-scale testing.

Carbon capture and utilisation in combination with hydrogen-based synthetic fuels made using renewable energy is expected to make CO_2 recycling possible. Hydrotreating CO_2 for methanol generation is a global research hotspot; catalysts and different technical approaches have been developed and a number of demonstration plants built. In addition, breakthroughs have been made in CO_2 and hydrogen synthesis for petrol production, and China is the first country to carry out pilot-scale mass production. Once renewables-based electrolytic hydrogen production that uses the CO_2 generated by fuel consumption and recycles it into synthetic fuels becomes a reality, it will reduce the time needed to achieve carbon neutrality.

11 A Three-point Framework for the New Energy System

The global energy transition towards a low-carbon, zero-carbon or even negative-carbon energy system is well under way. Renewables-based new energy is shifting rapidly from a supplementary energy form to the predominant energy form. China holds huge potential in making the switch to new energy carriers. In view of this, China needs to make long-term and short-term plans and endeavour to make new energy and the new economy the engine for building a modern socialist and carbon-neutral country.

11.1 Plan and Move Early

The goal of carbon neutrality is clear, but the pathway that leads to it is not. China should follow the principle of "long-term vision, medium- and long-term strategies and short-term plans" and devise concepts, blueprints and an implementation schedule to advance the energy transition and new energy development in a sound and orderly manner. First, the guidelines for achieving carbon neutrality and the energy transition should be studied. Second, China should implement its energy revolution strategy of "Four Revolutions and One Co-operation" in energy consumption, supply, technology, the energy system as a whole and international co-operation, and set targets to be achieved in each decade. Third, a "Five-year plan for energy development" should be created to identify targets, solutions and their implementation. Fourth, plans for new energy development should be drawn up for land use, power grids, energy storage, hydrogen, carbon capture (utilisation) and storage, and nuclear energy to provide guidance and stakeholder co-ordination.

11.2 Three Tactics

Energy efficiency and recycling. Energy efficiency is the "first fuel" and recycling the "first resource". Reducing energy demand is a prerequisite for carbon neutrality. It should be implemented across the economy, society and all sectors. Importance should be attached to reducing demand for energy and primary resources through the creation of a circular economy, especially in industries like steel, copper, aluminium and municipal waste. For example, a high proportion of electric furnace steel relies on the use of steel scrap.

Substitution of electricity for fossil fuels. Electrification is the key to a carbon-neutral world. China should accelerate the electrification of end uses across sectors. In transport, the country should continue its efforts to increase the uptake of electric vehicles, which will also help reduce dependence on imported oil. In buildings, China should promote electric cooking and heating, and consider exploring a new model of 100% electrification with distributed green electricity through pilot zero-carbon community projects. In industry, the deployment of electric boilers should be increased, and policies and actions should be tailored to the specific characteristics of each sector.

Energy decarbonisation. Decarbonisation is the starting point for carbon neutrality. Carbon neutrality is not just about renewable energy, but about removing carbon dioxide emissions and other greenhouse gases from the atmosphere. Nuclear power, bioenergy and thermal power with carbon capture, utilisation and storage (CCUS) are all possible alternatives, depending on cost, technology and supply security. Wind and solar power will likely dominate the power mix by 2030, but there are risks and challenges in relying on wind and solar power to achieve carbon neutrality. Resolving these risks and challenges requires the right types of technology, business model and infrastructure.

11.3 Five Enablers

Innovation. According to the International Energy Agency's 2021 report, Net Zero by 2050: A Roadmap for the Global Energy Sector, to achieve net-zero emissions by 2050, almost half the reductions will come from technologies that are currently at the demonstration or prototype phase. Countries are now in a race to achieve leadership in low-carbon technologies. China should also take part and define policies for these technologies. In those instances where China is behind the leaders—which includes nuclear energy, hydrogen, bioenergy, smart grids and CCUS—China should improve benchmarking and international co-operation to catch up as quickly as possible. For the technologies where China is abreast with or ahead of international competitors—information and communication technologies, solar photovoltaic power, batteries and buildings—it should roll out those technologies at scale and use the market to drive progress. For cutting-edge and conceptual technologies—such as hydrogen-based chemicals, the use of CO_2 as a raw material and hydrogen–carbon synthesis for methanol and ammonia—resources should be unlocked nationwide and scientific research intensified to aim for global leadership.

In addition, China should strengthen its planning of energy management and build a multi-level scientific research support system that identifies clear pathways forward. On the first level, China should rely mainly on colleges and universities to lay a solid foundation and focus on basic scientific research. On the second level, the country should consider how to engage national laboratories, focus on the translation of theories into reality, and seek breakthroughs in production processes, equipment design and manufacturing that result in new products and systems. On the third level, leading enterprises should establish innovation consortia and accelerate the application and industrial development of new technologies by initiating and implementing major scientific and technological research projects.

Reform. Electricity pricing and carbon pricing mechanisms should be supported. The key to stimulating the market lies in the design of the pricing mechanism.

First, China should: (i) deepen its reform of the power system and improve the electricity pricing mechanism; (ii) adjust and optimise feed-in tariffs, transmission and distribution fees, and selling prices; and (iii) create favourable conditions to accelerate electrification. The key tasks of the feed-in tariff reform include: establishing a thermal power capacity cost compensation mechanism and a new energy feed-in tariff adjustment mechanism; building an electricity spot market; and improving the auxiliary services market and storage pricing mechanism. The key priorities in the reform of transmission and distribution fees include refining cost management, gradual phase-out of cross-subsidies, introduction of time-of-use transmission and distribution fees and incentive pricing, and establishing a capacity price mechanism that allows interaction between large power grids and distributed power networks. The key points in the reform of electricity selling price include optimising the mechanisms for time-of-use electricity and tiered pricing for household electricity use; building a diverse electricity price system; allocating fairly the cost of electricity connection and consumption and refining the inter-provincial and inter-regional electricity pricing mechanism.

Second, China should step up efforts to improve the carbon market and encourage society to support carbon abatement. Conditions should be created to (i) shift gradually from a free carbon emissions allowance to competitive auctions to stimulate market vitality; and (ii) establish a carbon price stabilisation mechanism and a carbon credit storage mechanism to provide stable incentives for low-carbon investment. Efforts should be made to guide fair and international market-indexed carbon pricing and avoid possible barriers to green trade. As the carbon price becomes increasingly objective and reasonable, the option of combining carbon pricing with CCUS, green electricity certificates and other carbon reduction schemes should be considered.

Third, China should link carbon emission trading schemes with electricity trading. For thermal power projects, the carbon abatement cost of participating in carbon market trading should be reflected in the feed-in tariff, and the supply security provided by thermal power plants in the power system should be factored in through capacity pricing.

Investment. China should plan infrastructure investment and construction before they are needed. Reshaping the energy system will generate new value. Energy infrastructure construction should follow three principles.

First, it should facilitate a smooth energy transition. The continued availability of conventional energy sources such as coal, oil, natural gas, thermal power, hydropower and nuclear power needs to be planned to ensure supply security and reliability.

Second, plans for new investment should cover the new energy infrastructure needed, as well as milestones, investment cost and supporting policies. The required infrastructure includes wind and solar power, smart and resilient grids, distributed energy systems, energy storage facilities, hydrogen production from renewables, bioenergy, CCUS and electric vehicle charging systems. The whole of society should be encouraged to actively participate in the building of a new energy system and a new economy.

Third, social change is needed to achieve carbon neutrality. Revolutions in energy, buildings and transport, as well as changes in how society produces things and lives, are also needed. It is therefore necessary to speed up the transition to a low-carbon society and increase efforts to deploy buildings of ultralow energy consumption, zero-carbon factories and other new infrastructure of the future. China should continue to follow the "government-guided and market-led" approach and encourage the active participation of private capital.

Co-ordination. Better regional co-ordination and international co-operation are needed. To achieve carbon neutrality, China needs pacesetters and requires all regions to be aligned with the goal. Resource conditions decide how easy or difficult it will be for different regions to achieve carbon neutrality. The regional development of new energy and the new economy should be aligned with national strategies and plans, and should avoid rushing blindly at new energy projects and generating unreasonable competition between regions to attract investment. China should adapt new energy to local conditions and meet the basic requirement of ensuring energy security. Priority should be given to the development of large new energy power generation clusters and high-quality inter-regional transmission systems in robust power grids. Efforts should be made to develop distributed energy systems for local grid connection and for energy storage and use in communities and buildings. Because new energy systems are dominated by uncontrollable intermittent power sources, base load generating facilities should be retained to ensure basic supply security. Sudden permanent shutdowns of fossil energy generation should be avoided.

At the same time, China should use its market advantages and combine leading technologies with its Going Global Strategy to strengthen international co-operation on energy and industrial progress.

First, China should focus on technological co-operation with Europe, Japan and South Korea and promote industrial development by opening up its market to the outside world, strengthening patent protection and jointly building R&D platforms to progress technological development.

Second, China should advocate a green Belt and Road Initiative and help those countries and regions taking part to achieve green and lower-carbon development by using China's low-cost, low-carbon products, technologies and solutions. This would expand China's market share for low-carbon products and services while advancing the global energy transition.

Rule of law. Energy legislation should be improved. With control of carbon emissions as the goal, China should speed up the introduction of a regulatory framework to help achieve carbon neutrality.

First, China should enact laws as soon as possible to provide a legal foundation that furthers carbon neutrality. Those laws should be related to existing laws on energy conservation, renewable energy, coal and other associated legislation.

Second, research on carbon accounting, carbon footprints, carbon standards, carbon assessments and carbon certification should be conducted.

Third, using its expertise in energy efficiency, China should design a mechanism to set and cascade emission targets for the regions. Emissions which are already covered by the emissions trading system should not be included in the targets. Targets should be related to emissions and emissions intensity rather than energy use and energy intensity. To improve energy security the country needs to address all aspects of the value chain including resource acquisition, development, use and international trading.

Chapter 1: An Economic, Social and Energy Blueprint for Carbon Neutrality

Zhaoyuan Xu, Xiurong Hu
and Georgios Bonias

1 Introduction

China's economy is growing steadily. By 2035, the country's per capita GDP is expected to reach the level of a moderately developed country. By 2060, China will become a modern socialist country in all respects. Its economic growth will shift from capital-driven to innovation-driven and from investment-driven to consumption-driven. The economic structure will be continuously optimised, with the tertiary (service) sector accounting for about 70% of GDP by 2060.

China's total primary energy demand will plateau from 2030 onwards, peak around 2035 at about 6.1 billion tonnes of coal equivalent (tce) and then gradually decline to less than 5 billion tce by 2060. The country's energy intensity (energy consumption per unit of GDP) will drop by nearly 40% by 2035 and by more than 75% by 2060, compared with 2020.

A shift to clean energy supply will occur. The proportion of non-fossil energy in total primary energy demand will exceed 25% by 2030, more than 70% by 2050 and 82% by 2060. End users will switch to clean energy at a faster pace. In this setting, China's electrification rate is projected to exceed 30% by 2030, 50% by 2050 and 60% by 2060. Hydrogen will play an important role in decarbonising heavy industries and heavy-duty transport.

The goal of achieving peak carbon emissions and carbon neutrality is a major strategic decision made by the Central Committee of the Chinese Communist Party (CPC) and the State Council of the People's Republic of China after looking at the bigger picture of both China and the world. It is a solemn declaration that China is a responsible major country. Considering that China is embarking on a new journey of building a modern socialist country in all respects, the goal of carbon neutrality provides strategic guidance for accelerating economic development and restructuring. However, this is a very challenging goal, which requires systematic and long-term planning. In light of the short- and medium-term targets, transformation pathways for key sectors in different periods need to be identified to achieve green and sustainable development. Planning pathways to carbon neutrality, based on the medium- and long-term economic development targets, is therefore of importance when defining policies and measures.

Zh. Y. Xu (✉)
Department of Industry Economic Research,
Development Research Centre of the State Council,
Beijing, China

X. R. Hu
Nanjing University of Aeronautics and Astronautics,
Nanjing, China

G. Bonias
Shell International Limited, The Hague,
The Netherlands

Shell International B.V. and Development Research Centre of the State Council (DRC) of People's Republic of China, *Embracing the Future, Powering Growth: An Energy System Renewed for China*, Advances in Oil and Gas Exploration & Production,
https://doi.org/10.1007/978-3-031-73075-7_2

Using China's current carbon neutrality policy measures and targets, this chapter mainly uses the dynamic stochastic general equilibrium model to simulate the economic development, industrial structure and new energy demand that form the basis of the Carbon-neutral Scenario.

2 The Relationship Between China's Economic Development, Energy Consumption and Carbon Emissions

2.1 China Has Entered a New Stage of Development

According to the Fifth Plenary Session of the 19th CPC Central Committee, China entered a new stage of development in 2021 in which the country marches towards the Second Centenary Goal of building China into a great modern socialist country in all respects. This follows on from China achieving its First Centenary Goal of building a moderately prosperous society in all respects.[1] In 2020, China's GDP exceeded RMB 100 trillion, with per capita income going beyond $10,000 (measured in 2010 prices). This indicated a further narrowing of the gap with high-income countries and enhanced national competitiveness in the global arena.

China has more or less achieved industrialisation. The country's industrial structure is continuously optimised—industry extends to medium- and high-end value chains; the service sector accounts for an increasing share of the economy; and agriculture is progressing impressively in its modernisation. In 2020, the primary, secondary and tertiary sectors contributed 7.7%, 37.8% and 54.5% of the country's total value added respectively. China entered the later stage of industrialisation in 2011, started the second half of the later stage in 2015 and basically realised full industrialisation in 2020. The high-end and green manufacturing sectors are now key pillars of growth in industry. In recent years, China has achieved strong momentum in high-end equipment manufacturing and is increasing investment in smart industries. As a result, many intelligent manufacturing companies have been fostered and specialised intelligent manufacturing industrial clusters formed.

China is at a key stage of industrial transformation and development. Supported by a series of supply-side structural reform measures, China's real economy is continuously improving and experiencing rapid growth. Moreover, the service sector represents a rising share of the economy, while manufacturing remains strong. Service-oriented manufacturing is an important channel for higher added value, and high-tech manufacturing is growing. In addition, strategically important emerging industries and producer services are also advancing.

2.2 The Link Between Economic Growth and Higher Carbon Emissions Is Weakening

Energy is essential to production and an increase in its use is often a measure of economic development. However, as the use of energy by industry is continually changing as technologies develop and energy efficiency improves, the decoupling of economic growth from higher energy consumption is drawing near. According to data from the National Bureau of Statistics of China, China's energy consumption per unit of GDP dropped from 1.4 tce/RMB 10,000 in 2005 to 0.8 tce/RMB 10,000 in 2020, a decrease of about 43%, as shown in Fig. 1.

China's reduction in energy intensity was mainly the result of the following two factors: (i) transformation of the economic structure due to lower output by some energy-intensive industries and increased production by the tertiary sector, which reduced energy consumption per unit of GDP; and ii) energy efficiency improvements, especially reductions in energy consumption by energy-intensive sectors. As reported in the white paper "Energy in China's

[1] The New Development Stage Is an Important Stage in China's Socialist Development Course, People's Daily, January 13, 2021, page 1.

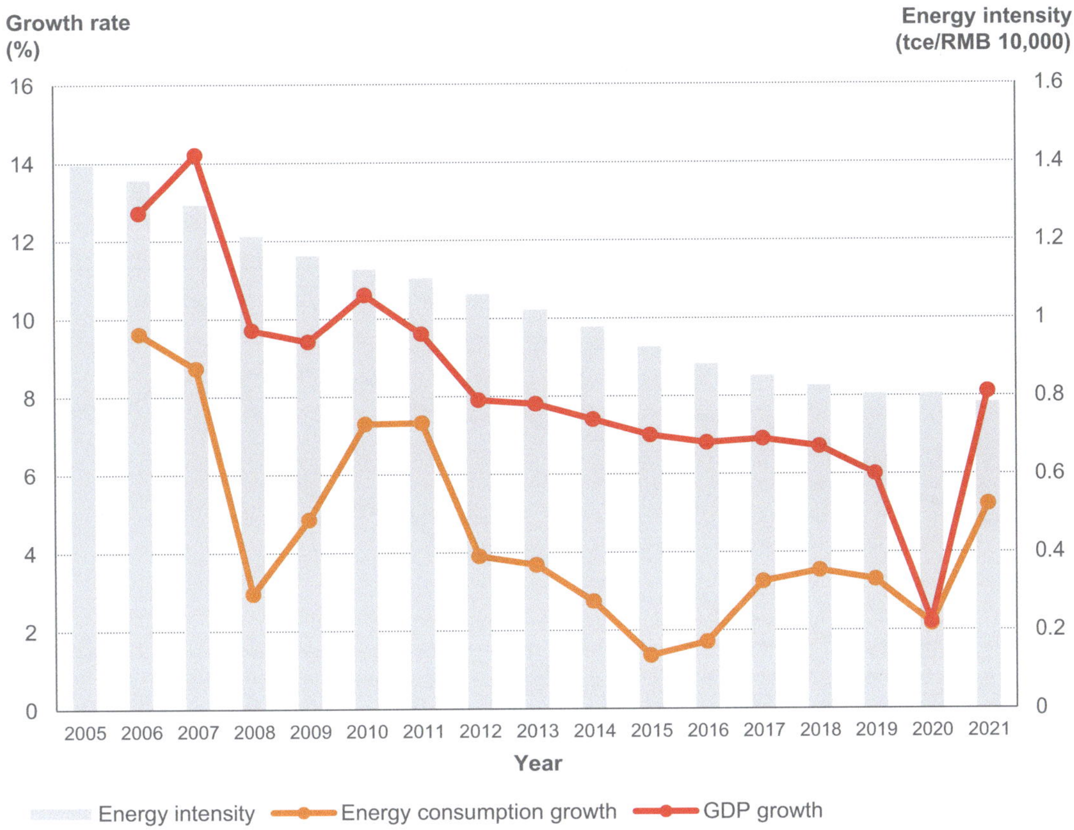

Fig. 1 China's energy consumption and GDP growth, 2005–21. *Source* Database of China Statistical Yearbook 2022

New Era", China's energy efficiency has improved substantially over the past few years. Between 2012 and 2019, average annual energy consumption grew 2.8%, which helped raise average economic growth to 7%.[2]

With the optimisation of the energy mix, China's economic and social development was significantly less dependent on carbon emissions. Between 2005 and 2019, China's GDP grew by a factor of four. In the same period, the CO_2 emissions per unit of GDP dropped by 48.1%, equivalent to a reduction of 5.6 billion tonnes of CO_2.

[2] The State Council Information Office of the People's Republic of China, Energy in China's New Era, December 2020.

2.3 Achieving Carbon Neutrality in the Post-industrialisation Era Is a Challenge

China, which is at a medium-to-high growth stage of its economy and society, is faced with a difficult challenge. China has already announced that it will achieve peak carbon emissions by 2030, then rapidly decouple carbon emissions from economic growth to reach carbon neutrality by 2060. This requires China to make greater efforts to achieve carbon neutrality than Europe and the USA (see Fig. 2). The European Union, as a well-established economy, achieved peak carbon emissions in the 1980s and 1990s, and its goal of carbon neutrality lies 50–70 years after its carbon emissions peak. China has only 30 years to transit from peak carbon emissions to carbon neutrality, which is comparatively short. This places huge demands on China for structural transformation

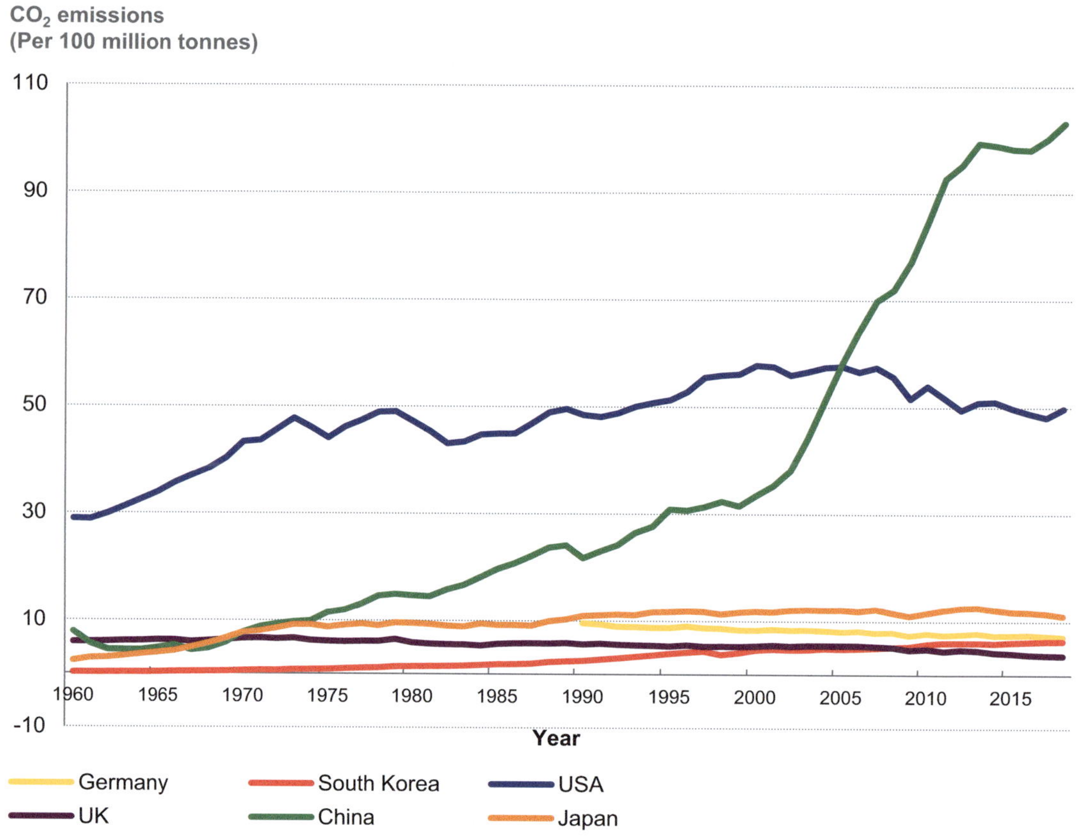

Fig. 2 CO_2 emissions in major countries, 1960–2018. *Source* China Stock Market & Accounting Research (CSMAR) Database

of its economy, technological innovation, capital investment and changes in consumption. In other words, it will be difficult for China to achieve carbon neutrality before 2060.

3 Methodology and Scenario Design

3.1 Methodology

Computable general equilibrium (CGE) models describe the interactions between energy, the environment and economy, and capture the direct and indirect effects of disruptions in the economic system. As a result, CGE has become an important modelling tool for the analysis of energy and environmental policies. In this study, the DRC–CGE model was used to simulate China's medium- and long-term economic and industrial development, reveal the interaction and feedback mechanisms between the economy and greenhouse gas emissions, and explore the new energy demand structure and possible pathways to carbon neutrality.

The DRC–CGE model is widely used in macro-policy analysis. It comprises three modules—economy, energy and carbon emissions—and analyses six types of economic participant: manufacturers, households, government, investors, exporters and inventories. It covers production factors such as capital, labour, land and other special resources. These factors are also subdivided—for example labour is subdivided into low-skilled, medium-skilled and highly skilled workers.

The model data are based on China's input-output tables in 2017 and other sources. After splitting and merging the data, the model covers 73 sectors (one for agriculture, 49 for industries, two for buildings and 21 for services). Power production is calculated for seven technologies: coal, hydropower, gas, solar photovoltaic, wind power, nuclear and biomass. The model gives a detailed description of four types of primary energy (coal, oil, natural gas and primary electricity) and eight types of end-use energy (coal, oil, natural gas, electricity, liquefied natural gas, oil products, coke and hydrogen). The relevant energy data are taken from the China Statistical Yearbook, the China Energy Statistical Yearbook and other sources. The emissions from burning fossil fuels are extracted from the International Energy Agency's World Energy Outlook 2021 and the Carbon Emission Accounts and Datasets.

3.2 Scenario Design

The goal of achieving peak carbon emissions by 2030 and carbon neutrality by 2060 has become the guiding principle of China's economic and social development. This chapter is therefore structured in line with the parameters and main policies of this dual-carbon strategy. The following parameters of China's medium- and long-term economic and industrial development are based on its economic development, changes in industrial structure and energy efficiency improvements of recent years.

First, population and labour. According to projections in the UN's World Population Prospects 2019, China's total population will peak in 2031 at about 1.464 billion and decrease to 1.333 billion by 2060. The country's workforce aged 15–64 peaked in 2013 at about 1.01 billion. An aging population will gradually reduce the workforce to about 987 million in 2030 and 749 million in 2060.

Second, total factor productivity (TFP). TFP refers to the additional output contributed by technological progress and its implementation, excluding the inputs of other factors such as capital and labour. If TFP makes a large contribution to the economic growth of a country or region, it means that the economic growth is mainly driven by technological progress. If TFP's contribution is low, it indicates that the economic growth of the country or region depends mainly on other factor inputs. Due to diminishing returns from marginal factors, technological advance is required to continuously improve factor utilisation efficiency and achieve sustained economic growth. Using the TFP survey data of the Vienna Institute for International Economic Studies, this chapter bases the TFP growth rate for 2020–60 on the TFP figures of the USA, Japan, the UK and France.

Third, household spending. Householder spending affects the demand for products and thus total energy demand. It is generally accepted that as incomes rise, expenditure on necessities like food decreases, while expenditure on entertainment, health care and other items increases.

Fourth, movements in the savings rate. At present, final consumption accounts for only about half of China's GDP, compared with 80% or more in developed economies. China is entering the later stage of industrialisation and the post-industrialisation era. The experience of developed countries shows that the proportion of final consumption in China is likely to rise gradually, while the contribution of investment to GDP will slowly dip. In the model, savings will be converted into investment and have a decisive effect on future investment and capital supply. Based on the relevant research[3] and on the savings rate of developed economies, this study assumes that the savings rate of urban residents in China will drop to 30% by 2050 and that of rural residents will fall to 30% by 2060.

Fifth, urbanisation. Urbanisation affects infrastructure investment and thus energy demand in heavy industries such as iron, steel and cement. Generally, an urbanisation rate of between 30% and 70% means rapidly increasing urbanisation. According to the findings of the

[3] Zhang, L., Brooks, R., Ding, D., Ding, H., He, H., Lu, J. and Mano, R., China's High Savings: Drivers, Prospects, and Policies, International Monetary Fund Working Paper WP/18/277, 2018.

seventh national population census of 2020, China's urbanisation rate was 63.9% at the end of 2020, up 14.2 percentage points from 49.7% in 2010; while the UN's World Population Prospects 2019 projects China's urbanisation rate to reach 83.4% by 2060. The present study predicts that China will start a new round of urbanisation in the 14th Five-year Plan for 2021–25 and reach about 73% by 2035. Growth in urbanisation will then slow down, reaching 78.2% by 2050 and more than 80% by 2060. The demand for steel and cement will contract with the slowing pace of urbanisation.

In addition to the above-mentioned variables, this chapter looks at carbon neutrality-related policies, which includes the following considerations.

China's efforts to achieve carbon neutrality are based mainly on policy and market instruments. Current measures focus primarily on the goal of achieving peak carbon emissions by 2030. The long-term objectives are as follows: by 2060, a green, low-carbon and circular economy and a clean, low-carbon, secure and efficient energy system will be fully established; energy efficiency will reach an internationally advanced level; and the proportion of non-fossil fuels in total energy use will be more than 80%.[4]

Actions to be taken include promoting a comprehensive green transformation of economic and social development; restructuring industry; accelerating the development of a clean, low-carbon, secure and efficient energy system; building a low-carbon transport system at pace; improving the quality of green and low-carbon urban and rural development; strengthening research efforts to achieve breakthroughs and wide deployment of green and low-carbon technologies; continuously consolidating and improving carbon sink capacity; opening up faster to the outside world; and taking green and low-carbon development to the next level.

The policy and mechanism instruments should achieve the following goals. First, improve investment policies. Investment in high-carbon projects, such as coal power, steel, aluminium, cement and petrochemicals will be strictly controlled and greater support will be provided to energy efficiency and environmental protection, new energy and low-carbon transport. Second, actively develop green finance. Green finance products and services and a green finance standard system will be developed. Third, refine fiscal, tax and pricing policies. Tax policies related to carbon abatement will be studied and a pricing mechanism that drives the large-scale development of renewable energy will be established. The green electricity pricing policy will be refined. Fourth, improve and optimise market-oriented mechanisms. The supporting systems for the national carbon emissions trading scheme will be further improved.

The present chapter outlines a pathway that combines the driving role of policies and market mechanisms to achieve China's dual-carbon goal.

First, energy efficiency improvements. As many energy efficiency measures in industry, buildings and transport are implemented and scaled up, overall energy efficiency will improve. The speed of improvement in energy efficiency depends on research in related fields and the costs of implementing those improvements. It is assumed that the cost of energy efficiency improvements are reflected in increased non-energy inputs.

Second, energy substitution. Substituting clean energy in energy development and electricity in energy use (the "two substitutions") at a faster pace is fundamental to achieving carbon neutrality. Clean energy substitution means replacing fossil fuels with clean energy such as solar, wind, nuclear and biomass on the energy supply side to reduce carbon emissions at the source of energy production. In this case, non-fossil energy will represent more than 80% of total energy use by 2060. Electricity substitution means replacing other fossil fuels with clean and efficient electricity in energy use. A higher rate of electrification is the key to cutting total energy

[4] Working Guidance for Carbon Dioxide Peaking and Carbon Neutrality in Full and Faithful Implementation of the New Development Philosophy, website of the CPC Central Committee Bimonthly (qstheory.cn), October 24, 2021.

Table 1 Abatement potential of CCUS in all sectors, 2025–60 (in 100 million tonnes per year)

Sector	2025	2030	2035	2040	2050	2060
Coal power	0.06	0.2	0.5–1	2–5	2–10	2–10
Gas power	0.01	0.05	0.2–1	0.2–1	0.2–3.5	0.2–3.5
Iron and steel	0.01	0.02–0.05	0.1–0.2	0.2–0.3	0.5–2.1	0.9–2.9
Cement	0.001–0.17	0.1–1.52	0.2–0.8	0.3–1.5	0.8–1.8	1.9–2.1
BECCS	0.05	0.01	0.18	0.8–1	2–5	3–7
DACCS	0	0	0.01	0.15	0.5–1	0.3–3
Petrochemicals and chemicals	0.05	0.1–0.5	0.1–0.8	–	1–1.5	1.5–2
Average	0.09–0.3	0.2–4.08	1.19–8.5	3.7–13	6–14.5	10–18.2

Source Annual report of the Chinese Academy of Environmental Planning, Ministry of Ecology and Environment; forecasts of Tsinghua University's Carbon-neutral Scenario

consumption and carbon emissions. The present chapter will make a preliminary assessment of end-use energy demand trends in various sectors.

Third, negative emission technologies. They include agroforestry carbon sinks, carbon capture, utilisation and storage (CCUS), bioenergy with carbon capture and storage (BECCS), and direct air capture with carbon storage (DACCS). A study shows that by 2060, China's forest carbon sink capacity will be between 0.75 and 1.7 billion tonnes a year.[5] This chapter focuses on the paths of abating carbon dioxide emissions from energy combustion. Emissions from industrial processes and other greenhouse gases are expected to become carbon neutral through forest carbon sequestration by 2060. The abatement potential of CCUS technology in different sectors is shown in Table 1.

Fourth, carbon pricing policy. Carbon pricing is seen as an important means to achieve deep decarbonisation in a cost-effective way. On July 16, 2021, China's national emissions trading system (ETS) came into operation. It is the world's largest ETS, covering more than 4.5 billion tonnes of annual carbon dioxide emissions from 2,162 key emitters in the power sector. Due to different assumptions in the existing studies to date, the carbon price is estimated to range from $8 per tonne to $120 per tonne when carbon emissions peak in 2030. In "Global Warming of 1.5 °C", a special report prepared by the Intergovernmental Panel on Climate Change (IPCC), to limit the global temperature rise to 2 °C, the carbon price is expected to reach $15–22 per tonne in 2030, $45–1,050 per tonne in 2050 and $420–19,300 per tonne in 2070. According to Zhang Xiliang et al., to meet the goal of carbon neutrality, China's carbon price should not be less than RMB 70 per tonne in 2025, RMB 100 per tonne in 2030, RMB 180 per tonne in 2035 and RMB 700 per tonne in 2050.[6] The present study assumes a carbon price of RMB 700 per tonne in 2060.

4 China's Economic Growth and Low-carbon Industrial Development in the Carbon-neutral Scenario

China's medium- and long-term economic growth, industrial structure, population and urbanisation rates, energy and electricity consumption and other key indicators in the coming four decades are strongly linked to how carbon neutrality can be achieved. This section looks at the population and labour projections of various organisations in China and the world, and

[5] Sustainable Development Strategy Research Group, Chinese Academy of Sciences, China Sustainability Report 2020: Pathway to Carbon Neutrality, Science Press, 2021.

[6] Zhang Xiliang et al., Pathways and Policies for Energy and Economic Transition Under the Goal of Carbon Neutrality, Management World, Issue 1, 2022.

compares China with other developed economies in terms of total factor productivity, industrial structure and other parameters to project China's economic development in the medium and long runs.

4.1 Changes in China's Economy

4.1.1 Macroeconomic Forecasts

China will see steady economic growth and will continuously optimise its economic structure. The period 2021–35 will be critical, as China's economic growth drivers will change. Average economic growth is expected to reach 5.8% in the 14th Five-year Plan (2021–25) and gradually settle in the 15th Five-year Plan (2026–30) at about 4.5%. According to estimates, by 2035, China's GDP will be RMB 202.2 trillion (measured in 2020 constant prices). When the exchange rate, price changes and other factors are taken into consideration, the country's per capita GDP is expected to reach the level of moderately developed countries in 2035.

Between 2035 and 2060, China's high-quality approach to economic development will bear fruit and sustained economic growth will result. It is projected that by 2050 and 2060, China's GDP (measured in 2020 constant prices) will be RMB 323.1 trillion and RMB 409.6 trillion respectively, and China will grow into a great modern socialist country in all respects (see Table 2).

4.1.2 From Capital-driven to Innovation-driven Growth

China's current economic growth is mainly driven by capital accumulation, especially in 2020 when the country was hit by the COVID-19 pandemic and the contribution of total factor productivity (TFP) was negative. As the population declines and capital accumulation slows down, the "catch-up" effect of TFP growth will gradually disappear. However, new technologies and new knowledge will drive up efficiency and the contribution of TFP to economic growth will steadily increase. In the report of the 20th National Congress of the Communist Party of China in 2022, "raising total factor productivity" is one of the tasks to achieve high-quality development. The present study forecasts the contribution of capital, labour and TFP to GDP growth. As shown in Fig. 3, TFP's contribution to GDP growth will steadily increase.

4.1.3 From an Investment-driven to a Consumption-driven Economy

Currently, investment and consumption are the main drivers of China's economic growth. As the country makes efforts to expand domestic demand and refine the consumption mechanism, consumption will become the main enabler of high-quality economic development in the new era.

Table 2 China's economic growth in 2020–60 in the Carbon-neutral Scenario

Year	Nominal GDP (in RMB trillion)	Actual GDP (in RMB trillion)	Per capita GDP (in RMB 10,000)
2020	101.4	101.4	7.2
2025	147.8	133.7	9.4
2030	203.0	166.4	11.7
2035	272.4	202.2	14.3
2040	359.0	241.4	17.2
2045	462.7	281.8	20.3
2050	585.7	323.1	23.7
2055	732.5	366.0	27.5
2060	905.1	409.6	31.7

Note Actual GDP and per capita GDP in the table are measured in 2020 constant prices
Source Calculations based on the models used by the project team

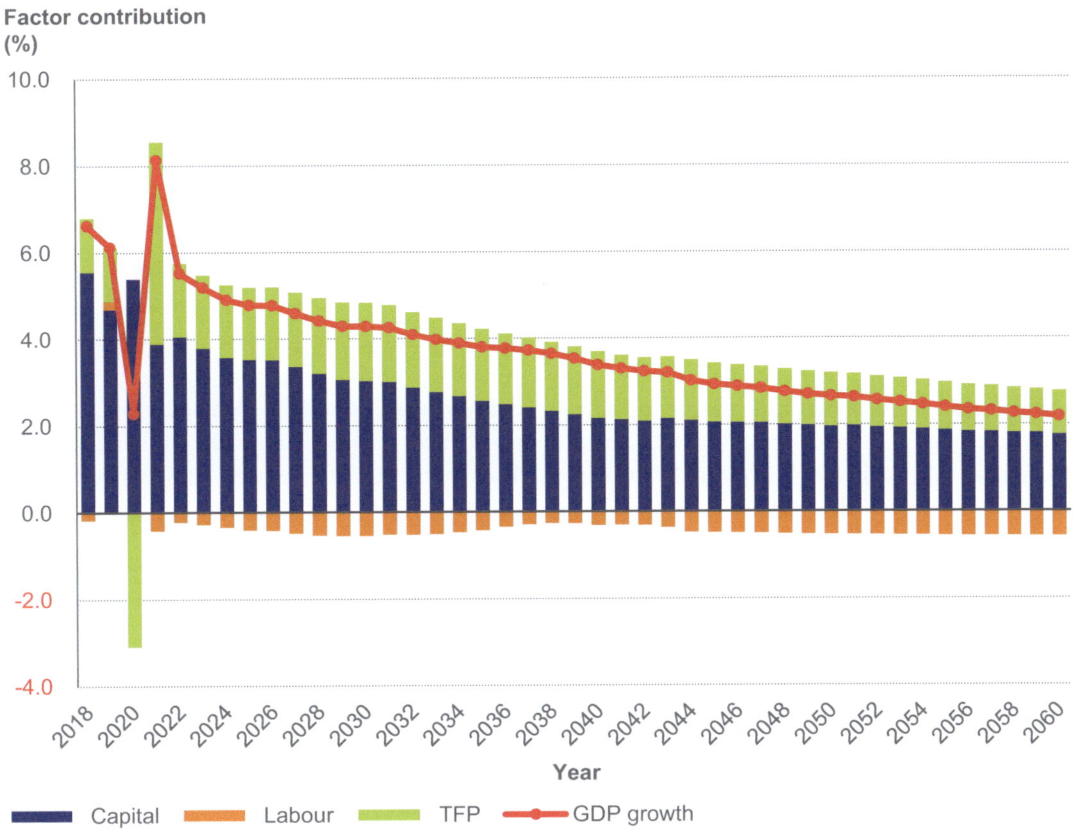

Fig. 3 Factor contribution to economic growth. *Source* Calculations based on the models used by the project team

Consumer spending represents an increasing share of China's GDP. In 2020, China's final consumption expenditure (the proportion of final consumption in GDP) registered 54.3%, far lower than that of developed countries such as the USA, the UK and France at more than 80%. China's people have a low purchasing power at present, with consumption accounting for less than 40% of the country's GDP. Hit by the COVID-19 pandemic, the proportion of consumer spending in China's GDP declined by 1.34 percentage points in 2020 from 39% in 2019.[7] In contrast, consumer spending in the USA contributed 69% to the GDP of the USA in 2020. In Japan and South Korea, the figure was around 60%. Looking ahead, China will continue to advance its new development paradigm centred on the domestic market. With further growth in average personal income and a gradual shift from an investment-driven to a consumption-driven economy, consumer spending will represent a significantly higher proportion of China's GDP, reaching an expected 44%, 47% and 50% by 2035, 2050 and 2060 respectively. The share of final consumption (including household consumption and government consumption) will be around 62% in 2060, as shown in Table 3.

The contribution of investment to economic growth is decreasing. Historically, the investment peak of industrialised countries has followed an upward trend. The USA had a peak investment rate of 20% in its initial industrialisation stage,

[7] Calculations based on the China Statistical Yearbook 2021.

Table 3 Proportion of consumption, investment and imports and exports in GDP, 2020–60 (%)

Year	2021–25	2026–30	2031–35	2036–40	2041–45	2046–50	2051–55	2056–60
Household consumption	40.9	42.7	43.9	45.1	46.2	47.4	48.6	49.8
Government consumption	14.4	13.8	13.5	13.2	12.9	12.6	12.3	12.0
Fixed capital formation	44.7	43.7	42.7	41.7	40.7	39.6	38.4	37.3
Exports	18.1	18.2	17.6	16.5	15.1	13.9	12.9	12.1
Imports	18.1	18.4	17.7	16.4	14.8	13.4	12.2	11.2

Source Calculations based on the models used by the project team

which increased to about 28% during its industrial revolution. After World War 2, Japan's investment rate peaked at 31.6% in 1973, while South Korea's peaked at 40.6% in 1996. China has been experiencing high savings for a long time, which inevitably leads to a high investment rate. In 2020, investment represented 45% of China's GDP, and in the coming decade increasing urbanisation will continue to fuel investment. As the population contracts and urbanisation slows down, long-term investment in new buildings and infrastructure will fall sharply to about 37% of GDP by 2060.

Imports and exports will make up a decreasing proportion of GDP. Export growth in the USA declined gradually after its peak in the 1970s, although it has rebounded in recent years. Japan's export growth peaked in the 1970s, then declined, but has picked up in recent years. Since China's accession to the World Trade Organization in 2001, it has maintained a high rate of export growth, although the international financial turmoil in 2008 caused this to dip significantly. It will be difficult for China to maintain its high rate of export growth due to the gradually weakening cost advantage of its manufacturing industry. As a result of higher personal income and rising consumption, China's demand for imported goods and services will keep growing steadily. The share of both imports and exports in the country's GDP will decline.

4.1.4 Significant Changes in Patterns of Consumption

Patterns of consumption are changing, shaped by rapid economic development and increasing personal income. As can be seen from the consumption data for 2013–18,[8] the proportion of food, tobacco, alcohol and clothing experienced the sharpest decline, with food, tobacco and alcohol decreasing from 31% in 2013 to 28% in 2019. Due to the lingering effects of the pandemic, spending on other categories also fell, while that on food, tobacco and alcohol rose again in 2020. In the long run, the proportion of spending on housing, transport and communications, and health care and medical services will gradually increase. By 2060, the share of food, tobacco and alcohol in total expenditure is expected to decline to 9.4%; the proportion of education, culture and recreation, health care and medical services, and personal retail financial services will gradually increase; and the share of housing (including rent, water and electricity bills, etc.) will dip slightly (see Fig. 4).

[8] The National Bureau of Statistics of China started to survey the income and consumption expenditure of urban and rural residents in 2013.

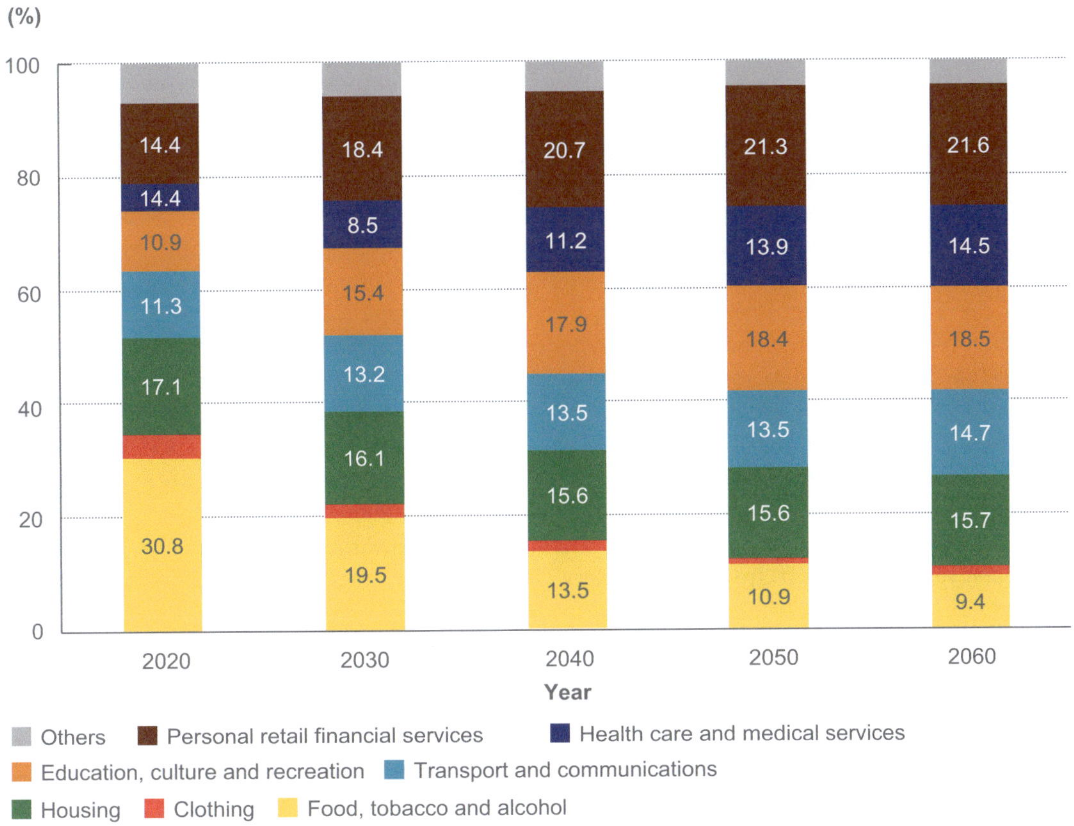

Fig. 4 Forecast of consumption patterns for 2020–60.[9] *Source* Calculations based on the models used by the project team

4.2 Optimising the Industrial Structure

China's industrial structure is continually changing. The tertiary sector is expanding, growing beyond that of the secondary sector in 2012 for the first time to become the largest economic sector. The tertiary sector made up more than half of the economy in 2015 and 54.5% of it in 2020.[10] In the secondary sector, heavy industries have shifted from extensive to intensive growth. Iron and steel, cement, building materials, chemicals, glass and other sectors have successively peaked in production. The output value of medium- and high-end manufacturing sectors represents an increasing share of the total. The contribution of China's manufacturing industry has declined since 2011, making up only 26.2% of the economy in 2020. Compared with other economies at a similar level of development, manufacturing in China accounts for a larger proportion of GDP. In recent years however, the relatively rapid decrease in manufacturing's share of GDP has become a national concern.

Between 2020 and 2035, China will be in a period of transition from late-stage industrialisation to post-industrialisation. Tertiary industry will make an increasing contribution to economic growth, of which producer services will extend to the high-end sector of the value chain and be deeply integrated with advanced manufacturing

[9] Definitions of consumption are based on the consumption expenditure categories of the National Bureau of Statistics of China.

[10] National Bureau of Statistics of China, Statistical Communiqué of the People's Republic of China on the 2020 National Economic and Social Development, 2021.

and agriculture. The tertiary sector is expected to account for 63% of the country's GDP from 2035 onwards. The share in GDP of secondary industries will steadily decrease. Manufactured goods used in buildings and service industries will decline, and personal expenditure on manufactured products will also decrease. The export of manufactured goods will likely account for a decreasing share of total final demand. By the end of the 14th Five-year Plan (2021–25), the proportion of China's manufacturing industry will decline by about 1.4 percentage points to around 24.8% compared with 2020.

Between 2035 and 2060, China will enter its post-industrialisation stage, when a transition to a service-oriented and knowledge-based economy will occur. The tertiary sector will dominate. The secondary sector will shift from a machinery-driven to an intelligent technology-driven economy, and will play a leading role in the fourth industrial revolution. Digital industries, such as intelligent manufacturing and the industrial internet, will evolve and mature. Based on the previous experience of developed countries, the primary, secondary and tertiary sectors are projected to make up 3.0%, 25.9% and 71.1% of China's GDP respectively in 2056–60, as shown in Table 4.

When per capita GDP exceeds $20,000, the share of GDP of China's secondary sector will be higher than that of Japan, the USA and Germany, but lower than that of South Korea, as shown in Fig. 5. The share of the tertiary industry shows a contrary trend.

Manufacturing technologies constantly advance. In 2021, the added value of China's high-tech manufacturing sector climbed by 18.2% on a year-on-year basis, and investment in high-tech manufacturing rose by 22.2% year-on-year. Thanks to continuous improvement in China's industrial technology and use of intelligent manufacturing, manufacturing undergoes continuous optimisation—the proportion of high-tech and medium-high-tech manufacturing sectors will continue to rise, while the share of resource-oriented, low-tech manufacturing and medium-low-tech manufacturing will gradually decline (see Fig. 6).

The output value of energy-intensive sectors will decrease (see Fig. 7). By 2060, the total output value of China's industries will increase by a factor of 3.5, compared with 2020. The output value of services, electrical equipment manufacturing, transport equipment manufacturing, commercial transport and the electric power and heat industries will more than triple by 2060. Even larger, the output value of new energy-based power generation and new-energy vehicle manufacturing will be more than 10 times higher than in 2020, representing the highest growth. Coal, oil and gas extraction, refining and cement will dip in output value. Other manufacturing sectors will double or triple their output value by 2060.

5 Pathways to Carbon Neutrality

5.1 Changes in Energy Demand and Energy Mix

China's total primary energy demand is expected to peak around 2035 at about 6.1 billion tonnes of coal equivalent (tce) and fall below 5 billion tce by 2060. Energy intensity will decrease by 75% by 2060, compared with 2020. The demand for coal, oil and gas will peak successively, and non-fossil energy will represent about 82% of

Table 4 Share of GDP of the primary, secondary and tertiary sectors in China (%)

Sector	2021–25	2026–30	2031–35	2036–40	2041–45	2046–50	2051–55	2056–60
Primary	6.1	5.8	5.2	4.7	4.1	3.6	3.4	3.0
Secondary	38.2	35.5	33.2	31.3	29.8	28.5	27.1	25.9
Tertiary	55.8	58.8	61.7	64.1	66.1	67.9	69.5	71.1

Source Calculations based on the models used by the project team

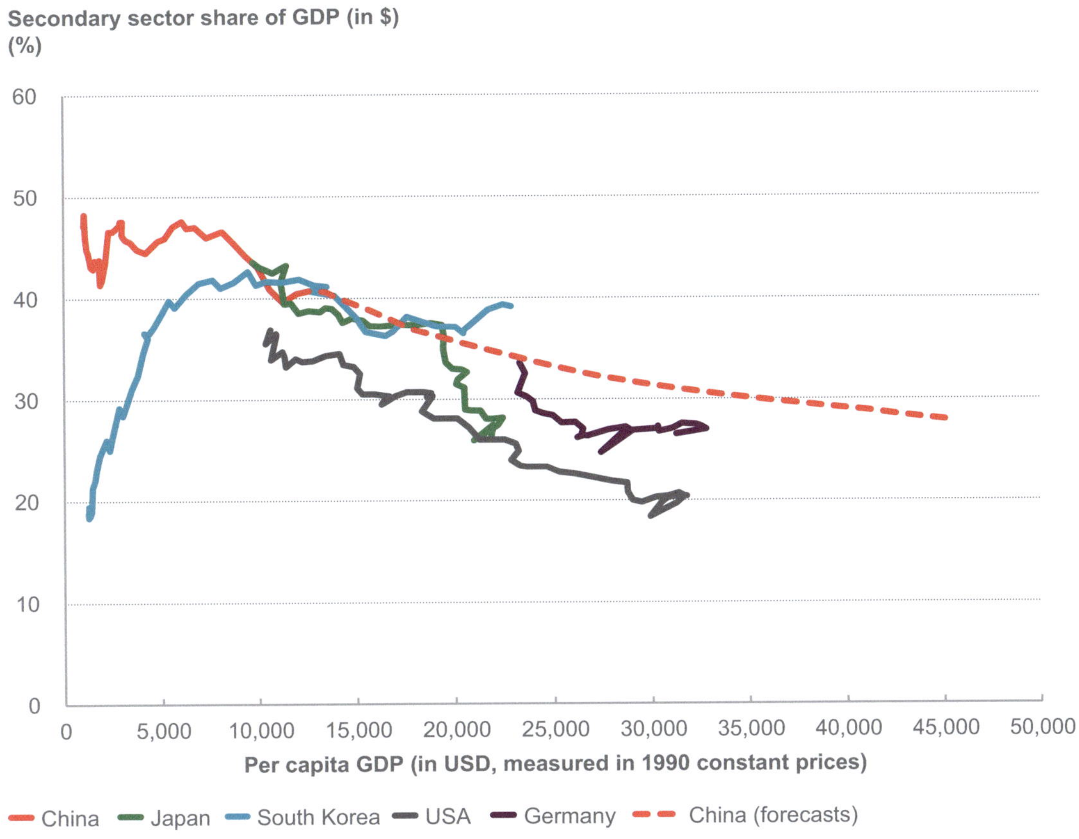

Fig. 5 Secondary sector share of GDP in China and other countries. *Source* China's forecasts are calculations based on the models used by the project team; the data of other countries are from the World Bank

total energy demand. A cocktail of intensified carbon constraints, improved energy efficiency and increased electrification will cause China's total primary energy demand to plateau from 2030 and peak around 2035. From 2035 onwards, economic growth and energy consumption will gradually decouple, and by 2060, the country's total primary energy demand will be about 4.92 billion tce. Compared with 2020, energy consumption per unit of GDP will decrease by 15% by 2025, by 68% by 2050 and by 75% by 2060, which equates to an average decline of about 3.5% a year over the four decades.

The transition to a renewables-dominated energy mix will accelerate. China's coal consumption is estimated to peak around 2025, before plunging to about 400 million tce by 2060. As electrification of the transport sector increases, oil consumption will peak between 2025 and 2030 at about 1.05 billion tce, and then fall to around 200 million tce by 2060. Natural gas consumption will peak around 2035 at 800 million tce, and then decline to about 200 million tce by 2060 (see Fig. 8). Non-fossil energy will develop rapidly, with its share of the energy mix climbing from 25.4% in 2030 to around 85% in 2060.

5.2 End-user Energy Demand

Final energy use in China will peak around 2030 at 4.3 billion tce, which will then drop to 2.8 billion tce in 2060, a decrease of 35%. End-user energy demand will gradually shift to clean energy carriers, achieving an electrification rate of more than 60% by 2060, of which about 12% will be hydrogen.

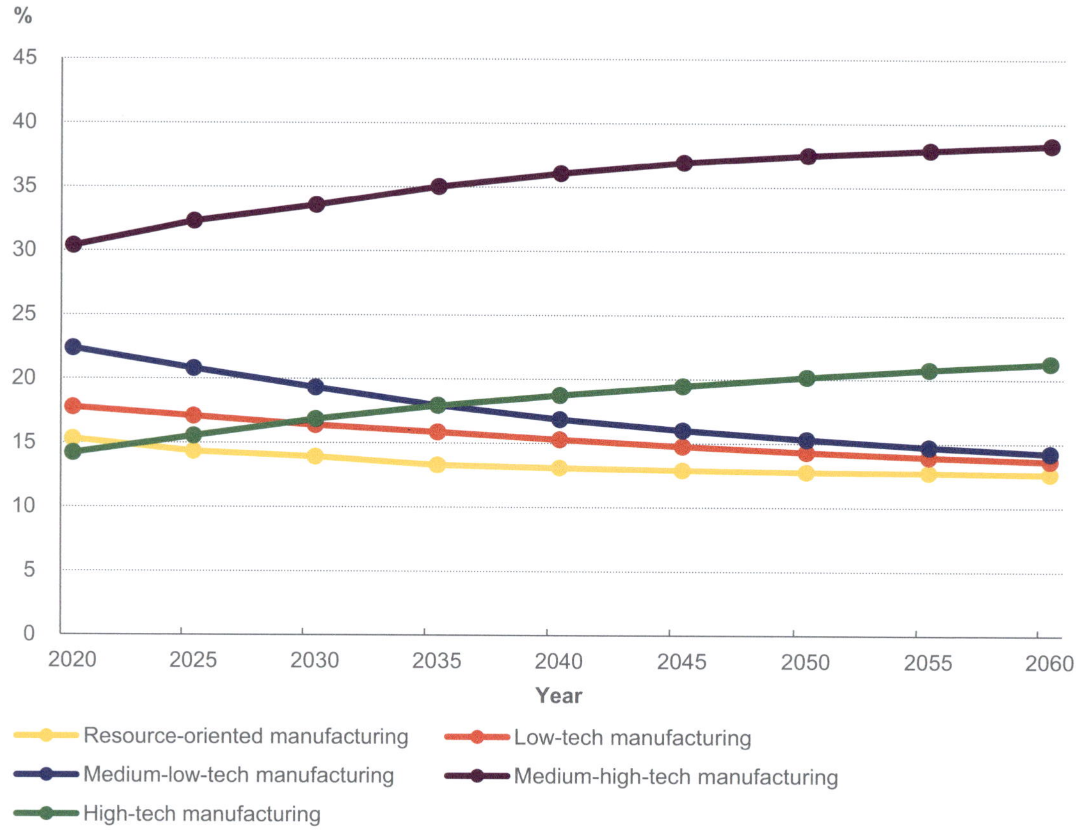

Fig. 6 Share of different sectors in the manufacturing industry, 2020–60. *Source* Calculations based on the models used by the project team

Final energy use occurs mainly in the industry, buildings and transport sectors. By 2060, industry will remain the largest energy consumer in China, representing 52% of final energy use, a decrease of 18.3 percentage points from 70.3% in 2020. Buildings' and transport's share of total final energy use will rise, with the former increasing from 20.3% in 2020 to about 30% in 2060, and the latter from 9.5% in 2020 to around 18% in 2060.

The shift by end users to clean energy will be rapid. By 2060, the average electrification rate of end-user energy demand will exceed 60% (see Fig. 9). In 2020, electricity accounted for 26.5% of China's total final energy use, with the electrification rate of four energy-intensive industries (chemicals, building materials, iron and steel, and non-ferrous metals) at 17.8%. Along with improved energy efficiency and greater use of alternative energy, 56% of industry's energy demand is expected to be electrified by 2060. The electrification rate of buildings is rapidly improving as well, reaching 44.1% during the 13th Five-year Plan (2016–20), and is expected to go beyond 70% by 2060. Transport currently has an electrification rate of about 3.7%, which is projected to exceed 60% by 2060, indicating the huge potential for electrification. Hydrogen will represent an increasing share of total energy demand as its use spreads to iron and steel, chemicals and heavy-duty road transport. The goal of carbon neutrality will drive the shift to a green and electrified modern energy system.

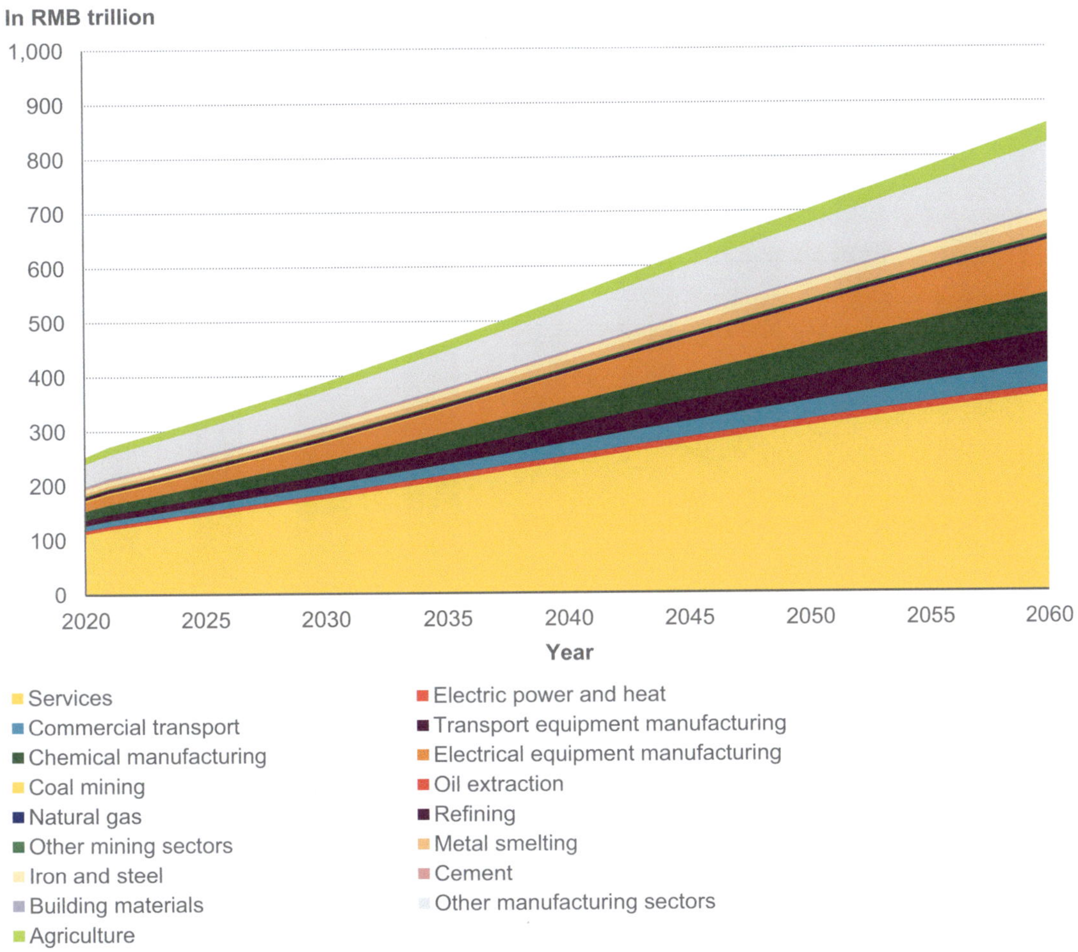

Fig. 7 Changes in total output value by sector. *Source* Calculations based on the models used by the project team

5.3 The Potential of Carbon Capture, Utilisation and Storage

Carbon capture, utilisation and storage (CCUS) will play a key role in achieving carbon neutrality. CCUS demonstration facilities are deployed mainly in thermal power. Emission reductions by CCUS in coal and gas power generation are expected to reach 14 million tonnes per year by 2030 and about 300 million tonnes per year by 2050. CCUS emissions abatement in the chemical industry is expected to grow from 89 million tonnes per year in 2035 to 225 million tonnes per year by 2060. From 2035 onwards, CCUS demonstration facilities will be deployed in the iron and steel sector, contributing an emissions abatement potential of 35 million tonnes per year, which is expected to reach about 130 million tonnes per year by 2060. CCUS in the cement sector will grow from 41 million tonnes per year by 2035 to 144 million tonnes per year by 2060. In addition, bioenergy with carbon capture and storage (BECCS) and direct air capture with carbon storage (DACCS) will need to provide about 500 million tonnes per year of emissions abatement for China to achieve carbon neutrality by 2060, as shown in Table 5.

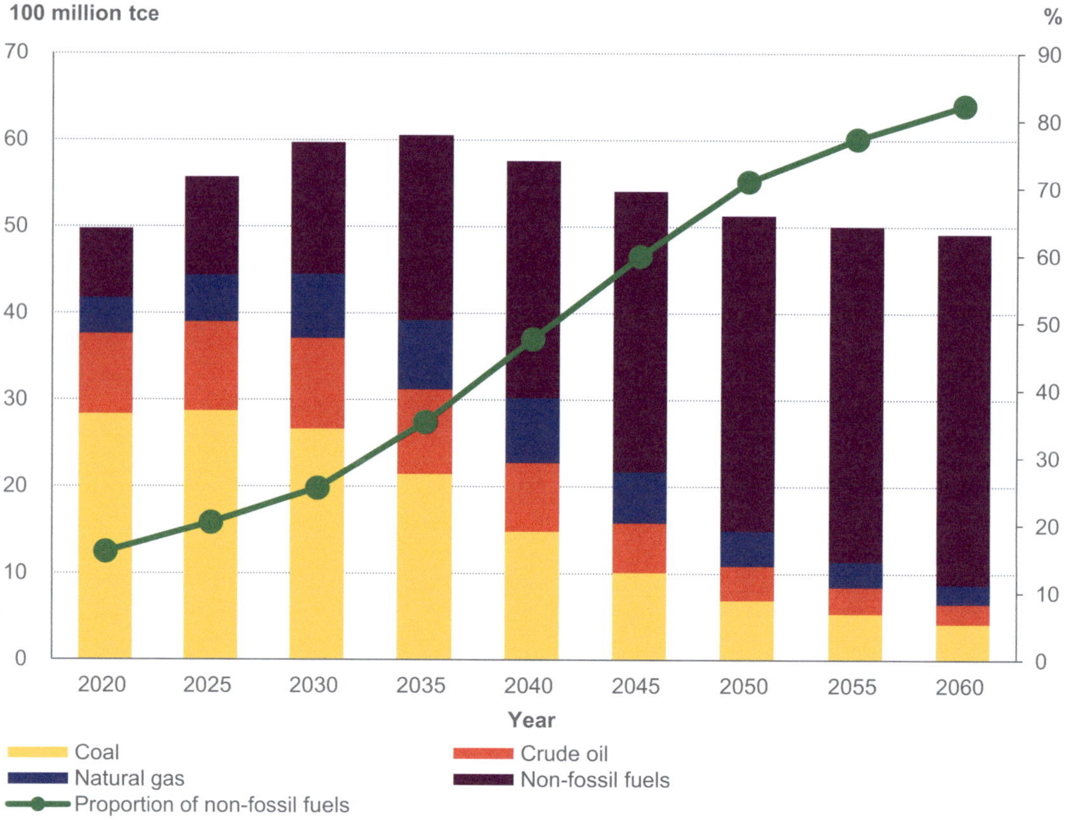

Fig. 8 Primary energy demand, 2020–60. *Source* Calculations based on the models used by the project team

6 Pathways to Carbon Neutrality in Key Sectors

6.1 Electricity

Decarbonisation of the electricity sector is the key to achieving the goals of peak carbon emissions by 2030 and carbon neutrality by 2060. To build a carbon-neutral country, the power generation mix needs to shift from high-carbon fossil fuels to zero-carbon renewables.

Electricity demand will gradually grow with rising living standards, economic growth and the electrification of industries. According to our models, electricity demand will be more than 9 trillion kilowatt-hours (kWh) by 2025, 10.5 trillion kWh by 2030 and around 18 trillion kWh by 2060 (including electricity use in hydrogen production).

Renewable energy will play a dominant role in the power mix. The proportion of renewable energy, mainly wind and solar, in China's power output will increase from 34% in 2020 to more than 90% in 2060. In view of the intermittency of renewables, it is necessary to develop complementary multi-energy and generation–grid–load–storage integration projects, gradually building a secure, cost-effective and low-carbon power system. On the one hand, the electricity sector needs to address the issue of carbon emission transfer resulting from the electrification of other sectors and applications. On the other hand, it is difficult for additional power demand to be met by non-fossil energy generation at a time of peak carbon emissions. Therefore, the electricity sector may experience a later carbon emissions peak, compared with other sectors.

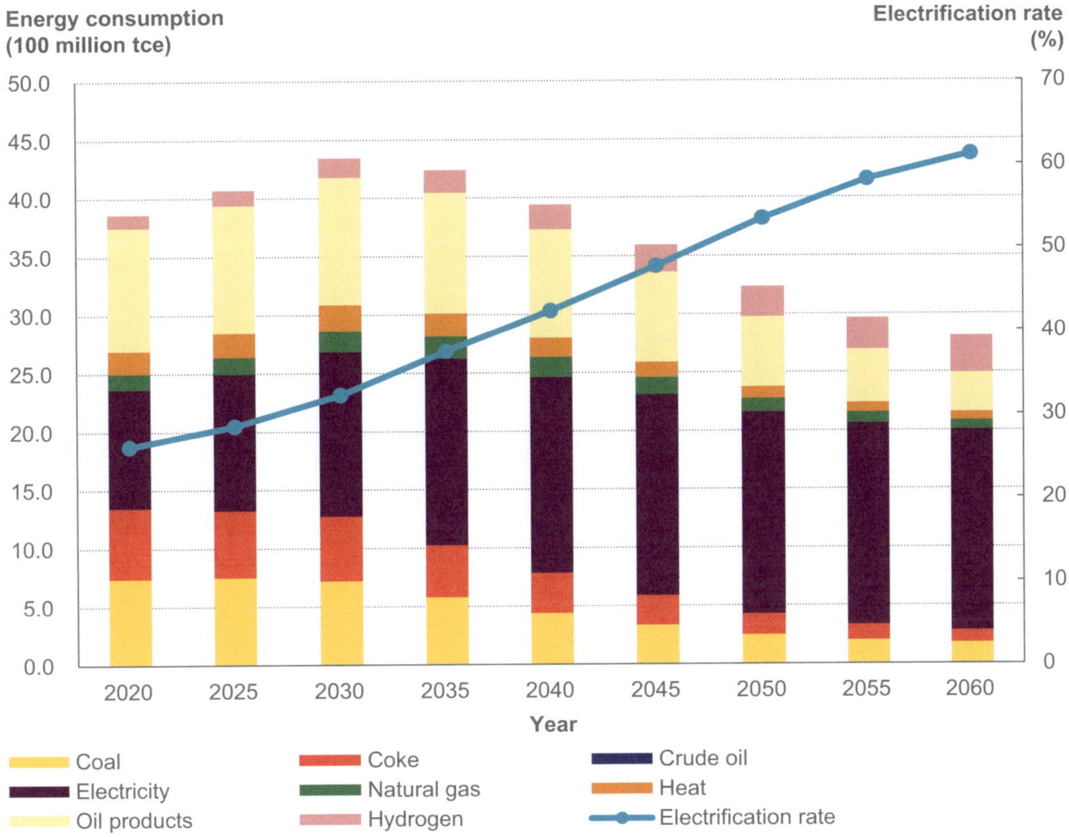

Fig. 9 End-user energy demand, 2020–60. *Source* Calculations based on the models used by the project team

Table 5 Carbon removal by CCUS in the Carbon-neutral Scenario (in million tonnes per year)

	2030	2035	2050	2060
Chemicals	0.65	0.89	1.84	2.25
Iron and steel	0.24	0.35	0.82	1.30
Cement	0.27	0.41	0.96	1.44
Electricity and heat	0.14	1.31	3.04	3.03
BECCS+DACCS	0.01	0.18	4.50	5.00

Source Calculations based on the models used by the project team

6.2 Industry

Energy consumption in industry is mainly driven by the energy-intensive sectors, including ferrous metal smelting and pressing, chemical feedstocks and chemical products, non-metallic mineral products, oil, coal and other fuel processing, and non-ferrous metal smelting and pressing. The output of the energy-intensive industries is peaking. For example, cement production has already peaked, and steel output is approaching a peak or may even have reached it. Final energy use in industry is forecast to plateau between 2028 and 2035 at 2.33 billion tce, before plunging to 1.46 billion tce by 2060, as shown in Fig. 10.

To achieve a higher electrification rate in industry, technological innovation and the electrification of more industrial processes are

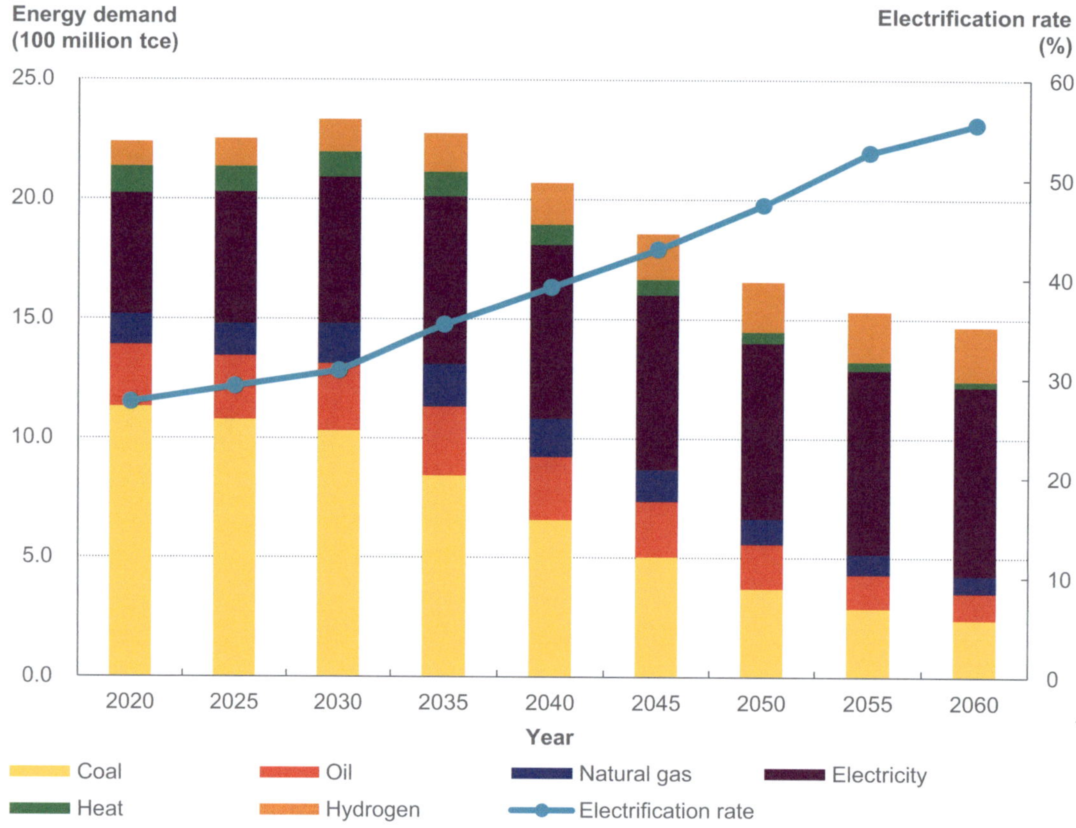

Fig. 10 Energy demand mix in industry, 2020–60. *Source* Calculations based on the models used by the project team

needed. Because the substitution of electricity for fossil fuels at scale in steel, cement, petrochemicals, glass and textiles faces many challenges, the electrification rate in industry will increase at a slower pace than other sectors, reaching about 40% by 2040. With the increasing use of hydrogen, electric boilers and other technologies, coupled with the sharp decline in demand for steel and cement and the rise of short-flow steelmaking, the electrification rate in industry will climb at a faster pace after 2040, reaching 56% by 2060 (see Fig. 10).

The paths to carbon neutrality vary across industrial sectors.

The key to decarbonising petrochemicals lies in hydrogen uptake and the substitution of electricity for fossil fuels. The energy consumption of this sector is expected to peak around 2035 at 720 million tce. Hydrogen demand in the petrochemical sector will gradually grow to 39 million tonnes by 2050 and 42 million tonnes by 2060, which is close to half of total hydrogen demand.

Short-flow processes and hydrogen-based iron reduction will play a pivotal role in decarbonising the iron and steel sector. These technologies will significantly reduce dependence on coal and coke and increase demand for electricity. According to our model projections, by 2050, steel output from electric furnaces will create demand for zero-carbon electricity of 1.9 trillion kWh, which will rise to 2.4 trillion kWh by 2060. More than half of iron and steel production will be electrified by 2060. Remaining long-flow steelmaking processes could adopt direct iron reduction by using hydrogen and carbon capture and storage. By 2060, the iron and steel sector will need about 26 million tonnes of hydrogen to support zero-carbon steel production.

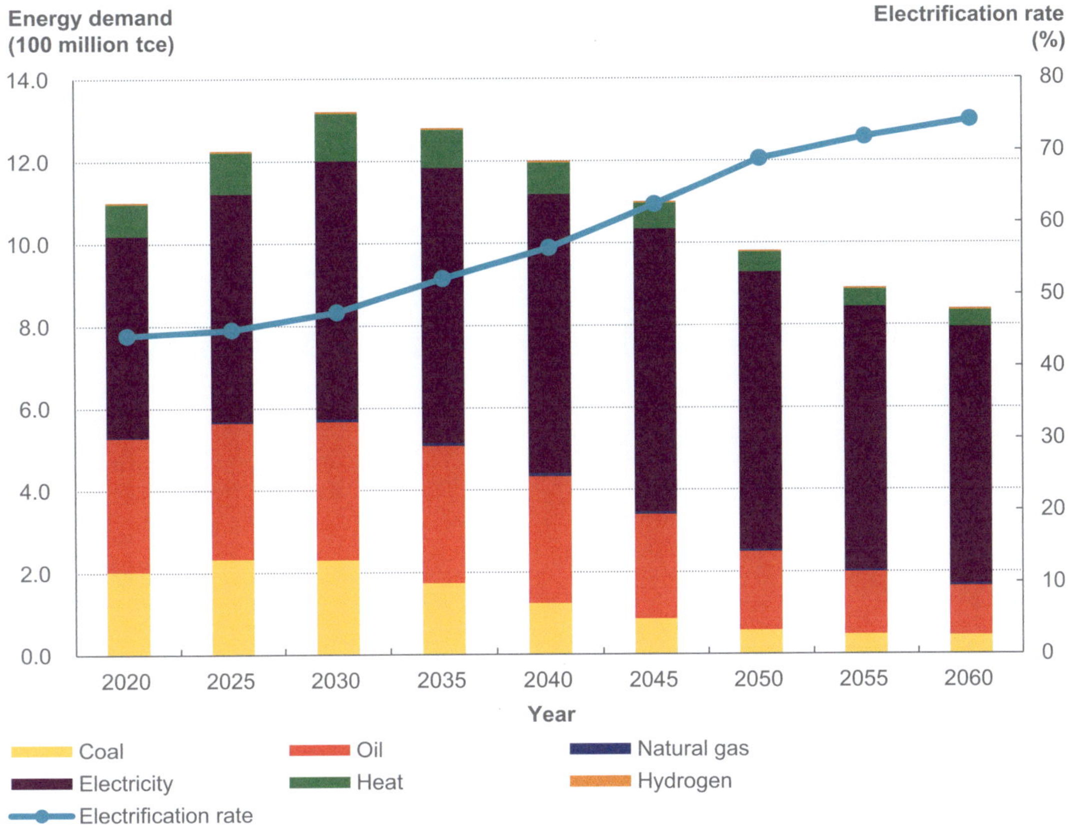

Fig. 11 Energy demand mix of the buildings sector, 2020–60.[11] *Source* Calculations based on the models used by the project team

Higher energy efficiency, carbon capture and storage (CCS), and biomass technology will be key instruments to decarbonise the cement sector. As infrastructure construction cools down, the demand for concrete and cement will gradually decline. After 40 years of technological innovation, China has become a forerunner of high levels of energy efficiency in cement production. By 2060, energy efficiency is expected to improve further and energy demand dip. According to our model analysis, energy demand in the cement sector will be 250 million, 140 million and 120 million tce by 2035, 2050 and 2060 respectively. That said, the sector will contribute 75% of China's total carbon emissions from industrial processes. CCS will therefore be needed to decarbonise the cement sector.

6.3 Buildings

Electrification and energy efficiency will shape the pace of decarbonisation of buildings. China's total building floor area will continue to grow. The floor area of urban residential buildings, rural residential buildings and public buildings is expected to increase to 47.8 billion square metres (m^2), 9.7 billion m^2 and 21.3 billion m^2 respectively by 2060. Of the total floor area of 78.8 billion m^2, about 24.2 billion m^2 has heating. Thanks to improvements in energy efficiency, the total energy consumption of buildings

[11] Building energy consumption is generally accepted internationally to mean the energy consumed during the use of civic buildings, including heating, air conditioning, hot water, cooking, lighting and household appliances. Civic buildings are divided into residential and public buildings (including commercial buildings)

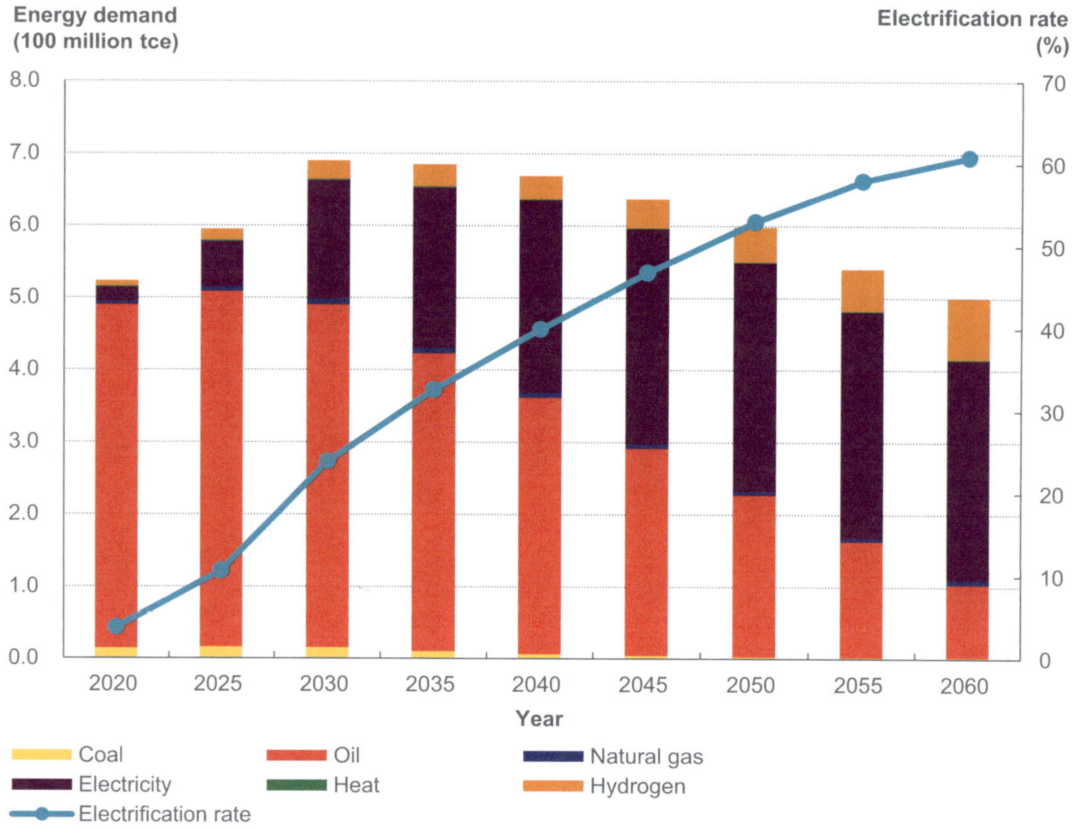

Fig. 12 Energy demand in the transport sector, 2020–60. *Source* Calculations based on the models used by the project team

in China will peak around 2030 at 1.32 billion tce and continue to decline to about 840 million tce by 2060.

The electrification of buildings will increase rapidly, reaching 73.3% by 2060 (see Fig. 11). In North China, electricity's share of urban heating, urban residential buildings (excluding heating), public buildings (excluding heating) and rural residential buildings will increase to 65%, 78%, 80% and 70% respectively.

6.4 Transport

The shift to electricity and hydrogen will be the key to decarbonising the transport sector. The measures needed include wider uptake of electric vehicles, electric buses and hydrogen-powered heavy-duty trucks, more electrified railways and the use of low-carbon alternative fuels like electricity and hydrogen in aviation and shipping. Rising numbers of new energy vehicle ownership will create additional electricity demand. As a result, total final energy use in the transport sector will peak at about 700 million tce in the period 2030–35, then gradually decrease to about 500 million tce by 2060 (see Fig. 12).

Total use of oil, currently the dominant energy type in transport, will peak between 2025 and 2030, then gradually decline. As the number of electric vehicles increases, electricity's share of total demand in transport will soar from 3.7% today to about 50% by 2050 and more than 60% by 2060, making electricity the dominant energy form in this sector. For long-distance heavy-duty road transport, hydrogen-powered trucks will be an important supplement.

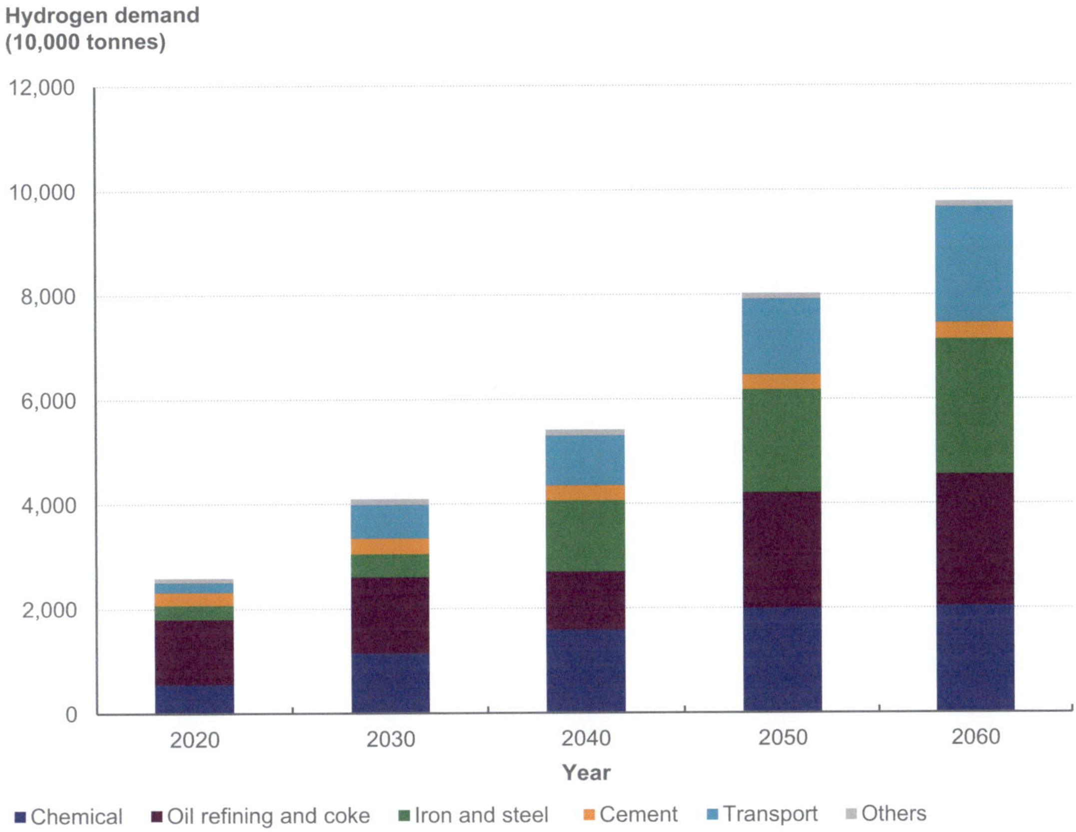

Fig. 13 Hydrogen demand of key sectors, 2020–60. *Source* Calculations based on the models used by the project team

6.5 Hydrogen

Hydrogen will play an indispensable role in hard-to-decarbonise sectors. To achieve carbon neutrality, China needs to increase annual hydrogen production from 25 million tonnes in 2020 to about 120 million tonnes by 2060.

Hydrogen will be essential to the decarbonisation of heavy industry and heavy-duty transport. It can provide a direct heat source for sectors like cement, and can be used as a reducing agent in the direct reduction technology of zero-carbon steel production. Moreover, hydrogen can be used as a raw material to produce almost all major chemical products. As the technology and input costs fall over time, hydrogen will gradually achieve cost competitiveness comparable to existing fuels. Hydrogen demand will grow significantly in the period 2040–60 (see Fig. 13). By 2060, hydrogen will account for about 12% of total final energy use (excluding non-combustible uses of fuels), about 18% of the total energy demand of industry and about 18% of transport's energy mix.

7 Recommendations

Improve Guidance on Pathways. China should implement General Secretary Xi Jinping's directives on combating climate change and promoting energy development. The country should strengthen national strategies and clear the path for controlling total carbon emissions.

The importance of improving energy efficiency should be recognised. Carbon neutrality hinges on making industrial process improvements and energy efficiency gains. Compared with 2020, China's energy consumption per unit of GDP will drop by about 40% by 2035, by 68% by 2050 and by more than 75% by 2060. Driven by higher energy efficiency, total primary energy demand will peak between 2030 and 2035, and decline to less than 5 billion tce by 2060.

Develop New Types of Energy. China should continue to substitute new energy for conventional energy sources, and push the proportion of non-fossil energy in total primary energy demand to more than 25% by 2030, more than 70% by 2050 and to 82% by 2060.

Accelerate the Substitution of Electricity for Fossil Fuels. Higher electrification of end-user energy demand is essential for the development of non-fossil energy and progress towards carbon neutrality. Efforts to replace fossil fuels with electricity in all sectors should be stepped up. China's overall electrification rate could then reach more than 30%, 50% and 60% by 2035, 2050 and 2060 respectively.

Support the Development of Hydrogen Technologies. China should deploy hydrogen and fuel cell technologies, which will direct renewable electricity into end-use sectors such as transport, industry and buildings for deep decarbonisation. By 2060, hydrogen supply is expected to exceed 94 million tonnes, mainly of decarbonised hydrogen.

Invest in CCUS and Other Carbon-removal Technologies. Carbon capture, utilisation and storage (CCUS) allows fossil fuels to remain an important energy option. It can help the deep decarbonisation of such sectors as cement and iron and steel. According to our models, in 2060, some carbon emissions will still need to be removed by CCUS and other technologies. For this reason, China should act early in the research and development, demonstration and large-scale roll-out of such technologies.

Keep Abreast of Game-changing Technologies. According to Net Zero by 2050: A Roadmap for the Global Energy Sector, a report released by the International Energy Agency in 2021, to achieve net-zero emissions by 2050, almost half the reductions will come from technologies that are currently at the demonstration or prototype phase. China should therefore focus on innovation and R&D in these technologies, especially small controlled nuclear fusion plants, CCUS and new techniques in production and energy efficiency.

Chapter 2: Industry

Ling Cheng and Georgios Bonias

1 Introduction

Coal dominates the industrial fuel mix and is the largest emitter of greenhouse gases. Currently, industry's share of total energy use and the proportion of coal in the industrial energy mix are decreasing, but their dominance in energy remains. Industry contributes 62% of China's total greenhouse gas emissions. If the electricity sector is excluded, iron and steel, building materials and chemicals represent the three main industry emitters.

Carbon capture and storage is still technologically immature. To decarbonise industry, the most important enablers include energy efficiency and the shift from fossil fuels to electricity, hydrogen, sustainably sourced biomass and other low-carbon alternative fuels. Industrial electrification will be the key to decarbonisation.

The low-carbon transformation of industry needs to be strongly supported by government policies. Despite the barriers to industrial decarbonisation—such as technological immaturity, excessively high costs and long payback period—China's large domestic market provides opportunities to build low-carbon competitiveness through economies of scale. The government should therefore take multi-level policy measures and promote the development and uptake of decarbonisation technologies at scale.

Industry, as a pillar of the national economy and a major driver of growth, represents the largest energy consumer and greenhouse gas emitter in China. Due to technological immaturity, high costs, imperfect policies, absent technology standards and other barriers, it is difficult to decarbonise industry. To help China meet the goal of carbon neutrality by 2060, focus should be directed at research on industrial decarbonisation technologies and the shift to clean and low-carbon electrification.

2 Industrial Decarbonisation

2.1 Industrial Carbon Emissions

Industry emitted 6.236 billion tonnes or 62%[1] —the largest proportion—of the total carbon emissions of China in 2021. Industry is the hardest-to-abate sector, and most countries have achieved limited reductions in industrial emissions. Between 1990 and 2018, the carbon intensity of industrial GDP in the USA, the UK, France and

L. Cheng (✉)
China Electric Power Research Institute, Beijing, China

G. Bonias
Shell International Limited, London, UK

[1] Shi Dan, and Li Peng, Simulation of industrial carbon emissions structure under the "dual carbon" goal, and policy implications, in Reform, 12, 2021.

Shell International B.V. and Development Research Centre of the State Council (DRC) of People's Republic of China, *Embracing the Future, Powering Growth: An Energy System Renewed for China*, Advances in Oil and Gas Exploration & Production,
https://doi.org/10.1007/978-3-031-73075-7_3

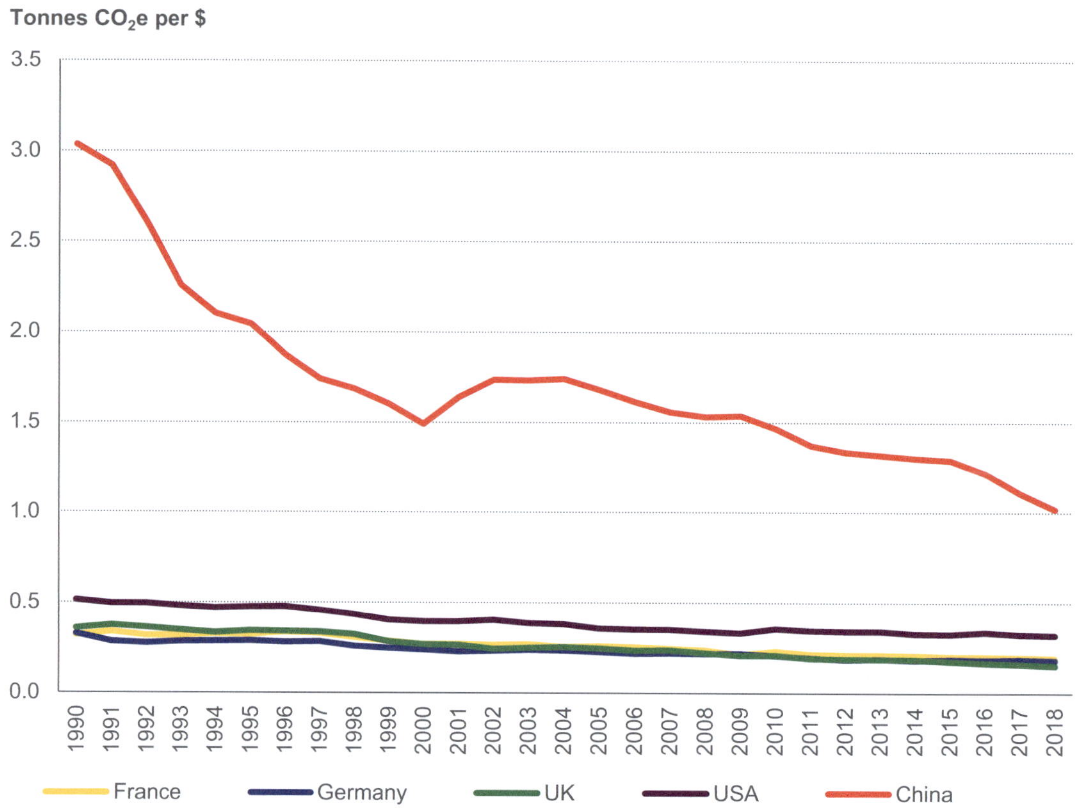

Fig. 1 Carbon intensity of industrial GDP in major countries. *Source* United Nations Framework Convention on Climate Change; World Bank

Germany did not decline greatly. In contrast, the carbon intensity of industrial GDP in China decreased by two-thirds (see Fig. 1). Despite the drop in carbon intensity between 1990 and 2018, because of the high growth rate of China's industrial production, total carbon emissions increased by 3.4 billion tonnes. Reaching net-zero emissions will require a much faster reduction in the carbon intensity of industrial GDP.

Iron and steel, cement and chemicals are important to China's economy, but they are also responsible for a large amount of industrial emissions. These three energy-intensive sectors emit high levels of pollution. In 2019, they were responsible for 12%, 11% and 6% of China's total carbon emissions respectively. In iron and steel, for example, coal dominated the energy mix at 87%. This excessive dependence on a fossil fuel led to a low level of energy efficiency in steelmaking. In 2018, China's major steelmakers used 555 kg of coal equivalent per tonne of steel produced, far higher than that of Germany at 251 kg and the USA at 276 kg.

International examples show that deep decarbonisation of industry is not yet happening at scale anywhere. Major economies have decreased their industrial emissions somewhat since 1990, but this has largely been the result of efficiency gains and switching fuels from coal to natural gas. The UK, for instance, which has achieved the greatest decrease in industrial emissions of the countries mentioned, has reduced the share of coal in industrial energy demand from 68% in 1990 to 7% in 2018, while natural gas increased from 0% to 38%.[2] Deep

[2] International Energy Agency, World Energy Balances, 2021.

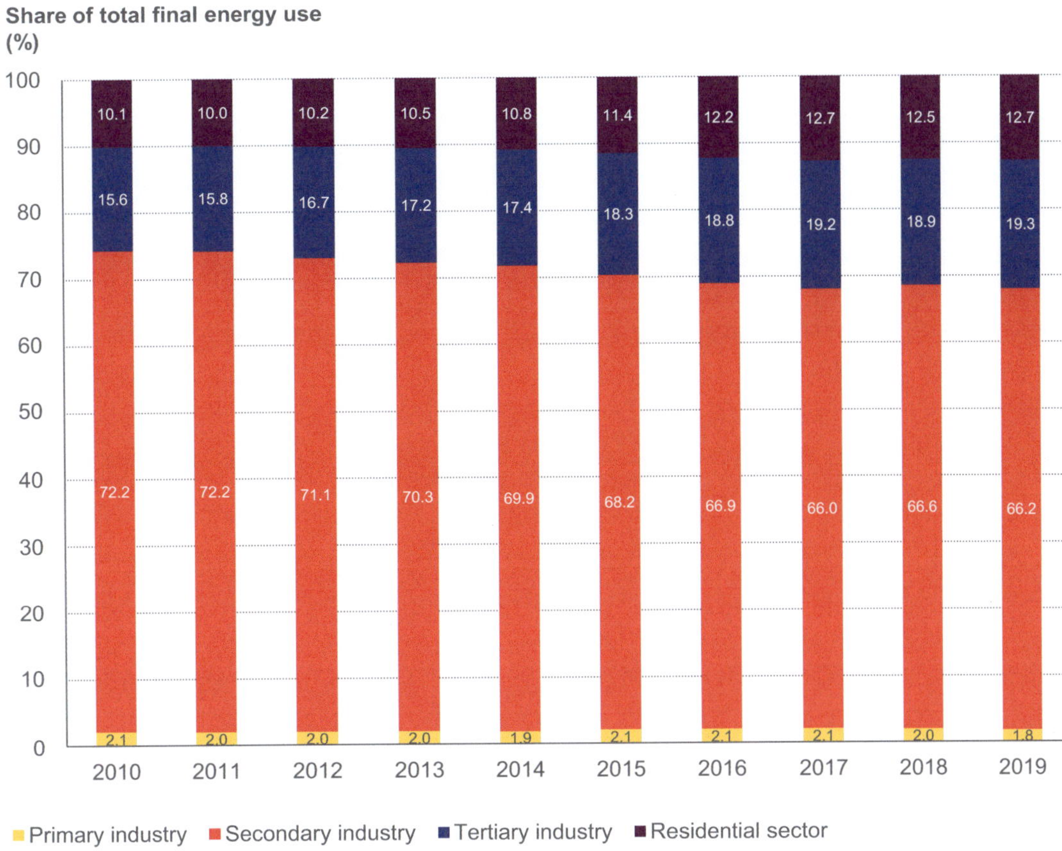

Fig. 2 China's final energy use mix, 2010–19. *Source* China Energy Statistical Yearbook 2020

decarbonisation technologies, such as carbon capture and storage and hydrogen, have not been implemented at scale.

2.2 Industrial Energy Mix

Industry accounts for the largest share of China's total final energy use. In recent years, driven by the shift to clean and low-carbon energy carriers and the energy efficiency gains from technological advances, industrial energy use fell from 72.2% in 2010 to 66.2% in 2019, a decrease of 6 percentage points (see Fig. 2). Energy use in transport and agriculture grew significantly, but from a small base. In buildings, commerce and other sectors, average annual growth in energy use remained level with that of total final energy consumption.

Coal is the dominant energy form in China's industrial fuel mix, followed by electricity. The proportion of coal in the industrial energy demand mix dipped from 54.6% in 2010 to 46.4% in 2019, indicating a gradual shift to clean energy. Electricity has become the second largest contributor to the industrial fuel mix at 25.5% in 2019. In recent years, oil use in industry declined, from 15.3% in 2010 to 13.4% in 2019. Petroleum and petroleum products are used for various purposes, mainly in chemical fibre manufacturing, petrochemical processing and transport. Demand for heat increased, mainly in textiles, chemicals and other sectors (see Fig. 3).

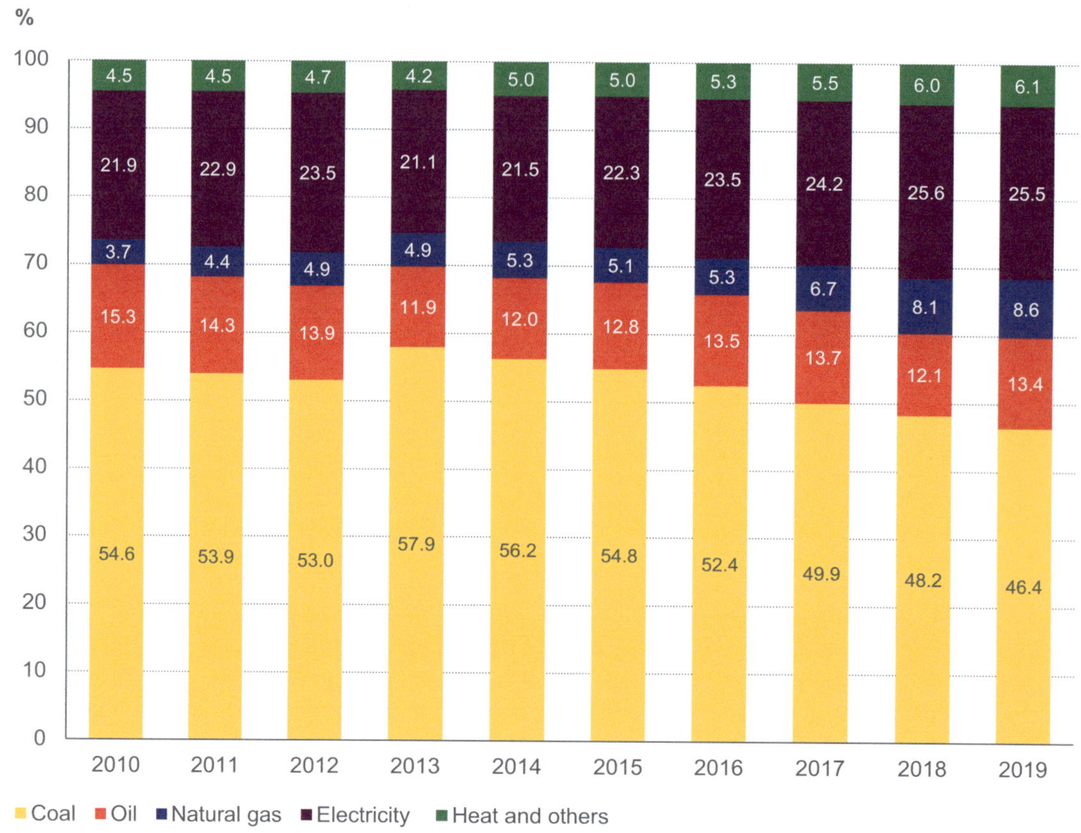

Fig. 3 China's industrial energy use mix, 2010–19. *Source* China Energy Statistical Yearbook 2020

2.3 Electrification Trends in Industry

Developed countries in Europe, America and other regions were the earliest to achieve industrialisation with a high electrification rate in the 20th century (industrial electrification rate is the proportion of electricity in the country's total industrial energy use). Developing countries are still moving towards industrialisation, while underdeveloped economies are in the early stage of this journey. In 2018, 30% of China's industrial energy demand was electrified, higher than the global average but lower than South Korea, South Africa, Japan, France, Germany and the UK. In fact, China's electrification rate was 19 percentage points lower than South Korea's, which had the highest level (see Fig. 4).

In 2019, electricity represented 19.7% of global final energy use, an average increase of 4.2 percentage points compared with 2000.[3] The global electrification rate has advanced steadily over the past 20 years after a rapid rise before 2000. The increase was generally higher in developed economies than in developing countries, but the combined effect of slowing economic growth, competition from alternative fuels, less reliance on nuclear energy and other factors drove down the growth rate.

Ferrous and non-ferrous metal smelting and pressing, chemical feedstocks and chemical product manufacturing, non-metallic mineral products, petroleum, coal and other fuel processing accounted for 76.8% of China's total industrial final energy use in 2019. These energy-

[3] International Energy Agency.

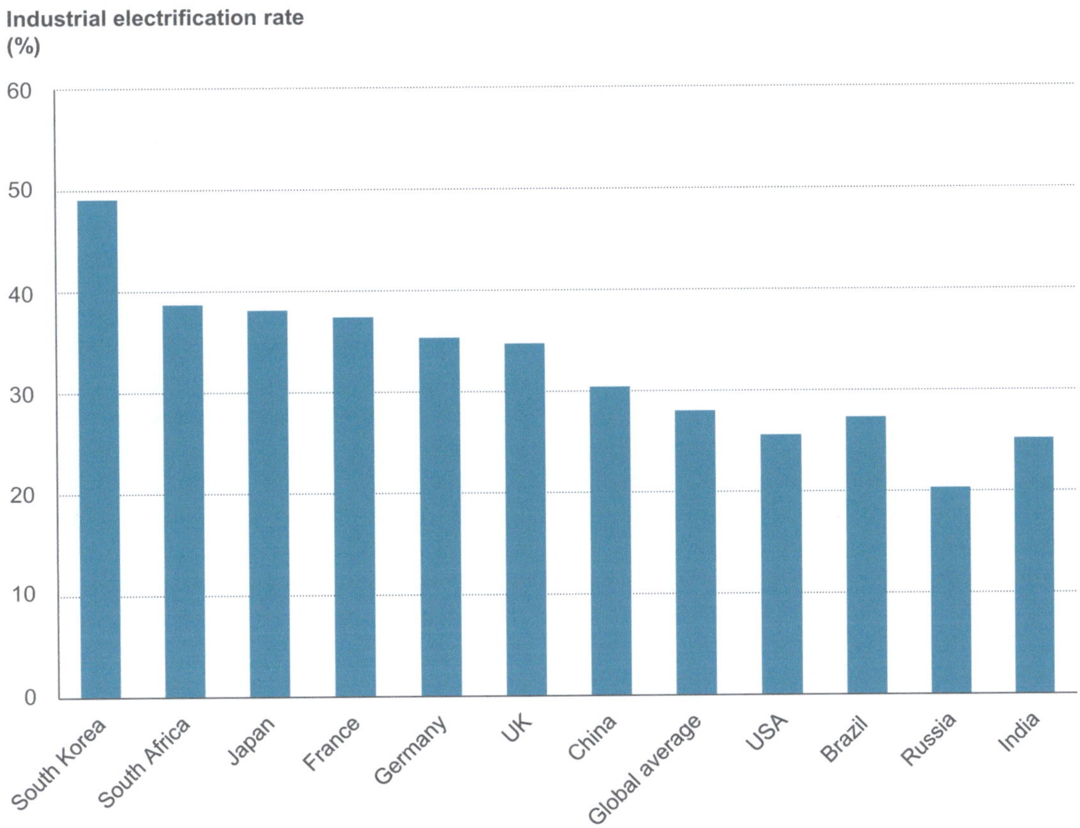

Fig. 4 Industrial electrification rate of selected countries in 2018. *Source* International Energy Agency

intensive industrial sectors, except non-ferrous metal smelting and pressing (thanks to the use of aluminium electrolysis), registered a lower electrification rate than China's industrial average.

Ferrous metal smelting and pressing increased its electrification rate from 10% in 2010 to 11.2% in 2019 as a result of environmental policies and industrial restructuring. Chemical feedstocks and chemical product manufacturing increased its energy use and electrification rate from 13.3% in 2010 to 15.9% in 2019. In the non-metallic mineral products sector, total energy consumption declined, but the electrification rate increased from 10.9% in 2010 to 16.7% in 2019. Petroleum, coal and other fuel processing industries gradually switched to clean and low-carbon energy and increased their electrification rate from 5.4% in 2010 to 6.6% in 2019. Electricity accounted for the biggest share of total energy use in non-ferrous metal smelting and pressing, rising from 53.5% in 2010 to 65.6% in 2019.[4]

3 Pathways, Challenges and Opportunities in Industrial Decarbonisation

Carbon capture and storage technology remains immature. To decarbonise industry, it is essential to improve energy efficiency and adjust the energy mix from fossil fuels to electricity, hydrogen, sustainably sourced biomass and other low-carbon fuels. Deep decarbonisation of industry continues to be a challenge, due to a combination of barriers such as technological immaturity, high costs, long payback periods and difficult supply chain co-ordination. However,

[4] China Energy Statistical Yearbook 2020.

China's large domestic market offers opportunities to build low-carbon competitiveness through economies of scale.

3.1 Technological Pathways to Industrial Decarbonisation

Industry, as the largest energy consumer, accounts for 62% of China's total greenhouse gas emissions. That in itself presents a difficult task for China's industrial decarbonisation.

Most decarbonisation technologies are commercially immature. The International Energy Agency estimates that less than 20% of emissions reduction globally by 2050 will be contributed by technologies that are already mature today.

Decarbonising China's industry poses enormous challenges for existing policies, which make the development and wider uptake of decarbonisation technologies urgent. These technologies will need to be deployed at scale in the 2030s, which will require policy changes in the short term. By 2035, 24 million tonnes of coal equivalent in the industrial energy mix should come from hydrogen.[5] This corresponds to more than 100 million tonnes of steel production or around 10% of current Chinese capacity.[6] Similarly, by 2035, 7% of the industrial production stock should be equipped with carbon capture and storage (CCS) infrastructure. This would correspond to more than 75 million tonnes of steel, which is around the current combined production capacity of Germany and Italy, the two largest European producers.[7] Immediate policy changes are necessary to reach this target, especially as the Chinese emissions trading system will not cover industrial sectors for some years. Efforts should therefore be increased to develop and widely deploy clean energy like hydrogen and CCS infrastructure.

[5] Shell Scenarios Sketch, Achieving a Carbon-neutral Energy System in China by 2060, 2021.

[6] World Steel Association, 2020 Steel Statistical Yearbook.

[7] Global Energy Monitor, Global Steel Plant Tracker, 2021.

3.2 Case Study 1 | UK Investment in Supporting Infrastructure

Public funding is needed to build the supporting infrastructure required to make private investment in CCS and hydrogen viable (see Fig. 5).

Public funding should support the development of CCS and hydrogen infrastructure to improve the business case for investment in facilities. CCS and hydrogen are among the most important technologies to support industrial decarbonisation, representing 37% and 11% respectively of emission reductions by 2050 in the International Energy Agency's Net Zero Scenario. Both require extensive supporting infrastructure to be viable, including storage and transport infrastructure for CCS and transport infrastructure for hydrogen. Private investment in supporting infrastructure is not viable unless it will be used by industrial companies. Similarly, investment in CCS- or hydrogen-based production by industrial companies is not viable unless supporting infrastructure exists. There is, therefore, a strong case for public support.

The UK's recent Industrial Decarbonisation Strategy announced significant investment in supporting infrastructure through several mechanisms.

First, the Carbon Capture and Storage Infrastructure Fund and the Net Zero Hydrogen Fund are to build up supporting infrastructure at a cost of GBP 1 billion and GBP 240 million respectively.

Second, the Clean Steel Fund, worth GBP 250 million, aims to support early adopters of clean steel technologies. The CCS Infrastructure Fund and Net Zero Hydrogen Fund will serve a similar purpose across other sectors, but no funding level has been indicated yet.

3.3 Case Study 2 | Carbon Contracts for Difference

Carbon contracts for difference (CfDs) complement carbon prices to support investment in low-carbon production, despite the long payback periods associated with decarbonisation technologies (see Fig. 6).

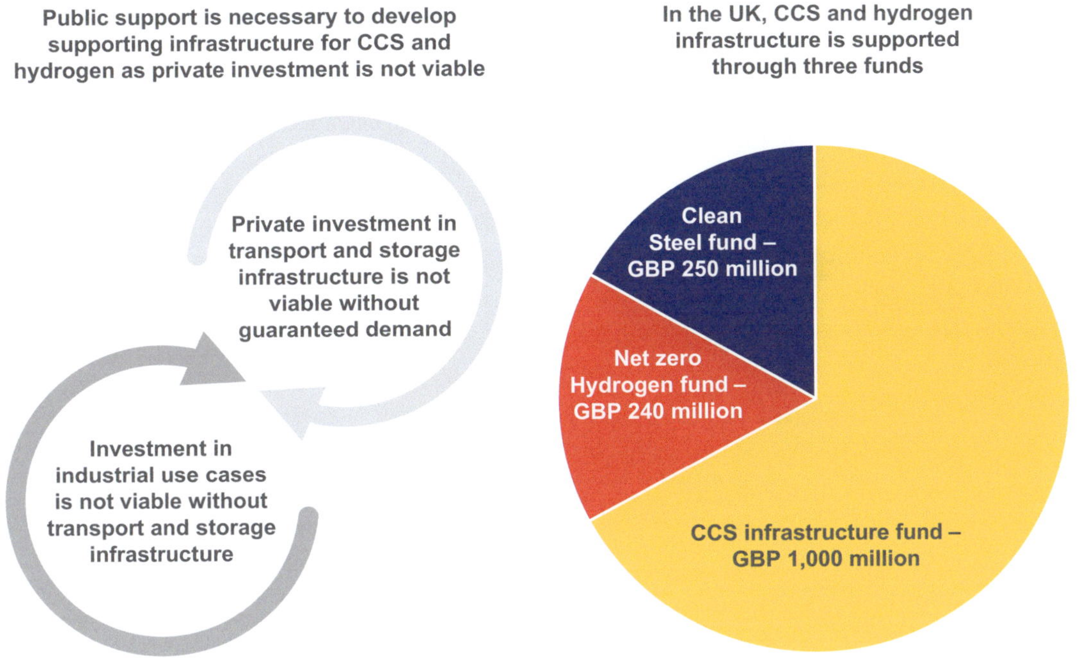

Fig. 5 Planned UK investment in supporting infrastructure for industrial decarbonisation policies. *Source* Shell–DRC project team

Carbon contracts for difference can alleviate two limitations of emission trading systems (ETS) in incentivising low-carbon industrial production.[8]

Carbon prices are too low to ensure that low-carbon production can break even with traditional methods. Production assets which are nearing the end of their useful life in Europe and the USA need to be replaced in the next decade with low-carbon assets; even the EU ETS price is projected to be too low to incentivise this by 2030. In China, where production assets are much younger, the ETS price would only need to be high enough in the next 20 years, but there is still a possibility that the price level will be too low to incentivise production on its own. CfDs can alleviate this issue by setting the price directly at the level which will ensure that companies break even from low-carbon production.

Carbon prices fluctuate, which can disincentivise investment given the long payback periods from investing in low-carbon production capacity. CfDs, in contrast, are fixed over a long term.

Carbon CfDs are gaining momentum across the European Union. The Netherlands is already trialling a form of CfD through its Sustainable Energy Production and Climate Transition Incentive Scheme (SDE++), extended from the previous SDE+ programme. The scheme has recently committed EUR 2 billion of funding for the Porthos CCS project, which will be provided as a subsidy to compensate companies for the difference between the cost of removing each tonne of CO_2 and the average EU ETS price in that year. The EU is considering introducing a CfD throughout the union to support industrial decarbonisation. This is driven by the realisation that given the need to retire many European industrial facilities, incentives for low-carbon production need to be in place as soon as possible.

CCS will probably be the key to decarbonising iron and steel, cement, chemicals and other industrial sectors, which is why it needs to be

[8] Sartor, O. and Chris Bataille, Decarbonising basic materials in Europe: How Carbon Contracts-for-Difference could help bring breakthrough technologies to market, 2019.

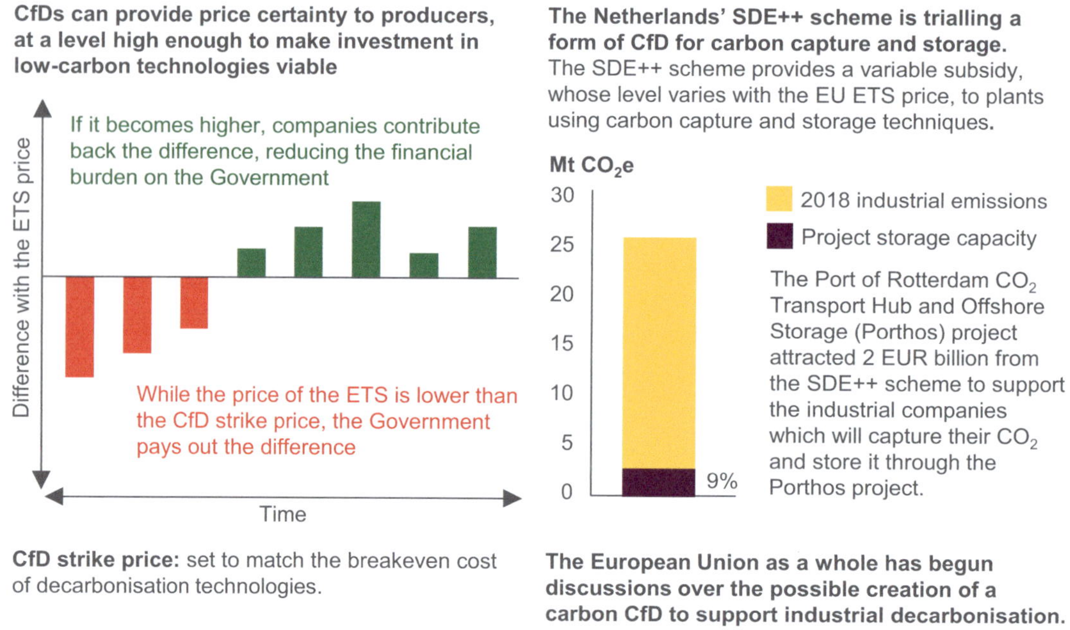

Fig. 6 Carbon contracts for difference. *Source* Shell–DRC project team

commercialised at scale as soon as possible. It is particularly important for China given the suitability of CCS for existing assets, as Chinese steel plants are only between 8 and 12 years old on average, relative to a typical 30-year lifespan, and are expected to continue to operate into the 2040s. CCS is well suited for retrofitting to existing blast furnace plants, and is currently the only option to abate process emissions from the cement and chemical sectors. Deployment should increase slowly in the 2020s, while pilot CCS plants are being monitored for technological maturity and commercialisation. The pace should then pick up in the 2030s. By 2060, almost 60% of fossil fuel-based heavy industry facilities need to be fitted with CCS (see Fig. 7).

Energy-efficient technologies can reduce carbon emissions at source. They are the most immediate and cost-effective way to abate emissions in the short term. As the world's largest developing economy, China is still progressing its industrialisation, which means demand for energy continues to grow. Energy use, especially fossil fuels, is the largest source of carbon dioxide emissions in China. Energy efficiency gains can effectively reduce industrial energy use at source, especially the consumption of fossil fuels in energy-intensive sectors, acting as a control valve for carbon emissions.

3.4 Case Study 3 | Industrial Energy Efficiency Accelerator

Public funding of innovative technologies to demonstrate their potential at scale is crucial, as high costs and technological immaturity may prevent developers from accessing private finance.

Innovation funding of low-carbon technologies has increased in the past 20 years, but more is needed given the early stage that many of these technologies are in. The International Energy Agency (IEA) estimates that public funding of low-carbon energy innovation has increased by 110% since 2000. Further innovation is necessary to achieve industrial decarbonisation, as the IEA estimates that only 17% of technologies required to achieve net-zero emissions globally in the industrial sector are currently mature. Funding

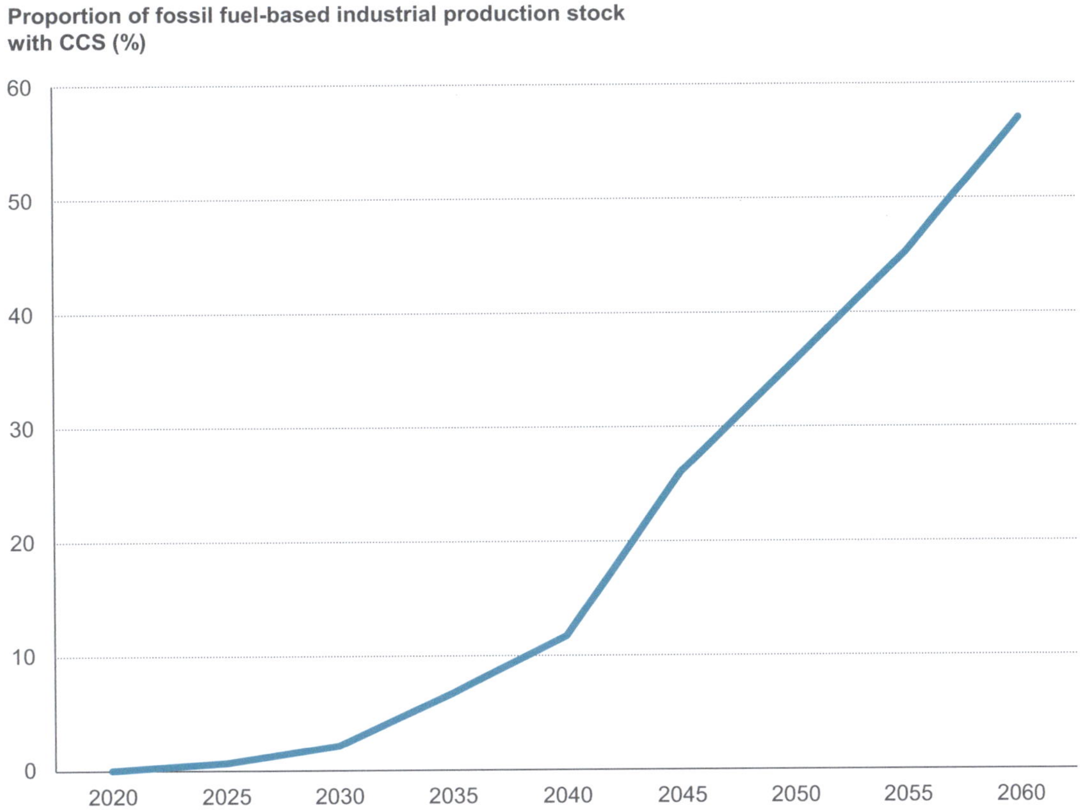

Fig. 7 CCS penetration in industry. *Source* Shell Scenarios Sketch, Achieving a Carbon-neutral Energy System in China by 2060, 2021

should support technologies through the innovation cycle to ensure that the most promising reach the market as quickly as possible.

The UK government supports the uptake of energy-efficient technologies by UK industry by enabling their demonstration at scale. Projects bring together a technology developer, who has already developed a technology, and an industrial company to test it at its production site. Such funding serves to demonstrate the suitability of technologies for specific industrial processes, as well as potentially inspiring uses in other sectors.

The programme, called the Industrial Energy Efficiency Accelerator, was launched in 2019. It supports projects in a wide range of industrial sectors.

One such project in the cement industry aims to demonstrate that multi-component cement, using a mixture of clinker, ground-granulated blast furnace slag and limestone, can achieve a higher rate of clinker substitution than single-component cements. The Mineral Products Association hopes to show that this multi-component cement could meet rigorous safety specifications and be suitable for full-scale production.[9]

A second project in fertiliser manufacturing uses a technology to produce feedstock by processing ammonia and organic waste streams from the waste water industry. The process also uses waste heat from combined heat and power plants to save energy.[10]

[9] The Industrial Energy Efficiency Accelerator, Low carbon multi-component cements for UK concrete application, 2021.

[10] The Industrial Energy Efficiency Accelerator, Improving energy efficiency in fertiliser production through wastewater treatment resource recovery, 2021.

3.5 Case Study 4 | ENERGY STAR

It is not easy for buyers of products to identify producers who have above-average energy efficiency performance. Performance labelling and certification, such as that of the ENERGY STAR programme in the USA, is one solution.

The U.S. Environmental Protection Agency (EPA) runs an energy efficiency performance certification programme for manufacturing plants as part of the wider ENERGY STAR programme. ENERGY STAR ranks production facilities using an industry-specific energy performance indicator. The top 25% of plants in each sector are awarded the certification. Plants in the cement, steel and some chemical industries are eligible.

The standard is intended to help buyers of industrial products identify the best-performing suppliers. Several public procurement programmes across the USA have committed to using ENERGY STAR qualified products. These include Washington, D.C. and Phoenix, Arizona.

The certification programme is accompanied by a partnership programme through which manufacturing companies can work with advisors to improve their energy performance. This programme also enables participating companies to share best practice and experiences, facilitating knowledge-sharing. Overall, the energy efficiency of plants in the cement sector has increased dramatically over the past 30 years, with the worst-performing plants improving the most.

End-use industrial sectors will be required to adopt decarbonisation measures to achieve the emission reductions required. These measures include material efficiency in end-use sectors through precise specification of requirements; the use of alternative, lower-carbon materials like timber in construction; and increased use of products and materials through reuse and recycling. Further improvements in energy efficiency and input use efficiency in production would contribute to emission reductions, particularly in older production facilities. In the IEA's Net Zero Scenario, material efficiency supports 20% of the total emissions reduction by 2050 (see Fig. 8).

3.6 Case Study 5 | EU Circular Economy Action Plan

Circular economy principles need to be embedded within policies in all sectors to overcome cross-sectoral barriers. Circularity requires co-ordination across supply chains to ensure that products are designed in ways that make recycling or reuse possible. As a result, policies to support circularity must touch on all stages of the supply chain and give consumers the tools and information they need to choose products according to the principles of circularity. Major economies have not yet implemented such wide-ranging policy packages, but the EU's recently published Circular Economy Action Plan is an example of how circular economy principles can be embedded across sectors. "High impact intermediary products" like steel, cement and chemicals are a focus area of the action plan.

The proposal most relevant to industrial producers is the application of circular economy principles in the assessment of permit applications for industrial facilities. The EU Industrial Emissions Directive sets the levels of emissions of pollutants and greenhouse gases for industrial facilities according to best available technology (BAT) documents, which are regularly reviewed by sectoral experts. Industrial facilities are assessed according to these BAT documents to obtain production permits.

Another proposal in the action plan is to facilitate industry-led reporting and certification systems to aid the implementation of industrial symbiosis (industrial symbiosis is the sharing and repurposing of secondary resources and by-products of industrial production). The plan underlines the possibility of companies saving on material costs by being in symbiosis with other sectors. For instance, the pharmaceutical industry produces by-products which can be useful to car

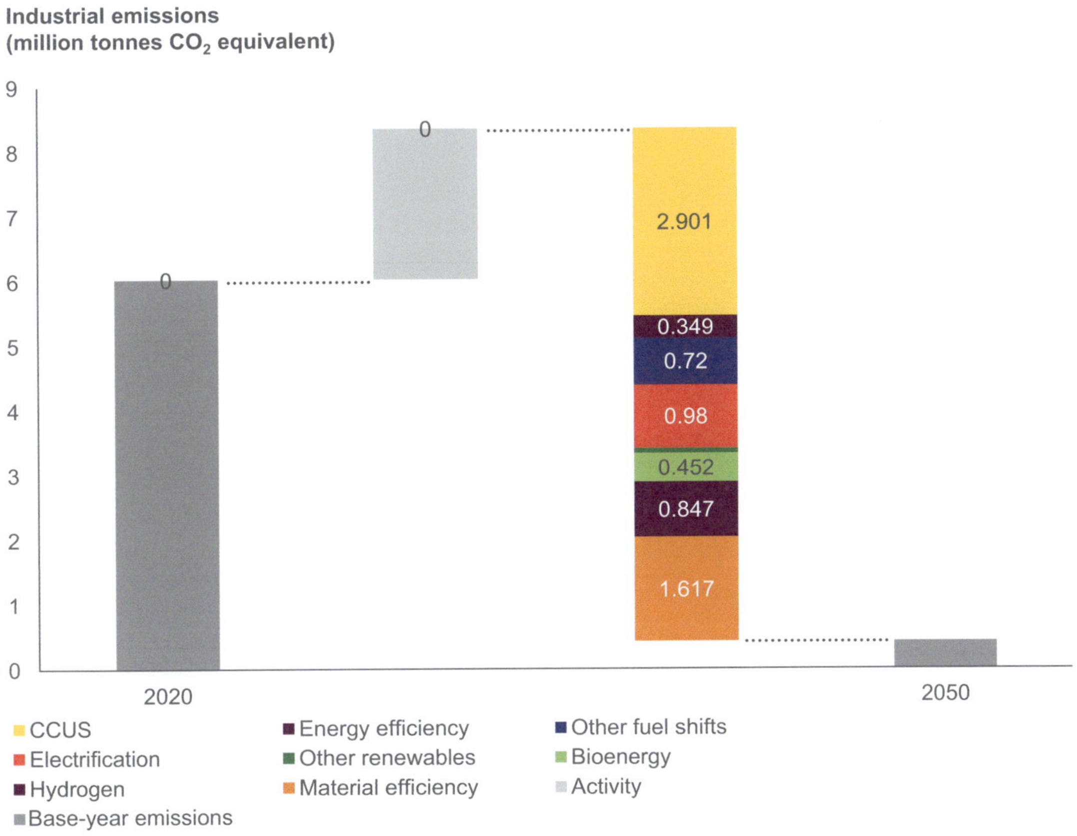

Fig. 8 Mitigation measures required to achieve net zero globally by 2050 in industry. *Source* International Energy Agency, Net Zero by 2050: A Roadmap for the Global Energy Sector, 2021

manufacturing. The plan suggests that the first step to encouraging this is better data sharing, hence the proposal to support industry-led reporting.

3.7 Changing the Industrial Fuel Mix Is the Key to Decarbonising Industry

3.7.1 Total Industrial Energy Use Is Projected to Decline, but Changes in the Industrial Fuel Mix Are Still Needed

To help achieve the dual-carbon goal, industry in China will optimise its internal structure and technologies and gradually deploy electric, digital and intelligent equipment to improve energy efficiency. As a result, total energy use in industry will steadily decrease, and the share of energy-intensive sectors in total energy use will decline. Between 2028 and 2035, industry's final energy use is expected to peak at about 3.9 billion tonnes of coal equivalent (tce), falling to about 3 billion tce by 2050 and to 1.46 billion tce by 2060.[11]

Energy transition will be the most important path to industrial decarbonisation. To achieve carbon neutrality, it is necessary to change how industrial products are manufactured and used. This applies particularly to the industrial fuel mix, such as reducing reliance on solid hydrocarbon fuels (mainly coal) and increasing the proportion of electricity use.

[11] State Grid Corporation of China, China's Energy and Electricity Outlook 2020.

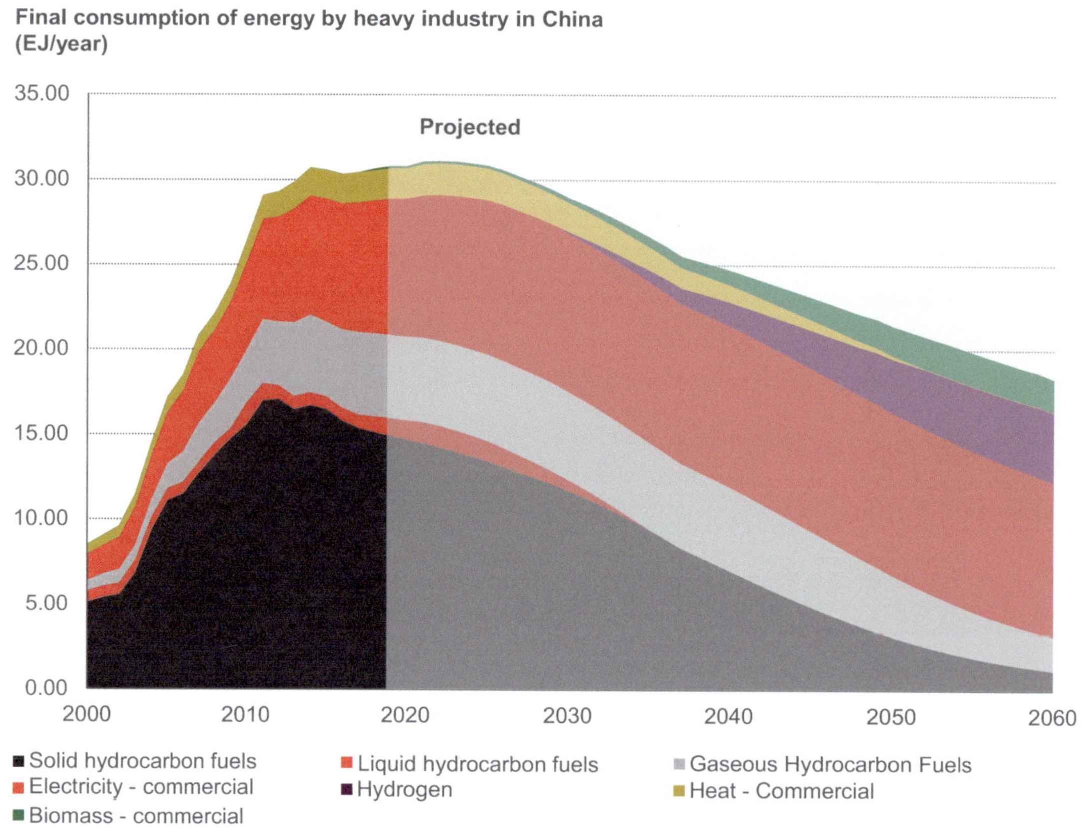

Fig. 9 Transformation of the industrial fuel mix in China. *Source* Shell–DRC project team based on Shell Scenarios Sketch, Achieving a Carbon-neutral Energy System in China by 2060, 2021

In the coming decades, most fossil fuels in the current industrial energy mix in China and globally need to be replaced with low-carbon fuels, including electricity, hydrogen and sustainable biomass. This will be made possible by the use of new fuels (see Fig. 9).

Electricity can be used as fuel. Some industrial processes can be electrified to a larger extent than they currently are, including low- to medium-temperature heat and motors. As the stock of scrap steel in China grows, the possibility of electrifying steel production through electric arc furnaces (also known as short-process steelmaking), made possible by the use of scrap rather than virgin steel, will also grow. The importance of electricity in the heavy industry fuel mix has increased from 18% of energy demand in 2000 to 26% today, but it will need to continue increasing to 50% by 2060.

Electrolysis-derived hydrogen can be used as a feedstock or as a fuel to produce, for instance, steel or ammonia. Hydrogen-based production is not mature in most industrial subsectors; hydrogen is not expected to start playing a role in industrial energy demand until 2030. By 2060, it should meet around 25% of fuel demand. By 2035, hydrogen use in industry must increase by 45% year-on-year to meet the net-zero target.

Sustainable biomass and biogas could also be used as fuel in some industrial processes. Biomass should start playing an important role from the 2040s and account for around 10% of total fuel demand by 2060. Biomass used in industrial production needs to increase by 11% year-on-year by 2030 to meet the net-zero target.

3.8 Case Study 6 | Low-carbon Steel Production in Sweden

Industrial production clusters favour co-operation between production and end-use companies and the adoption of circular economy principles.

The Norrbotten region of Sweden has two major projects under way to produce fossil-free steel at commercial scale by the mid-2020s:

- SSAB, LKAB and Vattenfall's HYBRIT initiative, which is supported by automotive manufacturer Volvo, plans to produce its first commercial shipment of fossil-free steel by 2026; and
- H2 Green Steel — a consortium of steel production (Bilstein Group and SMS Group), automotive (Mercedes and Scania), hydrogen production (EIT InnoEnergy) and building (Kingspan) companies — aims to create a fully integrated and digitalised greenfield fossil-free steel production plant.

Both these initiatives intend to make hydrogen-based steel production a reality at scale by the mid-2020s.

The two projects exemplify the importance of collaboration across the supply chain in several ways.

First, their location in the north of Sweden means they are close to cheap sources of hydroelectricity and good quality iron ore. Cheap renewable electricity makes the use of hydrogen for steel production much more cost-efficient, while local supplies of iron ore reduce transport costs.

Second, the participation of both production and end-use companies makes the project viable as it ensures that the steel meets specifications required by different end-use sectors, while financial support and assured demand from end-use companies makes investment viable for steel manufacturers.

Circular economy principles are employed in both projects. For instance, Volvo's electric vehicles are used in Hybrit's operations, and the waste heat generated by H2 Green Steel's production plant will be used by the local district heating system. The close co-operation between steel manufacturing and end-use sectors ensures that a large proportion of scrap steel will be returned to manufacturers for recycling.

3.8.1 Improving the Rate of Electrification

Although industry's share of total electricity consumption will eventually decrease, industry will remain China's largest electricity consumer. Electricity is already widely used in industrial processes, including electric motors for power, low- to medium-temperature heat pumps and electric boilers, electric furnaces for steelmaking, and to produce a small amount of hydrogen by electrolysis. As measures to reach the dual-carbon goal intensify, China's industrial structure will gradually adjust over time. Industrial electricity use will first increase, peak around 2040, and then decline (see Fig. 10). The main reason for this will be the large-scale roll-out of technologies, such as high temperature heat pumps and electric boilers in industrial processes. As the quantity of scrap steel used rises, electrified steel production will account for a growing share, and increased use of, electricity. Thanks to continuous improvements in energy efficiency, electricity use will then gradually decline.

According to China's Energy and Electricity Outlook 2020, published by the State Grid Energy Research Institute, the electrification of energy-intensive sectors will continue to increase up to 2040 (see Fig. 11).

The non-ferrous metals sector will expand at scale in the short and medium terms, driven by strong demand from segments like semiconductors, power grids, transport equipment, machinery, electronics, aerospace, building materials, weapons and equipment manufacturing and others. In addition, the non-ferrous metals sector will see remarkable energy efficiency gains, thanks to wider uptake of crude copper self-oxidation-reduction technology, high-current-density zinc electrolysis, and enhanced parallel flow electrolysis.

New building materials will drive up electricity demand steadily. On the one hand, the

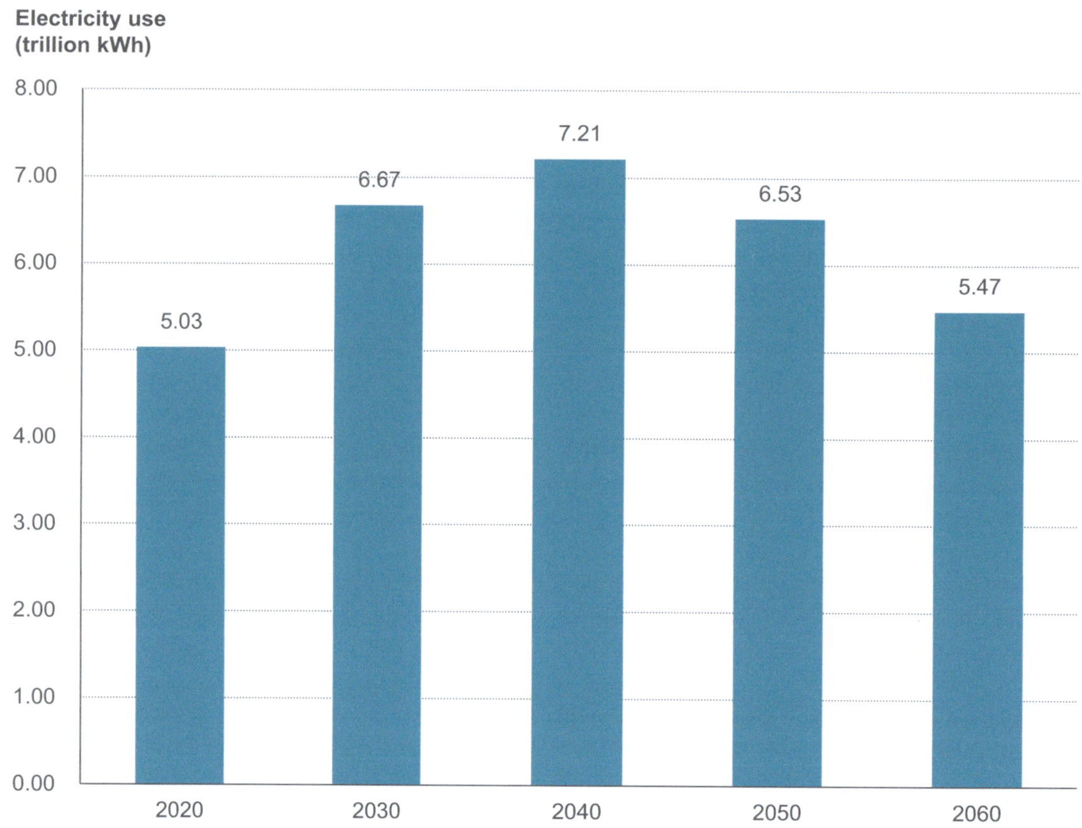

Fig. 10 Industrial electricity demand forecast. *Source* Research results of the project team

scope of expansion for conventional building materials is narrowing. The demand for cement and wall materials, for example, peaked in the 13th Five-year Plan period (2016–20), which implies a limited shift to electricity in cement clinker calcination and other processes. On the other hand, demand for renewable and high-performance materials will grow slightly.

The shift to electricity in the chemical sector will gain momentum in the medium and long terms. Nitrogen fertiliser (synthetic ammonia), methanol, soda ash, calcium carbide, olefins, caustic soda and other sub-sectors will continue to grow moderately, before peaking in 2025–30. As China's chemical sector makes structural adjustments and introduces new technologies, the total energy consumption of this sector will slow down, as the use of electricity instead of fossil fuels gradually yields results.

The ferrous metals sector is moving towards high-quality development. Stimulated by the country's efforts to promote supply-side structural reform, the sector will maintain stable progress over the long run. The deployment of new nanomaterial-coated heat transfer technology, intelligent burner control systems for industrial boilers and other innovative technologies, alongside greater production of steel from electric furnaces, will push energy efficiency to the next level.

Other manufacturing sectors such as those for transport equipment, communication devices and electrical equipment will see a rapid transition to the high-tech, high-end manufacture of new products, which will replace conventional production processes and products. The total electricity demand in these sectors is expected to peak around 2040.

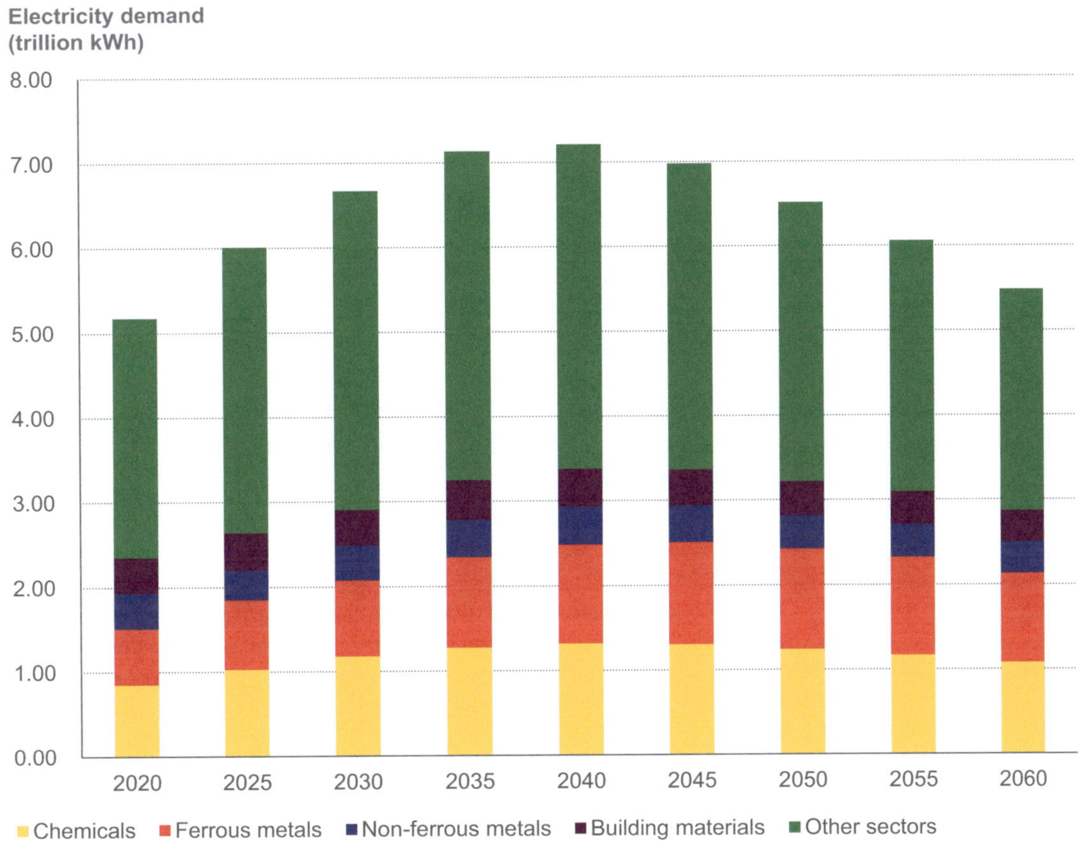

Fig. 11 Electricity demand forecasts for key energy-intensive industries in China. *Source* China's Energy and Electricity Outlook 2020; China Energy Statistical Yearbook 2020

3.9 Challenges and Opportunities in China's Industrial Decarbonisation

3.9.1 Barriers to Industrial Decarbonisation

The deep decarbonisation of industry faces challenges that other sectors do not. Several barriers stand in its way, all of which have the potential to halt or slow the uptake of decarbonisation technologies well below the required level to reach net-zero emissions by 2060.

First, technological immaturity and the need for expensive supporting infrastructure. Many technologies required to achieve decarbonisation are highly immature today. For example, carbon capture and storage (CCS) for industrial facilities is at a lower level of technological maturity than that for power plants, and costs are still very high. Investment in CCS by industrial facilities will not be viable without strong supporting infrastructure, particularly CO_2 distribution networks and storage facilities. Similarly, investment in supporting infrastructure is not viable if facilities do not intend to install CCS.

Second, high costs and lack of competitiveness. For example, hydrogen made by electrolysis is technologically immature and the need for supporting infrastructure like storage and distribution networks makes early-stage investment unviable. Decarbonised hydrogen remains much more expensive than fossil fuels in most applications. Switching to sustainably sourced biomass, biogas or other alternatives is also more expensive than fossil fuels. Competing demand for biomass from other sectors, such as power generation and construction, impacts supply potential. In addition, producing low-carbon materials is significantly more expensive than traditional production.

Third, long payback periods. Investment in some decarbonisation technologies may yield benefits over the long term, but these cannot be capitalised upon quickly enough to support the business case. Energy efficiency improvements are low-hanging fruit given the higher energy savings that can be realised by technologically advanced facilities; but energy efficiency measures often require high upfront investment while incurring long payback periods, which weakens the investment case.

Fourth, difficulties in supply chain co-ordination. Decarbonisation measures in downstream sectors, especially end-use ones, require co-ordination across sectors and along value chains. Co-ordination between the design, production, disposal and recycling stages is necessary to ensure products are used optimally. This is currently compromised by industry fragmentation. The environmental and resource costs of products are not properly priced into each stage of the supply chain, meaning price signals are not effective at encouraging optimal use.

3.9.2 Economies of Scale in China's Industrial Decarbonisation

China is uniquely placed to benefit from economies of scale in decarbonisation technologies. While industrial decarbonisation comes with costs, China's large domestic market provides opportunities to build low-carbon competitiveness. Key technologies to support industrial decarbonisation, such as electrolysers and carbon capture and storage (CCS), have large cost reduction potential through economies of scale. Leveraging these economies of scale can yield significant cost reductions in, for instance, low-carbon steel production. This reduction can range from 11% for blast furnaces with CCS to 20% for hydrogen-based direct reduced iron. This would lower the green premium on steel and make it more competitive with traditional steelmaking options.

China is also uniquely placed to benefit from economies of scale, given its vast domestic market. This is because huge scale means wider sharing of the costs of CCS or hydrogen technology development and infrastructure deployment. The country accounts for close to 50% of global steel capacity, mostly in blast furnace–basic oxygen furnace production, as shown in Fig. 12.

3.10 Case Study 7 | Public Procurement in the Netherlands

Green public procurement programmes supported by information can help build demand for low-carbon materials.

Public procurement programmes can be an important source of demand for industrial producers seeking to develop low-carbon products. Large investments are required to validate and bring to scale breakthrough technologies in low-carbon industrial production. Certainty of demand for low-carbon products is therefore critical to companies considering these investments. Across the OECD, public procurement accounts for around 12% of GDP and for more than one-third of construction and infrastructure projects that use large quantities of industrial materials.[12] Public procurement is, therefore, an important tool to provide demand certainty to companies considering investment in low-carbon production.

The Netherlands has focused on green public procurement since 2008, and has developed a world-leading system to ensure that environmental performance and innovation are rewarded. The system is characterised by:

- setting minimum performance standards for eligibility for each product group and regularly

[12] OECD, Going Green: Best Practices for Sustainable Procurement, 2015.

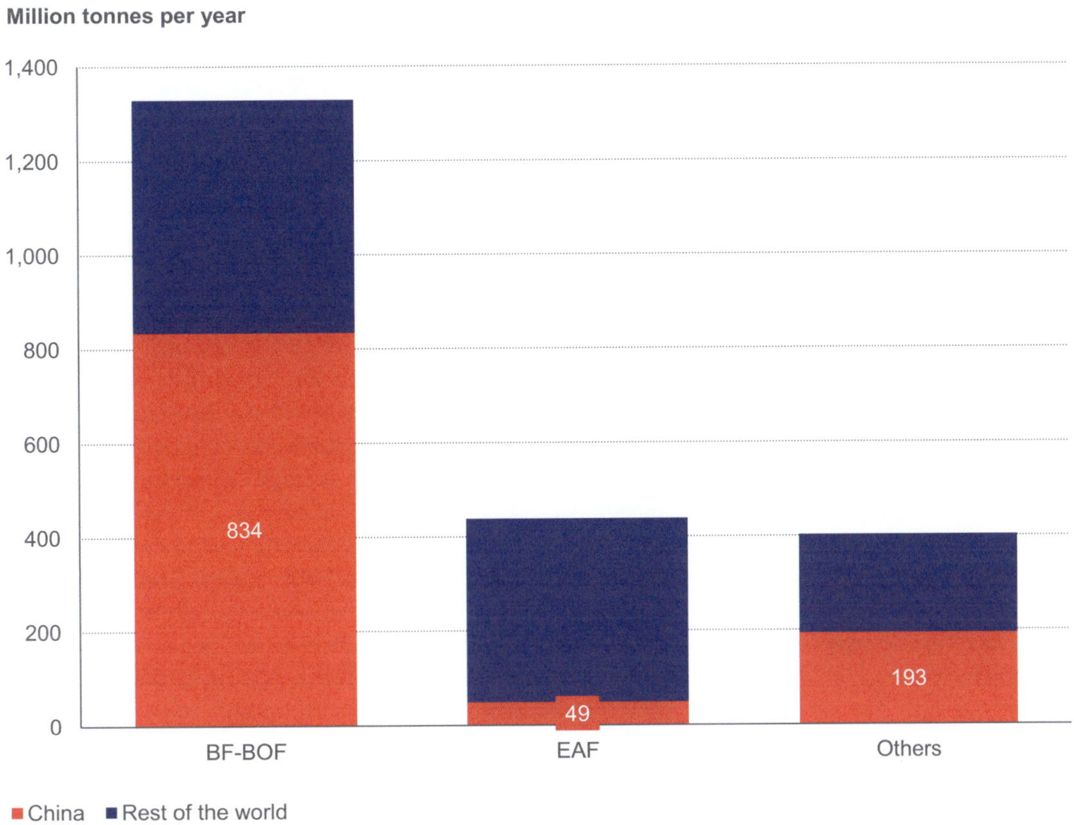

Fig. 12 Global steel production capacity, by type of production. *Note* BF–BOF = blast furnace–basic oxygen furnace production; EAF = electric arc furnaces. *Source* Shell–DRC project team

updating them to ensure that best-in-class environmental performance is encouraged[13]; and

- assessing the overall environmental performance of bids through the DuboCalc software tool, which is accessible to both procurers and bidders. The assessment of bids is based on multiple criteria in which environmental performance plays a key role, and the higher cost of more environmentally conscious or innovative bids can be offset by their better environmental performance.[14]

[13] Hasanbeigi, A., Nilsson, A., Mete, G., Fontenit, G. and Shi, D., Fostering Industry Transition through Green Public Procurement: A "How to" Guide for the Cement & Steel Sectors, 2021.

[14] Netherlands Enterprise Agency, Stimulation of sustainable energy production and climate transition.

3.11 Case Study 8 | Industry-led Initiatives

Industry-led initiatives contribute to building a credible business case for investment in low-carbon production by providing certainty over future demand.

Downstream sector-led initiatives to build demand for low-carbon products, such as SteelZero, could incentivise investment in low-carbon production across supply chains. Building demand for low-carbon materials, despite their higher cost, is critical to incentivise investment, as the higher cost of production for low-carbon materials means their competitiveness with traditional materials is not ensured. The SteelZero commitment to procure 100% net-zero steel by 2050 and 50% by 2030 is an example of such an industry-led commitment, relying on an independent standard. Companies from across steel-using sectors, particularly construction and infrastructure, have joined the initiative, putting pressure on steel providers to meet their needs.

Producers that are active across the supply chain are well placed to invest in circularity measures. For instance, cement producers that are active in the construction supply chain are in a good position to trial decarbonisation measures based on circularity. Holcim developed a range of cement with low embodied emissions, relying on low clinker specifications, optimised mix design and the use of construction waste as additives.[15] The firm's activity in both cement production and construction waste management gives it a strong position to adopt innovative circularity measures.

These industry-led initiatives across the supply chain reflect several key trends of market and regulatory pressure:[16]

- growing awareness of the current and future impact of carbon pricing: announcements about expanding carbon markets, including in China, and the threat of imposing future regulation on the emission intensity of key building materials;
- the desire to retain brand loyalty, particularly among end-customer-facing brands, by reducing emissions across supply chains; and
- the need to remain competitive as decarbonisation gains ground.

4 Decarbonisation Pathways for Key Industrial Sectors

Industry is the largest energy consumer and carbon emitter in China. When the electricity sector is excluded, energy-intensive iron and steel, building materials and chemicals are the three largest emission sources. To achieve the dual-carbon goal, efforts must be made to improve energy efficiency and reduce emissions in those three sectors.

4.1 Iron and Steel

4.1.1 Decarbonisation Prospects

The shift towards carbon neutrality will have a profound impact on the global iron and steel industry. According to the International Energy Agency's Sustainable Development Scenario, to keep the global average temperature rise below 2 °C, total direct carbon emissions from the global steel industry will have to drop by 55% by 2050, compared with 2019. Current forecasts expect global steel demand to grow by only 10% by 2050. International steel industry associations and companies have presented their abatement schedules: the European Steel Association announced that the carbon emissions from the European steel industry will drop by 30% by 2030, compared with 2018, and decrease by between 80% and 95% by 2050, compared with 1990; the Japan Iron and Steel Federation aims to achieve zero greenhouse gas emissions from iron-making processes and reduce total carbon emissions by 30% by 2050, and produce zero-carbon steel by 2100; and the Korea Iron & Steel Association aims to reduce carbon emissions from 135.7 million tonnes to 127.1 million tonnes by 2030.

In China, iron and steel is one of the biggest carbon emitters. Steel demand is mainly driven by urbanisation and industrialisation. Almost half (48%) of China's steel products are used in infrastructure such as buildings, railways and roads. In manufacturing, the automotive sector is the largest source of demand. More than 90% of steel production facilities use blast furnace–basic oxygen furnace production technology to make long-process steel. The plants that use scrap steel and electric arc furnaces account for only 9%, much lower than the international figure of 22%.

If the current production and consumption growth rate continues, China is expected to

[15] Holcim, Breaking New Ground in Sustainable Construction, 2020.

[16] The World Business Council for Sustainable Development and CLG Europe, Tomorrow's Markets Today: Scaling up demand for climate neutral basic materials and products, 2021.

become a moderately developed country in 10–15 years. China's urbanisation rate is expected to rise from 62% in 2019 to around 70% in 2030, when the demand for steel used to construct buildings and new infrastructure will decline. China's annual steel demand is projected to shrink to around 475 million tonnes by 2050, with an increasing proportion of steel made from scrap steel. As more and more buildings, vehicles and equipment reach the end of their service life, the supply of scrap steel is expected to grow by 10% annually, with the price falling accordingly. To meet the 2 °C temperature rise target of the Paris Agreement, China's short-process steel made in electric arc furnaces is expected to reach 60% of the country's total steel production.

4.1.2 Technological Pathways to Decarbonisation

Electric arc furnace (EAF) technology is the main pathway to iron and steel decarbonisation. According to the International Energy Agency's Iron and Steel Technology Roadmap of 2020, greater use of scrap steel in EAF and changes in fuel mix will be key to decarbonising the sector. The use of blast furnace–basic oxygen furnaces production (BF–BOF) in global steel production will have to drop from 70% in 2019 to 30% by 2050; the share of smelting reduction + BOF + carbon capture, utilisation and storage will have to increase to 10%; the share of EAF using scrap steel should rise from 22% in 2019 to 38% by 2050; and EAF using hydrogen-based direct reduced iron should reach 8%. Since the carbon intensity of EAF production is much lower than that of BF–BOF, a higher share of EAF will reduce the average carbon intensity of steel production. As the electric power system is decarbonised, the carbon intensity of EAF steelmaking will gradually decrease to zero.

Switching to hydrogen-based reduction is the main pathway to decarbonise existing long-process steelmaking plants. The use of decarbonised hydrogen as the reducing agent in steelmaking can help achieve zero-carbon steel production. Sweden-based SSAB is building a pilot plant using hydrogen-based direct-reduced iron and plans to achieve zero-carbon steel production in the early 2040s. German steel company Salzgitter is developing a pilot project, and Arcelor Mittal, one of the world's largest steelmakers, is also considering this technology. In China, China Baowu Steel Group partnered with China National Nuclear Corporation and Tsinghua University to use hydrogen in steelmaking in 2019, and is also working with Rio Tinto to decarbonise the steel value chain.

4.1.3 Roadmap for a Low-carbon Transition

To decarbonise the iron and steel sector, the following three factors need to be in place.

First, build a scrap steel recycling system suitable for the iron and steel sector. National industry standards for scrap steel recycling, classification, quality control, and testing and inspection should be improved at a faster pace. The scrap steel trading mechanism should be optimised; scrap steel recycling, processing and distribution should be more advanced; and in-depth co-operation between the upstream and downstream parts of the value chain should take place to unlock the potential of scrap resources in China. Policies on curbing exports of scrap steel should be formulated to avoid the outflow of large quantities of high-quality scrap at low prices; and policies to encourage the import of high-quality scrap steel should be drawn up to meet China's EAF steelmaking requirements.

Second, research and a broader uptake of EAF technology are required to improve cost-effectiveness. Efforts should be focused on R&D and on encouraging the uptake of advanced EAF equipment and processes, including large-capacity, high-power and intelligent electric arc furnaces and to improve the energy efficiency and economy of the technology. Policies should be formulated to increase the flow of scrap steel and encourage large steelmakers to scale up EAF production to achieve economies of scale.

Third, increase research and the deployment of hydrogen-based reduction in long-process steelmaking. The focus should be on researching the in-furnace reaction mechanisms and changes in the characteristics of furnace charge, promoting innovation in blast furnace steelmaking using hydrogen, developing safety materials that are resistant to hydrogen and high temperature and which prevent explosion and leakage, and increasing the capacity and output of steelmaking equipment using hydrogen. The development of clean power generation, hydrogen production from electrolysis and hydrogen-based steelmaking should be co-ordinated; and hydrogen production, transport and storage should be optimised to drive down costs and increase efficiency. Efforts should be made to encourage steelmakers to adopt hydrogen-based steelmaking instead of BF–BOF. Supporting policy mechanisms like subsidies for decarbonised hydrogen should be introduced, and co-ordination between hydrogen-based steelmaking and the hydrogen value chain should be strengthened.

By 2030, China's scrap steel resources will reach 450 million tonnes. EAF will be more cost-effective than BF–BOF, and its output will double to 400 million tonnes or 34% of total steel production. The cost of clean power generation and hydrogen production from electrolysis will fall sharply, with output reaching 57 million tonnes or 5% of total steel production. By 2050, China's scrap steel resources will exceed 550 million tonnes, and the annual output of EAF steel will reach 500 million tonnes or 65% of total steel production. The supply of scrap steel will increase and EAF steelmaking technology will improve steadily, and an electricity-centred fuel system for steelmaking will take shape. Clean hydrogen production from clean energy will make breakthroughs, driving down the cost of decarbonised hydrogen production. Steelmaking using hydrogen will be rolled out at scale instead of blast furnaces, with output reaching 230 million tonnes or 30% of the total. By 2060, the scrap steel stock will maintain a level of more than 500 million tonnes and annual output from EAFs will stabilise at about 500 million tonnes or 67% of the total. Steel production from hydrogen will reach 240 million tonnes or 32% of the total.

4.2 Cement

4.2.1 Decarbonisation Prospects

In 2018, China's share of total global cement production was 53%, higher than the rest of the world combined. After years of growth, the country's cement production peaked at 2.48 billion tonnes in 2014, dropped to 2.40 billion tonnes in 2016 and to 2.18 billion tonnes in 2018. The country's cement output is about 10 times higher than India's, the world's second-largest cement producer. Due to high transport costs relative to production expenses, cement is seldom exported or imported. The demand for cement for infrastructure and buildings accounts for 55% of the total.

Carbon dioxide emissions from the cement sector include those from burning fuel to heat kilns to more than 1,600 degrees Celsius, and those from limestone decomposition in the production process, which is responsible for about 60% of the total CO_2 emissions.

After 40 years of continuous technological innovation, China has become a forerunner in energy-efficient cement production. Most cement plants deploy advanced dry process kilns, rather than wet process kilns that use more energy. The heat consumption intensity of clinker is 3,600 kilojoules per kilogram in China, 15% lower than the world average. The intensity of electricity consumption per tonne of cement produced is less than 90 kWh. The country's cement sector aims to further improve its energy efficiency by between 13% and 16% by 2050. While higher energy efficiency can save energy and reduce CO_2 emissions, China still needs to take further steps to fully decarbonise the cement sector. Possible options include using zero-carbon fuels such as biomass or clean electricity to provide heat input, and deploying carbon capture and storage (CCS).

4.2.2 Technological Pathways to Decarbonisation

Cement is produced by chemical reactions that emit carbon dioxide (about 330 kilograms of carbon dioxide per tonne of Portland cement). Given this, the use of alternative feedstocks or the deployment of CCS will be necessary to decarbonise cement production.

Use of Alternative Feedstocks. Limestone decomposition is responsible for 60% of the total CO_2 emissions from cement production. Using alternative minerals instead of limestone or clinker can help reduce the inherent CO_2 emissions in chemical processes. Alternatives such as fly ash and slag are already widely used. However, in a zero-carbon economy, the generation of fly ash and slag will decline due to the decreasing number of steel production facilities using BOFs and coal-fired power plants.

Biomass Heating with CCS. Analysis shows that the use of biomass heating with CCS to abate process emissions from the cement sector may be the lowest-cost route to decarbonise cement production. In China, it also has the most policy support. In 2015, six ministries and commissions, including the Ministry of Industry and Information Technology, announced pilot projects to co-ordinate cement production with municipal waste disposal (China's municipal waste is increasing by 8–10% annually). The use of biomass is restricted by resource shortage and local feasibility.

Electric Heating with CCS. The required temperature for clinker calcination is between 1,000 and 1,450 degrees Celsius (°C). Electric heating using metallic and non-metallic components can provide a temperature of 1,000–1,500 °C and 1,500–1,700 °C respectively. This indicates that electric heating for cement production is viable. Moreover, a large number of cement plants in China are located in areas rich in renewable energy, which partly explains why electric heating with CCS is a feasible decarbonisation pathway for cement. However, electric kiln technology has not yet been commercialised and the cost of electricity for heating is about 1.5 times and 1.3 times that of coal and natural gas respectively. Furthermore, the cost of the metallic and non-metallic materials needed to manufacture electric heaters for cement production is far higher than that for rotary kilns. Low cost-effectiveness is the main reason why electric kilns have not been commercialised. In time, electric kilns could be used as an auxiliary technique to decarbonise the cement sector.

4.2.3 Roadmap for a Low-carbon Transition

To decarbonise the cement sector, efforts should be made in the following four areas.

First, intensify the research and development of low-carbon alternative feedstocks. Research on alternative raw materials for cement production should be intensified to develop low-carbon feedstocks as a replacement for existing conventional high-carbon ones, cutting down CO_2 emissions at source.

Second, increase the use of CCS for cement kilns. Innovative kiln designs and associated technologies should be adopted to increase the CO_2 concentration at the outlet of the kiln and reduce the cost of carbon capture. CCS should be widely deployed to treat the exhaust gas from the kilns, avoiding CO_2 emissions from fuel burning and limestone calcination. Carbon sequestration facilities should be built near large cement plants.

Third, build a biomass fuel collection and use system suitable for the cement sector. The national and industry standards for collection, quality control and testing and inspection of biomass for heating in cement production should be improved at a faster pace. The biomass trading mechanism should be optimised, and efforts should be made to promote in-depth co-operation between municipal waste, agriculture and the cement industry to maximise the potential of biomass resources in China.

Fourth, improve the cost-effectiveness of electric kilns. Further breakthroughs should be made in the design, technology and refractory materials for electric kilns to improve the energy efficiency of the equipment and enhance the cost-effectiveness of electric kilns at a faster pace. More policy support should be provided, such as subsidies for equipment or lower electricity prices for electric kilns. Rotary kilns that have high energy consumption and high emissions should be gradually phased out. The cost of electric kiln equipment should be reduced by economies of scale.

By 2030, cement produced from biomass-fired kilns in China will account for 10% of the total. CCS for cement production and electric kiln technologies will both achieve technical breakthroughs and pilot deployment will start. By 2050, cement production supported by biomass heating will reach 45% of the total; CCS will be more cost-effective and will be widely used in the cement sector; and the cost of clean electricity will fall sharply, with 15% of total cement production made in electric kilns. By 2060, biomass will become the dominant energy source for heating and will be used to produce half of all cement in China; CCS will be widely used and electric kilns will account for 20% of total cement production.

4.3 Chemicals

4.3.1 Decarbonisation Prospects

In 2016, China's petrochemical and chemical sector consumed 28% of total industrial energy. Of the thousands of chemical products made, the three basic chemical product categories of ammonia, methanol and high-value chemicals (including light olefins and aromatics) account for about three-quarters of total energy consumption in China's chemical sector. According to the International Energy Agency's projections in its 2 °C Scenario, this will still be as much as 23% in 2050. The potential for reducing demand for feedstocks in the chemical industry can be greatly improved by developing a circular economy and increasing the efficiency and recovery rate of materials and products. However, zero-carbon production is essential to fully decarbonise this sector.

Coal is a key energy source and one of the main raw materials in China's chemical production. In the absence of major technological breakthroughs, coal-based production is expected to continue to play an important role in 2050. Among others, Power-to-X (P2X) and biomass-based processes will increase in importance.

Power-to-X means that electricity is used to reduce and recombine carbon from carbon dioxide, hydrogen from water, and nitrogen from the air to generate usable organic or inorganic raw materials. For example, electrolysis-based hydrogen can be used to reduce nitrogen for ammonia generation. Hydrogen can be used to reduce carbon dioxide to produce organic matter such as methane and methanol, which can be further synthesised into ethylene, propylene and benzene, etc.

New technical pathways (such as Power-to-X and biomass-based solutions) are more expensive than those based on conventional fossil fuels. Strong policies will therefore be needed to support the scale-up of these new pathways. In China, coal-based chemical production will continue to dominate in the years to come, unless Power-to-X and biomass-based solutions gain significant financial advantages.

4.3.2 Technological Pathways to Decarbonisation

Power-to-X. Chemical production is the organic conversion of hydrogen and carbon. Synthesis reactions using hydrogen, carbon monoxide and carbon dioxide as feedstocks can generate many products in the chemical value chain. The Power-to-X pathway to decarbonisation uses hydrogen from zero-carbon electricity-based electrolysis and carbon dioxide as the main feedstocks.

Synthetic Ammonia. Decarbonised hydrogen from zero-carbon electricity-based electrolysis can be used in the Haber process for ammonia synthesis using hydrogen and nitrogen as feedstocks. According to the calculations of Deutsche Gesellschaft für chemisches Apparatewesen (DECHEMA), which operates an international chemical database, the production of 1 tonne of synthetic ammonia requires 178 kg of hydrogen and 9.1 megawatt-hours (MWh) of electricity. In addition, compression and other steps in producing 1 tonne of synthetic ammonia consume 1.4 MWh and 0.33 MWh of electricity respectively.

Methanol. The catalyst for the reaction of carbon dioxide and hydrogen in methanol production has been commercially produced, and some pilot plants have started operation. Iceland-based Carbon Recycling International currently produces 4,000 tonnes of green methanol a year using this technology, and plans to scale up to 40,000 tonnes per year. The carbon dioxide feedstock comes from a geothermal power plant and the 5 megawatt electrolyser for hydrogen generation is powered by geothermal energy. Japan-based Mitsui Chemicals has also built a pilot plant that synthesises methanol using hydrogen and carbon dioxide, and has carried out a feasibility study for its industrial-scale production.

High-value Chemicals. Although the technologies for directly converting hydrogen and carbon dioxide into olefins and aromatics (benzene, toluene and xylene, etc.) are still relatively immature, the commercial production of light hydrocarbons using methanol as feedstock is already taking place. If the methanol feedstock is derived from zero-carbon hydrogen or produced using carbon dioxide, the decarbonisation of high-value chemical production will be greatly advanced. The plants deploying the best processes now produce 1 tonne of light hydrocarbons from 5 gigajoules of energy.

4.3.3 Roadmap for a Low-carbon Transition

Most electricity-based feedstock technologies are still undergoing laboratory tests. Some technologies such as electricity-to-hydrogen and methanol production, have demonstration projects in place, but they remain uncompetitive in terms of cost-effectiveness when compared with the conventional technologies that use coal and petroleum. To reach the next stage, efforts should be made in the following three areas.

First, optimise technological processes to sharpen cost-effectiveness. Research on process reaction mechanisms and dynamics should be strengthened and new catalysts developed to improve the conversion efficiency of electricity-based feedstocks. The technological process, reaction conditions and reactor design should be optimised to reduce process energy consumption and costs.

Second, actively promote technology demonstration. In areas abundant in clean energy and carbon dioxide resources, integrated technology demonstration projects, such as efficient electrolysis-based hydrogen production and large-capacity methane or methanol production from carbon dioxide hydrotreating, should be started to lay a foundation for the commercialisation of electricity-based feedstock technology.

Third, improve industry standards and the value chain. Standards for electricity-to-feedstock technologies should be introduced. The upstream electricity-based hydrogen production and carbon capture value chains should be improved, and large-scale manufacturing of energy-efficient reactor equipment should be promoted to drive down the cost of raw materials and plants. Together, these steps will pave the way for large-scale commercial use of electricity-based feedstocks.

By 2030, some relatively cost-effective electricity-based feedstock technologies will be demonstrated or commercialised, of which electricity-based ammonia will be deployed at scale. By 2050, electricity-based feedstocks will be more cost-effective than fossil-based technologies and will be widely deployed in commercial operation. The cost of clean power generation will further decline, and the production capacity of electricity-to-ammonia and electricity-to-methanol will reach 16 million tonnes and 39 million tonnes respectively. By 2060, electricity-based feedstocks will be the main source of raw materials for the chemical sector. The output of electricity-to-ammonia and electricity-to-methanol will be 30 million tonnes and 48 million tonnes respectively.

5 Recommendations for Decarbonising Industry

Industry is a priority but also a tough challenge for China in its shift towards peak carbon emissions and carbon neutrality. Constrained by immature technologies and excessively high costs, industrial decarbonisation requires government policies to advance the R&D and wide uptake needed to progress at pace.

First, optimise the industrial structure faster. Structure is the key factor influencing the low-carbon development of industry. China's industry is still dominated by labour- and resource-intensive heavy sectors and relatively few sectors that are technology- and knowledge-intensive. To achieve carbon neutrality, China should fast-track the development of emerging and green industries, promote the low-carbon transformation of conventional energy-intensive and high-pollution sectors, improve quality and efficiency, sharpen competitiveness and build modern industrial and economic systems. In this way, green, low-carbon and sustainable economic growth can become a reality.

Build an industrial circularity system. China should tighten its management of constraint targets on resources and the environment, build a society-wide system for resource recycling to support abatement, and promote the reduction and reuse of resources in production and consumption. China should shift production and consumption from an inefficient, high-carbon and high-pollution approach to one that is efficient, intelligent, clean and low carbon. This will co-ordinate the development of the economy, society and resources.

Foster strategic emerging industries. China should develop knowledge- and technology-intensive industries, such as next-generation information technology, high-end equipment manufacturing, new materials, new bioenergy and new energy vehicles. The country should promote the efficient and co-ordinated development of strategic emerging industries and new infrastructure such as big data centres and electric vehicle charging networks.

Encourage leapfrog development of green industries. China should establish quantifiable, verifiable and reportable green industry development indicators and reinforce green product life-cycle management. It should build a green technology innovation system, speed up research on crucial technologies for green manufacturing and achieve breakthroughs in new and leading green technologies.

Transform and modernise energy-intensive and high-emission sectors. China should implement further improvements in energy-intensive sectors, step up efforts to eliminate outdated production facilities and gradually reduce their total production capacity. The country should explore the potential for shifting energy-intensive sectors to electricity and low-carbon alternatives.

Second, improve industrial energy efficiency. China's industries are energy inefficient. To address this, energy should be conserved to force industries to make energy efficiency improvements. Energy use should be strictly controlled and existing demand optimised to help transform and modernise industrial sectors.

Focus on energy efficiency and carbon reduction in key sectors. China should promote technological innovation in energy efficiency and carbon reduction in coal power generation, iron and steel, non-ferrous metals, building materials, and petrochemicals and chemicals. It should accelerate the uptake of advanced and energy-efficient products and equipment, replace outdated and inefficient equipment, and develop a green manufacturing system. The country should enhance the energy efficiency of new infrastructure, such as green data centres and 5G infrastructure.

Improve laws, regulations and standards for energy efficiency. China should review the Energy Conservation Law, energy conservation review procedures and other laws and regulations. It should improve standards for energy efficiency and carbon reduction in industry, and revise the mandatory national standards for energy consumption and the energy efficiency of products and equipment. It should encourage local government to implement energy use budget management.

Improve energy conservation. China should improve energy metering and monitoring, optimise data collection and analysis of energy use in industry, and gradually apply big data resources in energy efficiency and carbon reduction.

Third, advance industrial decarbonisation through electrification. In the short term, China should make industrial electrification one of the key enablers of the dual-carbon goal. China should refine the policy-guided, enterprise-oriented and market-driven approach to one in which the whole of society works towards industrial electrification in a scientific way.

Electrify industrial heat. To meet the need for hot water, steam and heat in industrial production, China should encourage industry to replace coal-fired boilers and kilns with electric alternatives. In ceramics, cement, glass, lime, gypsum, rock wool, bricks, floorboards and other building materials, electric kilns should be deployed widely. In metals, wider uptake of electric resistance furnaces, electric arc furnaces, medium- and high-frequency induction furnaces and other electro-metallurgical technologies should be encouraged.

Electrify industrial equipment. To meet the power needs of industrial production, electric blowers, electric air compressors, electric excavators, electro-hydraulic hammers and electric crushers should be widely used; electric heavy-duty trucks instead of fuel-powered vehicles should be widely deployed for industrial short-distance transport; electric drilling units and other electrical devices should be widely adopted in oilfield production; and electric shovels and conveyor systems should be widely deployed in mining.

Use Power-to-X to indirectly electrify industry. China should use decarbonised hydrogen as a reducing agent in steelmaking and to reduce carbon emissions in the blast furnace–basic oxygen furnace process. The country should promote large-scale deployment of Power-to-X in the chemical sector to meet the production requirements of basic chemical products such as ammonia, methanol and high-value chemicals, and help the chemical sector reduce carbon emissions from feedstocks.

Explore business models for industrial electrification. China should emphasise the role of the market and explore business models for industrial electrification that adapt to market dynamics. The country should adapt models such as energy management contracts, equipment leasing, build-operate-transfer, engineering, procurement and construction, and public-private partnerships to local demand. It should actively encourage emerging business models such as shared alternatives, integrated electricity substitution, internet + energy contracting, and new energy e-commerce. China should promote the wide uptake of integrated energy services for industrial systems, explore flexibility in electrical loads in industry, and encourage users to actively participate in demand-side response and use more green electricity.

Fourth, develop technologies like carbon capture and storage (CCS) to support industrial decarbonisation. To remove the barrier of technological immaturity in CCS, hydrogen and other decarbonisation technologies, the government should increase support for R&D and pilot deployment. It should promote deep co-operation between enterprises, universities, research institutions and users, and encourage relevant industry associations, social groups and leading enterprises to build innovation bases, joint laboratories and other collaborative platforms. This will help improve R&D efficiency and drive down the costs of decarbonisation technologies. The country should create favourable conditions for the development of technology demonstration projects that are expected to play a leading role in industrial decarbonisation.

Fifth, develop a consumer market that supports low-carbon industrial products. China should stimulate demand for low-carbon industrial products and create an environment conducive to the decarbonisation of industry. To this end, China should define labels and standards to ensure that all customers are able to identify low-carbon industrial producers. Reliable, independently verified information is the key to stimulating demand for low-carbon products. The aim of labelling and standards should distinguish best-in-class performance and/or innovation. The country should actively stimulate demand by setting standards. For example, the government can refine energy efficiency and low-carbon standards to reduce emissions in industrial production and across supply chains.

Sixth, improve trading systems and mechanisms and the role of the market. China should: (i) reform electricity pricing, set power transmission tariffs and distribution fees in an evidence-based manner; (ii) support investment in the power grid expansion needed for electrification and allow investment and operational costs to be included in the cost of power transmission and distribution; (iii) improve the time-of-use price mechanism and introduce peak and off-peak pricing that links supply with demand and increases the price difference between them; (iv) develop the power trading market and support the participation of renewable energy projects that replace fossil fuels in medium-term, long-term and spot trading; (v) establish a green bond market and fund issuance, attract private capital through multiple channels and promote the electrification of industry; and (vi) build a national carbon market at a faster pace, refine carbon accounting methods in key industrial sectors and carry out pilot trading in an emissions trading system to drive the decarbonisation of heavy industry.

Chapter 3: Buildings: Controlling Carbon Emissions Effectively

Han Wu and Georgios Bonias

1 Introduction

In 2018, the energy consumption of China's buildings sector by category and proportion of the total was urban residential buildings (24%), rural residential buildings (22%), public buildings (33%) and heating in the northern part of the country (21%).

Electrification and energy efficiency improvements are the most important ways to decarbonise buildings. Given this, China should encourage clean heating in northern China and some areas of southern China, switch to low-energy residential buildings and energy-efficient home appliances, accelerate the use of renewable energy in buildings, and advance the digital energy market.

To meet the goal of carbon neutrality, the proportion of electricity in the final energy use of China's buildings sector needs to increase from 25.4% in 2018 to 39.8% in 2030 and 73.3% in 2060, as proposed in this chapter.

Based on research on the electrification and decarbonisation of buildings, six technical pathways are described. These are low-carbon new buildings, digital integrated energy stations, clean heating, low-carbon heat sources, electric cookware and energy-efficient home appliances, and renewable energy use in rural areas.

Buildings are responsible for about 21% of China's total energy consumption and carbon emissions. As the country continues to urbanise, the total gross floor space of buildings will increase. As prosperity and living standards rise, China's total energy consumption from buildings will continue on its upward trend. For China to meet its dual-carbon goal, it needs to electrify the buildings sector, improve energy efficiency and reduce emissions at a faster pace. It also needs to make progress in its energy production and consumption revolutions.

2 Electrification in the Buildings Sector

2.1 China's Buildings Sector

In 2020, China's urban population was 902 million living in 310 million households, and the rural population was 510 million in 184 million households. The urbanisation rate, driven by mass migration from rural to urban areas, increased from 37.7% in 2001 to 63.9%.[1] Since 2014, the floor area of new civic buildings in China has been stable at about 2.5 billion square

H. Wu (✉)
State Power Rixin Technology Co., Ltd., Beijing, China

G. Bonias
Shell International Limited, London, UK

[1] Communiqué of the Seventh National Population Census.

Shell International B.V. and Development Research Centre of the State Council (DRC) of People's Republic of China, *Embracing the Future, Powering Growth: An Energy System Renewed for China*, Advances in Oil and Gas Exploration & Production,
https://doi.org/10.1007/978-3-031-73075-7_4

metres (m^2) a year, while the floor area of demolished buildings has remained steady at about 1.5 billion m^2, resulting in a growing total floor area.[2]

In 2018, China's total building area was about 60.1 billion m^2, of which 24.4 billion m^2 were in urban residential buildings, 22.9 billion m^2 were in rural residential buildings and 12.8 billion m^2 were in public buildings. In northern China, the urban floor area requiring heating was 14.7 billion square metres.[3]

China's average residential building area per capita was 33.9 m^2 in 2018, close to that of developed countries like the UK and Japan. However, the average building area per capita in public buildings was only 9.2 m^2, about one-third that of the USA and half that of Japan.[4] Shopping centres, hospitals, schools, transport hubs, cultural and sports buildings, and community centres all accounted for a small proportion of the total building area of public buildings.

2.2 Energy Consumption and Greenhouse Gas Emissions in Buildings

Energy use and greenhouse gas emissions in the buildings sector cover the entire life cycle from construction to operation and demolition, as shown in Fig. 1. This chapter focuses on energy consumption and emissions in the operational stage of buildings. Energy consumption and emissions in the production, transport and use of building materials in the construction stage come within the industry and transport sectors. Energy use and emissions during demolition will not be discussed here, as they are relatively low.

Energy use during building operation refers to the energy used to provide occupants or users with heating, ventilation, air conditioning, lighting, cooking, hot water and other services in residential and office buildings, schools, shopping centres, hotels, transport hubs, cultural, sports and recreation facilities and other buildings. Greenhouse gas emissions during operation include direct and indirect carbon dioxide emissions associated with energy use as well as other greenhouse gases.

According to an International Energy Agency report, final energy use and CO_2 emissions during building operations contributed 30% and 28% respectively to the global total in 2018.

2.3 Energy Use and Carbon Emissions in China's Buildings Sector

There are four types of building in terms of energy consumption in China: urban residential buildings, rural residential buildings, commercial and public buildings (referred to as "public buildings") and northern urban heating (referred to as "northern heating").

Northern heating refers to the use of energy for winter heating in urban areas.[5] It includes central and decentralised heating systems. Central heating systems predominate in North China. There are many central and district heating networks which use various technologies such as combined heat and power, coal-fired boilers, gas-fired boilers and large central heat pumps. Decentralised heating systems are more common in areas that cannot be reached by heating networks; these systems include household gas-fired heaters, household coal-fired heaters, heat pumps, air conditioners and direct electric heaters. The energy carriers used in these systems are mainly coal, gas and electricity.

Urban residential buildings (excluding urban heating in northern China) refers to the energy consumed by urban residential buildings for home appliances, air conditioning, lighting, cooking, hot water and for winter heating in non-northern urban areas. The main energy types are

[2] Research Report on Energy Efficiency Development in Buildings in China, 2021.

[3] China Urban-Rural Construction Statistical Yearbook 2018.

[4] International Energy Agency, World Energy Balances 2020; World Bank, World Development Indicators.

[5] All urban areas in Beijing, Tianjin, Hebei, Shanxi, Inner Mongolia, Liaoning, Jilin, Heilongjiang, Shandong, Henan, Shaanxi, Gansu, Qinghai, Ningxia and Xinjiang, as well as some urban areas in Sichuan. Tibet, western Sichuan and some urban areas in Guizhou are excluded.

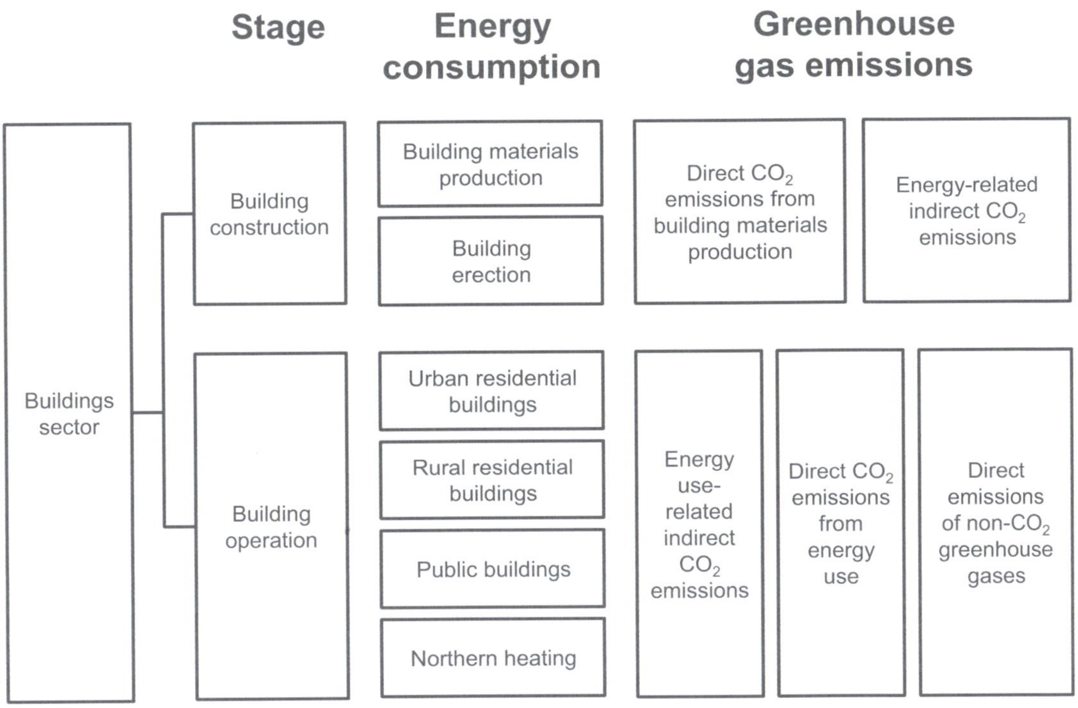

Fig. 1 Energy consumption and carbon emissions in the buildings sector. *Source* Research Report on Energy Efficiency Development in Buildings in China, 2021

electricity, coal, natural gas, liquefied petroleum gas and city (natural) gas.

Public buildings (excluding urban heating in northern China) refers to the energy consumed by public buildings. It includes buildings used as offices and for commerce, tourism, science, education, culture, health, communications and transport in non-northern urban and rural areas. The energy is used for air conditioning, lighting, electrical outlets, lifts, cooking, service facilities and heating. The predominant types of energy are electricity, gas, oil and coal.

Rural residential buildings refers to the energy use of rural families for cooking, heating, cooling, lighting, hot water and appliances. The energy types are mainly electricity, coal, biomass, liquefied petroleum gas and gas. Biomass is not included in national energy use statistics.

According to the China Energy Statistical Yearbook, the China Urban Construction Statistical Yearbook and other sources, total commercial energy consumption from China's building operations was about 1 billion tonnes of coal equivalent (tce)[6] in 2018, which amounts to 22% of the country's total energy use. The data for the four building energy use categories is given in Table 1.

Tree branches, straw, cow dung and other types of biomass are important sources of rural energy for cooking and heating. Total rural biomass energy consumption was about 91 million tce. This part of biomass energy is non-commercial energy, so it is not included in national data. Figure 2 shows the building area and energy intensity of the four types of building. The horizontal axis represents the building area, the vertical axis represents the energy intensity per unit of building area, and the figure in each of the four tiles is the corresponding total building energy consumption. Urban residential buildings and rural residential buildings accounted for the largest share of total building area. Northern

[6] One tonne of coal equivalent is an index for converting various energy forms to the calorific value of standard coal.

Table 1 Energy consumption from building operations in 2018

Energy use category	Area (billion m^2)	Households (million)	Energy consumption (billion tce)	Electricity use[7] (billion kWh)	Energy consumption per unit of building area ($kgce/m^2$)
Northern heating	14.7	–	0.212	57.1	14.4
Urban residential buildings (excluding heating)	24.4	298	0.241	540.4	9.9
Public buildings (excluding heating)	12.8	–	0.332	809.9	26.0
Rural residential buildings	22.9	148	0.216	262.3	9.4
Total	60.1[a]	446	1.001	1669.7	16.7[b]

Note [a]The total floor area of urban residential buildings, public buildings and rural residential buildings
[b]The overall mean
$kgce/m^2$ = kilograms of coal equivalent per square metre
Source China Energy Statistical Yearbook 2019; China Urban and Rural Construction Statistical Yearbook 2018; Research Report on Building Energy Efficiency Development in China 2021

heating and public buildings represented about one-quarter and one-fifth of total building area respectively. Public buildings (excluding northern heating) have the highest energy intensity, followed by northern heating. The energy intensity of urban residential buildings (excluding northern heating) and rural residential buildings were similar. The four categories each accounted for roughly 25% of total energy consumption, with public buildings (excluding northern heating) accounting for the most.

In 2018, carbon emissions from fossil energy use by China's building operations were 2.1 billion tonnes, of which direct carbon emissions, electricity-related indirect emissions and heat-related indirect emissions from combined heat and power accounted for 50%, 42% and 8% respectively. China's carbon emissions from building operations were 1.5 tonnes per person and 35 kg/m^2 of building area.

Electricity accounted for 70% of the total energy consumption of urban residential buildings and public buildings, meaning that indirect CO_2 emissions predominated. The heat from combined heat and power plants used by urban residential buildings in North China includes indirect CO_2 emissions. Coal made up 80% in the northern heating category and 60% in the rural residential buildings category, resulting in a large amount of direct CO_2 emissions. Northern heating, urban residential buildings, public buildings and rural residential buildings were responsible for 26%, 21%, 30% and 23% respectively of total carbon emissions in the buildings sector.

Figure 3 shows the building area, energy intensity and total carbon emissions of the four types of building. The horizontal axis represents building area, the vertical axis represents the carbon emissions intensity per unit of building area, and the figure in each of the tiles is the corresponding total carbon emissions. It can be seen from Figs. 2 and 3 that the carbon emissions and energy consumption of the four types of building differ: northern heating had a carbon intensity of 37.3 kg CO_2/m^2 due to substantial coal use; public buildings (excluding northern heating) had the highest carbon intensity per unit of area at 49.7 kg CO_2/m^2, which was 1.33 times that of the northern heating category; the carbon intensity of urban residential buildings (excluding northern heating) was 17.5 kg CO_2/m^2; and rural residential buildings had a carbon intensity of 21.0 kg CO_2/m^2 due to a low electrification rate and large coal consumption.

[7] Based on the annual average coal consumption of thermal power generation in China. Power consumption is converted into primary energy consumption expressed in standard coal. In this book, the average coal consumption of thermal power supply in 2018 was 308 grams of coal equivalent per kilowatt-hour (gce/kWh).

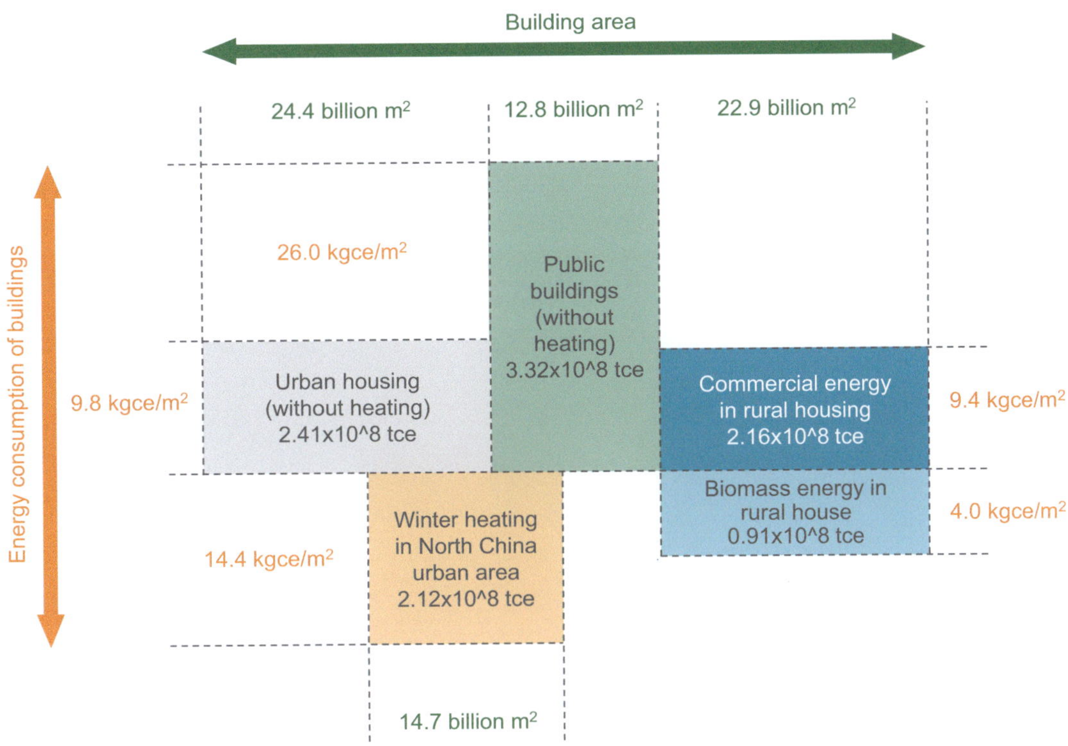

Fig. 2 Energy consumption from building operations in 2018. *Note* tce = tonnes of coal equivalent; kgce/m^2 = kilograms of coal equivalent per square metre. *Source* China Energy Statistical Yearbook 2019; China Urban and Rural Construction Statistical Yearbook 2018; and Research Report on Building Energy Efficiency Development in China, 2021

Energy consumption per unit of building area was lower in China than in developed countries; it was one-third that of the USA and half that of South Korea, Japan, the UK and Germany. Energy consumption per capita was far lower in China than in the USA, and was about half that of Germany, South Korea and Japan (see Fig. 4). However, due to the rapid growth of China's total building area and the high proportion of fossil fuels in the energy mix, total carbon emissions from buildings increased rapidly, including both direct and indirect carbon emissions. By 2020, the total carbon emissions from buildings exceeded those of the USA (see Fig. 5). To meet the goals of carbon neutrality, energy conservation and emissions reduction, China needs to take a development pathway that is different to other countries. This will undoubtedly pose challenges.

2.4 Electrification in China's Buildings Sector

Average coal consumption for thermal power supply was used to calculate the amount of energy generated for the buildings sector and the carbon emissions per unit of energy consumption. The results are given in Table 2.

Electricity's share of the final energy consumption of public buildings (excluding northern heating) and urban residential buildings was higher at 37.2% and 34.2% respectively than for rural residential buildings (18.5%) and northern heating (4.1%). The carbon emissions per unit of energy consumption is negatively correlated with the electrification rate. Due to the low electrification rate of northern heating and the proportion of coal used (up to 80%), the carbon emissions per unit of energy consumption were 2.6 kg/kgce,

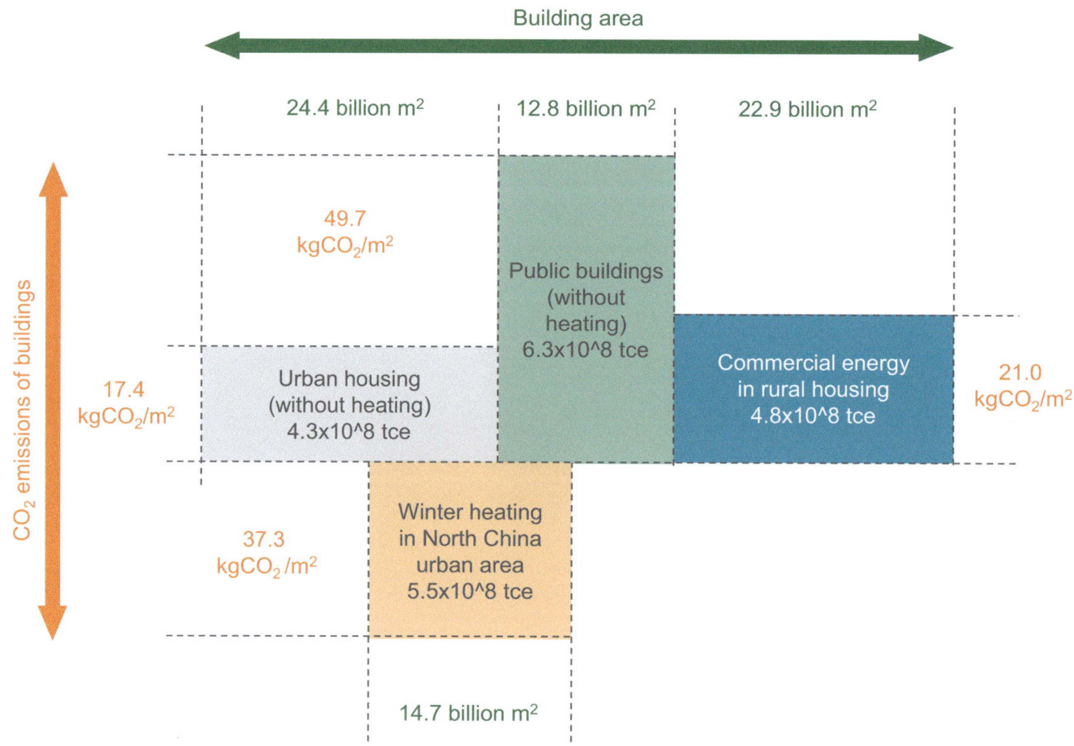

Fig. 3 Carbon dioxide emissions from building operations in China in 2018. *Source* China Energy Statistical Yearbook 2019; China Urban and Rural Construction Statistical Yearbook 2018; Research Report on Building Energy Efficiency Development in China, 2021

which is significantly higher than for other types of building.

A comparison of the electrification rate of the buildings sector in China with that of other countries shows that total energy consumption was 24.7% in China, which was about half that of Japan (52.3%) and the USA (50.5%). The share of commercial electricity use was 36.8%, about two-fifths that of Argentina (92.6%), half that of Australia (71.8%) and South Korea (67.5%), and much less than the USA (56.1%) and Japan (54.3%) (Figs. 6 and 7).

2.5 Electrification Policies for China's Buildings Sector

2.5.1 Clean Heating in Northern China

In the 13th Five-year Plan for 2016–20, China underscored the need for clean heating initiatives to encourage the shift from coal and gas to electricity in the northern part of the country.

At the national level, ministries and commissions issued policies, such as the Clean Winter Heating Plan for North China (2017–21) and the Notice on Pilot Fiscal Support for Winter Clean Heating in North China, to encourage clean heating in the north by means of central and local subsidies. At the local level, the governments of seven provinces in the Beijing–Tianjin–Hebei region and the Fenwei Plain issued 94 policies to support the shift from coal to electricity (14 at provincial level and 80 at prefectural city and county levels). These policies are mainly of the following types:

End-user equipment purchase policies in Beijing, Tianjin, Hebei, Shanxi, Shandong, Henan and Shaanxi. For example, the Tianjin Municipal People's Government covers 100% of the purchase and installation costs of low-

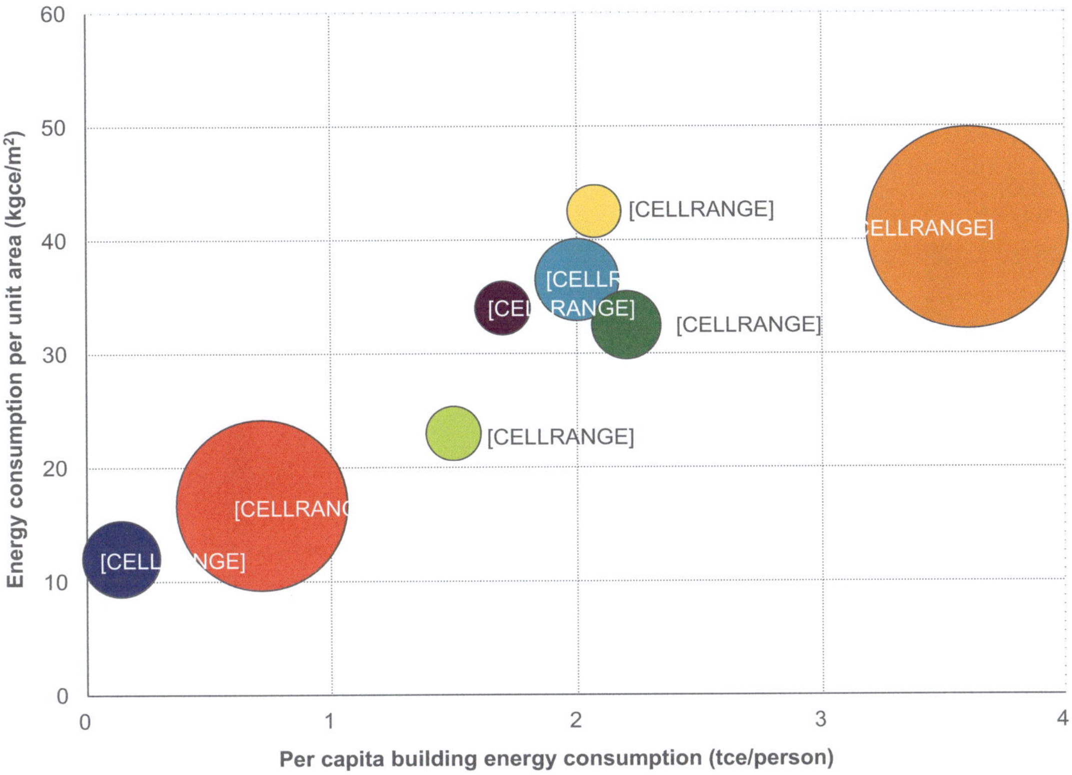

Fig. 4 Building energy consumption in China and other countries, 2017. *Note* tce = tonnes of coal equivalent; kgce/m^2 = kilograms of coal equivalent per square metre. *Source* International Energy Agency, World Energy Balances 2020; World Bank, World Development Indicators

temperature air-source heat pumps under the coal-to-electricity initiative and provides a subsidy of up to RMB 29,000 per household.

Heating fee (electricity price) policies in Beijing, Tianjin, Hebei, Shanxi, Shandong, Henan and Shaanxi. For instance, the Beijing municipal government gives a subsidy of RMB 0.2 per kWh to electric heating users during off-peak hours.

House retrofit policies. The People's Government of Beijing Municipality provides an energy efficiency and thermal insulation subsidy of RMB 20,000 per household for new builds and an energy efficiency and thermal insulation retrofit subsidy of RMB 10,000 per household for existing houses.

The above-mentioned policies have effectively driven the implementation of clean heating under the coal-to-electricity initiative in the seven provinces.

Thanks to strong policy support in Beijing, the number of users shifting from coal to electricity is more than 1.25 million, with electric heating the second largest source of heating in Beijing. The plains around Beijing have become more or less coal-free, significantly reducing air pollution in the capital.

By the end of 2020, 10.63 million households in the Beijing–Tianjin–Hebei region and surrounding areas, as well as the Fenwei Plain and Xinjiang, had switched to clean electric heating. As a result, the total building area of homes using electric heating was 826 million m^2, mainly in rural areas. During the 13th Five-year Plan for 2016–20, total electricity use for heating exceeded 50 billion kWh, equivalent to a reduction of 28 million tonnes in scattered (poor quality) coal burning and 50 million tonnes in CO_2 emissions.[8]

[8] Yearbook of the State Grid Corporation of China 2021.

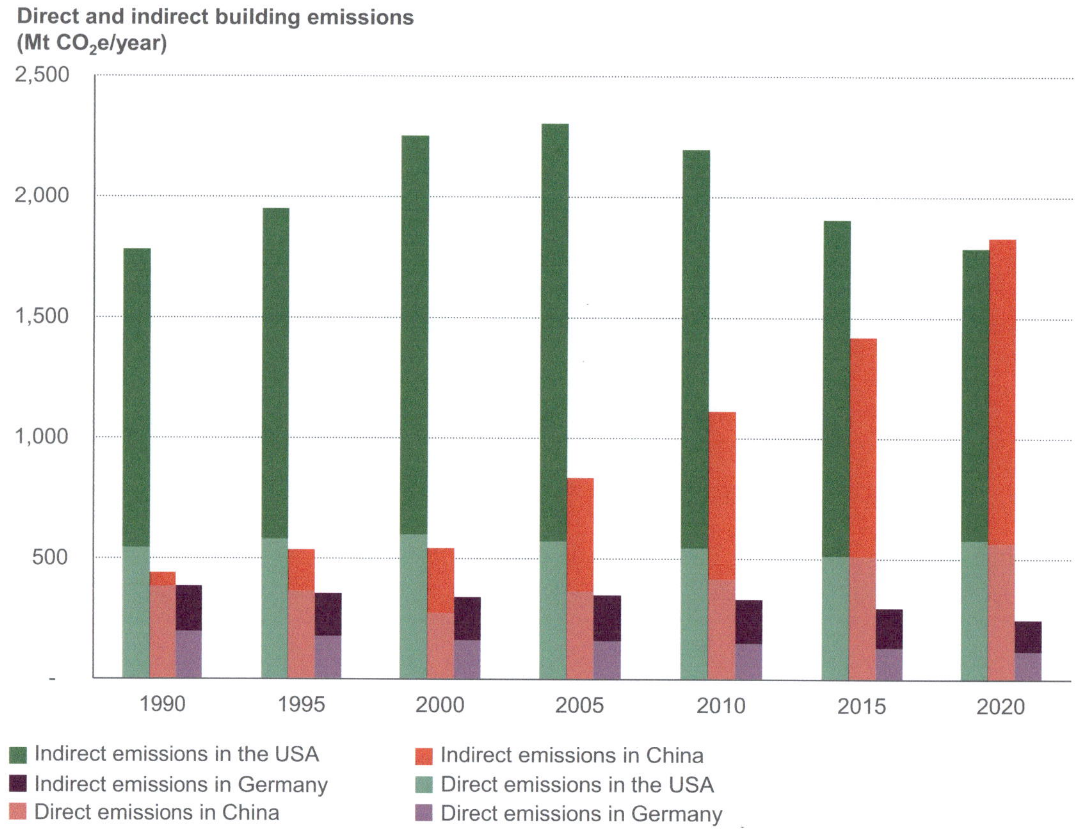

Fig. 5 Direct and indirect emissions from buildings. *Note* Total emissions are the sum of direct and indirect emissions. Indirect emissions are caused by the use of electricity and centrally produced heat in the buildings sector. *Source* Shell–DRC project team

2.5.2 Efforts to Shift Home Heating to Electricity

In May 2016, eight central departments, including the National Development and Reform Commission (NDRC) and the National Energy Administration (NEA), jointly issued a policy to promote the shift to electricity from fossil fuels. In the 13th Five-year Plan (2016–20), the initiative was comprehensively promoted in four key areas: residential heating in northern China, manufacturing, transport, and power (supply and consumption). As a result, scattered coal and fuel oil consumption of about 130 million tonnes of coal equivalent were replaced by electricity in final energy use. The share of thermal coal in total coal consumption increased by about 1.9 percentage points, and the share of electricity rose by about 1.5 percentage points to 27%.

Encouraged by the policy, as well as by economic growth and improvement in living standards, a growing number of urban and rural households bought home appliances. In the 13th Five-year Plan period, 2.2 billion home appliances were sold, effectively driving up the electrification rate in the buildings sector.[9]

2.5.3 Efforts to Control Air Pollution in Autumn and Winter

Since 2017, China has made far-reaching efforts to control air pollution in autumn and winter in 11 provinces, including Beijing, Tianjin, Hebei, Shanxi, Shandong, Henan, Shaanxi, Shanghai, Jiangsu, Zhejiang and Anhui. Coal- and oil-fired boilers with a capacity of 35 steam tonnes per

[9] Ibid.

Table 2 Electrification level of China's buildings sector in 2018

Energy use category	Area (billion m^2)	Households (million)	Energy consumption (billion tce)	Electricity use (billion kWh)	Proportion of electricity in final energy use (%)	Carbon emissions per unit of energy consumption (kg/kgce)
Northern heating	14.7	–	0.212	57.1	4.1	2.6
Urban residential buildings (excluding heating)	24.4	298	0.241	540.4	34.2	1.8
Public buildings (excluding heating)	12.8	–	0.332	809.9	37.2	1.9
Rural residential buildings	22.9	148	0.216	262.3	18.5	2.2

Source Research Report on the Energy Efficiency of Buildings in China 2021; China Energy Statistical Yearbook 2019; China Urban and Rural Construction Statistical Yearbook 2021

hour or less, combined heat and power plants with a capacity of 300 megawatts, and outmoded small coal-fired thermal power plants were shut down or replaced with electric boilers, accelerating the electrification of urban heating in northern China.

3 The Electrification and Decarbonisation Potential of Buildings

3.1 Electrification and Energy Efficiency Improvements Are Key to Decarbonising Buildings

According to the International Energy Agency,[10] the first step to reduce the carbon intensity of buildings is to increase the efficiency of building insulation (roofs, walls, floors, etc.) and appliances. Switching fuel to electricity, hydrogen, bioenergy or other renewables, especially low-carbon electricity, is an essential complement to greater efficiency. Fuel switching involves adopting new appliances by, for instance, changing to a highly efficient heat pump from a traditional boiler that runs on fossil fuel. Consumer behaviour also needs to change to minimise energy consumption. This includes closing windows when cooling or heating appliances are in use or only using a cooling device when a room is occupied.

[10] International Energy Agency, Net Zero by 2050: A Roadmap for the Global Energy Sector, 2021.

3.2 Factors Influencing the Electrification of Buildings

To better estimate the electrification potential of buildings, it is necessary to analyse the key factors that influence the electrification of buildings in China. These are principally the economy, policy and technology.

3.2.1 Economy

First, the urbanisation rate. Urbanisation is a process in which people move from single-storey rural houses to multi-storey urban residential buildings. This significantly raises the building electrification rate. Second, per capita income. Higher per capita income and living standards result in people buying more household, digital

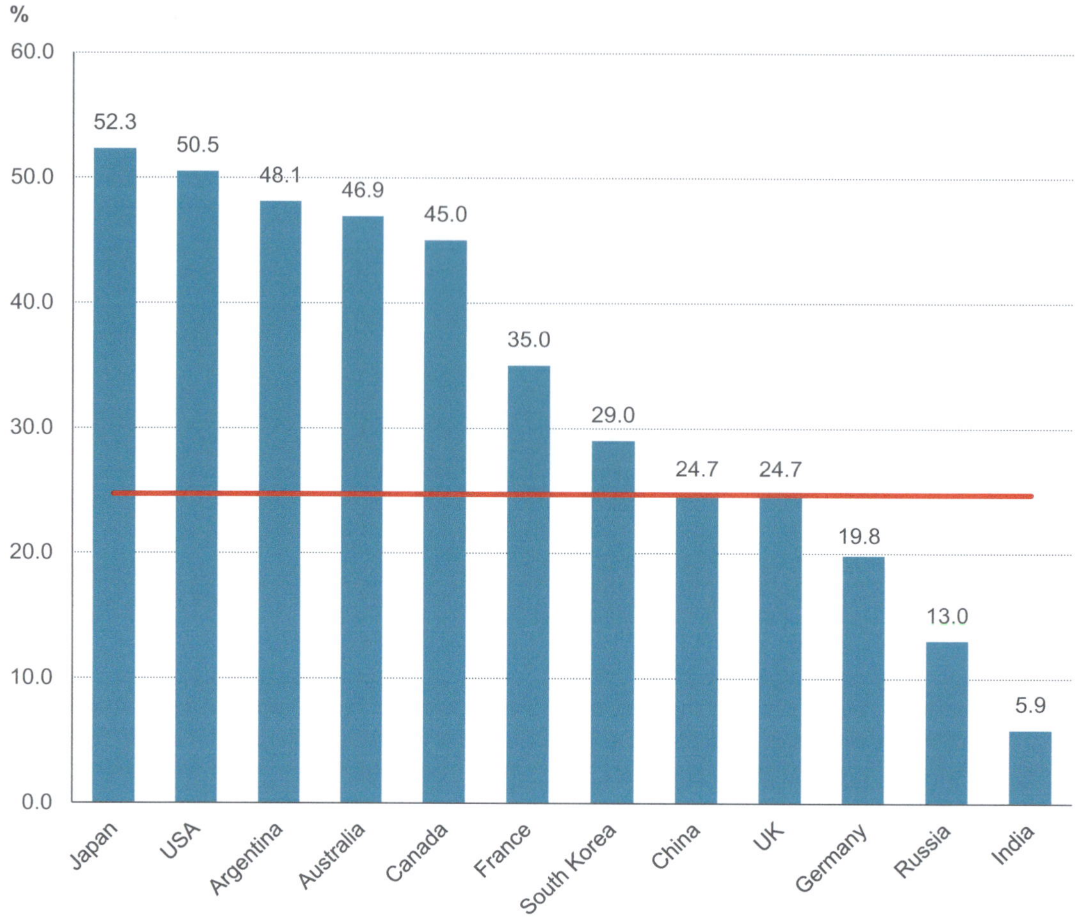

Fig. 6 Proportion of residential electricity use in final energy consumption. *Source* International Energy Agency, World Energy Balances 2020

and smart home appliances, which also significantly increases the electrification rate. Third, the transfer of the national economy's engine room to South China. Air conditioners are used for cooling in summer and heating in winter in southern China, which also improves the overall electrification rate of the buildings sector.

3.2.2 Policy

First, national polices on energy. These policies will fundamentally change how energy is produced and consumed. They support China's revolutions in energy production and consumption and advance the electrification and decarbonisation of buildings. Second, policies on energy efficiency. Electrification is the key to making buildings energy efficient and zero carbon. Third, policies to control air pollution. Replacing coal-fired boilers and scattered coal burning is the key to controlling air pollution. Substituting electricity for coal and oil will increase the use of electric heating and electric cooking technologies.

3.2.3 Technology

First, new energy development. The rapid development of solar photovoltaic and wind power will help form an electricity-centred energy consumption system. This will significantly increase the electrification rate and the use

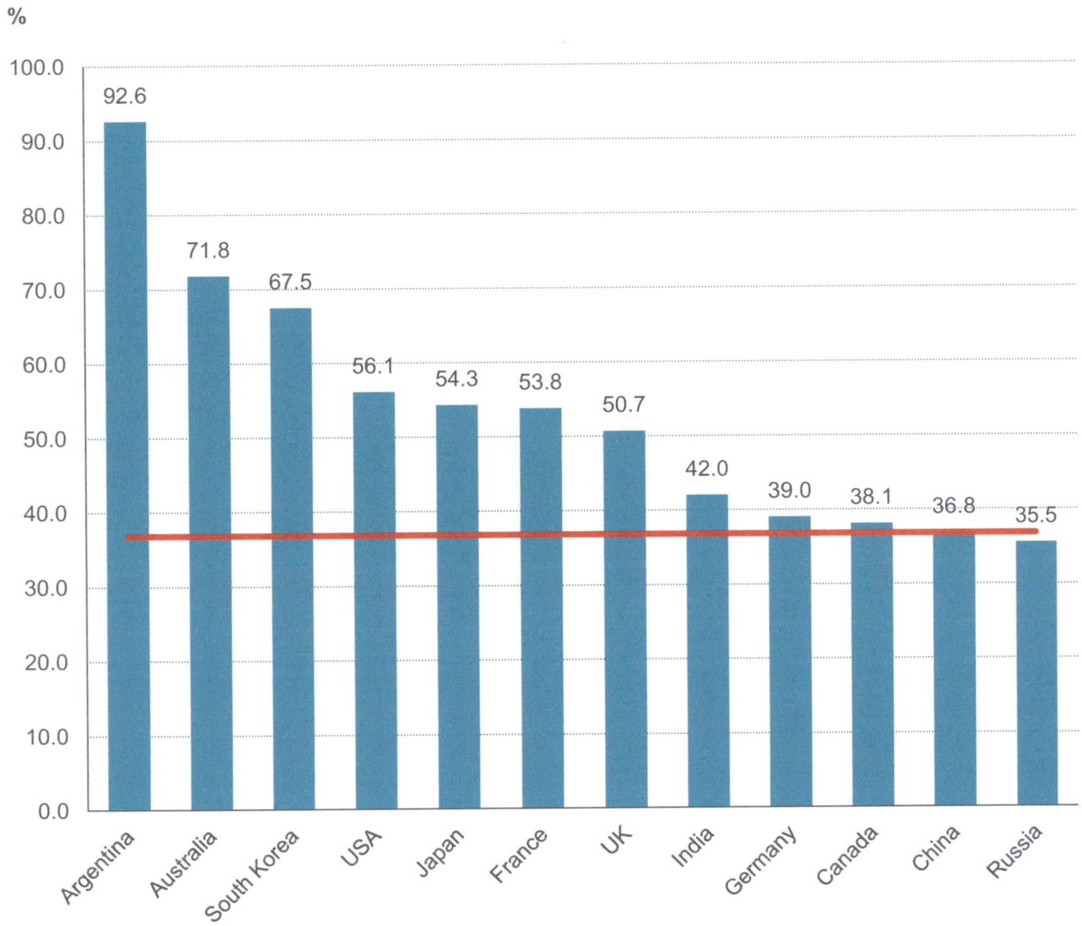

Fig. 7 Proportion of commercial electricity use in final energy consumption. *Source* International Energy Agency, World Energy Balances 2020

of clean and low-carbon energy in the buildings sector. Second, the development of new building energy technologies. With the gradual maturity and commercial deployment of new technologies such as building integrated photovoltaics, rooftop photovoltaics and passive (energy-efficient) buildings, the energy consumption per unit of building area will continue to decrease. Third, the development of rural renewable energy technologies. The evolution of distributed photovoltaic systems, small biomass power plants, solar electric water heaters and other small renewable-based devices will improve the electrification of energy demand in rural areas, especially in rural residential buildings.

3.3 The Electrification Potential of Buildings

In this section, we estimate the electrification potential of the buildings sector, taking into account the sector's current electrification rate and carbon emissions, the national dual-carbon goal, the level of urbanisation and the speed of new energy development.

3.3.1 Basic Data

First, China's urbanisation rate, total population and urban population. China is expected to have a total population of 1.44 billion and an urban population of 1.06 billion by 2030, which means an urbanisation rate of 73.6%. Reports published by the United Nations and other international organisations estimate China's urbanisation rate to be 83.2% by 2060, with a total population of 1.25 billion and an urban population of 1.04 billion.

Second, the annual building area of new builds and buildings demolished. The annual increase in building area of urban residential buildings, rural residential buildings and public buildings is estimated to be 1.2 billion m^2, 500 million m^2 and 500 million m^2 respectively. By 2030, the total building area of urban residential buildings, rural residential buildings and public buildings will be 38.8 billion m^2, 16.9 billion m^2 and 18.8 billion m^2 respectively, amounting to 74.5 billion m^2. By 2060, using the building area per capita in developed economies as a guide, the area of urban residential buildings, rural residential buildings and public buildings is estimated to reach 47.8 billion m^2, 9.7 billion m^2 and 21.3 billion m^2 respectively, totalling 78.8 billion m^2.

Third, the annual increase in urban heating area in northern China. If the urban heating area is assumed to increase by 500 million m^2 annually by 2025, and by 400 million m^2 annually by 2030, the urban heating area in northern China will reach 20.2 billion m^2 in 2030. If the proportion of urban heating in northern China in 2060 is the same as in 2030, the heating area will be 24.2 billion m^2 in 2060.

3.3.2 Energy Intensity and Total Energy Consumption

The energy intensity per unit area of the four building categories is based on the outcome of efforts to improve energy efficiency and reduce emissions, as well as on the increasing demand for energy due to a rising standard of living.

As living conditions improve and people's need for a better life is gradually met, the energy intensity of residential buildings in China will approach the level of developed countries. When carbon emissions peak around 2030, the energy intensity of urban residential buildings (excluding northern urban heating) is projected to reach 12.5 kilograms of coal equivalent per square metre (kgce/m^2), and that of rural residential buildings 12.7 kgce/m^2. Energy conservation, emission reductions and the implementation of the dual-carbon strategy are expected to lead to peak carbon emissions in public buildings before 2025 and lower their energy intensity to 22.0 kgce/m^2 by 2030. Thanks to improvements in insulation, high-efficiency windows, energy efficiency, heat metering and other measures, the energy intensity of urban heating in northern China will gradually decrease to 10.0 kgce/m^2 by 2030.

If China meets its carbon neutrality goal by 2060, the energy intensity of northern heating, urban residential buildings, public buildings and rural residential buildings is estimated to decrease to 6.0 kgce/m^2, 6.0 kgce/m^2, 14.0 kgce/m^2 and 10.0 kgce/m^2 respectively, with the average building energy intensity dropping to 0.5 kgce/m^2.

Figure 8 shows the energy intensity forecasts, which predict average building energy intensity to gradually decline after peak carbon emissions is reached around 2027.

A comparison of Figs. 8 and 9 shows that the total energy consumption of urban residential buildings and public buildings (excluding northern heating) will peak around 2030, although at a much higher level than in 2018. The main reasons for the increase are continuous growth in urban population and public building areas, as a result of which total energy consumption will keep rising after energy intensity peaks.

The total energy consumption of China's building operations will peak around 2030 at 1.32 billion tce, and then decline to about 840 million tce in 2060.

3.3.3 Electricity's Share of Final Energy Use and Total Electricity Consumption

Table 3 shows the proportion of electricity in final energy use in the buildings sector will continue to rise, reaching 39.8% by 2030. If the goal of carbon neutrality is met, the electrification rate will jump to 73.3% of final energy consumption in buildings.

Total electricity consumption in the buildings sector will be about 3.5 trillion kWh in 2030, more than double that of 2018. Total electricity use in the sector will be about 4.6 trillion kWh in 2060, 2.7 times higher than in 2018 (Table 4).

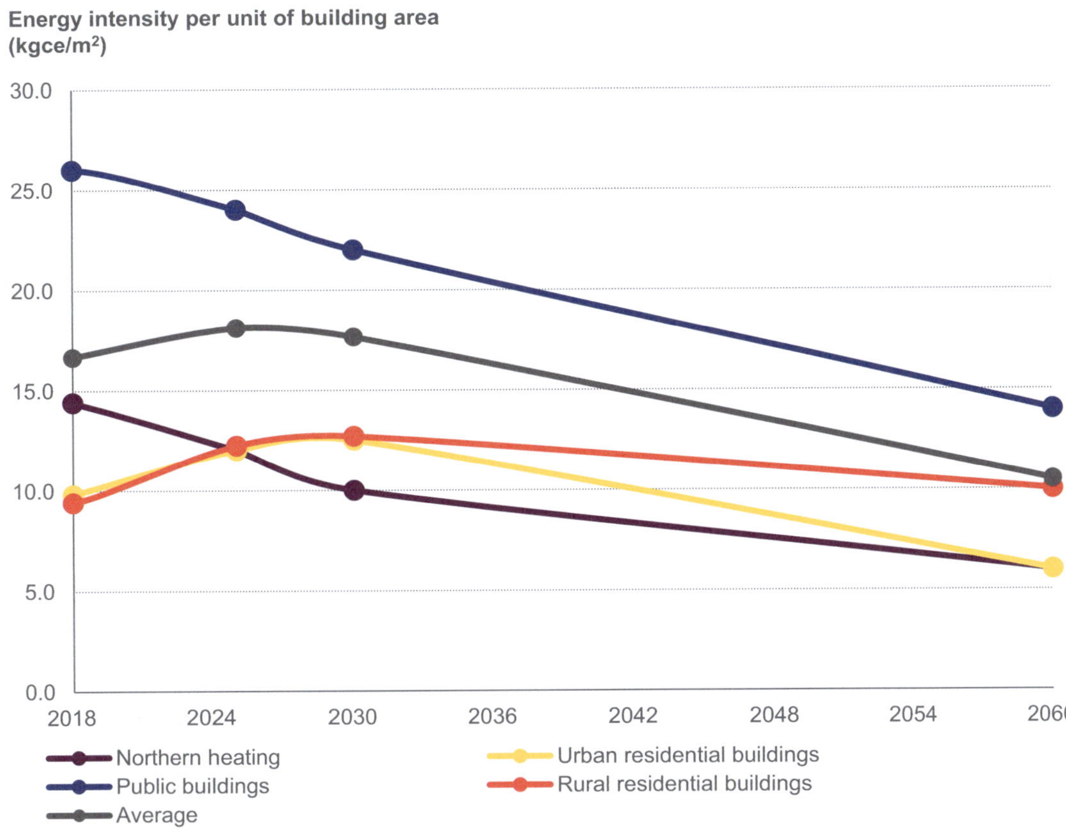

Fig. 8 Building energy intensity, 2018–60. *Source* Calculations based on the models used in this book

4 Technical Pathways to the Electrification and Decarbonisation of Buildings

This chapter describes six technical pathways to the electrification and decarbonisation of the buildings sector. They in turn depend on 12 key technologies: nearly zero-emission buildings, digital integrated energy stations, electric boilers with thermal storage for heating, heat pumps for cooling and heating, municipal central heating with supplementary electric heating, commercial electric cookware, household electric cookware, energy-efficient home appliances, solar water heaters, rooftop solar photovoltaics, solar-powered fish farms, and integrated biomass systems.

4.1 Low-carbon, Energy-efficient New Builds

A 2020 study found that implementing energy efficiency standards and installing low-carbon heating systems during construction adds only 3–5% to the total building cost. In comparison, retrofitting the same house 10 years later would incur up to 20% of the building cost. Ensuring new homes are low carbon and energy efficient represents a significant opportunity to decarbonise the sector.[11]

In China, the buildings sector still holds huge growth potential. The Chinese building stock is set to increase by almost 150% in some

[11] Element Energy, Development of trajectories for residential heat decarbonisation to inform the Sixth Carbon Budget: A study for the Climate Change Committee, 2021.

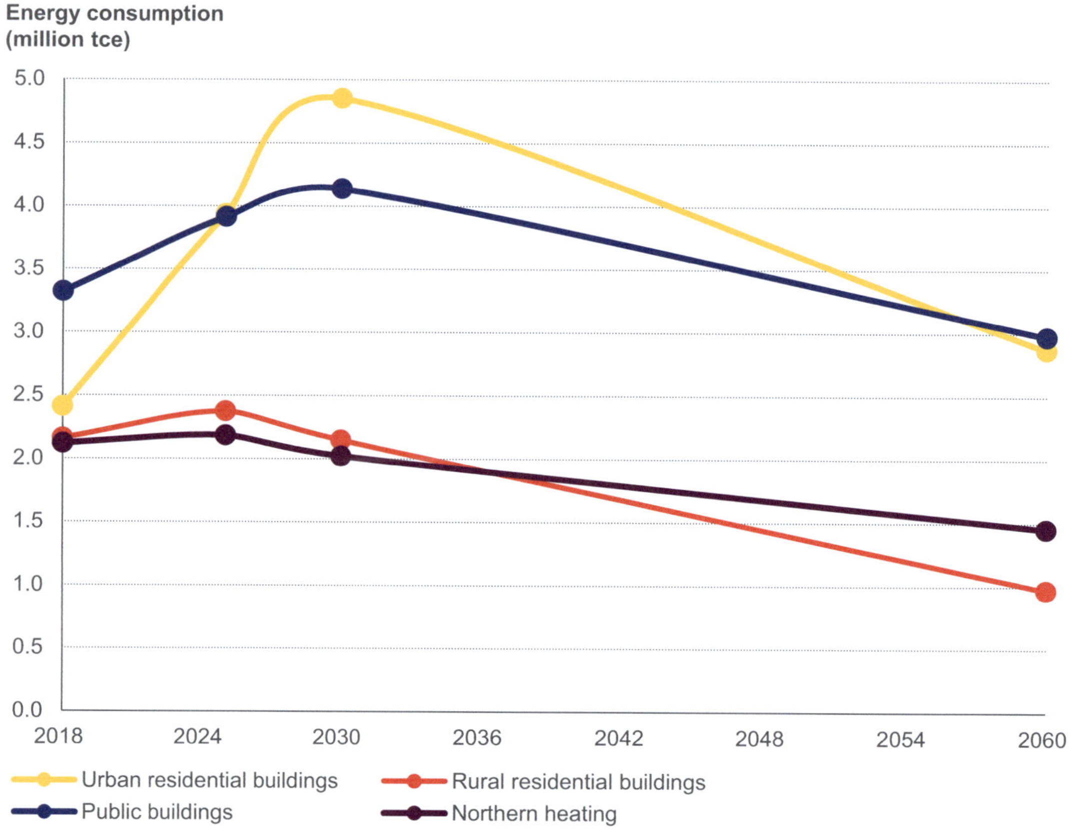

Fig. 9 Total energy consumption of buildings, 2018–60. *Source* Calculations based on the models used in this book

Table 3 Electricity's share of final energy use in the buildings sector

Year	Northern heating (%)	Urban residential buildings (excluding heating) (%)	Public buildings (excluding heating) (%)	Rural residential buildings (%)	Proportion (%)
2010	0.7	37.1	38.4	12.2	22.2
2011	0.7	35.9	40.5	12.1	21.7
2012	0.7	35.1	41.4	14.3	23.1
2013	0.8	44.1	46.1	13.9	27.1
2014	0.8	32.5	38.3	14.2	22.4
2015	2.2	32.9	38.1	14.7	23.2
2016	2.3	32.9	37.5	15.3	23.5
2017	3.9	34.2	38.7	14.4	24.2
2018	4.1	34.2	37.2	18.5	25.4
2025	8.0	40.0	43.0	27.0	33.3
2030	15.0	45.0	48.0	34.0	39.8
2060	65.0	78.0	80.0	70.0	73.3

Source Calculations based on the models used in this book

Table 4 Electricity use in the buildings sector

Year	Northern heating (billion kWh)	Urban residential buildings (excluding heating) (billion kWh)	Public buildings (excluding heating) (billion kWh)	Rural residential buildings (billion kWh)	Total (billion kWh)
2018	57.1	540.4	809.9	262.3	1669.7
2025	115.1	1037.4	1108.4	422.2	2683.1
2030	202.2	1456.3	1324.7	486.9	3470.1
2060	692.8	1642.4	1746.9	497.7	4579.8

Source Calculations based on the models used in this book

segments, meaning a large portion of the 2050 stock has yet to be built,[12] as shown in Fig. 10. Low-carbon new builds will therefore make a big difference.

The Technical Standard for Nearly Zero Energy Buildings divides energy-efficient buildings into three categories: ultralow energy buildings, nearly zero energy buildings, and zero energy buildings. Each category has different energy consumption reduction requirements. Ultralow energy buildings should have an energy consumption less than 50% of the relevant national energy efficiency standard. Nearly zero energy buildings should have an energy consumption that is 60–75% lower than the relevant national threshold. And zero energy buildings should use renewable energy resources in the building and from the surrounding area to produce more or the same amount of energy that they consume.

To meet the requirements of the nearly zero standards, buildings must follow the principle of climate adaptability and minimise energy consumption for heating and cooling, without compromising the building's function or damaging the environment. Passive building design measures—such as natural lighting, natural ventilation and insulation—should be maximised to reduce energy demand and improve resilience.

4.1.1 Case Study 1 | London

The London Energy Plan supports the city's ambition to become a zero-carbon city by 2050 by enforcing requirements on buildings to be zero-carbon ready (see Fig. 11). New developments are required to be net-zero carbon, meaning they should maximise opportunities to reduce the amount of energy used, both in construction and operation, as well as to use local, low-carbon sources of energy. Thresholds for energy performance were set which go beyond national standards, reflecting the higher level of ambition of the city. This level of ambition is enabled by the city's planning authority, as developers must detail how they intend to meet the requirements in their applications for planning permission, the success of which is conditional on their meeting the requirements.

New builds and retrofits are tackled together. Developers that cannot meet the standards must contribute to local carbon offset funds, which can be used to fund low-carbon retrofits. Some developers who are unable to meet the standards through energy savings or low-carbon energy production on-site may meet the standards by contributing to their local authority's carbon offset fund. The offset funds go toward funding low-carbon retrofits in the area through grants. The carbon offset fund for the London Borough of Camden, for instance, provides grants of up to 50% of the cost of insulation, heat pumps, solar panels or solar thermal systems in local homes and businesses.

In parallel, London encourages retrofits through additional measures such as the Solar Together initiative, which helps individuals and businesses identify certified providers to lower their costs. Solar Together allows Londoners to sign up to a group-buying scheme for solar panels. An auction is held to ensure that the most competitive providers are selected. This makes

[12] Lixuan Hong, Nan Zhou, David Fridley, Wei Feng and Nina Khanna, Modeling China's Building Floor-Area Growth and the Implications for Building Materials and Energy Demand, 2014.

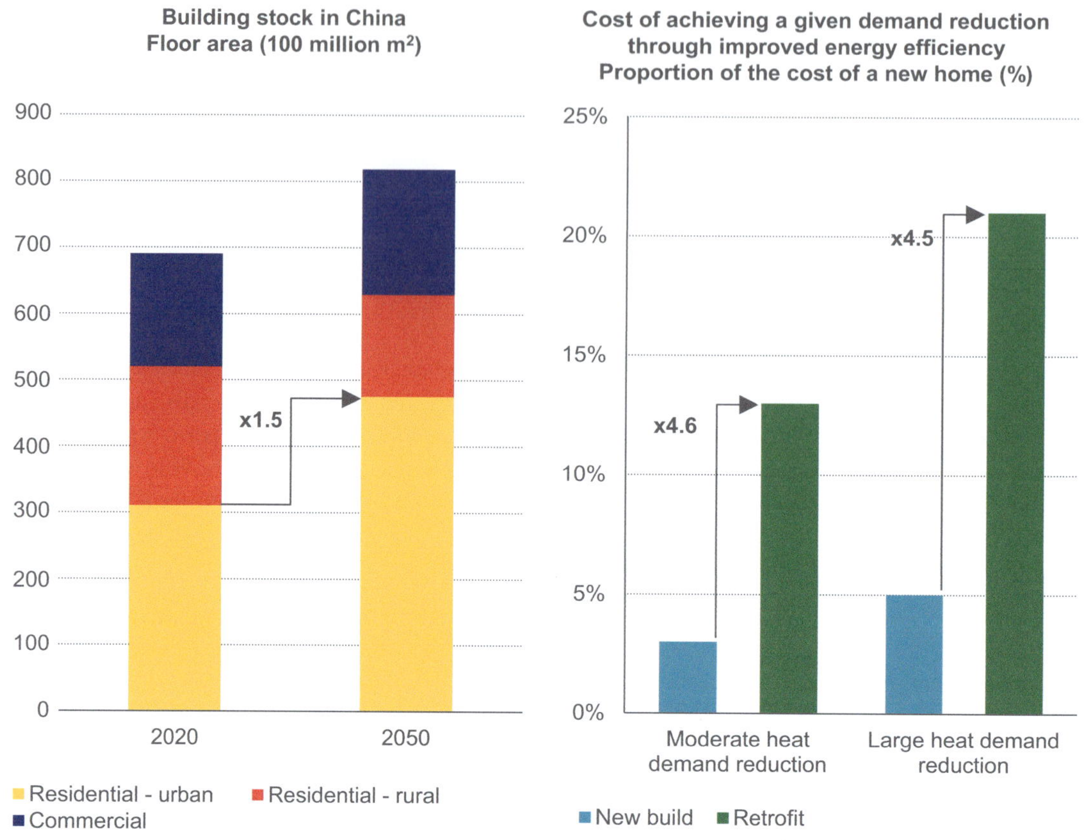

Fig. 10 The decarbonisation potential of new builds. *Note* Assuming a single-household home building cost of GBP 125,000 (or $175,000 at the average exchange rate for 2021). *Source* Element Energy and Hong et al. (see footnotes 11 and 12)

for high-quality installations, while lowering the cost for Londoners of looking for recommended providers. The scale achieved through group-buying means costs are reduced for each participant. The initiative also supports the development of the supply chain by providing large-scale orders, which incentivises investment in skills and service-building among providers.

4.2 Digital Energy Services for Large Public Buildings

The term large public buildings generally refers to premises that have a floor area of more than 20,000 m^2 and which are used for multiple purposes: as offices or for commerce, tourism, science, education, culture, health, communications or transport. Due to the size and use constraints of these buildings, their energy intensity from air conditioning, ventilation, lighting, lifts and elevators is higher than in other types of public building. This is an important reason for the increase in energy intensity of public buildings in China.

Large public buildings rely on various mechanical and electrical equipment systems. Air conditioning, water supply, waste water and lighting systems are typically responsible for 70% of the buildings' energy consumption, of which air conditioning alone accounts for more than 40%. The energy efficiency of mechanical and electrical equipment is, therefore, the key to reducing energy use in large public buildings.

Integrated energy services for large public buildings typically comprise digital energy

Fig. 11 London's ambition to become a zero-carbon city by 2050. *Source* Shell–DRC project team

technology and various other products and services such as an energy management system, energy efficiency analysis tools, energy contracting and power demand-response. By providing multi-energy system services, the energy intensity per unit of building area, as well as energy costs and carbon emissions, can be reduced. Measures include using low-grade heat and cooling (such as waste heat and natural cooling from the air, ground or water) and smart digital technologies.

4.2.1 Case Study 2 | Tokyo

Tokyo's cap and trade scheme encourages the efficient allocation of emission reductions across covered buildings. The scheme, whose first mandatory compliance period began in 2010, covers large commercial buildings (see Fig. 12), which together were responsible for 40% of all commercial building emissions in 2015. A mandatory cap on their emissions would impose larger costs on buildings that have fewer options to reduce emissions due to their age, construction materials or use patterns. The scheme incentivises emission reductions for those buildings that can achieve them cost effectively by purchasing credits from other covered facilities or from smaller commercial buildings and buildings outside Tokyo. By having five-year compliance periods, the scheme gives facilities time to plan their emission reduction measures in the medium term. The reduction targets have increased significantly for each compliance period, allowing for a progressive increase in ambition towards the Tokyo Metropolitan Government's longer-term emissions target.

By crediting facilities which procure or generate renewable electricity, the programme addresses all components of building emissions and bolsters local renewable generation. The programme is primarily focused on emission reductions through energy efficiency improvements, as these can be realised by individual facilities regardless of their source of energy. However, the scheme also supports more holistic approaches to emission reductions by providing renewable energy certificates to facilities which procure or generate renewable energy. Some 22% of emitted credits under the scheme have renewable energy certificates, meaning it has had an impact on local generation and wider electricity grid renewable penetration.

Public building retrofits provide an opportunity for energy-saving companies to build skills. It also places the responsibility for saving energy on them and limits the information burden on the authorities.

4.2.2 Case Study 3 | Paris

Retrofitting public buildings can support the development of low-carbon retrofit supply chains and have positive effects on the private sector (see Fig. 13). Low-carbon retrofits require additional skills compared to traditional construction, for instance in geothermal systems, air and vapour barriers and insulation. Public buildings make up a large share of the building stock and a large proportion of city emissions—on average, 19% of fixed assets are publicly owned in OECD countries. Retrofits for public buildings can support the development of skills in the supply chain by providing contracts on a large enough scale to make investments viable, while leading to a significant reduction in emissions in cities.[13]

The Paris scheme places the responsibility for achieving energy savings on covered buildings onto the company undertaking the renovation, which reduces the need for highly accurate standards.

Setting standards for construction methods and materials and appliances is a common method to achieve energy efficiency gains. This approach requires accurate information on the energy efficiency gains possible thanks to the standards. It requires monitoring and enforcement to ensure that the standards are respected.

The scheme in Paris circumvents these issues by giving the renovating company responsibility for the energy savings to be realised. Energy efficiency contracts are signed between the Paris municipality and an energy service company (ESCO), in which the company guarantees energy savings over 20 years. If the savings are not realised, the ESCO is liable for the remainder.

The energy savings target was set at an average of 30% across all the schools covered. This enabled improvements to be allocated more efficiently between buildings, depending on their condition and potential for improvement, compared to setting a target for each school.

[13] Hasanbeigi, A., Becqué, R., and Springer, C., Curbing Carbon from Consumption: The Role of Green Public Procurement, International Orginization of Pendamic Controls, 4, 2021, 1–104.

Holistic approaches to building retrofits help improve the acceptability of retrofits while improving their financial viability.

4.3 Clean Heating in Northern and Southern China

4.3.1 The Coal-to-Electricity Clean Heating Initiative in Northern Rural Areas

The shift from coal to electricity in northern China has yielded remarkable results: the widespread use of scattered (poor quality) coal has declined and air quality in Beijing and the surrounding areas has improved.

Rural residential buildings make up 95% of the buildings in the clean heating initiative. The policy plays an important role in encouraging rural residents to use electricity, which improves people's quality of life and speeds up the electrification of rural residential buildings.

A cost-effective and sustainable technical pathway should be selected that follows the principle of "adapting the use of electricity, natural gas, coal and geothermal energy for heating to local conditions" and takes the total operating cost into consideration. Table 5 is a comparative analysis by the State Grid Corporation of China of the annual life-cycle cost of electric heating per unit of building area in five locations in four provinces. Life-cycle costs include the construction of supporting power networks and the maintenance of electric heating equipment.

From the comparison of the above annual life-cycle costs, the following conclusions can be drawn:

First, central and decentralised electric heating have the same life-cycle economy. Taking reliability, power network, home comfort and regional differences into consideration, a central electric boiler with thermal storage is the best option. Among distributed technologies, air-source heat pumps are preferable.

Second, a thermal insulation retrofit can greatly reduce the investment, operating and life-cycle costs of electric heating.

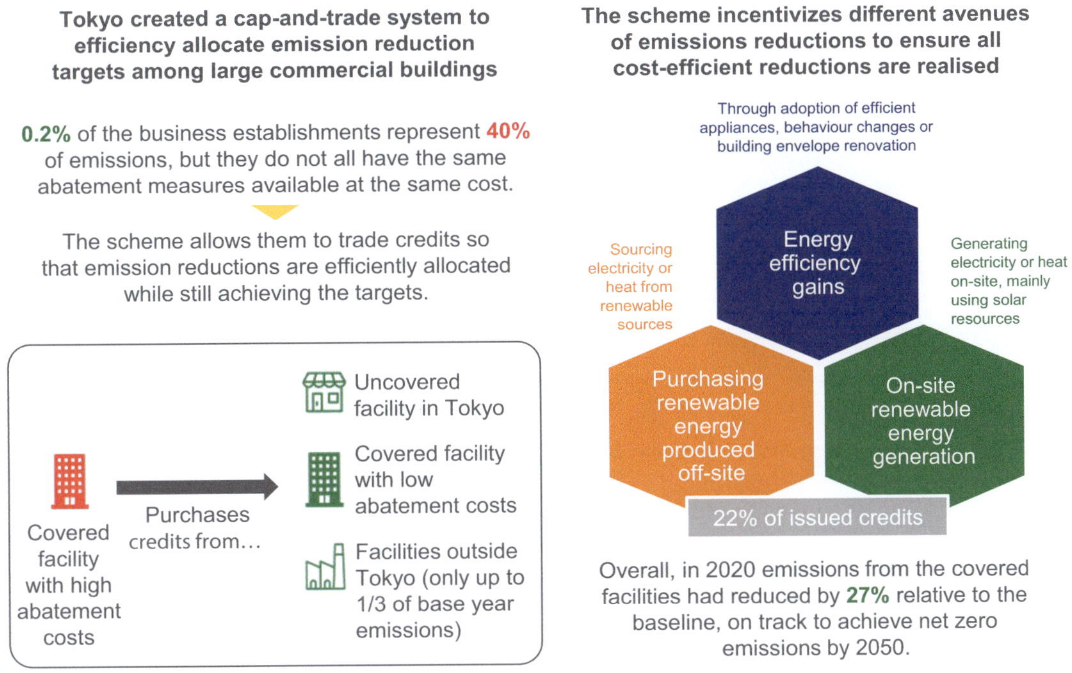

Fig. 12 Tokyo's cap and trade and incentive systems. *Source* Roppongi, H., Suwa, A., and Puppim De Oliveira, J. A., Innovating in sub-national climate policy: the mandatory emissions reduction scheme in Tokyo, in Climate Policy, 17 (4), 516–532; International Carbon Action Partnership, Japan–Tokyo Cap-and-Trade Program, May 2021,1–6

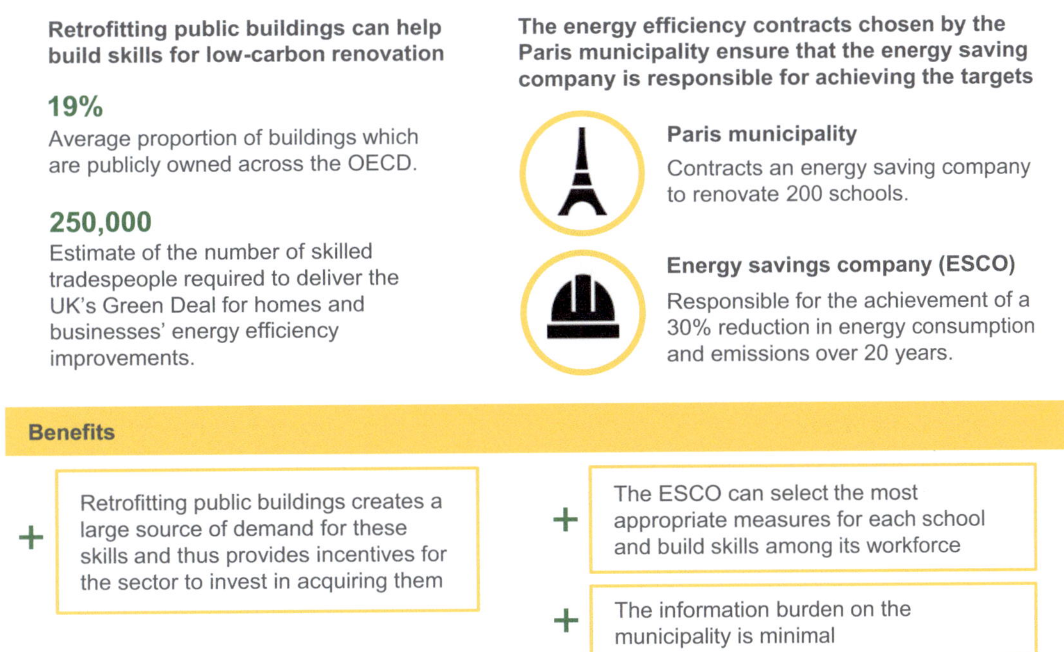

Fig. 13 Green and low-carbon retrofits of public schools in Paris. *Source* Canada Green Building Council, Trading Up: Equipping Ontario Trades with the Skills of the Future, 2019; C40 Cities, Case study—Paris Retrofit Project, 2014

Table 5 Comparative analysis of the annual life-cycle cost of electric heating per unit of building area in five villages

Category	Central heating RMB per m^2/year		Decentralised RMB per m^2 /year	
	Electric boiler with thermal storage (water, solid-based)	Air-source heat pump	Electric heater with thermal storage	Air-source heat pump
Changmenkou village, Beijing	41.53	32.78	45.41	25.29
Changmenkou village (incl. building insulation), Beijing	25.89	20.14	30.20	18.39
Dongliumu village, Tianjin	25.56	31.28	31.34	23.23
Qilindian village, Hebei	28.89	32.06	32.43	25.00
Luanzhuang village, Shandong	34.02	43.25	42.69	43.67

Source Marketing Department, State Grid Corporation of China

Third, heat pumps are particularly advantageous in buildings in areas with a long heating season as the operating and life-cycle costs of heat pumps are then far lower than those of other technologies.

Electric boilers with water-based thermal storage should be prioritised in clusters of villages, new rural communities and suburban residential communities. In areas with relatively high winter temperatures (above −10 °C), the use of central air-source heat pumps should be explored, and the cost effectiveness and energy efficiency of electric heating improved through optimised control. In villages and towns where households are concentrated, central and decentralised electric heating systems are recommended. In villages where households are dispersed, distributed electric heating systems are more suitable.

4.3.2 Heat Pumps for Southern Areas with Hot Summers and Cold Winters

As China builds a moderately prosperous society, people's hopes for a better life gradually increase. In recent years, people in southern China have voiced the need for clean heating in winter.

National support policies to promote heating in South China have yet to be introduced. The Ministry of Housing and Urban-Rural Development advocates the science-based selection of appropriate heating technology in southern provinces. In alignment with the "action plan to make our skies blue again" and the "plan for energy development", the governments of Hunan, Jiangxi, Hubei and other provinces have proposed pilot central heating schemes for key industrial parks and development zones. Before 2010, cities such as Nanjing, Hangzhou, Wuhan, Yangzhou and Nantong developed central heating projects in some communities. Both Wanlou and Xiangtan used industrial waste heat and water-source heat pumps to provide central heating for residential communities. Hefei used multiple technologies, including ground-source heat pumps and wastewater-source heat pumps, to provide heating for Hefei Binhu Science City. Wuhan built two large combined heat and power plants, which serve two districts and one development zone.

Our recommendations for suitable pathways forward are as follows. First, urban areas with a high level of economic development and a high living standard should be given priority, especially cities with hot summers and cold winters. Second, the selection of river water, air, ground and wastewater source heat pumps and other technologies should be adapted to local conditions. Thanks to their high energy efficiency ratio, heat pumps are both energy efficient and cost effective, and can provide heating in winter and cooling in summer, greatly improving the technology utilisation rate. Third, centralised systems are preferred as they avoid multiple installations and allow operation and maintenance services to be provided by the developer.

Case Study 4 | Copenhagen

The district heating system in Copenhagen needs to be progressively decarbonised to increase its contribution to municipal emission reduction goals. District heating systems are a highly efficient source of heating that can significantly reduce emissions relative to decentralised systems thanks to economies of scale. In Denmark, where district heating systems have been in operation since the 1960s, the emission intensity of heating is 60% lower than it would have been without district heating, holding the fuel mix constant.[14] To capitalise on the benefits of district heating, Copenhagen is working to change the fuel mix underlying its system by moving away from fossil fuels, which today make up more than 40% of generation, towards renewables and waste heat. Combined heat and power plants are switching from coal or natural gas to biomass, while waste heat from local waste incineration or industrial plants is used where possible.

National legislation supports the city in implementing these changes while ensuring that costs remain affordable to users. The Heat Supply Act enables municipalities to make their own local heat plans with district heating companies. In addition, it provides municipalities with the authority to mandate connections to the system, which facilitates investment in the system by providing certainty over future income. The connection rate in Copenhagen is currently more than 98% of buildings.[15] Overall, the district heating system provides cheap heating to households, which pay around 45–55% less than they would using oil or natural gas central heating (see Fig. 14).

[14] Application for the 'Global District Energy Climate Award', Copenhagen District Heating System, 2009.

[15] Chittum, A. and Østergaard, P. A., How Danish communal heat planning empowers municipalities and benefits individual consumers, 2014.

4.4 Decarbonising Northern Urban Heating

Central heating predominates in northern urban areas. The main heating sources include combined heat and power plants and coal-fired and gas-fired boilers. Coal-fired heating accounts for more than 70% of the heating mix. Improving the energy efficiency of heating systems and using clean electricity and low-grade waste heat are the key technical pathways to decarbonise heating in northern urban areas.

4.4.1 Improving the Energy Efficiency of Central Heating Systems

Retrofitting energy-efficient and low-carbon solutions to central heating systems should be encouraged. Energy-efficient equipment such as balance valves and variable frequency drives should be used. Hydraulic imbalances in the heat supply network should be countered to reduce heat loss and improve on-demand and precision heating. This will also contribute to the "Three Zeros" target of zero throttling, zero overflow and zero overheating, which will reduce the amount of energy consumed per unit of building area. Intelligent heating management systems should also be installed to provide automatic heat regulation, remote monitoring and automatic fault-finding to improve the heating experience of residents and enhance energy efficiency.

Old heat supply networks, heat exchanger stations and indoor heating systems should be upgraded or replaced at a faster pace. Heat supply networks that are more than 15 years old and have severe problems of heat escaping should be renewed. Safety equipment should be included in pipeline retrofits whenever old heating systems are modernised. District heating networks should be designed to meet future requirements for zero-carbon heating. Old heat exchanger stations should be replaced with modern unmanned stations with automated control.

4.4.2 Optimising the Heat Source Mix

The mix of heat sources used should be optimised. In the short and medium terms this includes improving the efficiency of clean coal,

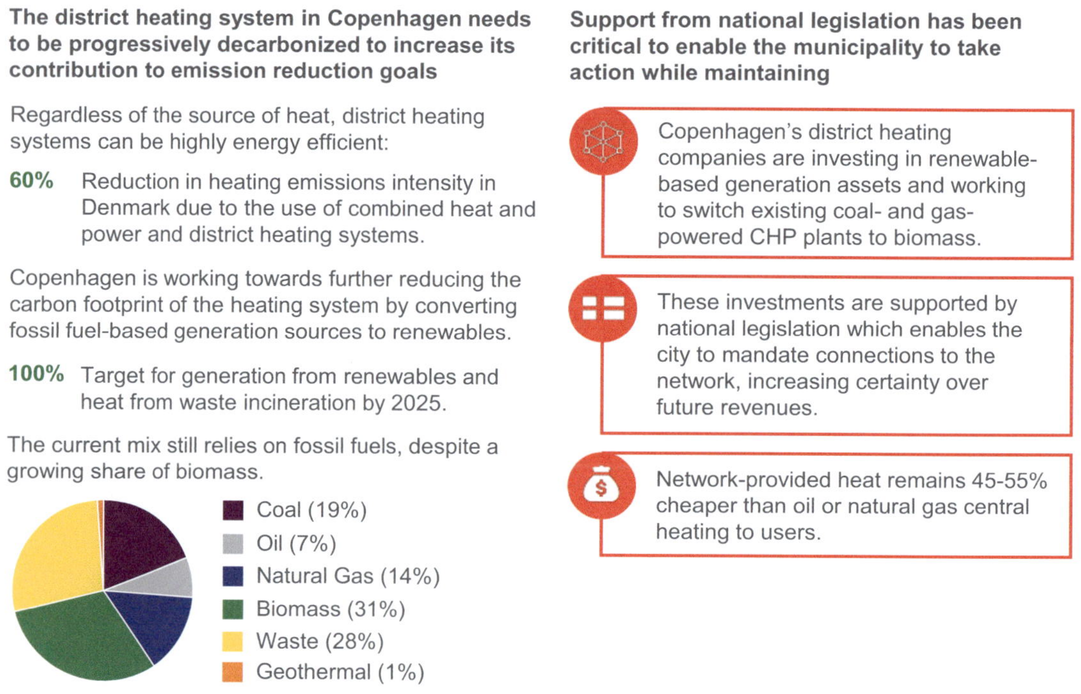

Fig. 14 Decarbonisation of the district heating network in Copenhagen. *Source* Shell–DRC project team

using natural gas and increasing the use of geothermal energy and waste heat. In the medium and long terms it means increasing the use of clean electricity and hydrogen.

Clean Coal.[16] Improve the efficiency. Combined heat and power (CHP) plants should be encouraged and new technologies deployed to reduce the return water temperature in heat supply networks and increase waste heat recovery. The heating potential of CHP units should be used to the full to enhance heating capacity.

Natural Gas. Gas-fired boilers should be prioritised as a peak load heat source for central heating systems, providing gas supplies are guaranteed. Coal-fired boilers should be gradually replaced by gas-fired boilers.

Geothermal Energy and Waste Heat. Standards and regulations for the development of geothermal energy should be introduced to ensure the orderly development of geothermal resources. Geothermal heating should be included in the scope of urban infrastructure development. Waste heat from data centres should be recovered at a faster pace, and the recovered waste heat should be used for heating in the cold season and for domestic hot water supply throughout the year.

Clean Electricity and Hydrogen. Priority should be given to the use of water, ground- and air-source heat pumps, gradually increasing the use of clean electricity for heating. Demonstration applications of hydrogen in district heating and CHP plants should be encouraged and advanced.

4.4.3 Case Study 5 | Stockholm

Stockholm's district heating system buys locally available waste heat on a trading platform that was established in 2014. This initiative has been particularly successful among the city's data centres, which generate large amounts of waste

[16] In China, clean coal is defined as the use of coal in an environmentally satisfactory and economically viable way.

Table 6 Comparison of commercial electric cooking technologies and conventional cookware

Benefits	Commercial electric cooking technologies	Conventional gas- and coal-fired cooking technologies
Energy efficiency and low carbon	Induction devices provide a thermal efficiency of more than 90%, triple that of conventional gas-fired stoves. They generate no exhaust fumes, greatly reducing carbon emissions	Average efficiency. CO_2 and other emissions are generated due to the use of fossil fuels
	No open flame or combustible gas, which eliminates secondary explosions. No unhygienic right angles where water and spillage can collect	Use of open flame; flammable and explosive
	Precision-control and easy to use and maintain; programmable control and timing; labour-efficient	Difficult to precision-control and labour intensive

Source Based on presentations of the State Grid Corporation of China

heat and have to spend money on cooling systems. More than 20,000 residential apartments can be heated from the waste heat of one 10 MW data centre. The additional revenue makes selling the heat to the networks significantly more attractive than other cooling solutions. In 2019, heat recovery reached 113 gigawatt-hours, or a 30% increase on the previous year.[17]

Regulatory changes in the EU would support the development of waste heat recovery initiatives throughout the union. The changes needed to increase the uptake of data centre waste heat recovery to decarbonise district heating systems include:

- using waste heat as a means to achieve renewable energy targets in heating and cooling systems: this would provide incentives for governments to support projects as part of their decarbonisation strategies;
- including waste heat from data centres in countries' analyses of their heating and cooling sectors, to inform their energy efficiency assessments; and
- encouraging the installation of new data centres in areas that have district heating systems to avoid waste heat in areas that cannot use it.

[17] Codema, From Data Centres to District Heating & Cooling: Boosting waste heat recovery to support decarbonisation, 2021.

4.5 Using Electric Cookware and Energy-efficient Home Appliances at Scale

4.5.1 Commercial Electric Cooking Technologies

Commercial electric cooking technologies are already widely used. Induction stoves and ranges, electric steam cabinets and induction steam cabinets, infrared cooktops and induction cooktops, and induction soup stoves can all replace gas- and coal-fired alternatives.

Compared with traditional appliances, commercial electric cooking technologies are energy efficient and low carbon, safe and hygienic, and they offer precision control and labour efficiency. The main barrier to the wide uptake of electric cookers is how to change habits and behaviours, which takes time. Table 6 compares commercial electric cooking technologies with traditional cookware.

In commercial catering, especially in buffets and open kitchens, an attractive cooking environment that allows the chefs to produce quality dishes in safe and hygienic surroundings is important. Commercial electric appliances meet the needs of restaurants and hotels, which typically choose electric appliances to improve the standard of food and drink they serve.

4.5.2 Household Electric Cooking Technologies

Household electric cooking technologies include induction cookers, rice cookers, microwave ovens and electric kettles. Electric cookware is safe and reliable. It has automatic control and a thermal efficiency of 80–92%, which means significant energy savings. With no fuel residues or exhaust fumes, household electric cookware is clean and hygienic. Electric kitchenware offers one-button operation and precision-control of the cooking temperature. Household electric cookware is especially suitable where natural gas is not available or open-flame cooking not allowed.

Nowadays, urban residential buildings, especially in new residential communities, have easy access to electricity and gas. The people who live in rural areas are mainly the elderly and children, which makes the safety and convenience of electric cookware important. The use of electric cooking technologies should therefore be encouraged in rural areas.

4.5.3 Energy-efficient Home Appliances

Energy-efficiency labels have become an important factor when people buy home appliances. Energy-efficient home appliances include refrigerators, washing machines, monitors, flat-panel TVs, electric water heaters, microwave ovens and copiers. They are technologically mature and can effectively reduce energy use in buildings and help households control their electricity bills.

4.6 Increasing the Uptake of Renewable Energy in Rural Areas

Thanks to the often vast areas of land at their disposal, rural areas are advantageously positioned to use renewable energy sources such as solar and biomass energy.

4.6.1 Solar Water Heaters

Solar water heaters convert solar energy into heat. They are used widely across much of China. In rural areas, the use of solar water heaters can significantly improve the quality of life and reduce household energy consumption.

4.6.2 Rooftop Solar Photovoltaic Systems

In June 2021, China's National Energy Administration announced it would start a national pilot programme to encourage the development of rooftop solar photovoltaic systems.

Because of the large number of rooftops in China, there is enormous potential to use them to generate solar energy. This in turn will reduce demand for grid electricity during peak periods, lower the need for investment in power distribution networks and encourage households to use renewable energy. It is an important measure to achieving peak carbon emissions and carbon neutrality, as well as to rejuvenating rural communities.

4.6.3 Solar-powered Fish Farms

Solar parks and fish farms complement one another. Solar module arrays are deployed above the surface of the ponds in which fish or prawn are farmed. Photovoltaic arrays also shield the fish from the sun.

4.6.4 Integrated Bioenergy Systems

China's rural areas are rich in biomass such as agricultural residues like crop stubble and animal manure. Previously, crop stubble was burned, which was wasteful and polluted the environment. Now it can be used in an integrated bioenergy station, which can produce energy from multiple sources, such as sewage, domestic waste and other agricultural residues. Such a station improves energy use efficiency and meets the energy needs of agriculture, rural areas and farmers.

5 Policy Recommendations for Decarbonising Buildings

To progress towards carbon neutrality, the buildings sector needs to decarbonise with the help of three different types of policy.

5.1 Support Clean Heating in Northern China

First, reform the billing system for heating. At present, heating is mainly charged in China by building area, which does not encourage households to conserve energy. The billing system needs to change and charge for actual heat consumption as soon as possible in the places where conditions permit. During this shift, attention should be focused on the impact of the new billing system on low-income families. For example, placing an additional financial burden on low-income families can be avoided by providing subsidies to those families or helping them with more efficient types of heating.

Second, determine the number of users who have shifted from coal to electricity. Local government should be encouraged to maintain accurate records of the number of users that have made the shift and remove any demolished or vacant houses from their records. Management of multi-household buildings should be improved to avoid homes acquiring electric heating equipment but not using it to ensure policies are implemented effectively.

Third, actively promote building insulation improvements. According to the findings of a programme in Beijing, the maximum heat load of buildings that had been improved fell by 23% and their electricity consumption decreased by 28%. This is equivalent to a reduction of about 15% in the total investment cost, effectively cutting life-cycle costs. However, provinces outside Beijing have not yet promoted strongly enough the renovation of buildings to reduce the need for heating. Local government should be urged to subsidise and make mandatory building insulation improvements for those shifting from coal to electricity, thus reducing the cost of living for residents and lowering the burden on government finances.

Fourth, improve oversight of the ban on coal. The tradition of burning coal or wood is deeply rooted in China. Because coal is cheaper than electricity, there is a risk that households using electric heating will revert to coal. In line with the national policies of controlling air pollution and making our skies blue again, local governments are urged to manage more effectively the ban on coal and prevent users from reverting to low-quality scattered coal.

5.2 Support the Use of Renewables in Rural Areas

Renewable energy is now firmly established in rural China. Barriers to its efficient use remain, however, such as insufficient support and lack of training, education and skills. The following recommendations would address these challenges.

1. **Improve Government Support**

Strong government support should be given to ensure the sustainable development of renewable energy in rural areas. Financial support policies should be introduced and incentive and reward mechanisms implemented to encourage the active participation of farmers in the deployment and use of renewables. Legislation on the development of renewable energy should be passed and the instance responsible for oversight identified.

2. **Accelerate R&D**

Efforts in research and development (R&D) should be increased to develop new energy technologies that best meet the economic conditions and development needs of rural areas. The principle of adapting to local conditions should always be followed and the advantages of wind, biomass, solar and geothermal energy should be harnessed to develop ecological types of agriculture.

3. **Improve Knowledge and Skills Among Farmers**

Develop a pool of highly skilled talent. Educate more farmers in clean energy development to build a talent pool for the development and use of renewables. Promotional campaigns on the benefits of rural energy technologies should be developed to encourage farmers to deploy and use clean energy.

5.2.1 Case Study 6 | UK Energy Efficiency Obligation

The UK requires large energy suppliers to support low-income households with measures to improve home energy efficiency (see Fig. 15). Households who either live in private housing and receive benefits or live in social housing are eligible for the scheme. Measures which can be funded through the scheme include loft, cavity wall and solid wall insulation as well as boiler replacement or repair. Fifteen per cent of the measures must be delivered in rural areas, and the amount each supplier has to contribute is based on its share of the UK's energy market. The UK energy regulator, Ofgem, implements the scheme.

Placing the obligation for improving energy efficiency in low-income homes on energy companies overcomes a number of barriers, including:

- overcoming the split-incentive problem in the rental sector: utility companies, rather than building owners, invest in the energy efficiency improvements, while households gain the benefit of lower energy bills;
- reducing the hassle cost of looking for adequate suppliers, as only accredited suppliers are eligible under the scheme, although households have to vacate their homes during renovation work; and
- addressing just transition concerns by offering energy efficiency improvements where they make the most difference to household energy bills: targeting the households whose energy bills have the greatest potential to be reduced means avoiding the maximum amount of unnecessary emissions, as well as impacting the most vulnerable households.

5.2.2 Case Study 7 | Warmer Kiwi Homes, New Zealand

The objective of the Warmer Kiwi Homes programme is to improve public health and reduce poverty, while also lowering emissions. The programme provides grants for insulation or low-carbon heater installations, worth up to 90% of the cost. They are offered to households who live in low-income areas or receive state benefits.

The objectives of the policy are to improve health among the populations targeted, through reductions in dampness and to achieve more comfortable living temperatures, and to reduce energy expenditure. Over two years, almost 25,000 households benefited from either or both grants, worth more than NZD 65[18] million in total. An impact assessment found the total benefit cost ratio of the scheme to be more than NZD 4.6 for every dollar invested, meaning it cost-effectively delivered the objectives set out, as well as providing improvements in energy efficiency and emission reductions.[19] This policy is an example of addressing just transition concerns by prioritising impact on low-income households in policy design. Warmer Kiwi Homes also provides an easy-to-use tool to apply for the grants and be connected with approved suppliers, which contributes to reducing search costs for households.

5.3 Long-term Action Plans to Improve Building Energy Efficiency

Energy efficiency improvement is the only way to electrify and decarbonise buildings. We recommend that long-term action plans be formulated to improve building energy efficiency.

First, develop prefabricated buildings and accelerate the construction of clean building projects. Guided by the philosophy of innovative,

[18] NZD 65 million was equal to around USD 45 million in average 2021 exchange rates. NZD 4.6 was equal to around USD 3.2.

[19] Arthur Grimes and Nick Preval, Warmer Kiwi Homes, Evaluation 2020: Phase 1.

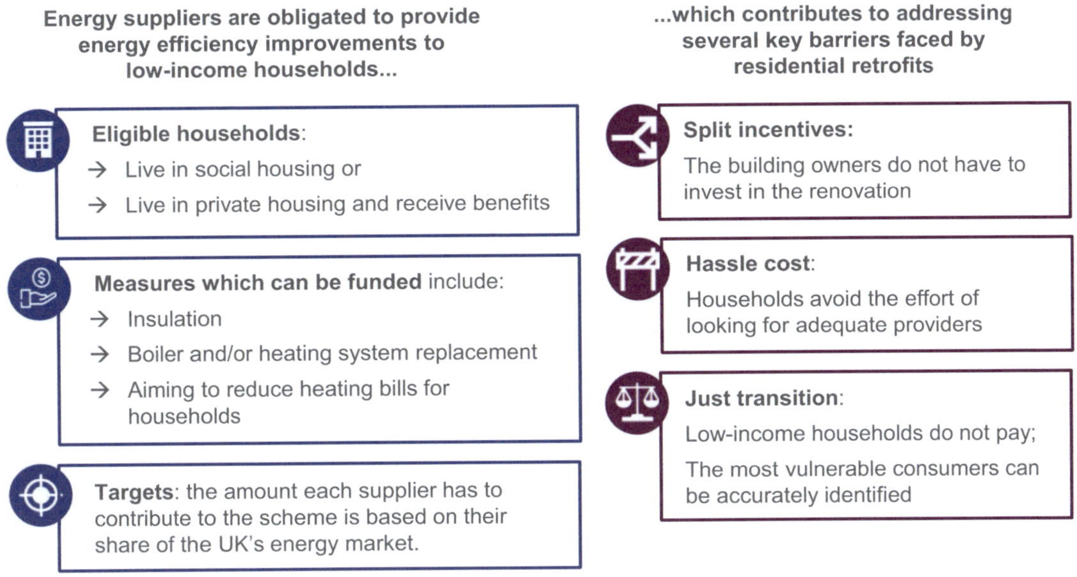

Fig. 15 UK energy efficiency obligation scheme. *Source* Ofgem, Energy Company Obligation, 2021

cost-effective, safe, green and beautiful buildings, the required design, production, construction and quality management systems should be established at a faster pace. The construction of prefabricated buildings should be accelerated in selected areas and the quality of the buildings improved.

Second, build high-quality properties to reduce energy consumption. China should use eco-friendly and energy-efficient technologies, materials and products to develop green buildings and build pilot or demonstration projects for large public buildings. The country should establish standards for energy-efficient buildings to improve quality and efficiency. It should renovate existing buildings in a planned manner and encourage the adoption of market-oriented models such as energy performance contracting and energy efficiency trading. Residential building retrofits are expensive and beyond the affordability of most users, especially in multi-occupant large buildings. Subsidies should be given to encourage energy efficiency retrofits in old buildings.

Third, actively develop clean energy-based heating and increase the proportion of clean energy in the building energy mix. Taking geological conditions, geothermal resources, energy demand and other factors into consideration, China should promote the use of clean and renewable energy in building projects, reduce the consumption of conventional energy for urban heating and domestic hot water, and increase the use of clean energy in heating. Heat pumps (sourced from river water, the air, ground or sewage) and other technologies should be adapted to local conditions.

Fourth, explore the use of carbon emission trading schemes to reduce emissions from commercial buildings. A cap-and-trade system on emissions can facilitate the development of locally appropriate solutions. This policy is ideal for large commercial or industrial buildings, where the energy consumption and emissions profile of each building can vary greatly.

5.3.1 Case Study 8 | Barcelona

Barcelona's 2000 Solar Thermal Ordinance was the first policy in Spain to mandate renewable energy consumption for new buildings. The solar ordinance was launched to increase the use of heat from renewable sources. It requires all new buildings to have at least 60% of their hot water demand provided by solar energy. The policy was successful, enabling the adoption of solar thermal installations able to produce hot water for up to 45,000 households. Given this success, the national government implemented a similar policy in 2006, showing the power of city-level policies to inspire change on a wider scale by demonstrating success and public endorsement.

Building on the ordinance, Barcelona now has a suite of policies in place to incentivise emission reductions across existing buildings. Measures to encourage energy savings and energy efficiency include subsidies available for up to 40% of the cost of a comprehensive renovation project, and 25% for partial ones. Grants are also available to install renewable electricity or heat generation in existing buildings.

In addition to these mandates and financial mechanisms, the city has focused on providing information, both to improve the uptake of voluntary measures and to facilitate the identification of relevant suppliers. Barcelona Energy Agency provides a catalogue of information to encourage energy saving measures, as well as to identify renewable energy providers. To incentivise investment in low-carbon retrofits or renewable energy generation systems beyond the grants provided, the agency also provides links to organisations able to provide information about accredited providers of energy renovation services or install renewable generation capacity.

To encourage adoption of these measures, the city also provides information on how public buildings are contributing to the generation of renewable energy. Municipal activities must incorporate renewable energy generation whenever feasible. The municipality publishes a map of renewable energy generation projects in the city. By setting an example and transparently sharing this information, the local government helps build public support for the policies and provides examples of successful low-carbon building projects (Fig. 16).

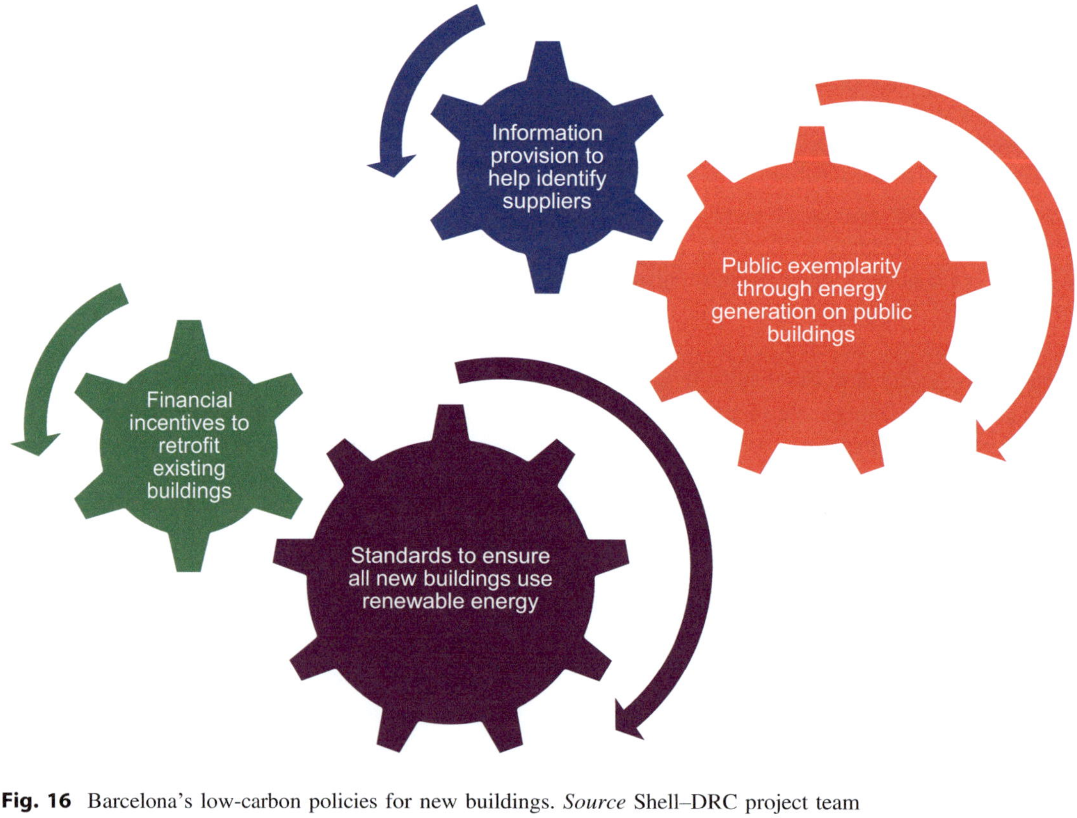

Fig. 16 Barcelona's low-carbon policies for new buildings. *Source* Shell–DRC project team

5.3.2 Case Study 9 | Energiesprong, The Netherlands

Energiesprong retrofits tackle the hassle cost to residents by involving them in the tendering, planning and implementation process, as well as by providing design improvements and energy savings.

Energiesprong is a whole-building retrofit standard and funding approach, which was first implemented in the Netherlands but has now been trialled in many European countries. The Energiesprong approach emphasises compromise-building among stakeholders, while seeking to make the retrofits attractive to the tenants. Tenants are involved in the decision-making process, from tendering for a contractor to following the work's progress.[20] In addition, the retrofits are designed and planned to improve the appearance of the houses or apartment blocks, which provides an extra incentive for tenants to take part.

The financing structure for Energiesprong is designed to ensure that all parties have incentives to take part and perform, thus overcoming the split-incentive problem. Energiesprong homes reach very high levels of energy self-sufficiency through the combination of effective insulation and on-site renewable energy generation. Tenants go from paying high energy bills to providers to paying a lower, fixed amount to their landlord, which is used to fund the renovation. As a result, the scheme benefits both the tenants, who pay lower bills and enjoy higher levels of comfort, and the landlords. In addition, the energy performance results are guaranteed by the contractors, who are liable to pay if the buildings underperform over the 20- or 30-year guarantee period. This provides strong incentives for contractors to perform optimally, rather than lower costs by making compromises during the renovation.

The Energiesprong approach is also intended to lower implementation costs as the retrofits are multiplied, through learning effects and economies of scale. Scale will help bring the costs down as more and more of the retrofit components can be designed and assembled off-site. For instance, a developer in the UK assembled solar panels onto the roof components off-site, but added pipework on site, whereas developers in the Netherlands, where more projects have already taken place, provide pre-assembled roof panels with pipework, which reduces assembly costs. Learning effects, both over the performance of manufactured components and the design process in collaboration with stakeholders, will support this cost reduction.

[20] CIBSE Journal, Energiesprong—the Dutch system that could rescue Britain's social housing, 2018.

Chapter 4: Transport: The Transition to Electric Vehicles

Yongwei Zhang, Jin Zhu, Jian Zhang and Georgios Bonias

1 Introduction

Transport, especially road transport, consumes large amounts of energy and emits large amounts of carbon. Electrification is the key to decarbonising the transport sector and to China achieving its energy security strategy and decarbonisation goal.

Thanks to advances in core technologies, lower costs, improved infrastructure and good market prospects, China is well positioned to decarbonise its transport sector. The country is already a global leader in batteries and is making breakthroughs in electric drivetrain and fuel cell technologies. New business models, such as high-power charging, smart charging and battery swapping should increase market penetration. New energy vehicles in the commercial passenger transport segment will become cost competitive earlier than private passenger cars and are expected to reach cost parity around 2025. Battery-electric vehicles are suitable for light-duty and short-distance freight and will be cost competitive by 2030. Hydrogen fuel cell technology should become cost competitive in time in medium- and heavy-duty freight transport.

Electric vehicles are projected to make up a rapidly increasing share of the transport fleet. In our Carbon-neutral Scenario, the penetration of battery-electric vehicles in passenger transport will increase to 25% by 2030 and be almost entirely electric by 2050. Nearly all new vehicles are projected to be electric by 2040 in passenger road transport and by 2045 in road freight transport. The entire fleet of road freight transport vehicles will be mostly electric by 2050.

The electrification of transport will lead to a decline in conventional energy use, with fuel consumption expected to peak around 2025. Electricity and hydrogen will grow rapidly as alternative fuels, with consumption reaching 1,000 terawatt-hours and 28 million tonnes respectively by 2050.

The electrification of transport will rely on multiple technical pathways and will be implemented at varying speeds in different regions. At the early stage of development, conventional fuel vehicles, hybrid-electric, battery-electric and hydrogen fuel cell vehicles will co-exist. At the transitional stage, alternative fuel vehicles, hybrid and plug-in hybrid electric vehicles will predominate. At the final stage, only zero-emission vehicles will exist in the form of battery-electric and hydrogen fuel cell vehicles. A differentiated approach will be adopted to electrify transport: economically developed regions with a strong industrial base will be

Y. W. Zhang (✉) · J. Zhu · J. Zhang
China EV100, Beijing, China

G. Bonias
Shell International Limited, The Hague, The Netherlands

Shell International B.V. and Development Research Centre of the State Council (DRC) of People's Republic of China, *Embracing the Future, Powering Growth: An Energy System Renewed for China*, Advances in Oil and Gas Exploration & Production,
https://doi.org/10.1007/978-3-031-73075-7_5

given priority; less-developed regions will adapt the pace of electrification to their local conditions.

Technological innovation, supply chain support and infrastructure development are priorities. In the short term, some products and segments will require more advanced technologies and customised business models. To scale up transport electrification, challenges need to be addressed in the medium and long terms in infrastructure development, aftersales, supply chain security and the supply of key natural resources.

2 Developments in Transport Decarbonisation

The large-scale and market-driven electrification of road transport is an important pathway towards decarbonisation. Following early policy incentives and other efforts to establish a market, the uptake of new energy vehicles in the private passenger car, electric taxi, electric bus and road freight transport segments has grown rapidly. However, the electrification of medium- and heavy-duty trucks is advancing relatively slowly. How to leverage China's first-mover advantages and achieve synergies with global efforts in electrification is the key to switching from policy-driven to market-driven electrification of road transport.

2.1 Electrification Is the Priority for Road Transport Decarbonisation

Electrification of road transport is the key to decarbonising the transport sector, which comprises four segments: road, rail, air and water. In 2019, transport represented about 10% of China's total carbon emissions of around 1.1 billion tonnes.[1] Road transport dominates China's transport sector, in terms of both energy consumption and CO_2 emissions. In 2019, road transport was responsible for 74% of carbon emissions in the transport sector (see Table 1). Road transport electrification is now at a crucial stage of transition from policy-driven to market-driven, in which electrification technologies and supporting market mechanisms are the main concern of the government, business and consumers alike. This chapter considers the electrification of road transport as the main issue in transport decarbonisation.

Cars are the main object of road transport electrification. In China, road transport vehicles comprise two-wheelers, three-wheelers and four-wheel motor vehicles. As there are no firm data on the number of two-wheelers and three-wheelers, and because their electrification technologies are simpler and their CO_2 emissions are relatively small, this chapter will focus on four-wheel cars, excluding low-speed electric variants.

2.2 Electrification Is More Advanced in Passenger Road Transport than in Road Freight Transport

2.2.1 The Electrification of Passenger Vehicles Advances Steadily, While that of Commercial Vehicles Progresses Slowly

In 2020, sales of new energy vehicles in China reached 1.367 million, accounting for 5.4% of all new vehicle sales. Of these, 1.115 million were battery-electric vehicles, 251,000 were plug-in hybrid electric vehicles, and 1,000 were fuel cell electric vehicles.[2] Sales of new energy passenger cars in 2020 were 1.246 million. About 79,000 commercial new energy passenger vehicles (including city buses and long-distance coaches) and about 42,000 new energy trucks were sold (see Fig. 1). By the end of 2020, the number of new energy vehicles reached 4.92 million, accounting for 1.75% of total car ownership.[3]

[1] "Five Optimizations" Help the Transport Sector Achieve the Dual-carbon Goal, www.rmzxb.com.cn/. March 15, 2022.

[2] China Association of Automobile Manufacturers.

[3] Estimates based on data from the Ministry of Public Security.

Table 1 Carbon emissions and the electrification of transport segments in China

Transport segment	Total carbon emissions (2019) (%)	Electrification rate (2020)	Stage of electrification	Scale
Road transport (cars)	74	1.75%	Start market-oriented and large-scale development	Large scale and high proportion of private vehicles
Rail transport	8	72.8%	Mature	–
Air transport	10	Almost 0 in commercial air transport	Nascent	–
Water transport	8	Almost 0 in some sightseeing/passenger and cargo ships	Nascent	–

Source National Development and Reform Commission; Ministry of Public Security

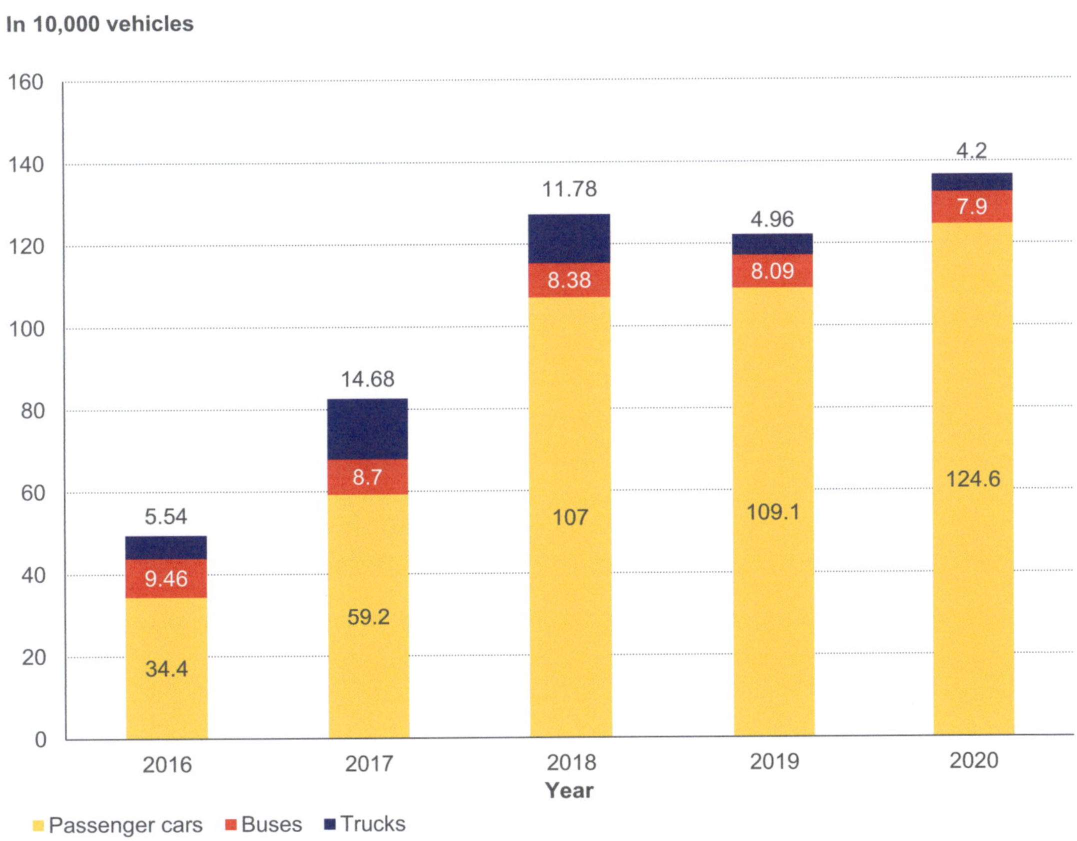

Fig. 1 Sales of new energy vehicles in China by segment. *Source* China Association of Automobile Manufacturers

2.2.2 More Consumers Are Buying Electric Passenger Cars and the Electrification of Taxis and Buses Is Advancing Rapidly

In 2014, sales of new energy passenger vehicles grew rapidly in China. Around 800,000 private electric passenger cars were sold, representing 70% of new energy vehicle sales in 2020 (see Fig. 2), which indicates a gradually improving acceptance of the vehicles by consumers. Driven by the national policy to electrify public transport, a strong economy in the commercial vehicle segment and other favourable factors, the total number of new energy taxis was 77,200 at the end of 2019, representing 5.5% of China's taxi fleet.[4] As their cost-effectiveness improves and if the battery-swapping business model succeeds, the shift to electric taxis will progress at an even faster pace.

New energy buses have achieved the highest uptake at 60% of the national fleet. Some city bus fleets, including those of Shenzhen and Changsha, are 100% electric. Other cities, like Foshan and Zhangjiakou, are actively promoting the demonstration or uptake of fuel cell buses.

2.2.3 The Electrification of Road Freight Transport Is Progressing Slowly

New energy vehicles in freight road transport have reached a bottleneck. In 2019, following a previously rapid uptake, sales of new energy goods vehicles in China declined by two-thirds to 49,600 compared with 2017 (see Fig. 3). The market penetration of new energy trucks was close to 5% in 2017, but then dropped to its current level of 2%.

New energy freight vehicles mainly comprise mini-trucks and light-duty trucks. In 2019, these two categories represented 56% and 44% respectively of the new energy freight transport fleet, while medium- and heavy-duty trucks accounted for almost none (see Fig. 4). The slow uptake of new energy heavy-duty trucks was constrained by factors such as the high cost of the battery and fuel cells, and insufficient support infrastructure.

[4] Dominance Shift from Slow NEV Charging to Fast Charging—A New Round of Industrial Plans Approved, www.cb.com.cn/. October 13, 2020.

2.3 China Has a First-mover Advantage in the New Energy Vehicle Market, While European Countries Are Also Gaining Ground

China is a global leader in new energy vehicle sales, but many countries have started to catch up. In 2020, global sales of new energy vehicles exceeded 3 million (see Fig. 5), with China topping the list for the sixth consecutive year with a share of 45%. Since 2019, major European countries have increased their support for new energy vehicles. Partly as a result of increasingly stringent emission standards, sales of new energy vehicles in Europe are gaining momentum. In 2020, Germany recorded sales growth of 254%, becoming the world's second largest consumer market for new energy vehicles. Norway recorded 70% market penetration for new energy vehicles, making it the world leader by uptake (see Fig. 6). Italy, Spain, the Netherlands and other countries also achieved significant sales growth.

3 Decarbonisation Potential in Transport

Electric vehicles are likely to increasingly dominate the passenger car market, as significant technological improvements and cost reductions take hold. The transition to electric vehicles in passenger transport is rapidly climbing the S-curve. Falling battery costs are reducing the purchase price, increasing battery capacity is providing longer-range vehicles, and expanding supply chains are gaining economies of scale in production. Fuel costs will also fall as electricity becomes cheaper than fossil fuels. As a result, electrification will remain the principal option for decarbonising China's transport sector.

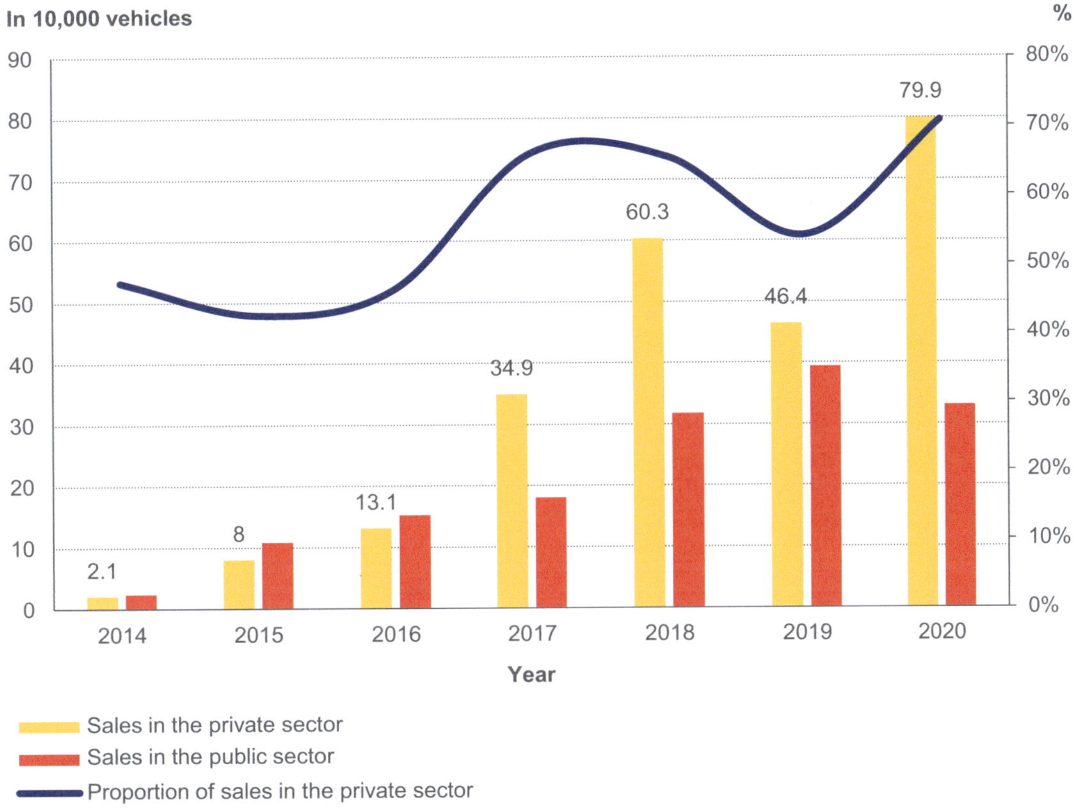

Fig. 2 New energy passenger vehicle sales in the public and private sectors. *Source* New energy vehicle insurance registration data

Hydrogen is expected to play a more prominent role in the freight sector. While electric vehicles will be used for urban freight and, to some extent, in medium-duty trucks, hydrogen fuel cells could be an alternative for long-haul heavy-goods vehicles. The combination of longer range and lighter weight, refuelling infrastructure that is similar to that for conventional trucks and refuelling times that are faster than for battery recharging, could coalesce into attractive business models for freight fleet operators.

3.1 Power Batteries and Hydrogen Fuel Cells Have Great Prospects

New energy vehicles are driven by power batteries, hydrogen, biofuels or solar power. Based on developments in the industry and on business surveys, this chapter analyses the potential of the core vehicle components and technologies.

3.1.1 China Leads the Global Power Battery Market

Power battery technology has made great progress, yet still holds huge potential for further advancement. As of 2020, the energy density of a single power battery in China was close to 300 watt-hours per kilogram (Wh/kg), and the cost of producing a battery dipped to RMB 0.8/Wh.

Chinese power battery companies are in the first rank thanks to their strength in the global arena. Innovation in the power battery field includes technical breakthroughs in materials, systems and structure. Material innovations include low-cobalt batteries, lithium-silicon batteries and semi-solid state batteries. For example,

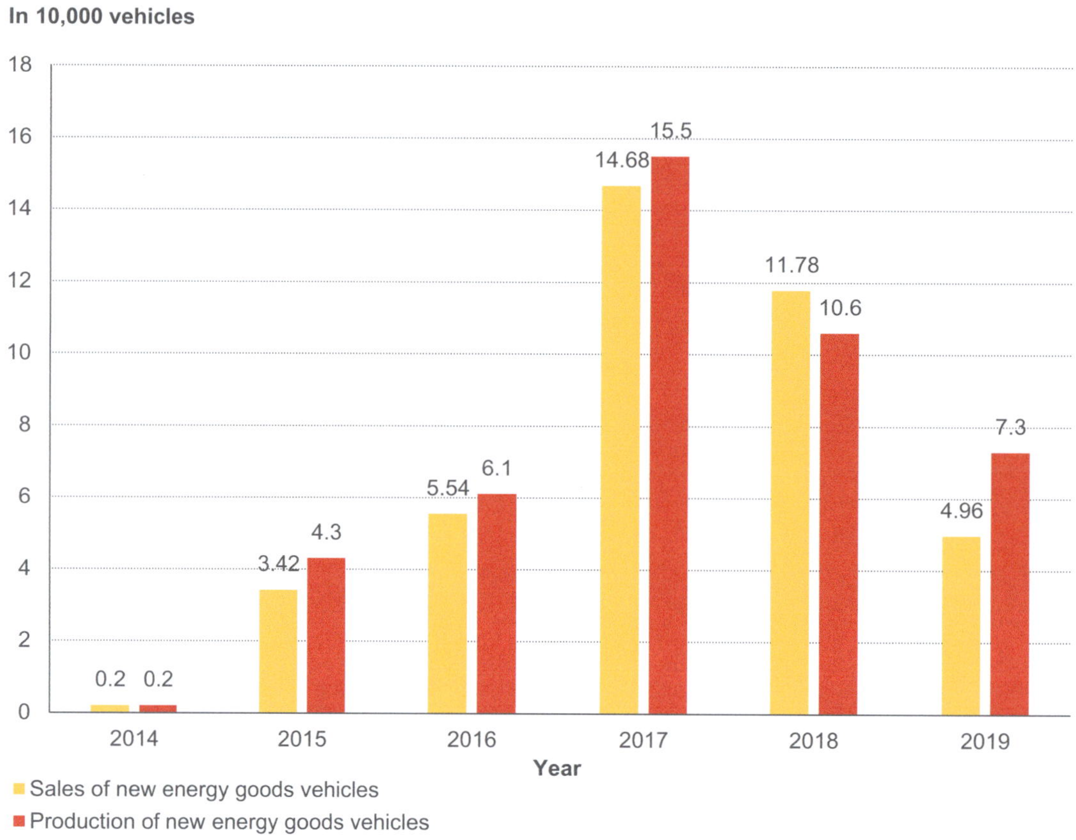

Fig. 3 Production and sales of new energy goods vehicles. *Source* New energy vehicle insurance registration and production data

a solid-liquid hybrid electrolyte pouch battery has an energy density of up to 360 Wh/kg, and a lithium-silicon iron phosphate battery has an energy density of 200 Wh/kg. Chinese companies are improving the specific energy of battery systems through structural innovations. Examples include cell to pack from CATL, blade batteries from BYD, jelly roll to module from Gotion High-Tech, and other concepts such as module-to-car and cell-to-chassis that are undergoing research. According to industry forecasts, from 2025 onwards the focus will shift to solid-state batteries, lithium-rich manganese-based solid solution batteries, lithium-sulphur batteries, metal-air batteries and metal-negative electrode batteries. There is still much scope for improvement in power battery performance.

A mixture of system reform, improved automation and increased economies of scale are driving down the cost of power batteries. The cost of mass-produced power battery systems is projected to decrease to about RMB 0.5 per Wh/kg by 2050 (see Fig. 7).

3.1.2 Creating a Fuel Cell Industrial Chain

In recent years, China's automotive fuel cell industry has made breakthroughs in fuel cell development. The power density, cold-start temperature, lifetime and maximum efficiency of fuel cells have been greatly improved. For example, in 2020, the power density of graphite bipolar plate stacks was 2.2 kilowatts per litre (kW/L), an improvement of about 47% compared with 1.5 kW/L in 2015; the lifetime of graphite bipolar plate stacks was 12,000 hours, an increase of 300% compared with 2015 (see Fig. 8). In addition, the fuel cell component

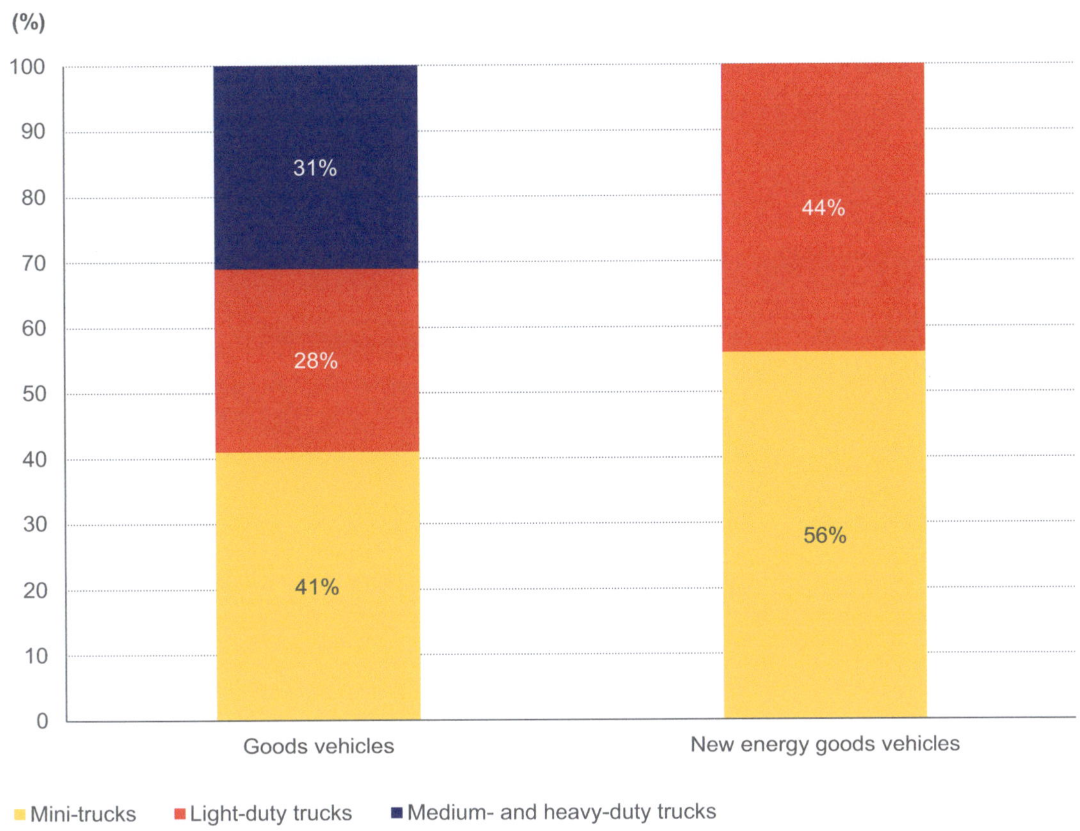

Fig. 4 Mix of goods vehicles and new energy freight vehicles in 2019. *Source* New energy vehicle insurance registration data

value chain has taken shape in China, and system integration has greatly improved.

Driven by the synergies of technological development and economies of scale in China, the cost of fuel cell systems will continue to decline. By 2035, the cost of a fuel cell system for passenger vehicles is expected to be RMB 455 per kilowatt, and that for commercial vehicles RMB 1,000 per kilowatt (see Fig. 9).

3.1.3 Motor Technology Is Continually Improving

In 2019, the power-to-weight ratio of mass-produced motors in China reached more than 4.0 kW/kg, an increase of more than 30% compared with 2016.[5] Many companies have launched their independently developed insulated gate bipolar transistor (IGBT), double-sided cooled module and high power density motor controller, as well as all-in-one drive modules, which are comparable to those of the global leaders. Future drive motors and controllers will see further improvements in efficiency, speed, density, noise suppression, energy consumption and costs. The evolution of new technologies and locally developed and produced critical materials, components and equipment should lead to further cost reductions (see Fig. 10).

3.1.4 Vehicle Safety Is Improving

New energy vehicles are less likely to catch fire than conventional vehicles, although there is still much room for improvement in safety (see Table 2). The fire risk of new energy vehicles is about 25% lower—in 2020 the risk was only

[5] China Society of Automotive Engineers, Annual Tracking Report of New Energy Vehicle Drive Motor Technology 2019, April 20, 2020.

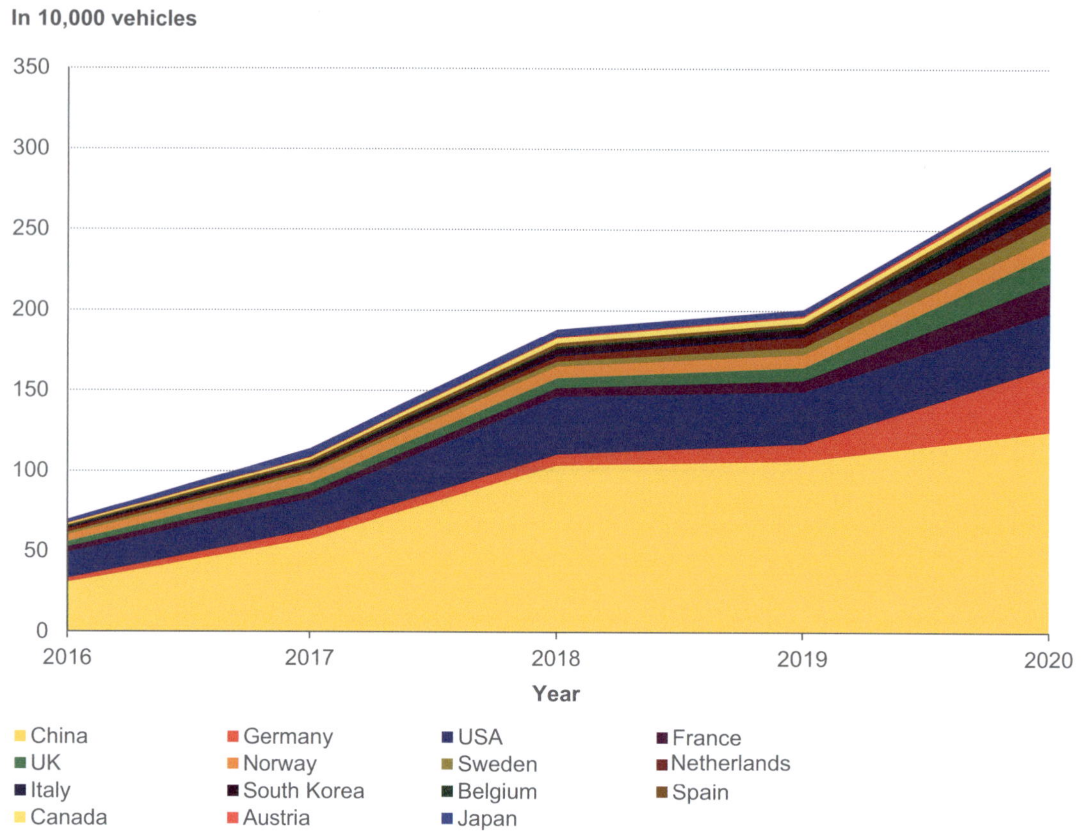

Fig. 5 New energy vehicle sales in major countries, 2016–20. *Source* International Energy Agency, Global EV Outlook 2021

2.6‱.[6] This will improve through the combined effects of research on the thermal runaway mechanism, improvements in materials, better use of big data, and more stringent standards and regulations. According to industry estimates, the fire risk of electric vehicles will be less than 0.1 per 10,000 vehicles by 2030.[7]

3.1.5 Vehicle Energy Consumption Will Decrease

Energy consumption is the chief indicator of vehicle economy, and an important factor determining carbon emissions. There is scope for further reductions in the energy consumption of battery-electric and hydrogen fuel cell vehicles (see Fig. 11).

[6] Data speaks louder—Academician of Chinese Academy of Engineering: The fire probability of NEVs is less than half that of ICEs, SOHU.com, October 10, 2020.

[7] China Association of Automobile Manufacturers, Energy-saving and New Energy Vehicle Technology Roadmap 2.0, 2020.

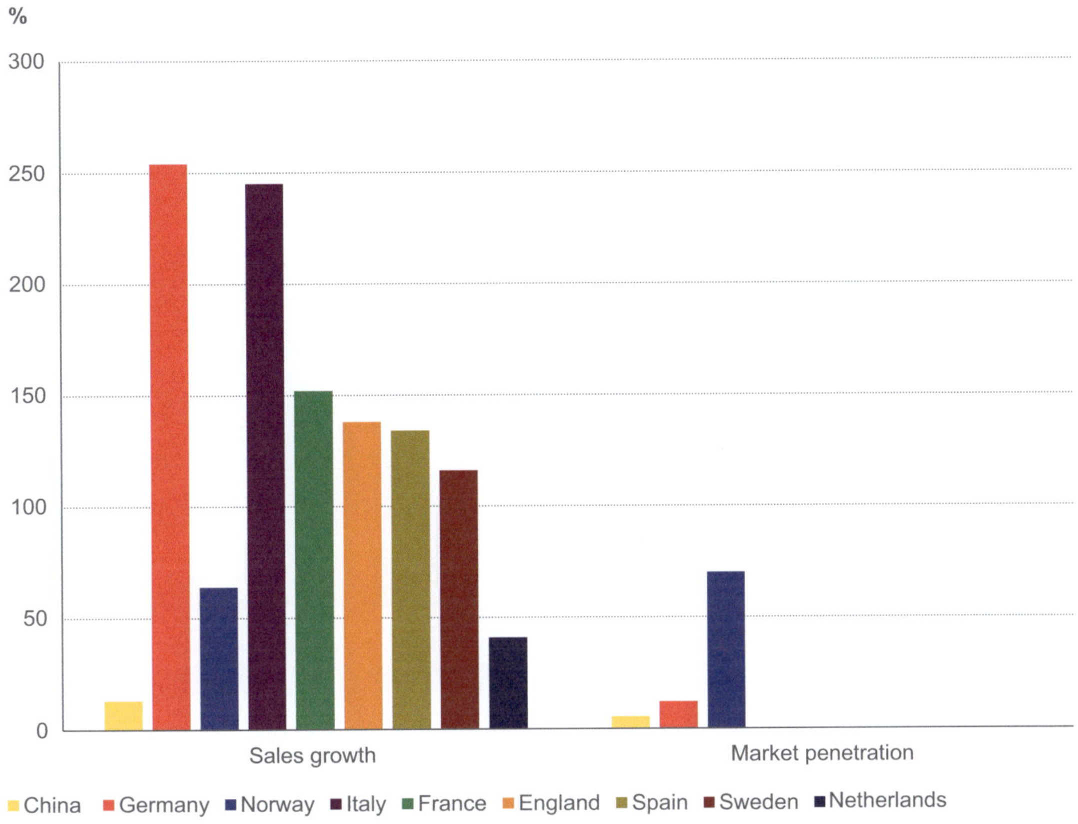

Fig. 6 New energy vehicle sales growth and market penetration in major countries in 2020. *Note* The market penetration figures for China, Germany and Norway only are available. *Source* International Energy Agency, Global EV Outlook 2021

3.2 There Is Room for Improvement in the Cost Efficiency of Electric Vehicles

3.2.1 The Cost Competitiveness of Battery-electric and Hydrogen Fuel Cell Vehicles Varies by Application

Small electric vehicles are cost effective compared with conventionally fuelled vehicles, thanks to the combination of small battery capacity and low running costs. Larger electric models have comparatively poor cost competitiveness, but are expected to achieve cost parity with internal combustion engine vehicles by around 2025. Overall, in the private passenger vehicle segment, battery-electric vehicles will be more cost-competitive than fuel cell vehicles in the short and medium terms (see Fig. 12).

Commercial electric passenger vehicles (taxis) have been given priority over the private segment; they achieved cost competitiveness with the internal combustion engine around 2022, thanks largely to their high frequency of use. Fuel cell commercial passenger vehicles are expected to be cost competitive around 2027 (see Fig. 13), but will not achieve parity with their battery-electric counterparts for some time.

Some electric buses with low battery capacity are already cost competitive with internal combustion engine buses, and most other bus types are expected to reach cost parity around 2025. Whether fuel cell buses are cost effective depends on how quickly the cost of core components and refuelling decreases. This is

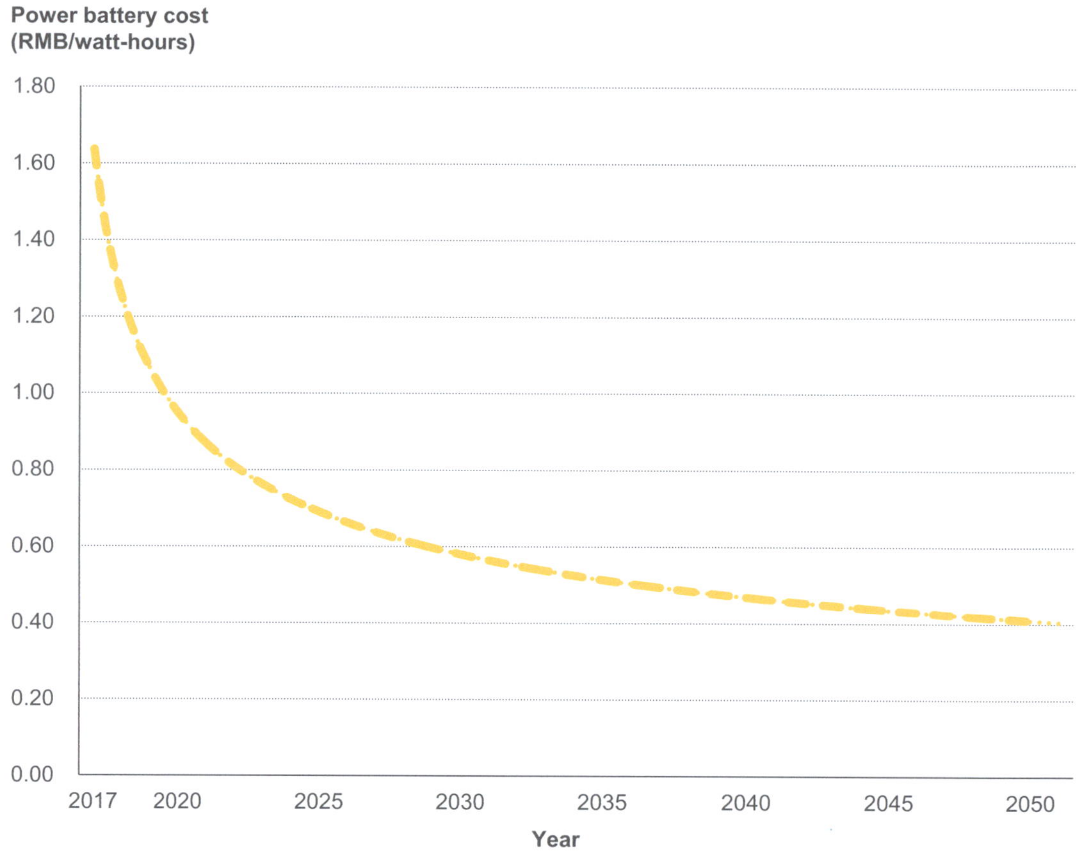

Fig. 7 Predicted movements in power battery system costs. *Source* China Automotive Technology & Research Center

expected to occur around 2030 (see Fig. 14). Fuel cell buses will be more cost effective than battery-electric buses for long-distance travel.

3.2.2 Battery-electric Vehicles Are Suitable for Short-distance, Light-duty Freight Transport; Fuel Cells for Long-haul, Heavy-duty Freight Transport

In short-distance, light-duty freight transport, battery-electric vehicles are forecast to be cost effective by 2030, fuel cell electric vehicles later. The heavy-duty freight transport segment will see a contrary trend, in which fuel cell vehicles achieve cost effectiveness before battery-electric (see Figs. 15 and 16).

3.3 Support Infrastructure—Such as Charge Points, Battery-swapping Stations and Hydrogen Refuelling Stations—Will Gradually Expand

3.3.1 Charging and Battery-swapping Stations

New business formats such as high-power orderly charging, intelligent charging and battery swapping will be rolled out at mass market scale. A landscape defined by slow charging in parking spaces for private electric vehicles and fast charging at public charging stations has already taken shape in China. In coming years, low-power DC charging will take over from AC

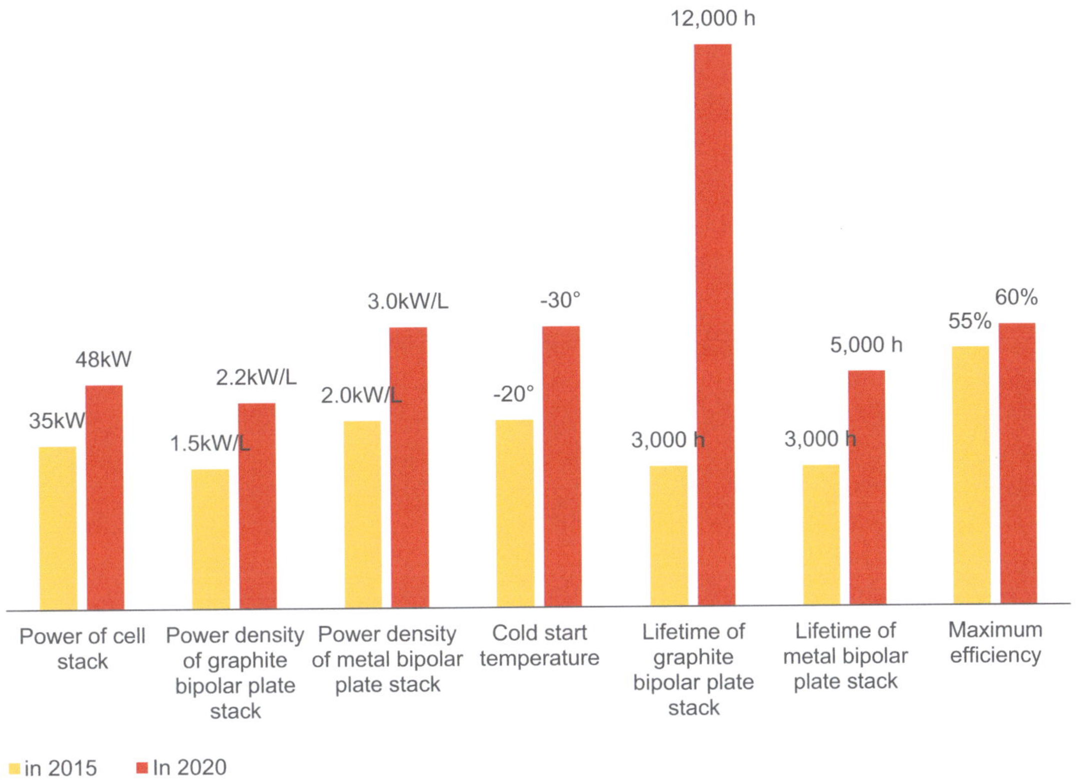

Fig. 8 Improvements in key technical indicators of fuel cells. *Source* Ouyang Minggao's speech at the China EV100 Forum 2021 in January 2021

charging in communities. Orderly and intelligent charging technologies will be deployed and private charger sharing and community-level public chargers will be more widely adopted. The "fast charging + quick battery-swapping" energy service network for electric vehicles will effectively remove pain points such as range anxiety and long recharging times. Considering the strong prospects that orderly charging, vehicle-to-grid and other technologies offer, efficient interaction between new energy vehicles and power grids will be possible in the future. Distributed solar photovoltaic power generation and storage systems, and integrated electric vehicle charging and discharging systems, will grow.

3.3.2 Hydrogen Refuelling Stations

Thanks to collaboration between multiple actors, hydrogen refuelling stations will open at a faster pace. At present, the development of hydrogen refuelling station infrastructure is at the initial stage, mainly as part of demonstration projects run by local government. Co-location with conventional oil and gas refuelling infrastructure is an important trend that will effectively contribute to the large-scale development of fuel cell electric vehicles. The multiple actors involved in building hydrogen refuelling stations will include energy, chemical and gas companies, specialised hydrogen refuelling station operators and equipment suppliers, and vehicle makers and vehicle operators.

3.4 Most Vehicles Will Be Electric by 2050

Thanks to government policies, electric vehicles and hydrogen fuel cell vehicles will become mainstream in the future. Section 2.4 focuses on

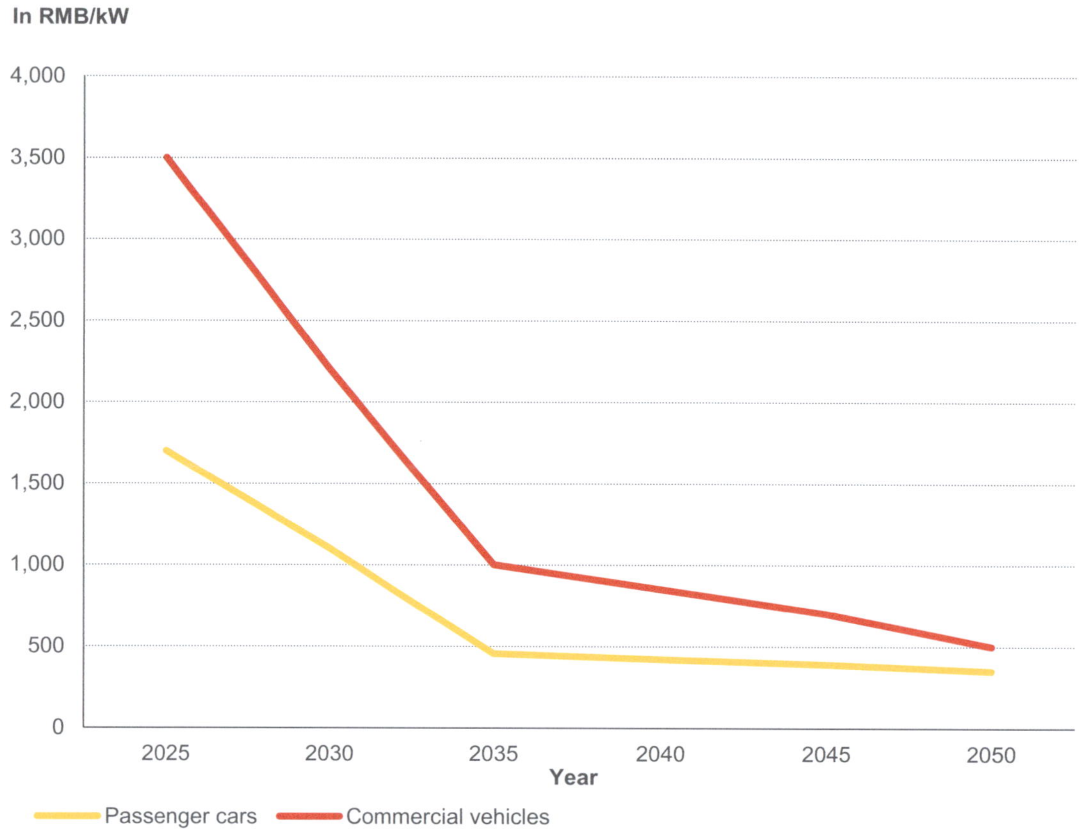

Fig. 9 Cost reductions for fuel cell systems. *Source* China Association of Automobile Manufacturers, Energy-saving and New Energy Vehicle Technology Roadmap 2.0, 2020

the changes in future market share of new energy vehicles and internal combustion vehicles.

The Development Scenario is based mainly on the new energy vehicle sales targets of the New Energy Vehicle Industry Development Plan (2021–35), which forecasts that new energy vehicles will account for about 20% of total vehicle sales by around 2025 and for about 35% in 2030.

The Carbon-neutral Scenario assumes faster electrification due to increasingly stringent carbon policies and competition from other countries. In that scenario, the market penetration of new energy vehicles will reach 30% in 2025 and about 50% in 2030.

3.4.1 In the Passenger Car Segment, Battery-electric Vehicles Will Predominate and the Internal Combustion Engine Will Almost Disappear by 2050

In the Development Scenario, the number of conventional internal combustion engine vehicles will peak around 2030, and then begin to decline to almost zero by 2050. The uptake of battery-electric vehicles in the passenger car segment will climb rapidly from 1.3 to 12.6% between 2020 and 2030, and reach 80% in 2050. Hybrid, plug-in hybrid and fuel cell electric vehicles will all hold market share in 2050 (see Fig. 17).

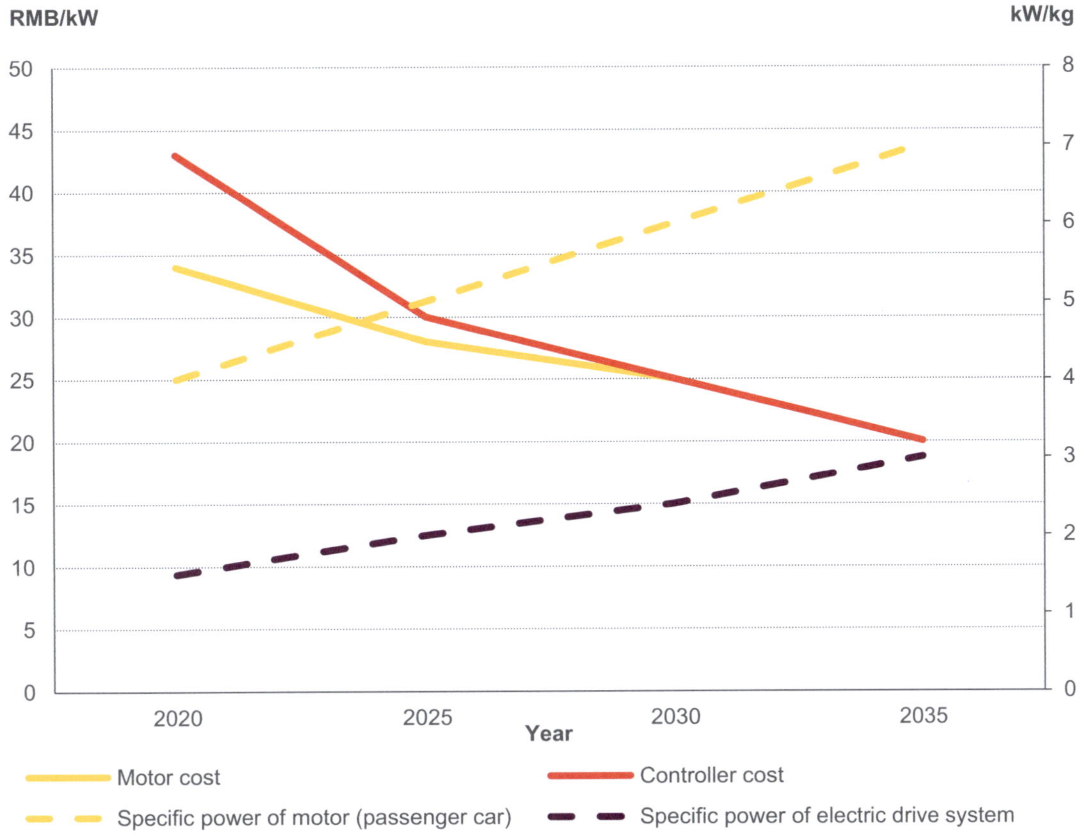

Fig. 10 Electric drive system improvements and the potential for cost reductions. *Source* China Association of Automobile Manufacturers, Energy-saving and New Energy Vehicle Technology Roadmap 2.0, 2020

Table 2 Safety features in selected Chinese batteries

Company	Technology	Industrial development status
BYD	Blade battery that gives off no smoke or fire	Mass production in March 2020
CATL	NCM 811 battery with self-isolation safety technology; material innovation, cell structure optimisation, battery thermal management improvement and system thermal diffusion control	The 100 kWh battery was the first to be put into mass production in late 2020
Farasis Energy	New thermal runaway technology	Deployed in the GAC Avion V in June 2020
Sunwoda	Battery solution that only gives off smoke, not fire	December 2020
SVOLT	A gel battery based on cobalt-free cathode and electrolyte materials, features high conductivity self-healing and flame retardance, and can prevent thermal diffusion without compromising battery performance	Launched in December 2020
GAC Avion	The magazine battery system safety technology can be used in lithium iron phosphate and ternary battery packs	Deployed in vehicles in 2021

Source Various publicly available sources

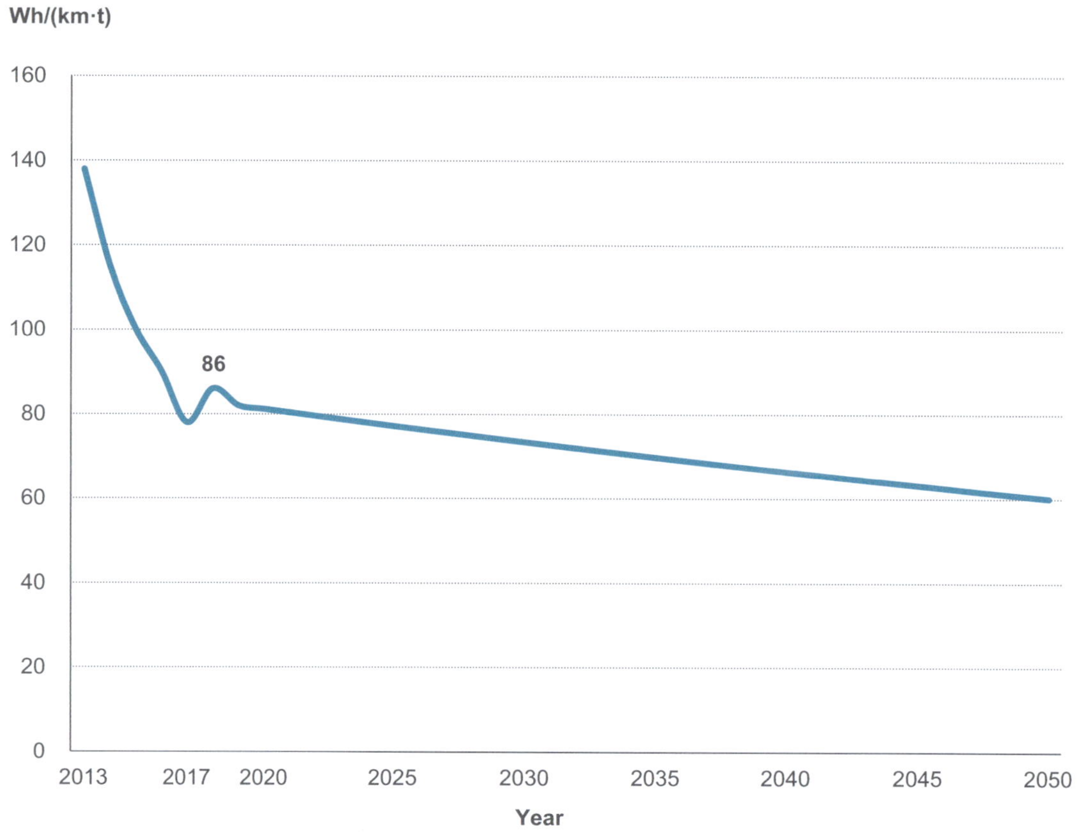

Fig. 11 Energy consumption potential of Class A battery-electric passenger vehicles. *Source* China Automotive Technology & Research Center

In the Carbon-neutral Scenario, the number of conventional fuel vehicles will decrease quickly to 2030, and then maintain the same rate of decline as in the Development Scenario. In the passenger vehicle segment, battery-electric vehicles will achieve a penetration rate of around 25% in 2030 and 80% in 2050. Hybrid and plug-in hybrid electric vehicles will have more or less disappeared by 2050. Fuel cell electric vehicles will have a 20% market share in 2050. A 100% shift to new energy vehicles will be possible by 2050, with battery-electric vehicles predominant (see Fig. 18).

3.4.2 Battery-electric and Fuel Cell Buses Will Predominate

In the Development Scenario, battery-electric buses will comprise more than 40% of the bus fleet in 2030 and 60% by 2050. The proportion of alternative fuel buses will peak at 15% in 2030 and gradually decrease to about 10% by 2050. Fuel cell buses will make up only 4% in 2030 but grow significantly to 30% by 2050. The share of petrol and diesel buses will steadily fall from 50% in 2020 to zero in 2050 (see Fig. 19).

In the Carbon-neutral Scenario, battery-electric and fuel cell buses will achieve faster

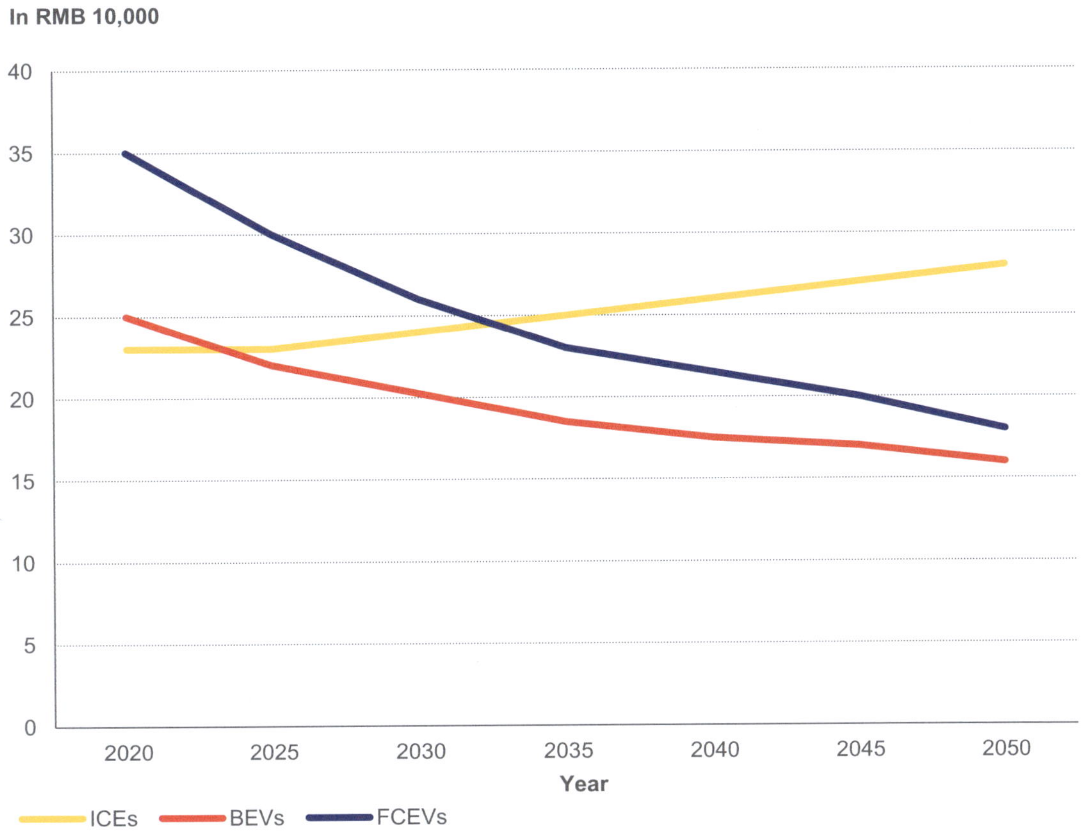

Fig. 12 The cost reduction curve of private electric vehicles. *Note* (1) Ease of recharging, range anxiety and purchase subsidies are not taken into account, but the cost of use is included. (2) The cost reduction potential is based on forecasts and research. (3) ICE = internal combustion engine; BEVs = battery-electric vehicles; FCEVs = fuel cell electric vehicles. *Source* Calculations made by the project team

penetration, whereas alternative fuel and plug-in hybrid buses will be more or less phased out by 2040. By 2050, all buses will be electric, with battery-electric comprising 60% of the fleet and fuel cell buses 40% (see Fig. 20).

3.4.3 Battery-electric and Fuel Cell Trucks in Light-duty Road Freight Transport

In the Development Scenario, the penetration rate of battery-electric light-duty trucks will steadily increase from 1.3% in 2020 to 10.5% in 2030, before escalating to 80% by 2050. The uptake of fuel cell light-duty trucks will rise rapidly after 2040, from 1.4% to 10% in 2050. The share of petrol and diesel trucks will drop by 87.2% between 2035 and 2050 (see Fig. 21).

In the Carbon-neutral Scenario, the uptake of battery-electric light-duty trucks will soar from 11.6% to 90.8% between 2030 and 2050. The penetration of fuel cell light-duty trucks will increase from 1.5% in 2040 to 9.2% in 2050. Petrol and diesel light-duty trucks will decline at a significantly faster pace, with most disappearing before 2045. Petrol and diesel light-duty trucks will hold only 10% of the market in 2045, and will be fully phased out by 2050 (see Fig. 22).

3.4.4 Hydrogen Is the Main Electric Vehicle Technology for Medium- and Heavy-duty Trucks

In the Development Scenario, the penetration of battery-electric medium- and heavy-duty trucks

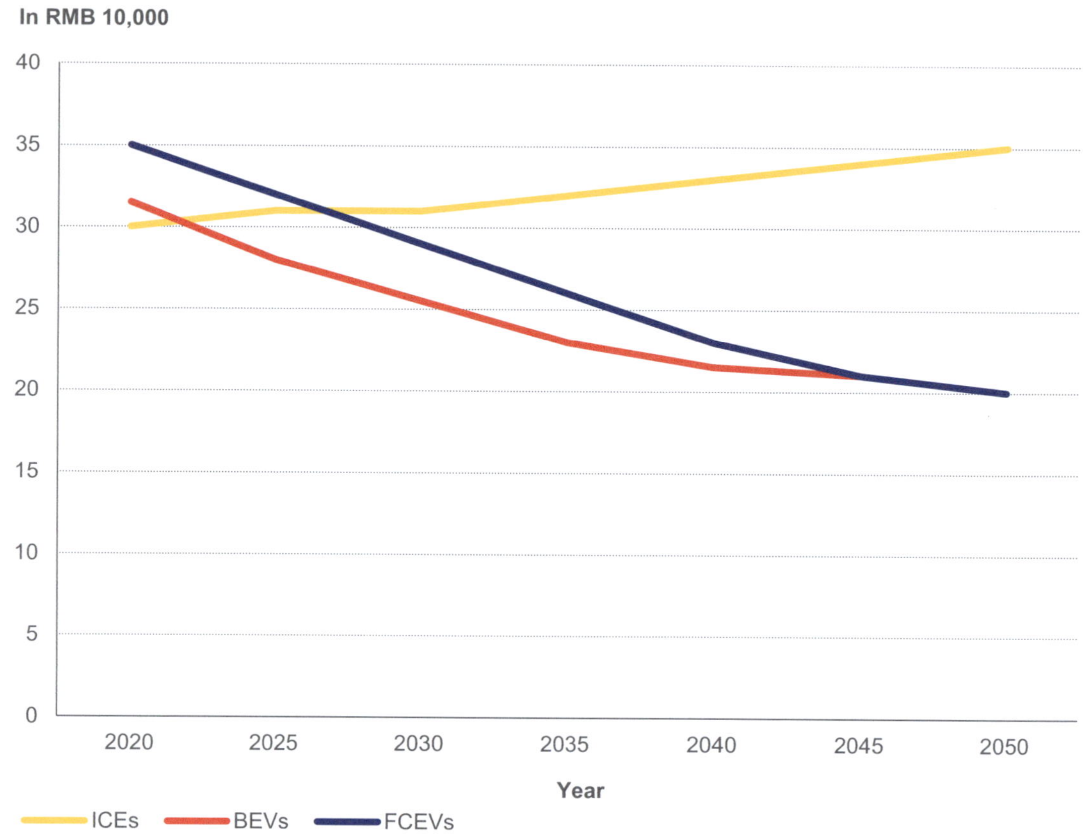

Fig. 13 The cost reduction potential of taxis. *Note* (1) Ease of refuelling, range anxiety and purchase subsidies are not taken into account, but the cost of use is included. (2) The cost reduction potential is based on estimates and research. (3) ICE = internal combustion engine; BEVs = battery-electric vehicles; FCEVs = fuel cell electric vehicles. *Source* Calculations made by the project team

will rise steadily from 0.7% in 2020 to 1.4% in 2035, before jumping to 31.8% by 2050. The uptake of fuel cell medium- and heavy-duty trucks is expected to increase rapidly after 2040, from 5.8% to 33.6% in 2050. Alternative fuel medium- and heavy-duty trucks will always account for a significant share, at 20% by 2050. The proportion of petrol and diesel trucks will decrease rapidly after 2035, from 85.6% to 14.6% in 2050 (see Fig. 23).

In the Carbon-neutral Scenario, medium- and heavy-duty trucks will be mainly fuel cell vehicles, which will rise rapidly to 66% by 2050. Battery-electric and alternative fuels will account for 20% and 10% respectively of the medium- and heavy-duty truck fleet in 2050. Petrol and diesel medium- and heavy-duty trucks will be phased out, especially after 2040 when their penetration will fall from 66.9% to less than 4% in 2050 (see Fig. 24).

3.4.5 Battery-electric Will Predominate in Short- and Medium-distance Light-duty Road Freight Transport, While Hydrogen Fuel Cell Trucks Will Lead in Long-haul Heavy-duty Freight

In 2020, light-duty trucks accounted for 68% of China's truck fleet, while medium- and heavy-duty trucks made up 32%.

In the Development Scenario, the number of battery-electric trucks will steadily increase from 1.1% in 2020 to 7.9% in 2030, before rising

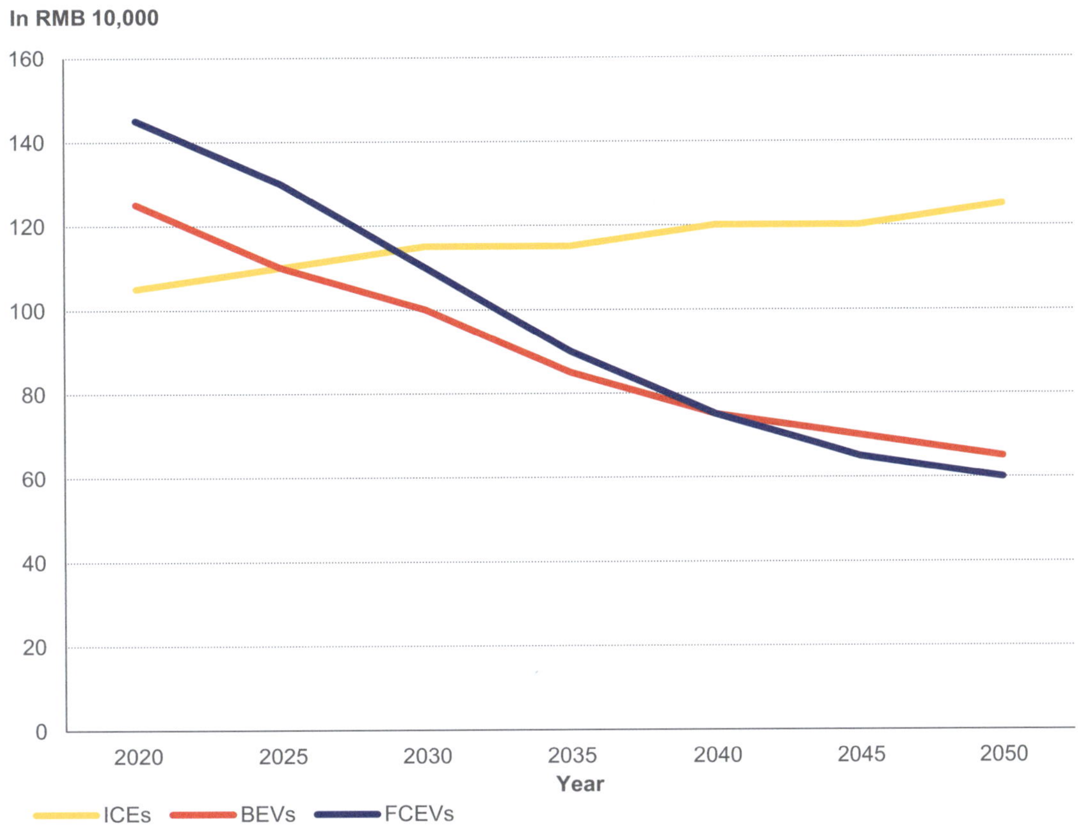

Fig. 14 The cost reduction potential of buses. *Note* ICE = internal combustion engine; BEVs = battery-electric vehicles; FCEVs = fuel cell electric vehicles. *Source* Calculations made by the project team

significantly to 64.7% in 2050. The uptake of fuel cell trucks will increase rapidly after 2040, from 2.8% to 17.5% in 2050. The number of petrol and diesel trucks will decline by more than 70% between 2035 and 2050 (see Fig. 25).

In the Carbon-neutral Scenario, the share of battery-electric trucks will increase rapidly around 2035, from 23.9% to 68.4% in 2050. Fuel cell trucks will grow quickly after 2040 from 5.6% to 27.2% in 2050. The phase-out of petrol and diesel trucks will accelerate; by 2050, their market share will be only 1.3% (see Fig. 26).

3.4.6 Conventional Fuel Vehicles Will Gradually Be Replaced by New Energy Models

In the Development Scenario, the proportion of electric vehicles will exceed 20% in 2030, while that of petrol and diesel vehicles will decrease to less than 80%. By 2040, more than 60% of vehicles will be electric, and by 2050 vehicle electrification will be more or less complete (see Fig. 27).

In the Carbon-neutral Scenario, electrification is consistent with that in the Development Scenario, but the penetration of electric vehicles will be higher. By 2050, battery-electric and fuel cell vehicles will account for 80% and 20% of the market respectively, while conventional fuel vehicles will be more or less phased out (see Fig. 28).

3.4.7 Demand for Conventional Fuels Will Gradually Decrease, While Demand for New Energy Will Grow

In the Development Scenario, fuel consumption for vehicles will peak at 277 million tonnes in 2028, then decline year by year to 19 million

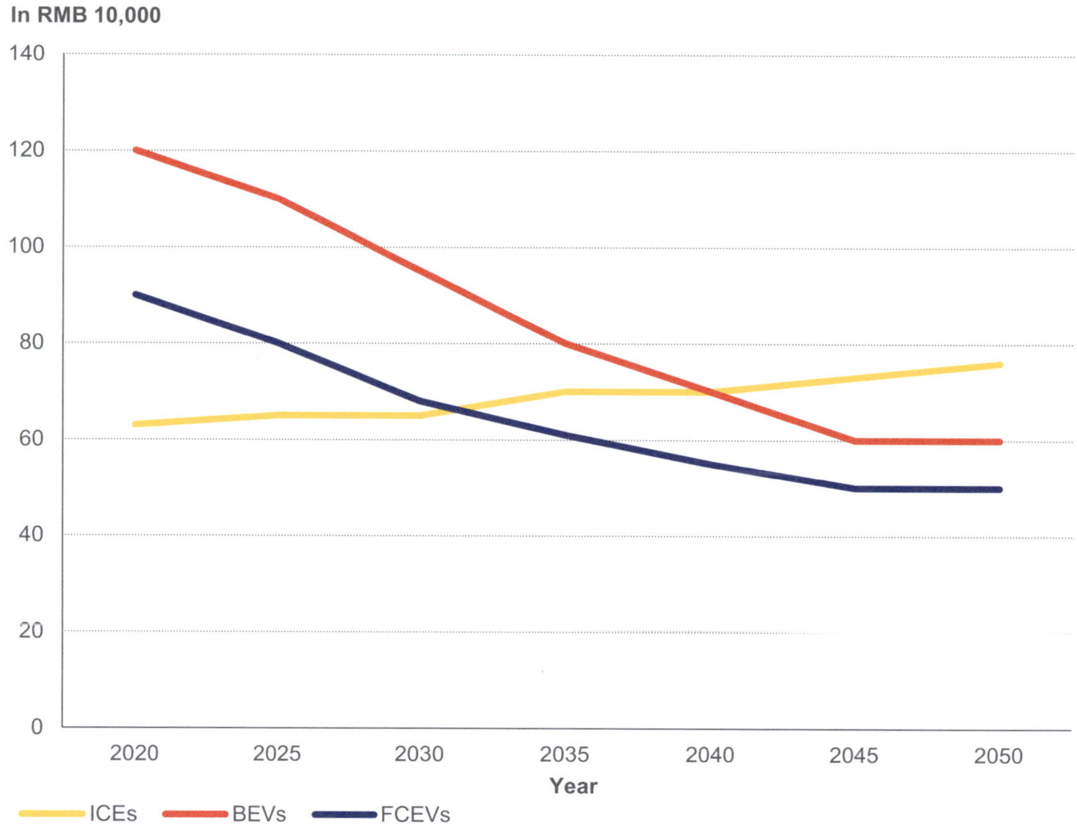

Fig. 15 The cost reduction potential of light-duty trucks. *Note* ICE = internal combustion engine; BEVs = battery-electric vehicles; FCEVs = fuel cell electric vehicles. *Source* Calculations made by the project team

tonnes in 2050. The demand for alternative fuels will peak at 18 million tonnes around 2030. Both electricity and hydrogen demand will grow. In 2050, demand for electricity will reach more than 1,000 terawatt-hours, while demand for hydrogen will be about 15 million tonnes, as shown in Table 3.

In the Carbon-neutral Scenario, vehicle fuel consumption will peak at 264 million tonnes in 2025 and then decline year by year to 4 million tonnes in 2050. The demand for alternative fuels will peak at 15 million tonnes around 2030. Both electricity and hydrogen demand will grow. In 2050, demand for electricity will reach more than 1,000 TWh, while that for hydrogen will be about 28 million tonnes, as shown in Table 4.

4 Challenges of Transport Decarbonisation

The priorities in the next stage of electrifying the transport sector are to solve technical weaknesses, secure the supply chain, improve the aftersales market, accelerate the construction of infrastructure and ensure large-scale resource supply.

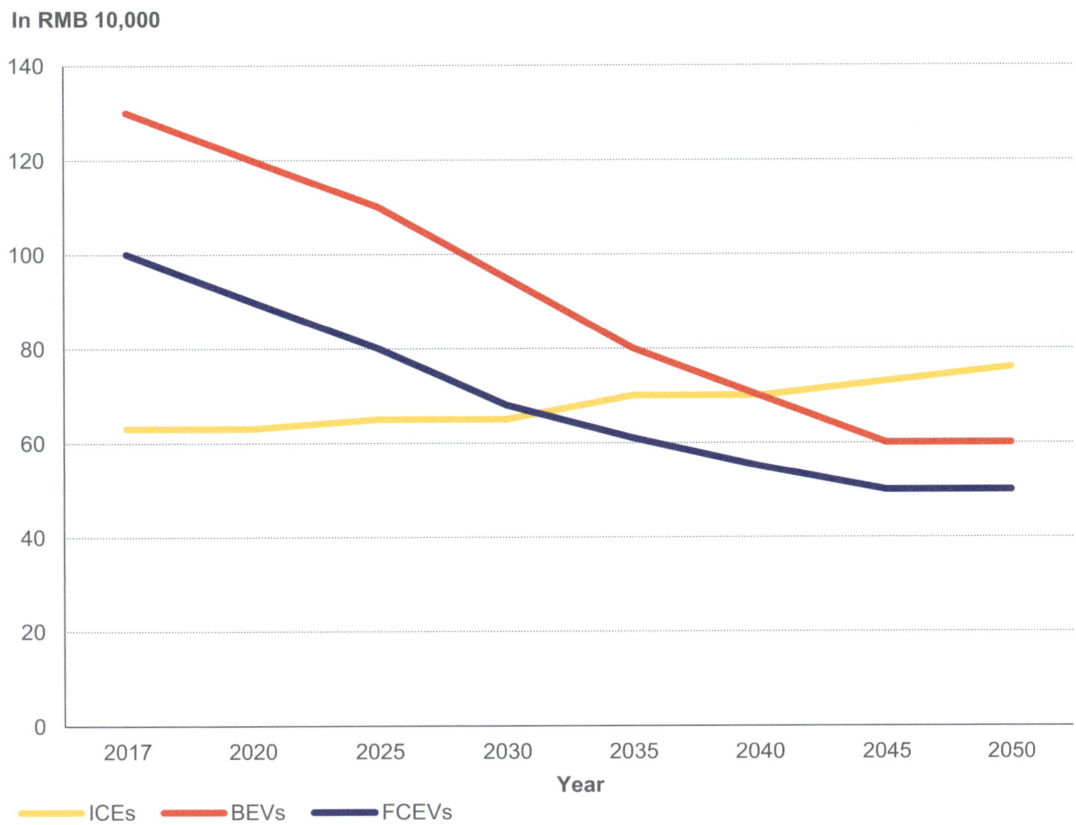

Fig. 16 The cost reduction potential of heavy-duty trucks. *Note* ICE = internal combustion engine; BEVs = battery-electric vehicles; FCEVs = fuel cell electric vehicles. *Source* Calculations made by the project team

4.1 Some Products Need to Be Improved

4.1.1 Insufficient Range and Poor Driving Experience

Temperatures are low in winter, making the comparatively short driving range of electric vehicles an issue for drivers. There are four main causes of this problem. First, the energy in a power battery decreases by 10–20% at low temperatures in winter. Second, the vehicle's energy consumption increases in winter and its regenerative braking function is more or less lost, reducing the total efficiency of the power system by 10–20%. Third, the power consumption of in-vehicle heating at low temperatures is higher than that of cooling in summer; as a result, in-vehicle electricity consumption is more than 10% of the energy used. Fourth, the vehicle's calculation of the range remaining is impaired in low temperatures and errors easily occur, which increases range anxiety and erodes the driving experience.[8]

4.1.2 High Purchase Cost and Low Residual Value

Although the life-cycle cost of electric vehicles is close to that of fuel vehicles, there is still a large difference in purchase price. Some electric vehicles cost 30–40% more and have a 30–40% lower residual value than their internal combustion engine equivalents (see Fig. 29). The market for used electric vehicles is therefore not large. The reasons for this include the difficulty of estimating the remaining life of the battery, depreciation in the value of the vehicle during

[8] Ouyang Minggao's speech at the Diaoyutai State Guesthouse in Beijing in January 2021.

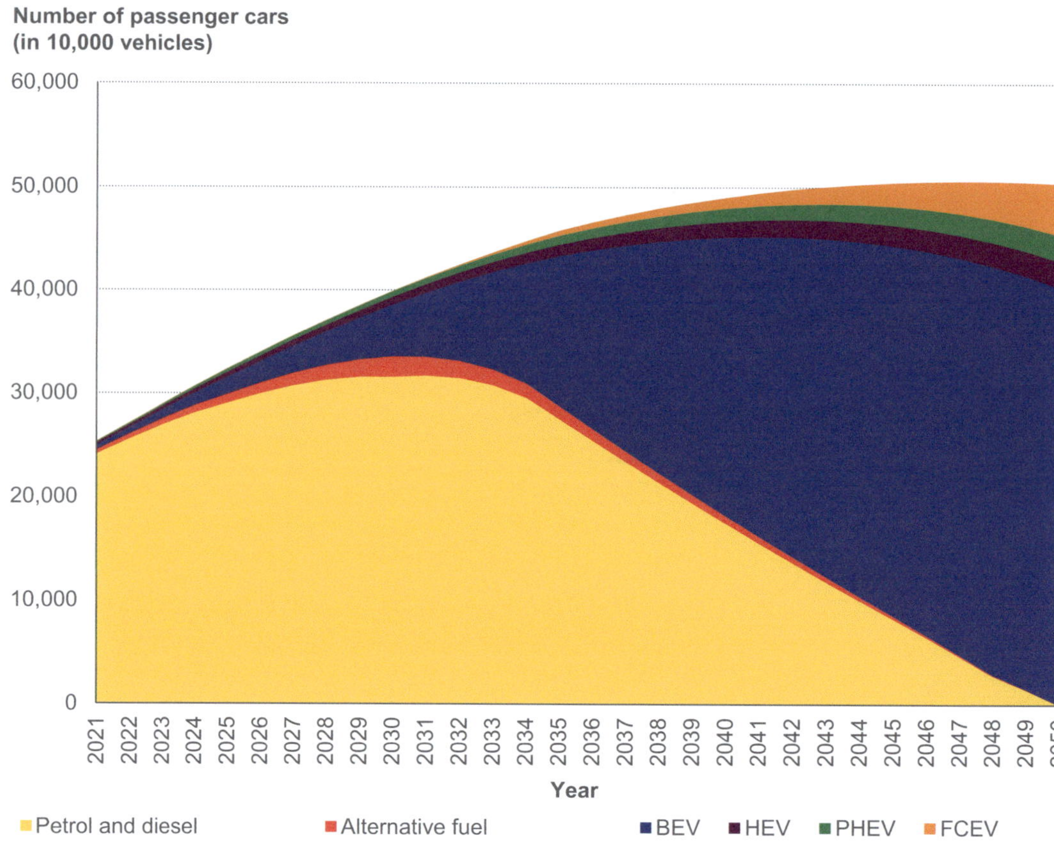

Fig. 17 Passenger vehicle electrification in the Development Scenario. *Note* BEV = battery-electric vehicles; HEV = hybrid electric vehicles; PHEV = plug-in hybrid electric vehicles; FCEV = fuel cell electric vehicles. *Source* Calculations made by the project team

this early stage of electric vehicle development, fragmentation of the available data, rapid development in vehicle technology and falling purchase prices.

4.2 There Is an Urgent Need for Cost Efficiency in Some Vehicle Segments

There is a need for cost-effective solutions in some segments due to the immaturity of certain vehicle technologies, the high cost of core components (see Table 5) and the lack of refuelling infrastructure and resultant range anxiety in drivers.

4.3 Supply Chains and R&D

4.3.1 Difficulties in Transforming the Supply Chain

The number of participants in the electric vehicle supply chain is increasing, which makes supply chains more fragmented and management more difficult, compared with internal combustion engine suppliers.

4.3.2 R&D Can Help Remove Reliance on Imports

The key technologies for some critical parts for electric vehicles have not yet been independently developed in China. Some products are highly dependent on imports, which are often at the high

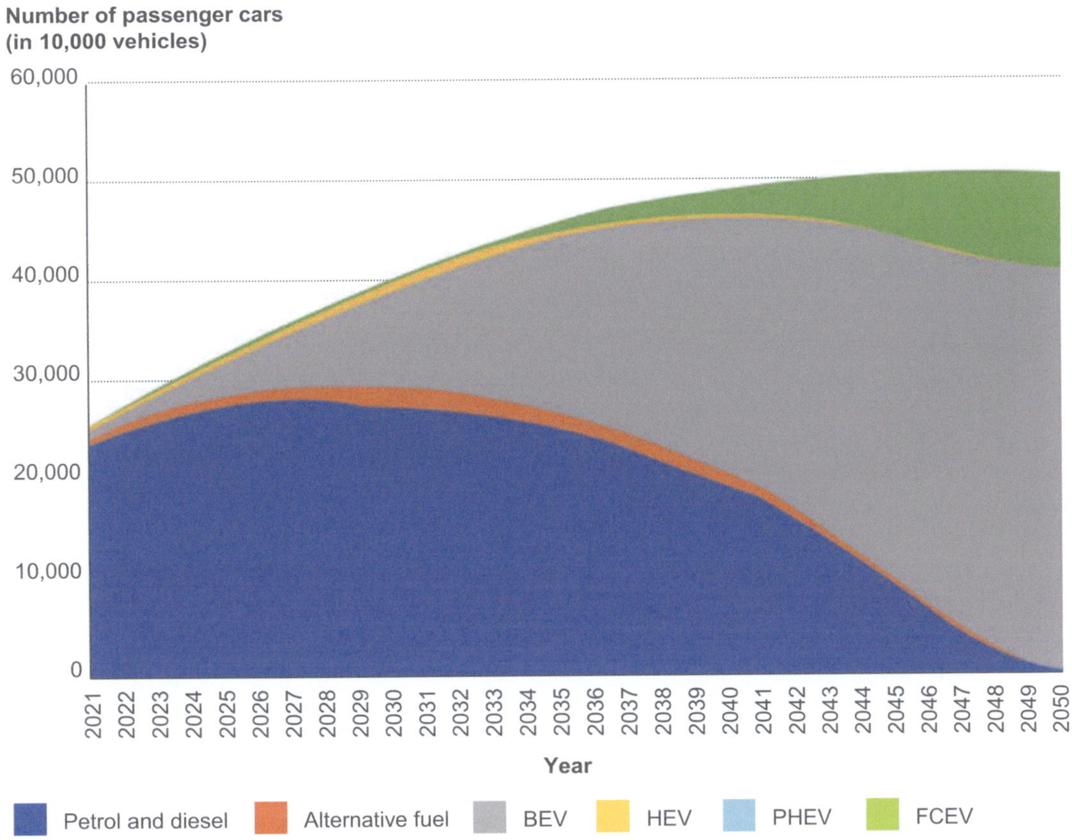

Fig. 18 Passenger vehicle electrification in the Carbon-neutral Scenario. *Note* In this scenario, technologies advance faster than expected, especially for hydrogen fuel cell electric vehicles and their support infrastructure. BEV = battery-electric vehicles; HEV = hybrid electric vehicles; PHEV = plug-in hybrid electric vehicles; FCEV = fuel cell electric vehicles. *Source* Calculations made by the project team

value-added end of the spectrum and crucial for next-generation vehicles. R&D efforts should therefore be intensified to master technologies in these high-value areas.

4.4 Improvements in the Aftersales Market Are Needed

4.4.1 Maintenance Mechanics

Because new energy vehicles are newcomers to the retail market and have a smaller customer base compared with internal combustion engine vehicles, there are relatively few sales outlets and service workshops. This is especially problematic in small and medium-sized cities and rural areas where car ownership is generally low. In addition, the workforce in new energy vehicles is small. According to government estimates, the number of people working in energy efficiency and new energy vehicles in China was only 170,000 in 2015. This is expected to increase to 1.2 million by 2025. However, due to the rapid development of the new energy vehicle sector, a large workforce gap of 1.03 million will remain.

4.4.2 Power Battery Recycling

There are urgent challenges that need to be addressed in the power battery recycling market. By 2020, decommissioned power batteries in China amounted to about 200,000 tonnes, many

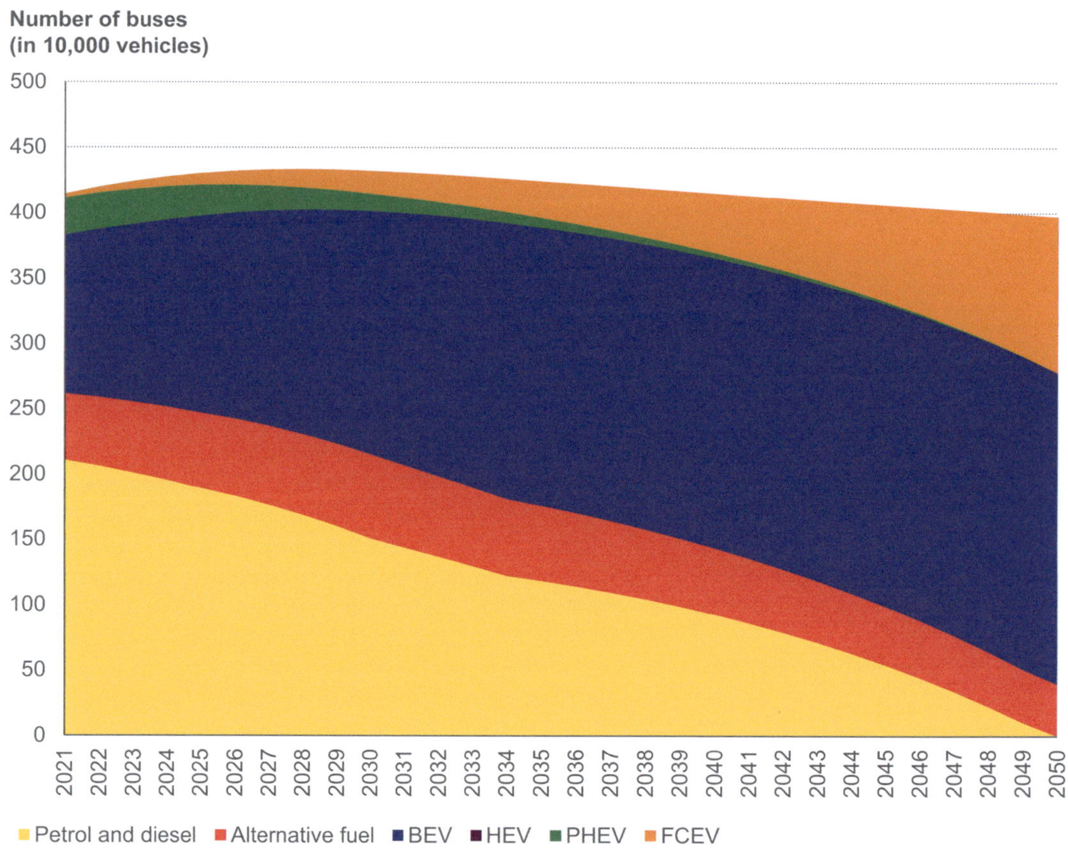

Fig. 19 Bus electrification in the Development Scenario. *Note* BEV = battery-electric vehicles; HEV = hybrid electric vehicles; PHEV = plug-in hybrid electric vehicles; FCEV = fuel cell electric vehicles. *Source* Calculations made by the project team

of which flowed into non-standard channels like small workshops, which resulted in safety and environmental hazards in battery recycling and disposal. Most of the recycled power batteries processed by these channels were re-assembled in new vehicles, posing new safety and environmental hazards when they re-entered the market. In addition, a cost-effective business model for recycling batteries has not yet been developed. It is difficult to make a profit recycling lithium iron phosphate batteries that are free of cobalt, nickel and other high-value metal materials. Accordingly, those in the recycling business face difficulties.

4.5 Infrastructure Construction Needs to Increase

4.5.1 There Are Insufficient Charging Facilities in Residential Areas

Due to constraints like limited power supply, insufficient parking spaces and uncooperative property management companies, there is now much resistance to the construction of charging infrastructure in residential blocks. This is especially the case in old residential communities where resistance restricts the uptake of electric vehicles.

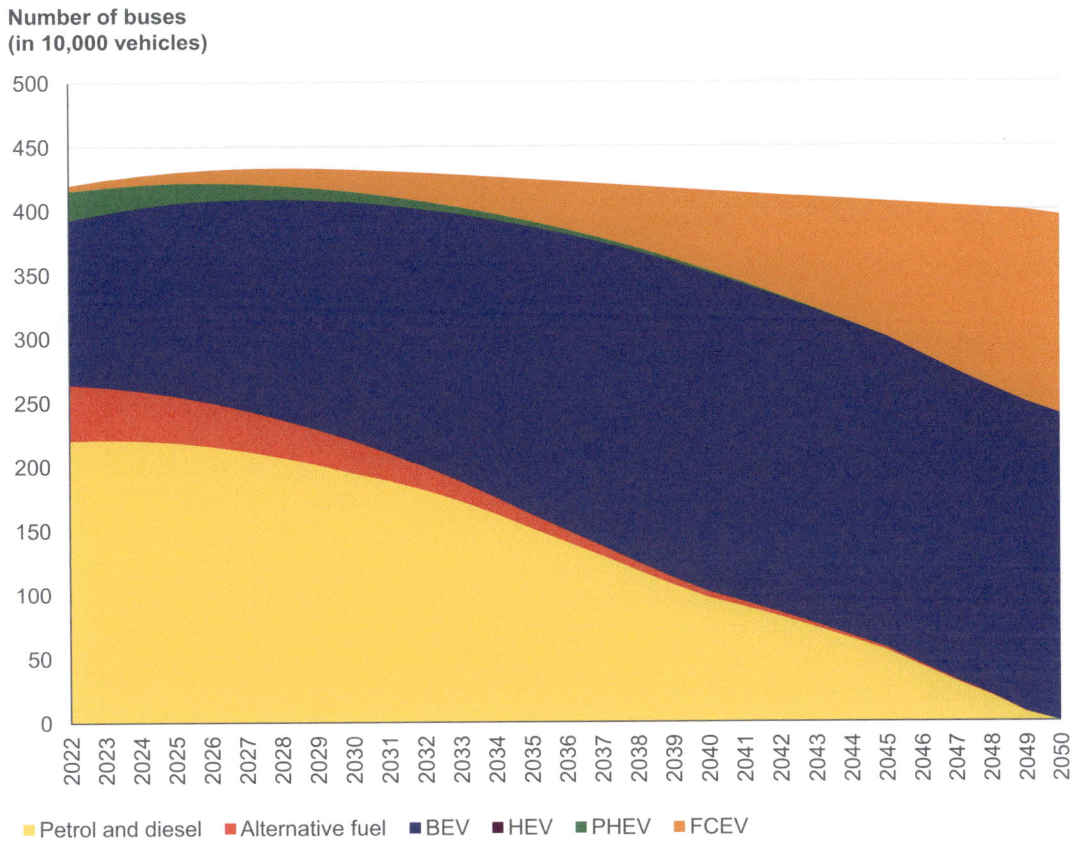

Fig. 20 Bus electrification in the Carbon-neutral Scenario. *Note* BEV = battery-electric vehicles; HEV = hybrid electric vehicles; PHEV = plug-in hybrid electric vehicles; FCEV = fuel cell electric vehicles. *Source* Calculations made by the project team

4.5.2 The Cost of Public Charging Is High in Some Areas

In some cities, especially in central areas of tier-1 cities (Beijing, Shanghai, Guangzhou and Shenzhen), the supply of parking spaces is tight and parking fees are even taken for charging. The high charging cost results in a low use rate of public charge points in central urban areas. If the charging price at peak hours in Beijing (10:00–15:00 and 18:00–21:00) is used, the refuelling cost difference between electric vehicles and fuel vehicles driving 300 kilometres is at least RMB 30 (see Fig. 30), which puts electric vehicles in a disadvantaged position. The reasons for the price difference include high industrial and commercial electricity tariffs and expensive parking in central urban areas.

4.5.3 Uneven Distribution of Chargers

The pace of charging infrastructure expansion varies greatly across provinces. Most provinces have a vehicle-to-charge-point ratio of about 3:1, and some remote provinces do not have charge point clusters at all. It is currently difficult for charge point operators to make a profit. Progress in the construction of public charge points in some cities is slow, which obviously leads to insufficient public chargers.

4.5.4 Lack of Integration Between Electric Vehicles and Power Grids

First, it is difficult to guarantee off-peak charging under the current market mechanisms. In China, there is a fixed pricing mechanism for residential

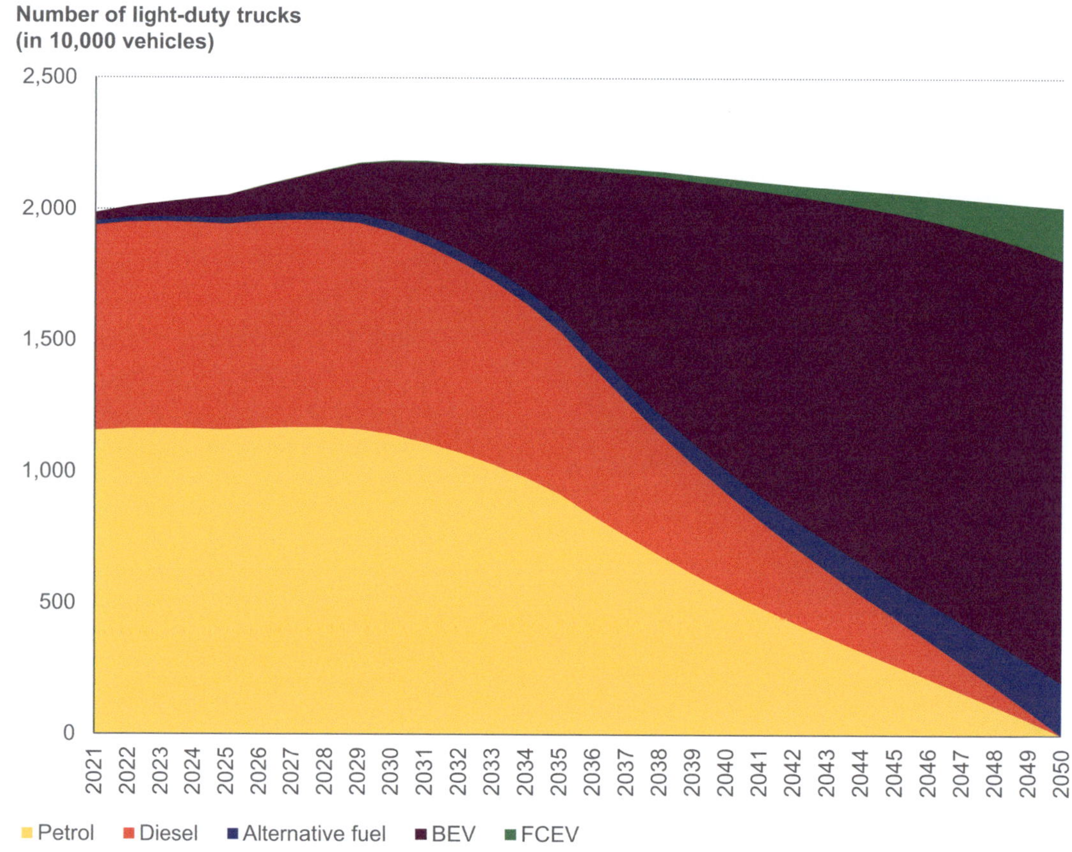

Fig. 21 Electrification of light-duty trucks in the Development Scenario. *Note* BEV = battery-electric vehicles; FCEV = fuel cell electric vehicles. *Source* Calculations made by the project team

electricity, in which electric vehicle chargers are not included. There are no incentives for users to charge their vehicles at off-peak hours, which makes it harder for grid operators to balance supply with demand.

Second, communication between power grids and vehicle chargers is currently not possible. As a result, vehicle-to-grid cannot be implemented, and the services that hold financial potential, such as vehicles providing auxiliary services to the grid, cannot be applied. China does not have the necessary power market regulation mechanisms in place to promote renewable energy consumption through large-scale mobile energy storage enabled by electric vehicles.

5 Case Studies: Best Practice from Around the World

This section presents case studies which illustrate how barriers to transport decarbonisation are being addressed internationally. Examples from around the world demonstrate how governments are enabling mass market uptake of new energy vehicles while municipalities attempt to reverse the traditional car-centric approach to urban development. The cases also present how the public and private sectors are developing innovative business models to support the business case for new energy vehicles.

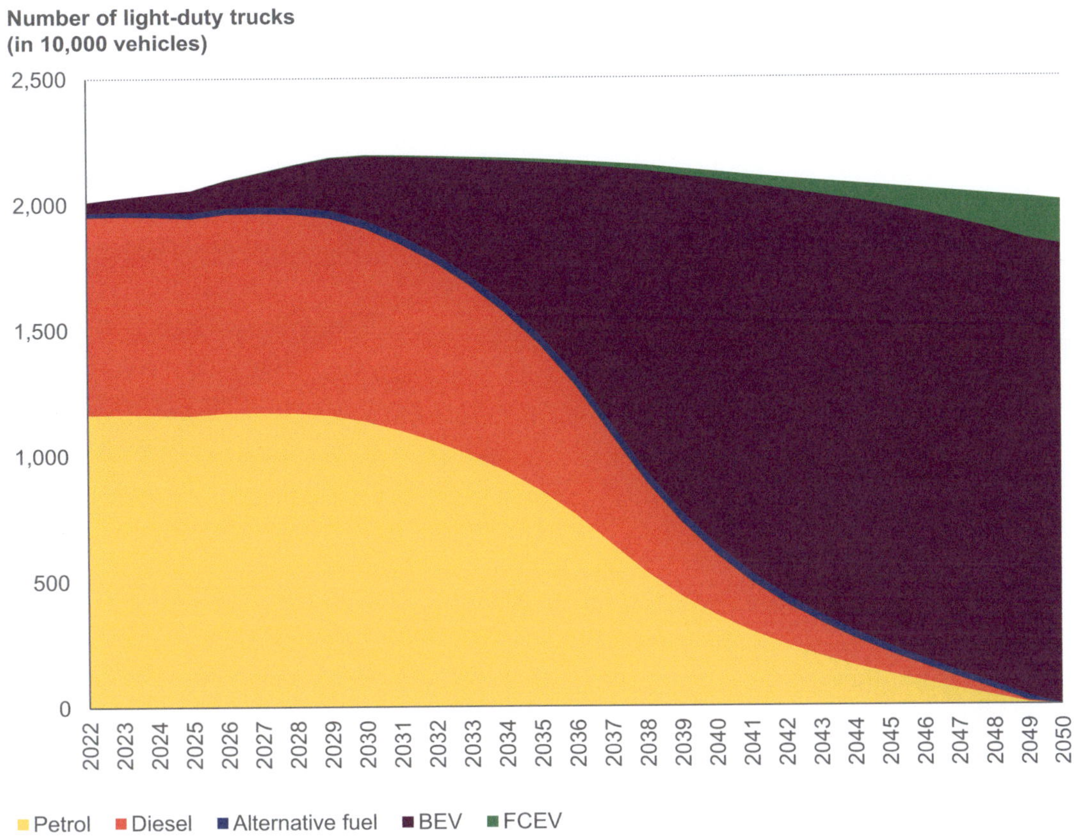

Fig. 22 Electrification of light-duty trucks in the Carbon-neutral Scenario. *Note* BEV = battery-electric vehicles; FCEV = fuel cell electric vehicles. *Source* Calculations made by the project team

5.1 How to Increase New Energy Vehicle Uptake

5.1.1 Transform the Fossil Fuel-based Automotive Industry

Politically driven targets and regulations for both supply and demand emit strong signals for a greener automotive future.

Case Study 1

The EU adopted stricter fleet-wide average emission targets of 95 grams of CO_2 per km for cars from 2021. As of then, the average emissions of all newly registered cars must be below the target. The penalty for exceeding the target each year is payable as an excess emissions premium for each car registered. Several studies estimate the penalties the targets are expected to generate in 2021, which range from €3.3 billion ($4 bn) to €34 bn ($41 bn) in Europe.

Several European countries have also introduced bans on the sale of internal combustion engine cars. A 10-point climate change plan by the UK government bans the sale of petrol and diesel cars from 2030, 10 years earlier than previously planned. While Norway targets 2025, neighbouring Sweden and Denmark have set the ban at 2030. France and Spain have targeted 2040.

The impact of emission standards and supply-side regulations on costs is often overestimated; in the EU, the impact on manufacturing and vehicle costs from such legislation has been minimal. The EU 2015 emission standards (phased in from 2012) did not cause an unusual increase in the average price of a passenger car between 2011 and

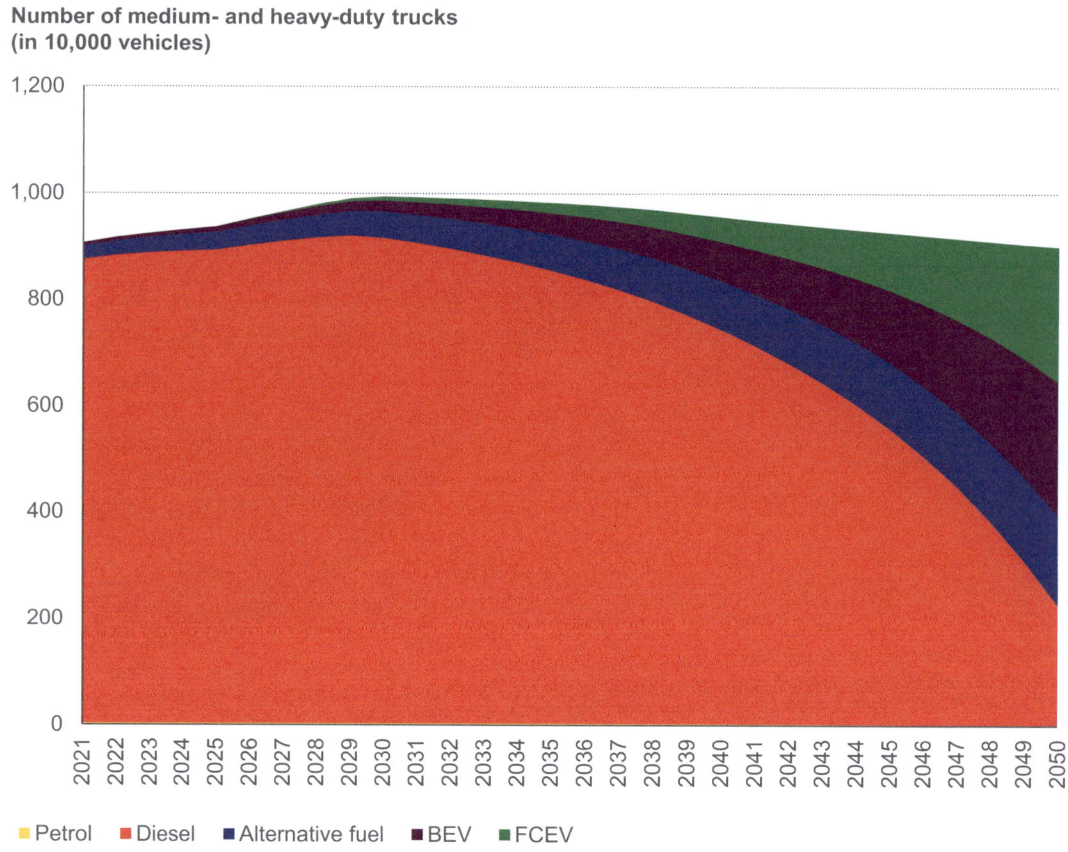

Fig. 23 Electrification of medium- and heavy-duty trucks in the Development Scenario. *Note* BEV = battery-electric vehicles; FCEV = fuel cell electric vehicles. *Source* Calculations made by the project team

2013, as shown in Fig. 31. On the contrary, average car prices fell in 2013 compared with 2012. Nor has the average car price followed the average emissions trend in the EU over the past 15 years. The increase in manufacturing costs has also been overestimated. As shown in Fig. 32, estimates suggested there would be an additional cost to manufacturing of between €1,000 and €4,000 ($1,200–4,870) per vehicle to meet an emissions cap of 120 grams per km. In effect, the cost increase was no more than about €200 ($243) per vehicle to comply with an emissions cap of 123 grams per km in 2014.

These policies have also encouraged innovation, offsetting some of the anticipated cost increases. For example, EU emission standards in the 2010s accelerated innovation and the development of new powertrain and fuel efficiency technologies, such as hybrid and ultralow-emission vehicles in the UK. There is also some evidence that little innovation would have taken place in the absence of such regulations.

The EU emissions target and bans by countries on the internal combustion engine in cars have provided a strong signal to manufacturers for systematic and permanent change. Original equipment manufacturers must initiate a long-term technical and portfolio shift to low-emission vehicles to avoid non-compliance and penalties. This requires significant investment in R&D to develop their own competitive electric vehicle models and adapt their supply chains with new equipment and parts to ensure a timely start to production. Regulatory visibility also reduces risk for the financial sector when allocating

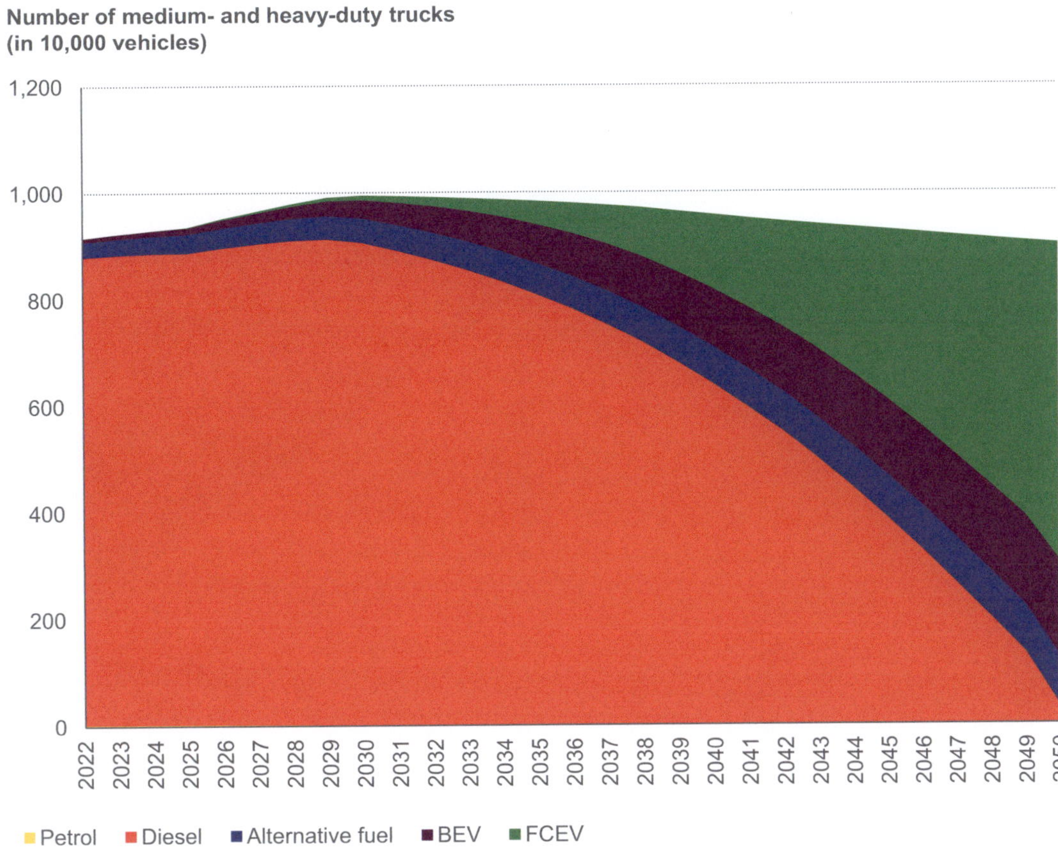

Fig. 24 Electrification of medium- and heavy-duty trucks in the Carbon-neutral Scenario. *Note* BEV = battery-electric vehicles; FCEV = fuel cell electric vehicles. *Source* Calculations made by the project team

capital to greener technologies and provides a firm timeline to the public and private sectors for developing support infrastructure like electric vehicle charging stations.

In response, European car manufacturers are investing heavily in R&D and are preparing to launch a greater number of electric vehicle models. European auto manufacturers spent the most on R&D in 2018–19, investing a total of £33.3 billion ($44 bn), 35% more than the £24.6 bn ($32.5 bn) they spent in 2014–15. All major manufacturers have announced plans to substantially electrify their fleets from 2021 onwards. For example, Ford has removed 2.0 and 1.5 litre petrol engines from the S-Max and Mondeo ranges and launched 1.0 litre 48 volt mild-hybrid Fiesta and Focus models instead. Similarly, Volkswagen and Skoda have dropped most 2.0 TSI petrol engines from their sport utility vehicles. The response to the UK sales ban on internal combustion engine cars has been positively welcomed by many fleet-related organisations.

5.1.2 Make Sure the Infrastructure Is Used

Creating anchor demand for new technologies through strict mandates and targets ensures greater use and therefore reduced costs for support infrastructure.

Case Study 2

California is leading the US market in zero-emission vehicle deployment with a market share more than four times the national average. The state was the first in the USA to establish a zero-

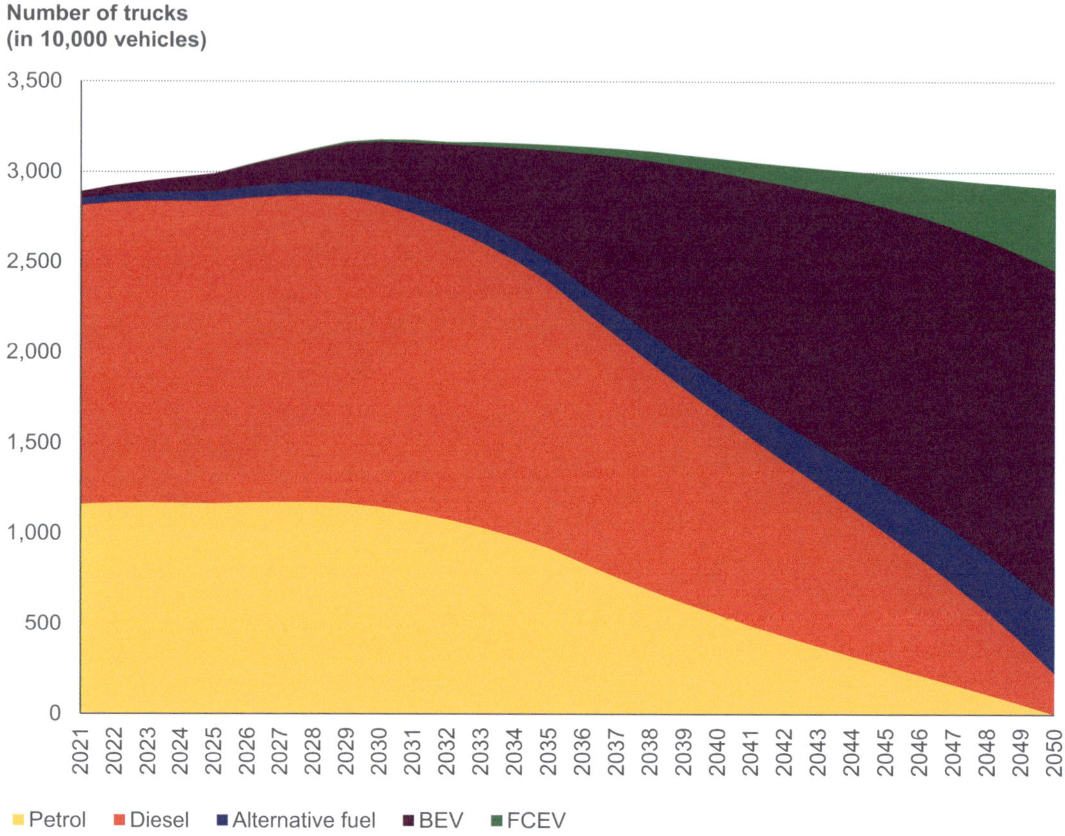

Fig. 25 Electrification of trucks in the Development Scenario. *Note* BEV = battery-electric vehicles; FCEV = fuel cell electric vehicles. *Source* Calculations made by the project team

emission vehicle mandate in 1990. The California Air Resources Board adopted a broader set of regulations for new cars sold between 2017 and 2025 known as the Advanced Clean Cars Program, where manufacturers are required to fulfil a percentage credit requirement for zero-emission vehicles.

California's recently announced mandates for zero-emission trucks are the first in the world, requiring truck manufacturers to start selling zero-emission alternatives from 2024 and that 40–70% of their sales should be zero emission by 2035. To achieve the 2030 target of a 40% reduction in greenhouse gas emissions, California has now extended its programme with various push and pull policies for commercial freight vehicles. In the pull against polluting vehicles, following the less stringent low-emission standards imposed by the federal government, California agreed a deal in August 2020 with several major automakers to increase fuel economy from 38 to 51 miles per gallon[9] by 2026. In the push for zero-emission vehicles, the manufacturers are to sell 100,000 zero-emission trucks by 2030 and 300,000 by 2035. The role of hydrogen fuel cell trucks will be critical, given their greater suitability for long-haul vehicles. A summary of California's targets and policies is presented in Fig. 33.

The mandates, coupled with public funding and sharing of infrastructure costs, assure the private sector that cost reductions will come with greater use. Hydrogen infrastructure faces significant costs, particularly when stations see little use. As the market complies with the mandates in California, infrastructure should see higher use

[9] 1 mile = 1.6 kilometres, 1 gallon = 3.8 litres.

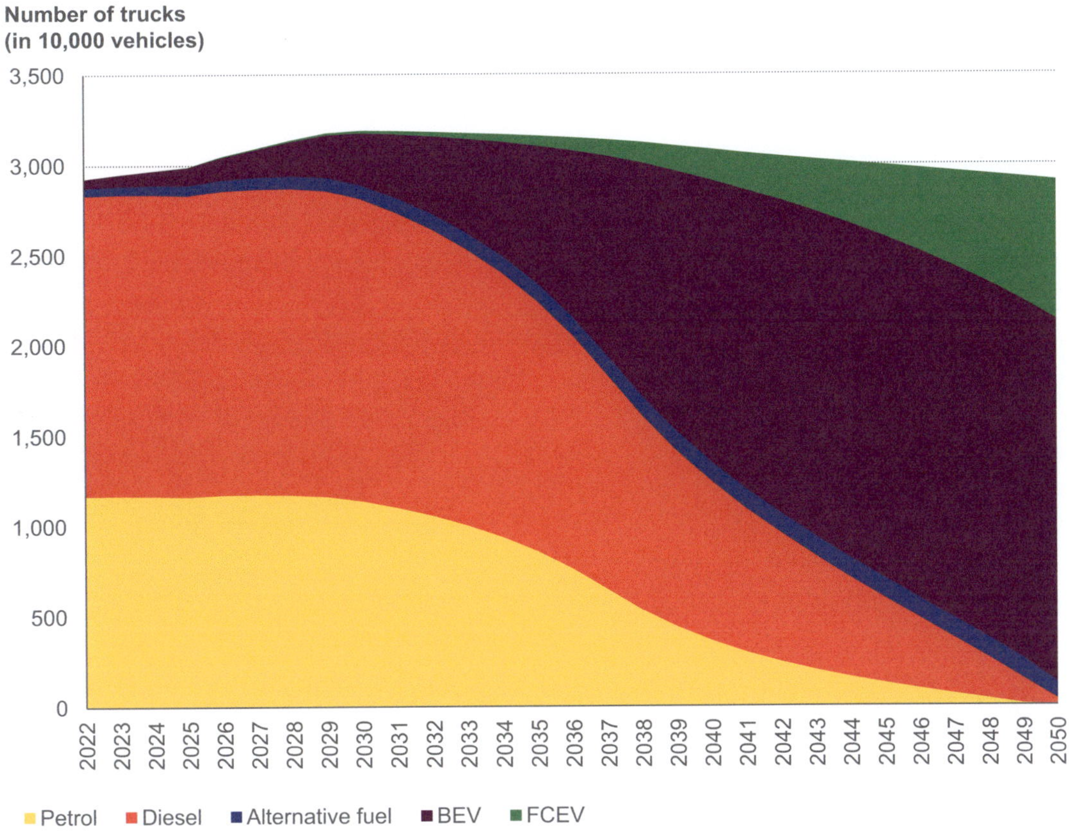

Fig. 26 Electrification of trucks in the Carbon-neutral Scenario. *Note* BEV = battery-electric vehicles; FCEV = fuel cell electric vehicles. *Source* Calculations made by the project team

rates, increased number of pumps per station and falling component costs with increasing scale. This suggests a business case for the construction and operation of refuelling networks, boosting private sector investment confidence in the technology.

The California Energy Commission's Clean Transportation Program announced competitive grants for hydrogen refuelling infrastructure projects in 2020. That same year, the commission provisionally approved a $70 million grant to FirstElement Fuel, Shell and Iwatani Corporation to build new hydrogen refuelling stations. These will add hydrogen refuelling stations to existing petrol stations and will contribute to California's Hydrogen Highway, which comprised 42 hydrogen stations in 2020.

5.1.3 Introduce Innovative Business Models

Innovative business models for new energy vehicles are helping consumers overcome the financing challenge.

Case Study 3

In the UK, car-sharing services such as Zipcar are making electric vehicles accessible to consumers without the need for upfront payments or car ownership. Car sharing is an environmentally and financially feasible option to use a car and is therefore a great way to make electric vehicles accessible to all. Zipcar reports that its electric vehicle fleet of 325 Volkswagen e-Golfs were popular during its first year of launch in 2018, when 22,000 unique members used the vehicles.

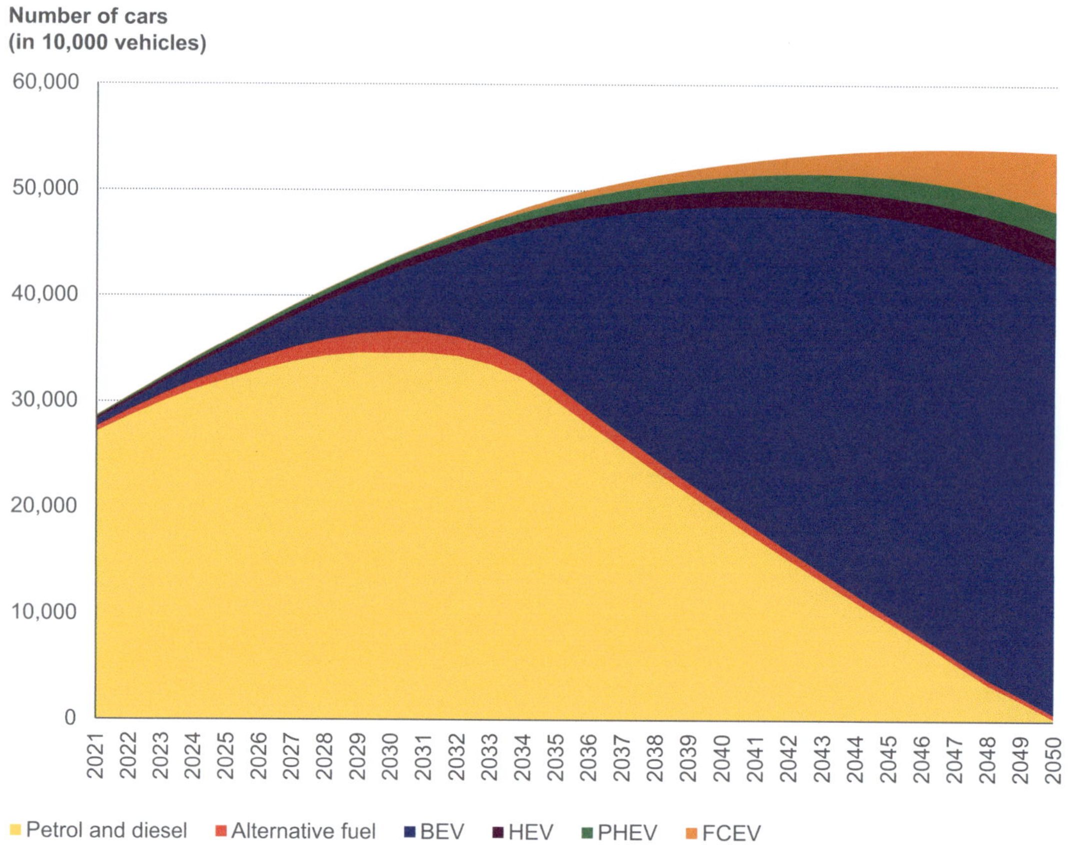

Fig. 27 Electrification of cars in the Development Scenario. *Note* BEV = battery-electric vehicles; HEV = hybrid electric vehicles; PHEV = plug-in electric vehicles; FCEV = fuel cell electric vehicles. *Source* Calculations made by the project team

Zipcar's 2025 vision is to be fully electric, which would deliver significant environmental benefits to London.

However, a lack of support infrastructure remains a barrier to the widespread use of electric vehicles. Figure 34 shows that of the respondents to a car club survey in London who had used a car-sharing battery-electric vehicle, 45% were not satisfied with vehicle charge points. They found charge points difficult to locate and often occupied when found. They also noted that the instructions on how to charge the vehicle were poor.

Zipcar's efforts to expand charging infrastructure in London and test new technologies like mobile charging strengthen the business case for private sector investment in electric vehicle infrastructure. Zipcar has partnered with five charging companies across London, and in the past year has undertaken 10,000 car charges to help lower the barriers to use. A new trial involving US-based FreeWire Technologies and Zipcar studied the use of mobile charging technology and compared it with traditional static charge points (in mobile charging, a charge point is brought to the car rather than the other way around). This helps reduce the cost of installing charge points in the ground and increases the options for those drivers who do not have access to off-street parking.

Case Study 4

In the Netherlands, vehicle-to-grid charging enabled by NewMotion (now Shell Recharge

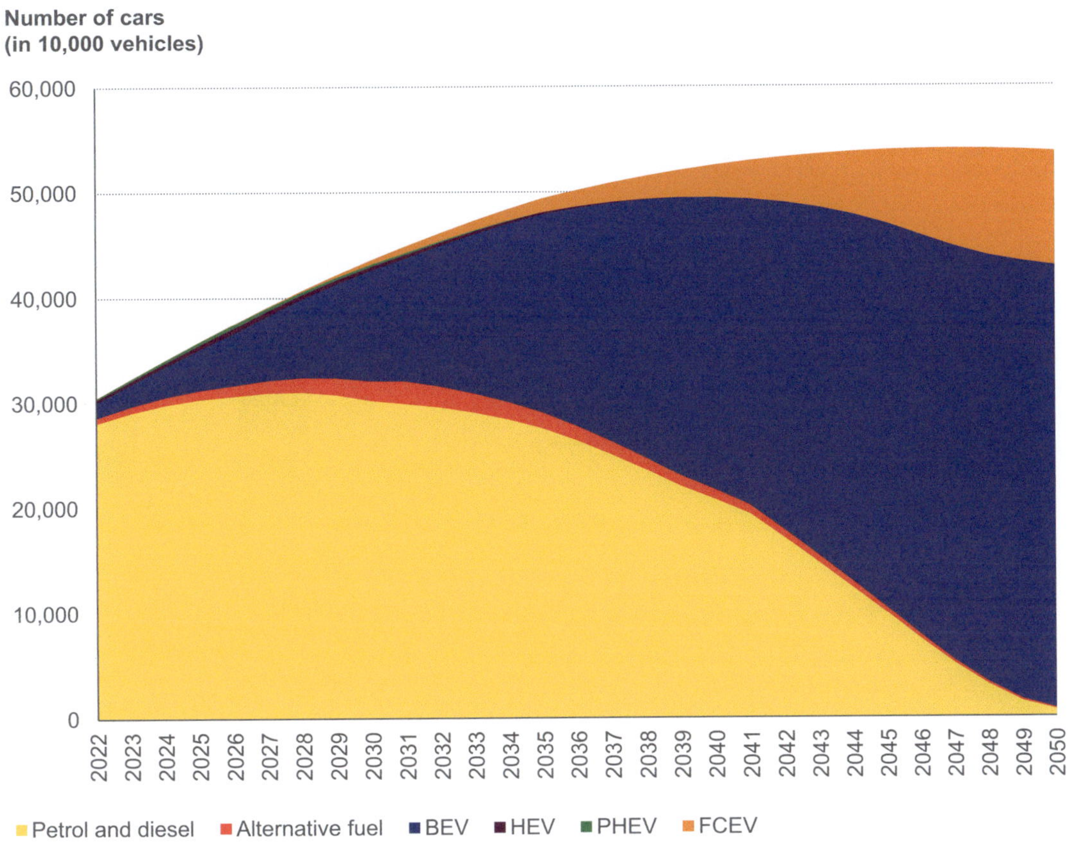

Fig. 28 Electrification of cars in the Carbon-neutral Scenario. *Note* BEV = battery-electric vehicles; HEV = hybrid electric vehicles; PHEV = plug-in electric vehicles; FCEV = fuel cell electric vehicles. *Source* Calculations made by the project team

Table 3 Energy consumption in the Development Scenario

Year	2025	2030	2035	2040	2045	2050
Annual consumption of fuel (million tonnes)	273	271	224	145	79	19
Annual consumption of electricity (TWh)	129.94	228.87	459.82	750.26	934.82	1,033.90
Annual consumption of hydrogen (million tonnes)	0.22	0.74	1.97	3.85	7.60	15.23
Annual consumption of alternative fuels (million tonnes)	11	18	14	13	12	14

Source Calculations made by the project team

Solutions), allows EV owners to supply electricity to the grid. Vehicle-to-grid charging operates through specially designed bi-directional charging stations. Electric vehicle owners can either charge their cars or sell the electricity in the battery to the grid, earning money and reducing their operating costs. The payments for vehicle-to-grid charging vary but can be significant. For example, a company in the UK pays a £30 ($40) fixed monthly fee to customers if they engage in at least 12 vehicle-to-grid sessions a month. This could amount to

Table 4 Energy consumption in the Carbon-neutral Scenario

Year	2025	2030	2035	2040	2045	2050
Total consumption of fuel (million tonnes)	264	247	204	135	62	4
Total electricity consumption (TWh)	159.88	314.53	537.96	808.12	958.62	1,011.15
Hydrogen consumption (million tonnes)	0.25	1.22	3.65	8.62	17.59	27.90
Consumption of alternative fuels (million tonnes)	9	15	12	9	7	4

Source Calculations made by the project team

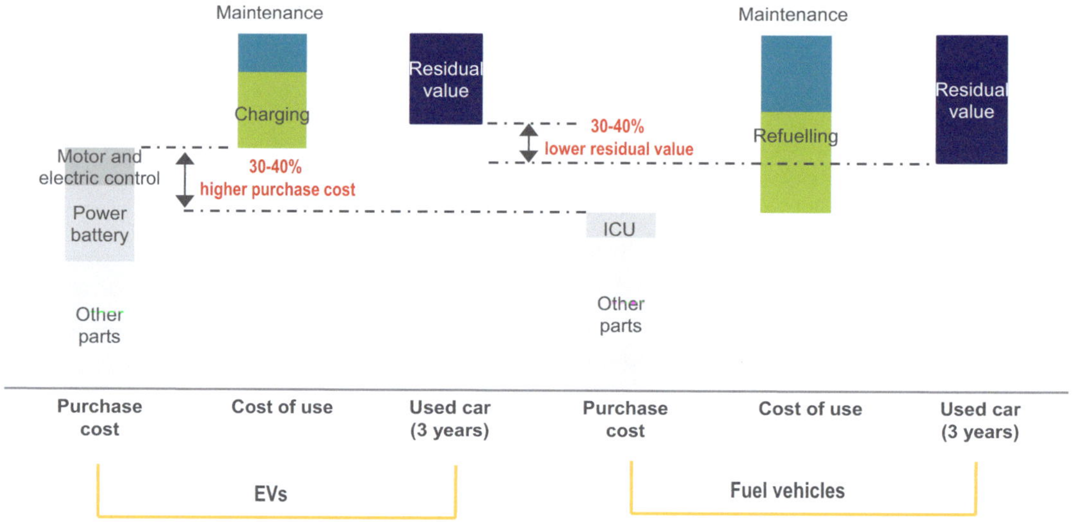

Fig. 29 Cost comparison between electric and fuel vehicles. *Note* ICE = internal combustion engine; EVs = electric vehicles. *Source* Based on research by the project team

Table 5 Barriers to the electrification of vehicles in some segments

	Heavy-duty trucks	Light- and medium-duty trucks
Battery cost	☆☆☆	☆
Fuel cell cost	☆☆☆	☆☆☆
Fuel cell cost	☆☆☆	☆☆☆
Cost-effective range	☆☆☆	☆
Vehicle cost	☆☆☆	☆
Infrastructure development	☆☆	☆☆☆
Fast-charging technology	☆☆☆	☆☆

Note ☆ Some barriers, ☆☆ Big barriers, ☆☆☆ Severe barriers
Source Based on research by the project team

£3,600 ($4,800) over a 10-year lifetime of the vehicle. While this does not directly address the upfront cost challenge, when combined with other business models such as car leasing it reduces both the operating and lifetime cost of the vehicle.

Through vehicle-to-grid charging, electric vehicles can also help balance electricity demand and supply, as shown in Fig. 35. The vehicles can absorb excess daytime electricity generation, particularly solar-generated power. They can in turn increase supply during the evening peak

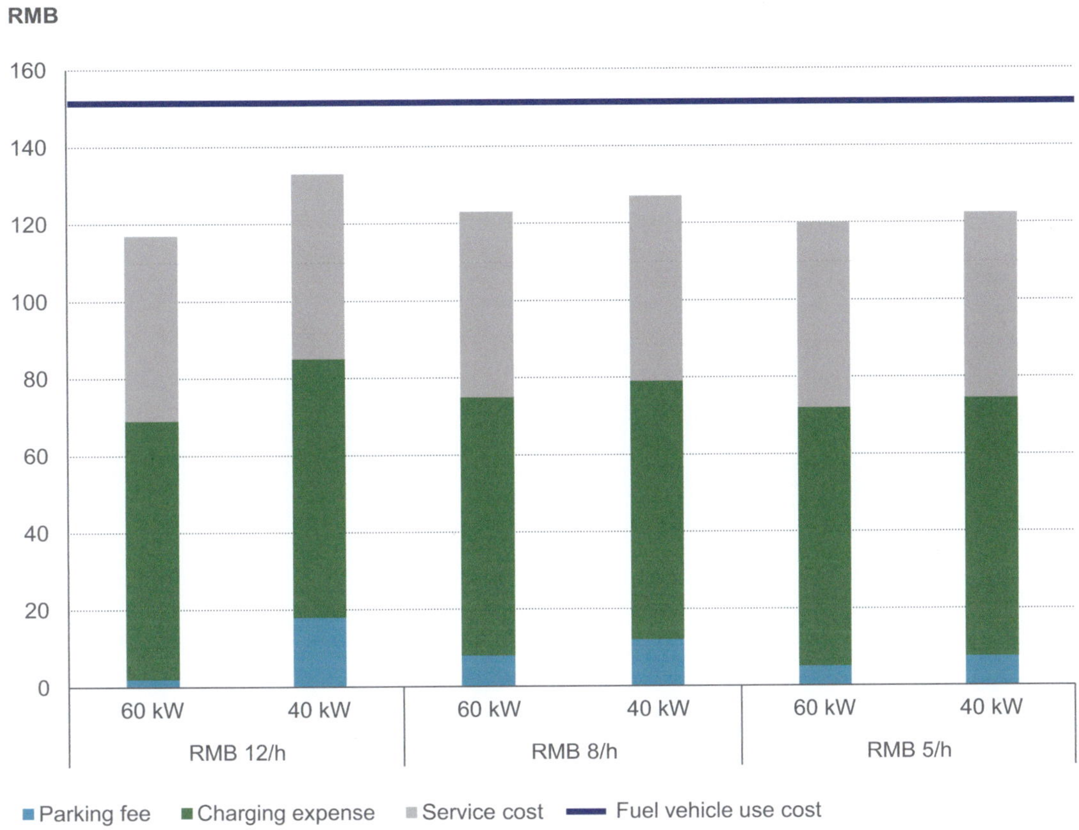

Fig. 30 Recharging cost for different power classes and at different parking fee rates for vehicles driving 300 km. *Source* Calculations made by the project team

period by returning power to the grid. Charging vehicles between 11:00 and 16:00 and discharging at peak times would help balance electricity supply with demand.

Case Study 5

In Switzerland, Hyundai Hydrogen Mobility, a joint venture between Hyundai Motor Company and H2 Energy, provides cost-effective fuel cell vehicle lease models. End users receive eight-year pilot contracts on a pay-per-use basis which gives them mileage, warranty, services, insurance and access to hydrogen. While Hyundai imports the hydrogen trucks, H2 Energy operates the platform for implementing the mobility system.

The joint venture has also partnered with Hydrospider and H2 Mobility Switzerland to promote a business model for fuel cell vehicles in tandem with green hydrogen production and more refuelling infrastructure. Hydrospider, a joint venture between H2 Energy, Alpiq and Linde, is responsible for decarbonised hydrogen production and delivery to the refuelling stations. Hyundai Hydrogen Mobility, Hydrospider and members of the H2 Mobility Switzerland Association are jointly developing the infrastructure for fuel cell truck refuelling using existing traditional stations. The partnership is also looking to set up an aftersales network to boost fleet operators' confidence in the viability of hydrogen technology.

The ecosystem allows fleet operators to avoid upfront initial investment costs, enables refuelling stations to make reasonable margins from the start, and gives fleets access to sufficient quantities of decarbonised hydrogen. The

Fig. 31 Average price and average CO_2 emissions of passenger cars. *Note* The figures in the left-hand axis represent the average price of passenger cars in the EU-27 after adjusting for inflation. *Source* Shell–DRC project team based on European Environment Agency, Average CO_2 Emissions of New Cars Registered in the EU, 2020; International Council on Clean Transportation (ICCT), CO_2 Emissions Target for New Passenger Cars and Light Commercial Vehicles in 2020–2030, 2016; ICCT, European Automobile Market Statistics in 2016–2017, 2017

partnerships in Switzerland have united various players in the energy and mobility industries in a joint, private sector system. Further partnerships are expected to be forged with logistics, retail and trade partners to build a European industrial hydrogen ecosystem.

Figure 36 summarises several factors that have made Switzerland's hydrogen freight ecosystem an attractive and sustainable business model for the private sector. A road tax exemption of up to €41,000 ($50,000) per year per vehicle for zero-emission trucks makes a clear financial case for freight operators. Switzerland generates more than half of its electricity from hydropower and has therefore the potential to produce local, large-scale decarbonised hydrogen from electrolysis. Local production means the country does not face import costs and the falling price of renewables is expected to make decarbonised hydrogen inexpensive in the long run. Without the use of capital subsidies, the private sector has already introduced business models that ensure affordability and the long-term viability of hydrogen trucks in Switzerland.

5.1.4 Provide Support Infrastructure

Large-scale deployment of support infrastructure is more important than fiscal mechanisms to

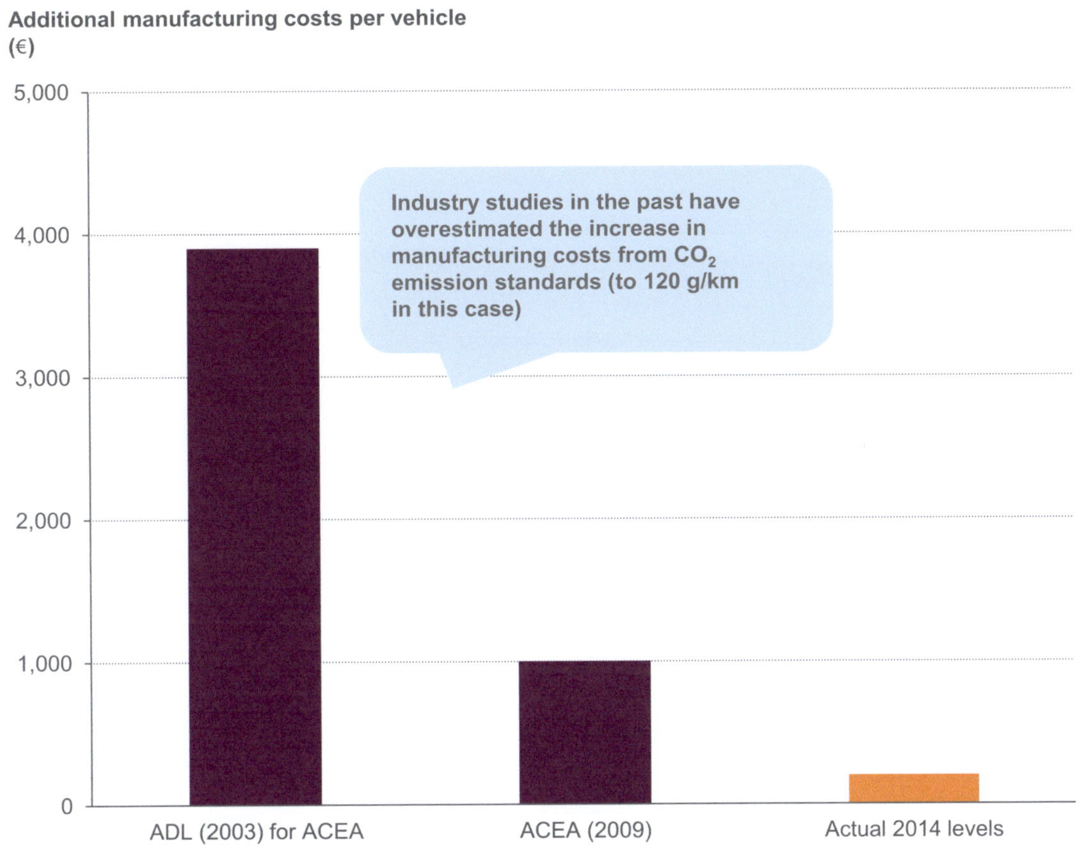

Fig. 32 Additional manufacturing costs and EU emission standards release times. *Note* The figures in the left-hand axis represent the average price of passenger cars in the EU-27 after adjusting for inflation. *Source* Shell–DRC project team based on European Environment Agency, Average CO_2 Emissions of New Cars Registered in the EU, 2020; International Council on Clean Transportation (ICCT), CO_2 Emissions Target for New Passenger Cars and Light Commercial Vehicles in 2020–2030, 2016; ICCT, European Automobile Market Statistics in 2016–2017, 2017

make new energy vehicles more competitive, helping to reduce range anxiety and increase consumer confidence.

Case Study 6

The Netherlands has both the densest and the fastest growing electric vehicle charging infrastructure network in the world. The country's leadership is partly due to its successful roll-out of charging infrastructure, achieving more than 160% growth between 2017 and 2020. The Netherlands provides by far the most charge points per 100 km globally, as shown in Figs. 37 and 39. As the electric vehicle fleet has grown, the number of charges per vehicle has either remained the same or grown year by year, boosting consumer confidence in charging reliability.

The Dutch government's approach has focused on offering charger incentives for businesses and on developing public charging stations. Businesses can receive a deduction of up to 36% of the amount invested in a charge point or the possibility to depreciate 75% of the investment cost of a charge point. For consumers, if there is not a charge point near where they live or work, they can request a free public charge point to be installed by the government. While consumers still need to pay for the energy consumed, they are not liable to pay for purchase, installation or use of the charge point.

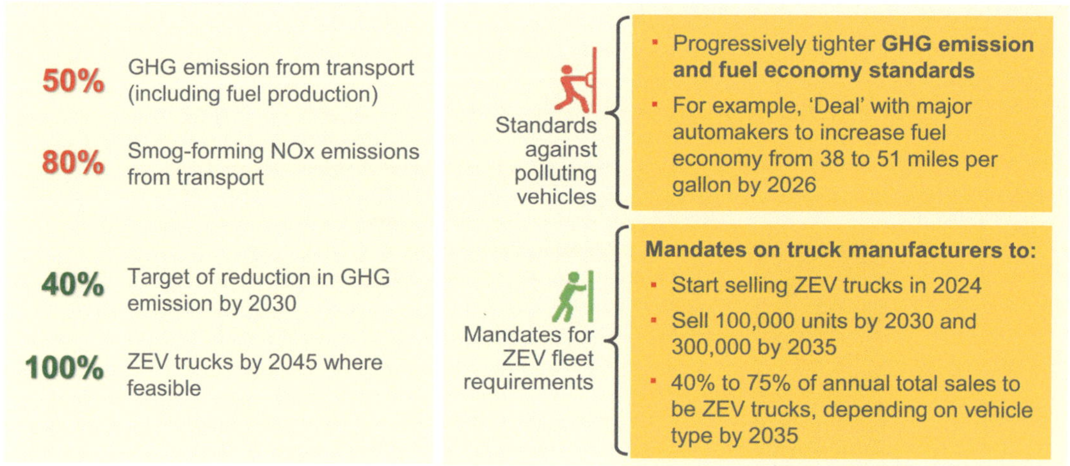

Fig. 33 California's zero-emission (hydrogen fuel cell) truck targets and support policies. *Note* GHG = greenhouse gas emissions; No_x = nitrogen oxide; ZEV = zero-emission vehicle. *Source* Shell–DRC project team based on the California Air Resources Board, the California Fuel Cell Partnership, the Center for Climate and Energy Solutions, and Transport and Environment

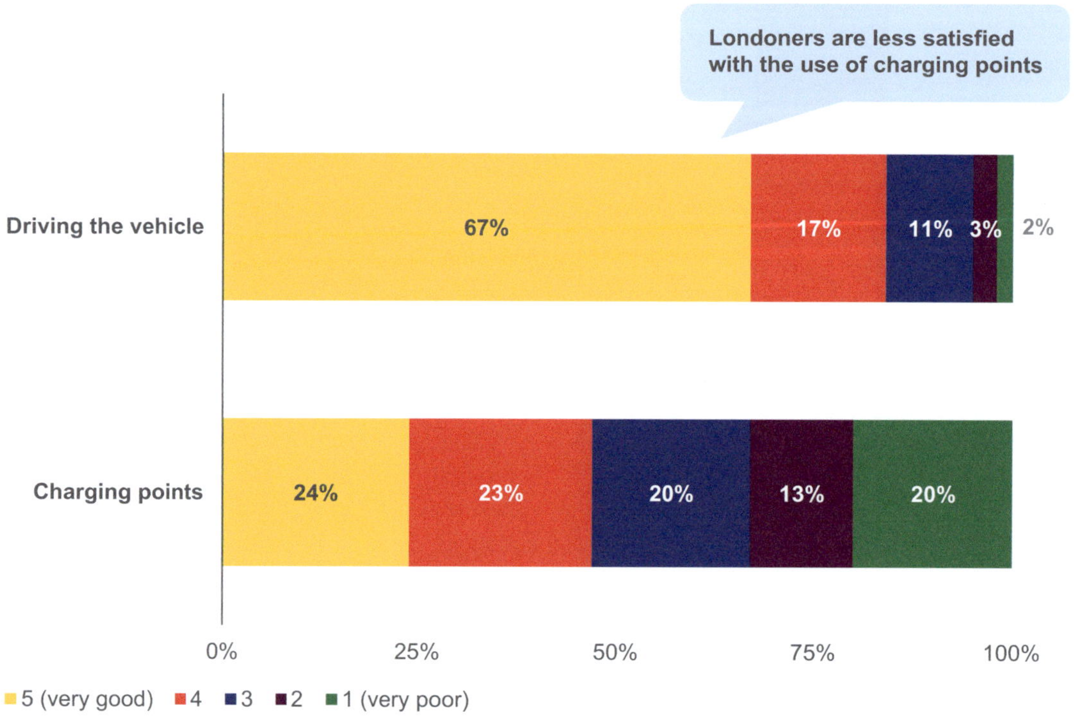

Fig. 34 Londoners' satisfaction with electric vehicle charge points. *Note* The survey was completed by almost 1,570 flexible car club members (including Zipcar), of which 49% had used a car-sharing battery-electric vehicle before. *Source* Shell–DRC project team based on Zipcar and CoMoUK's Car Club Annual Survey for London

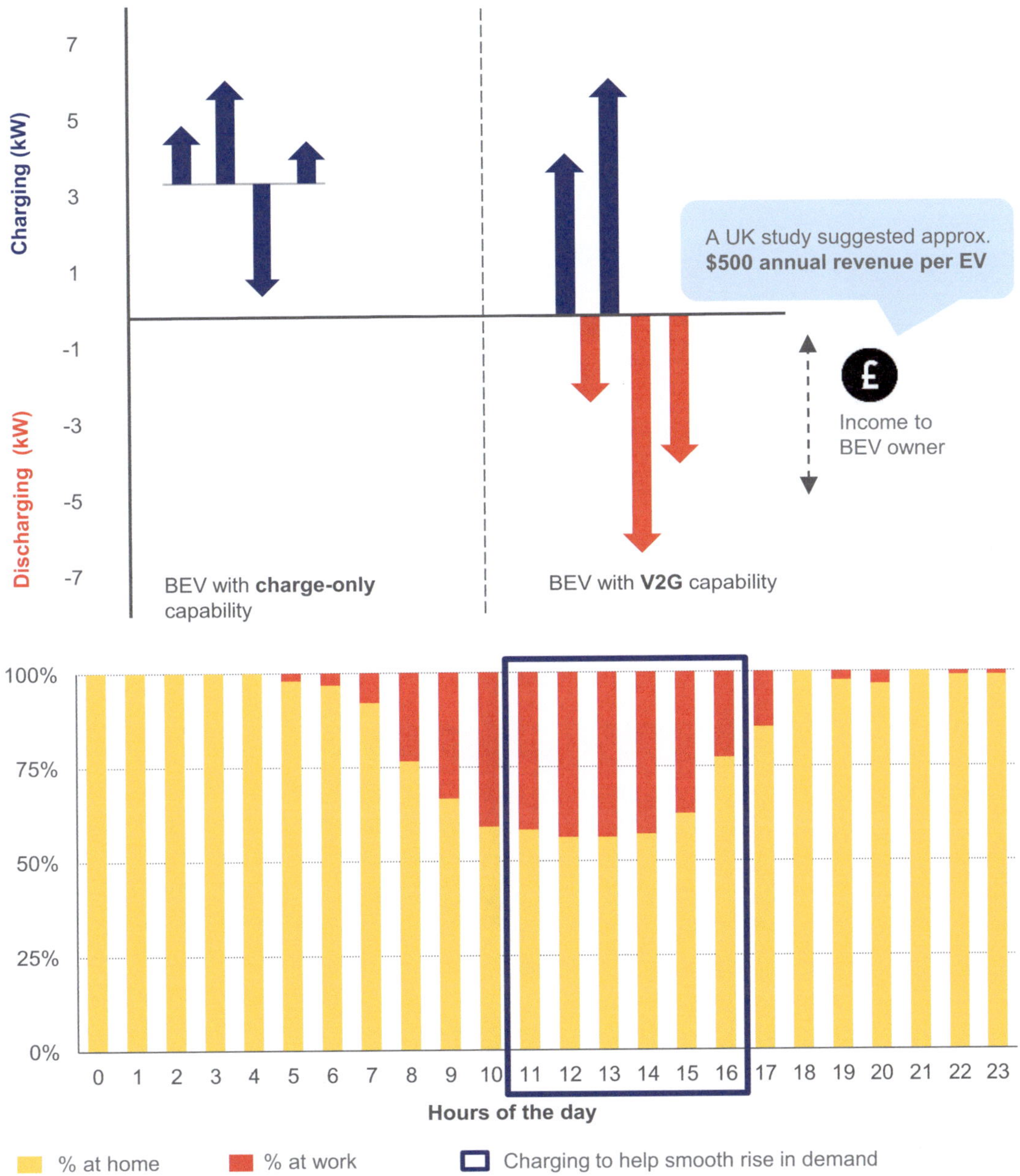

Fig. 35 Vehicle-to-grid charging helps regulate electricity supply and demand. *Source* Shell–DRC project team

Looking ahead, the government is working towards a position in which the business case for charging infrastructure is profitable. This will require the government to change its role in the provision of charging infrastructure. In 2015, the government, along with various municipalities and interest groups, signed the Green Deal for Publicly Available Charging Infrastructure. The deal includes achieving a further cost reduction in infrastructure development through innovation, efficiency and scaling up. The government ultimately hopes that charging infrastructure can be developed in coming years without government incentives.

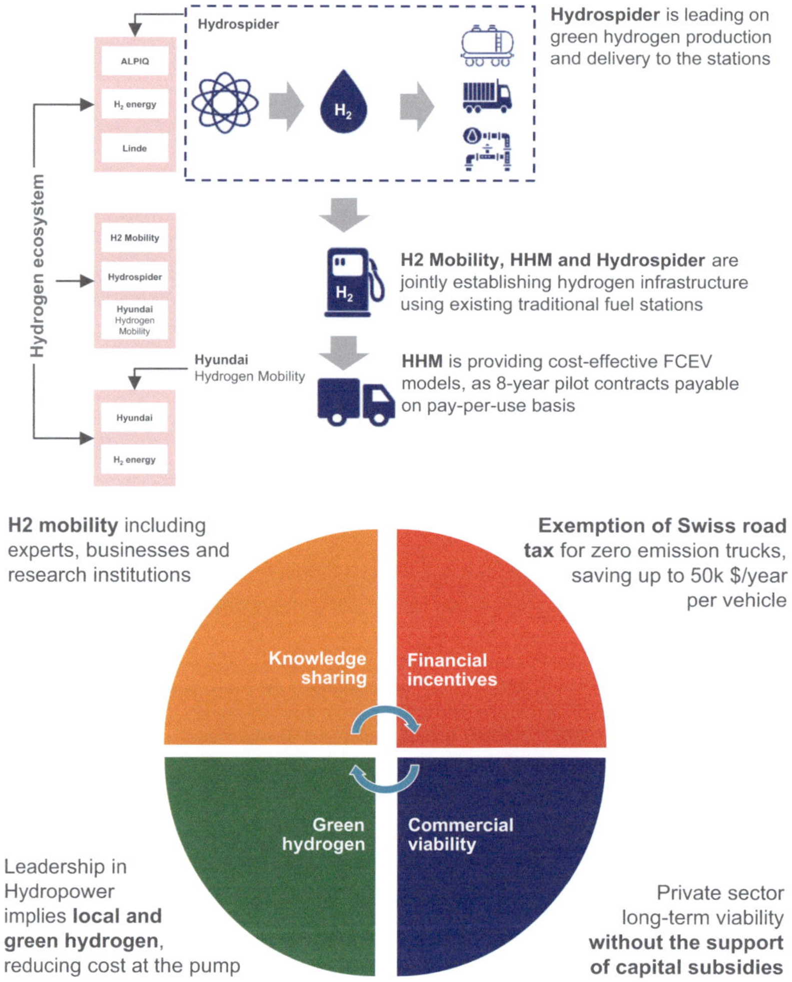

Fig. 36 Public–private partnership on hydrogen fuel cell trucks in Switzerland. *Source* Shell–DRC project team

The government is also taking steps to ensure interoperability in electric vehicle infrastructure. A network of publicly accessible private fast charging points has been set up by allowing market players to operate them on main roads. Given the public nature of this undertaking, the licensing conditions contain a provision stipulating that these charge points be interoperable. This means the charge points must be able to charge any car irrespective of make or service provider. The central government has also standardised the payment structure for charging using a single card or mobile application. The country is also pushing for interoperability throughout Europe, increasing the attractiveness of electric vehicles for longer journeys. For example, the Netherlands and Germany signed a co-operation agreement between eViolin and e-clearing.net on roaming in 2016.

Incentive mechanisms for electric vehicle purchases are progressively decreasing, where subsidies will be eliminated by 2025 at the latest and tax exemptions by 2030. The Netherlands' National Climate Agreement, published in June 2019, suggests further deployment of fast and public charging infrastructure will be prioritised over incentivisation mechanisms. As purchase subsidies are phased out, motor vehicle tax on

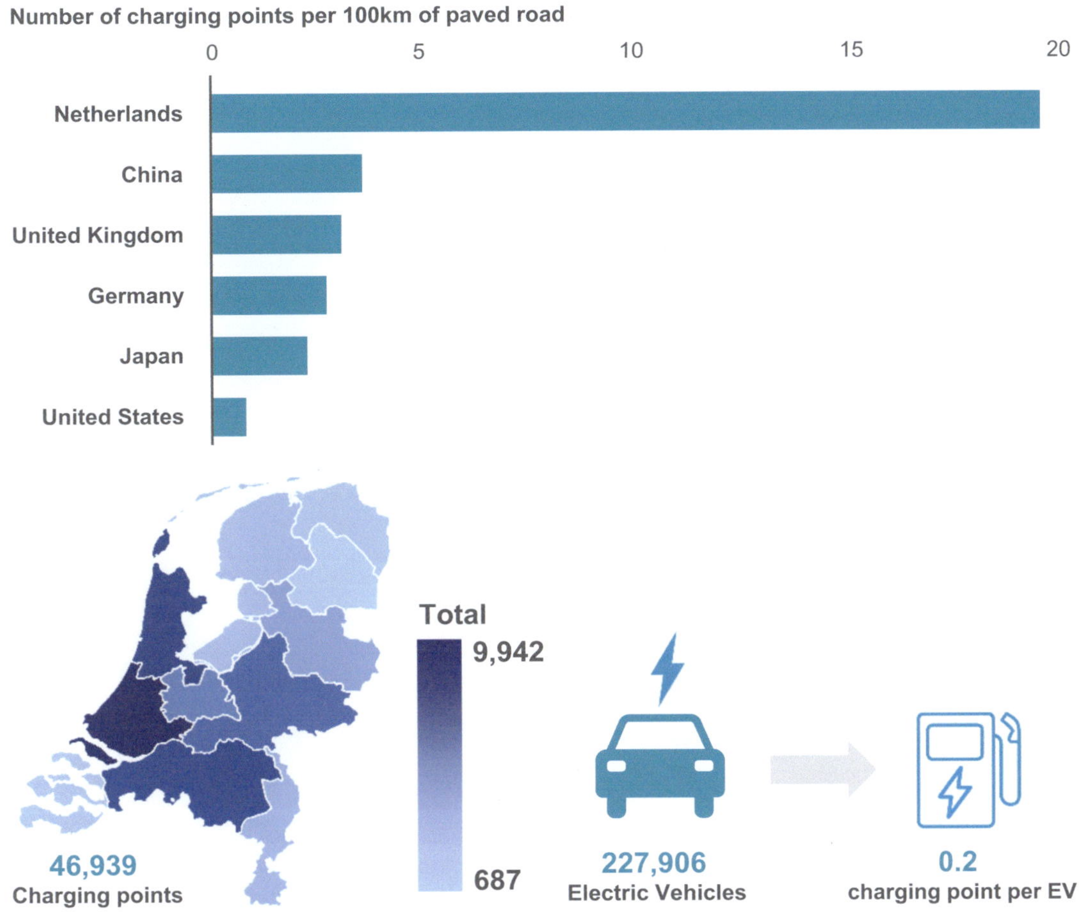

Fig. 37 Number of charge points in countries in 2018. *Source* Shell–DRC project team

electric vehicles will be phased in to offset the decline in government revenues. In order to foster a thriving second-hand market in electric cars for private individuals, the government will develop a scheme for the reimbursement of charge credit, purchase subsidies or battery guarantees.

Case Study 7

Australia is making hydrogen competitive with fossil fuels by making major investments in hydrogen production projects, limiting the need for fiscal incentives. In 2019, the country's National Hydrogen Strategy specified its goal of "H2 under $2" (Australian dollars). Consequently, the government launched an AUD 300 million ($215 m) fund to support hydrogen-powered projects in 2020. The reduced cost of hydrogen production will present a much cheaper way to fuel vehicles like Hyundai's NEXO. For example, if consumers were charged AUD 5 ($3.8) per kg at the pump, filling the NEXO's high-pressure tank would cost just AUD 30 ($22.6), which is enough to travel more than 650 km.

Given the strong commitment of the Australian government to hydrogen, several refuelling infrastructure projects have been initiated. Toyota and ARENA are developing a hydrogen refuelling station powered entirely by renewable energy in Melbourne's western suburbs. Haskel will supply ATCO and Fortescue with refuelling

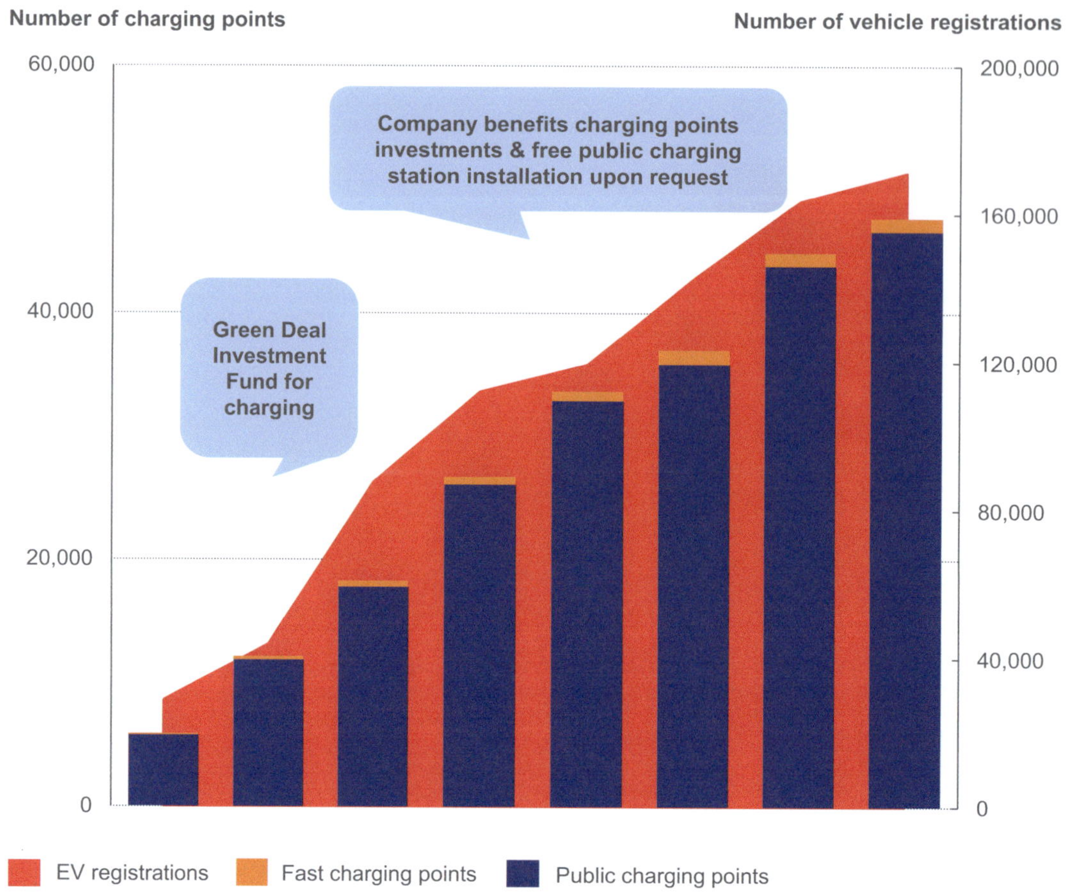

Fig. 38 Number of public charge points in the Netherlands in 2018. *Source* Shell–DRC project team

systems for their Hydrogen Refueller project. Hydrogen project developer Infinite Blue Energy has proposed partnering with electric vehicle charging provider NewVolt to integrate hydrogen refuelling stations at charging sites.

Low-cost decarbonised hydrogen, combined with emerging refuelling infrastructure, means Australia may only need limited public support for the widespread uptake of hydrogen-powered road freight transport. As shown in Figs. 39 and 40, in 2030, the net present value of hydrogen in long-haul freight in Australia is positive compared to diesel trucks, even if taxes on diesel are excluded. Hydrogen trucks are even more attractive if a small (AUD 40 per tonne, \$30/t) or large (AUD 130/t, \$100/t) carbon tax is included.

5.2 Other Measures

5.2.1 Integrate Alternative Modes of Transport

Integrating different modes of transport into a holistic system by providing infrastructure in a co-ordinated manner and ensuring ease of use can reduce the need for private road transport.

Case Study 8

Transport for London (TfL) is a statutory body responsible for various public modes of transport, enabling physical, institutional and operational integration across London. A key feature to achieving integration is the presence of a strong institutional framework which can plan,

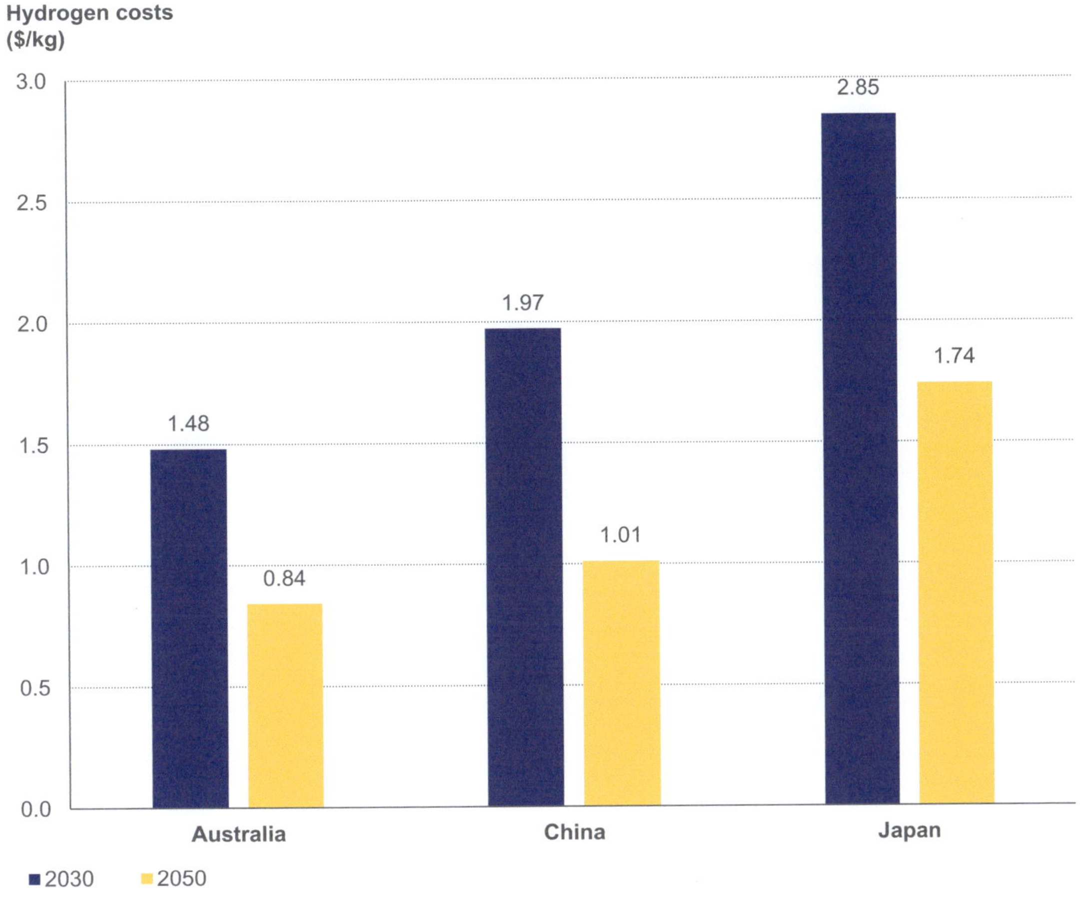

Fig. 39 Hydrogen cost forecasts. *Note* The 2030 and 2050 hydrogen cost includes production (of decarbonised hydrogen), transport (through pipelines) and storage. *Source* Shell–DRC project team

co-ordinate and monitor implementation. TfL has authority over a network of principal road routes, various rail networks including the London Underground and London Overground, Docklands Light Railway and TfL Rail. It also controls London's buses and taxis, cycling provision and river services.

TfL provides transport data information, which helps to improve both the operational integration of services and consumer experience with real-time, accurate information. By freely sharing transport data, TfL encourages the creation of new apps to make travel easier for customers and get more solutions to solve the city's transport challenges. Open Data by TfL is now generating annual economic benefits and savings of up to £130 million ($172 million) for travellers, TfL and the Greater London Authority, with 42% of Londoners reporting they use an application powered by TfL data and 83% saying they use its website.

Integrated payment solutions such as pay-as-you-go "Oyster cards" and mobile payments facilitate transferability and accessibility across transport modes. The Oyster smart card permits payments across buses, light rail, trams and the underground, making it easier to switch between transport modes during a single journey. Contactless cards and mobile payments are also possible in all these applications. Fares are differentiated among transport applications. However, "Hopper fares" also exist which allow passengers to use different modes of transport during a limited time, usually one hour. The key

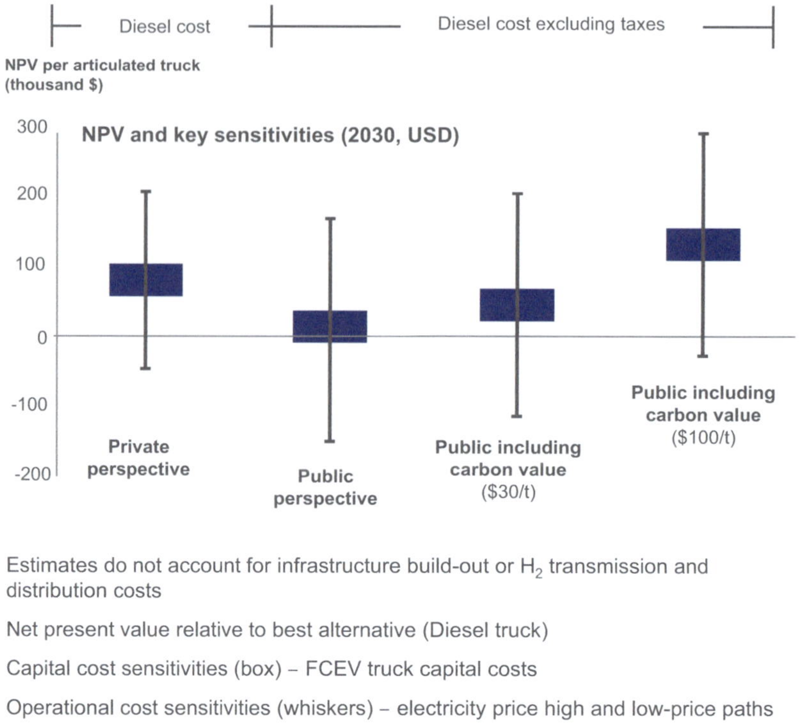

Fig. 40 Cost evaluation of delivered hydrogen. *Note* NPV = net present value; H_2 = hydrogen; FCEV = fuel cell electric vehicle. *Source* Shell–DRC project team

pillars of multi-modal integration are summarised in Fig. 41.

5.2.2 Rethink Urban Planning

Some cities are aiming for residents to have all the facilities and services they need (such as work, shopping, health and culture) within 15–20 minutes walking or cycling distance from their homes, thereby reducing transport demand.

Case Study 9

Paris is working towards building 15-minute neighbourhoods that encourage self-sufficient communities. It was a core component of Mayor of Paris Anne Hidalgo's 2020 re-election manifesto. The concept is bold and will require significant but careful changes to urban planning at the neighbourhood level. The aim is to encourage more self-sufficient communities within each arrondissement of the French capital, with grocery shops, parks, cafes, sports facilities, health centres, schools and even workplaces just a walk or bike ride away.

Such neighbourhoods are expected to bring significant benefits to residents, building a city intended for people, not cars. The implementation will require anti-zoning policies that undo decades of orthodox urban planning and industrial-era economic development that separated different activities into distinct parts of a city. These will reduce commuting and travel time for residents. They are also expected to improve health by reducing pollution, promoting mobility and increasing local consumption of goods and services. Less use of cars will also lead to safer streets for residents and visitors.

A range of policies have been proposed that co-ordinate related efforts by different government departments, as shown in Fig. 42. For example, measures include trimming car lanes to make space for pedestrians, cyclists, shops and restaurants or replacing car parking spaces with

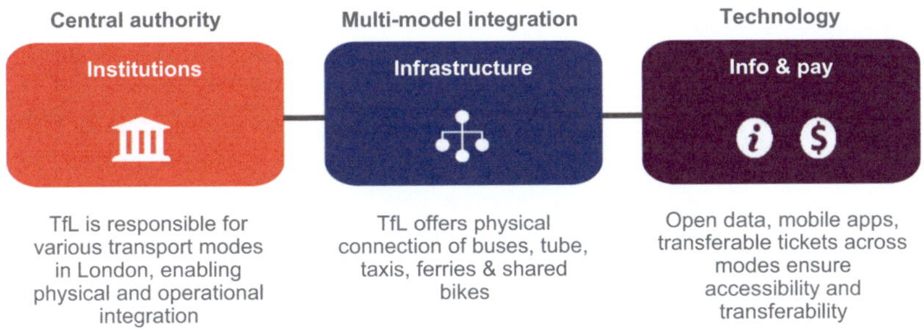

Fig. 41 The efficient integration of different modes of transport enables physical, operational and institutional co-ordination. *Source* Shell–DRC project team

gardens and playgrounds. Other measures include using civic facilities for multiple uses, greening the city by planting new trees and populating it with citizen kiosks for information and community cohesion.

In the COVID-era, where changing transport demand has reoriented urban lives, the 15-minute city comes at an opportune time. Post-COVID cities are expected to move away from their current role as large employment centres surrounded by dispersed residential communities. Employers are repurposing their offices and downsizing, and there is an overall shift towards city-centric living. With this trend, 15-minute neighbourhoods can reverse traditional urban development into car-free environments that offer improved health and a better quality of life for residents.

Case Study 10

The UK Government has announced the largest ever funding boost to create a "new era" for walking and cycling. The first stage of the £2 billion ($2.6 bn) investment will be a £250 million ($330 m) emergency active travel fund in response to COVID-19 to help councils reallocate road space for cyclists and pedestrians. This will include wider pavements, cycle- and bus-only corridors and safer junctions. During the pandemic, cycling and walking were not only good for exercise but also provided a safe, socially distanced way of commuting.

In addition to infrastructure provision, the government is providing incentives to residents to become more physically active. Vouchers are issued for cycle repairs and plans are proposed for greater availability of bike repair facilities. Employees are also encouraged to sign up to a cycle-to-work scheme, which gives employees a discount on new bikes. E-scooter trials have also been introduced to help people switch to greener alternatives than public transport. The government also intends to launch a campaign to encourage more people to take up alternative modes of transport.

More than 50 local councils are using the funds to set up low-traffic neighbourhoods. More than 200 such neighbourhoods have been formed, which limit motor vehicle access to certain areas and redirect through-traffic away from local streets and corridors. Typical measures include modal filters, speed humps, temporary cycling lanes, widened pavements, double yellow lines, cycle parking, bollards and 20 mph (32 km/h) speed limits.

5.2.3 Develop New Business Models

Flexible, shared mobility solutions in public transport can reduce the use of private cars, associated fuel demand and emissions, while incurring only limited infrastructure costs.

Case Study 11

In Switzerland, the City of Zurich has piloted an on-demand public transport service. Transit operator Zurich Public Transport (VBZ) partnered with the mobility firm ViaVan (now known as Via) to launch a service called Pikmi.

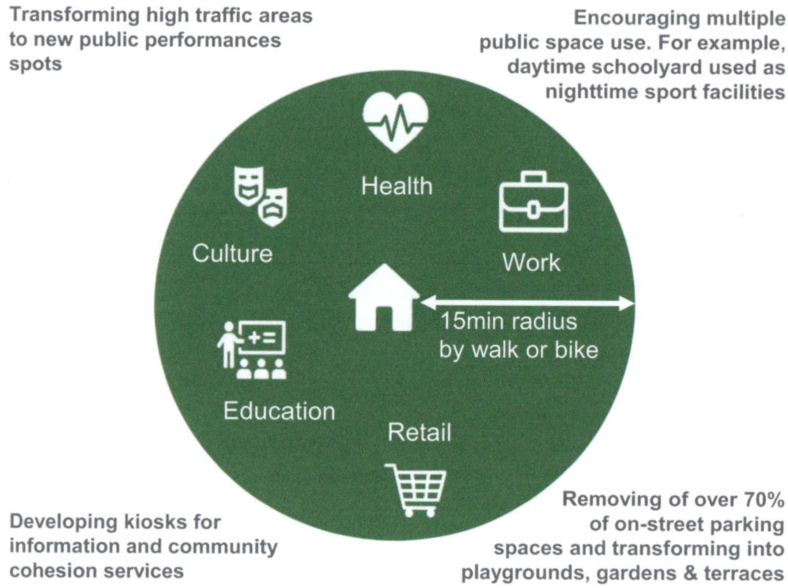

Fig. 42 Paris is reducing transport demand through efficient urban planning and design changes. *Source* Shell–DRC project team

The 18-month pilot, initiated in late 2020, will run from 20:00 to midnight every day, filling current gaps in public transport provision.

The business model has expanded access to integrated public transport with no additional infrastructure. Pikmi does not run on predefined routes or with a fixed timetable and has 150 additional virtual bus stops that did not require any infrastructure investment. The location of these additional stops is provided in detail on the app. Passengers can book rides through the app and pay by presenting a valid public transport ticket. These tickets are valid for two hours, so passengers can use the well-integrated service in conjunction with other forms of public transport.

The shared mobility solution provides fast and efficient public transport journeys that reduce the need for private transport. The algorithm underlying the app avoids unnecessary detours and long waiting times. The technology matches multiple passengers headed in a similar direction into a single vehicle, calculating the most efficient route in real time. The pilot is aiming to demonstrate how technology can increase efficiency in public transport, while reducing private vehicle traffic, fuel consumption and associated emissions.

Case Study 12

Warehouse on Wheels is a last-mile delivery partnership between Ford and Gnewt to trial a digital delivery service in London. Gnewt operates the largest fully electric delivery fleet in the UK with more than 70 electric vans (in 2019). Ford's cloud-based, multi-modal routing and logistics software MoDe:Link manages all aspects of parcel delivery from depot to doorstep. For example, the vans collect parcels from depots and stop at strategic locations most efficient for each batch of orders. Ford's software platform then co-ordinates with nearby foot couriers—or potentially with bicycle couriers, drones or autonomous robots in the future—to fulfil the last leg of each delivery.

The aim of the business model is to cut emissions, reduce congestion and improve delivery times. Businesses can optimise processes and increase van use, as well as improve customer experience by offering improved delivery windows at reduced cost. It will also help reduce traffic and congestion in cities, especially around kerb spaces where vans typically load and unload. The surge in online shopping caused by COVID-19 presents further opportunities to develop and use such innovative urban delivery models that also help reduce transport emissions.

6 Pathways to a Decarbonised Transport System

The electrification of transport in China should be planned and promoted according to the maturity of the electric vehicle technology and the stage of economic development that the region concerned has attained. Electrification of transport can be achieved by gradually switching from diverse vehicle technologies to zero-emission vehicles, including battery-electric and fuel cell vehicles; prioritising those regions that have a good foundation for electrification, as well as a high level of economic development and an ambitious approach to low-carbon development; and promoting the shift to new energy vehicles in less developed small and medium-sized cities and rural areas at a realistic pace.

6.1 The Transition to Zero-emission Vehicles

1. Early stage (electromotive vehicles, XEV-1): Different types of vehicle technology co-exist, including conventional internal combustion engine vehicles, energy-efficient vehicles, plug-in and extended-range hybrid electric vehicles, battery-electric vehicles and hydrogen fuel cell electric vehicles.
2. Transitional stage (electromotive vehicles, XEV-2): Alternative fuel, hybrid, plug-in hybrid, battery-electric and fuel cell electric vehicles.
   Category 1: Vehicles that use natural gas and other alternative fuels fully or partly, including natural gas vehicles and natural gas hybrid electric vehicles.
   Category 2: Vehicles that have both an internal combustion engine (fuelled by petrol or diesel) and an electric motor. These are mainly referred to as hybrid electric vehicles.
   Category 3: Vehicles that use petrol or diesel and an external electrical energy source. They include plug-in hybrid and extended-range electric vehicles.
   Category 4: Vehicles that use only electricity or hydrogen instead of petrol or diesel. They comprise battery-electric and hydrogen fuel cell electric vehicles.
3. Ultimate stage (zero-emission vehicles): Vehicles that have zero emissions are the ultimate goal as they reduce emissions and lower the automotive industry's dependence on fossil fuels. They include battery-electric and fuel cell electric vehicles.

6.2 Categories of Vehicle by Application

There are typically nine vehicle categories in passenger and road freight transport (Table 6).

6.3 Classification of City Tiers by Endowments and Local Conditions

Cities in China are divided into four tiers, as shown in Table 7.

Small and medium-sized cities and rural areas that have a relatively weak foundation, but see incremental opportunities for electric vehicle market development, are divided into four categories, as shown in Table 8.

6.4 Urban Transport Electrification

6.4.1 Advancing Urban Transport Towards Complete Electrification by 2050

In the Carbon-neutral Scenario, new passenger road transport vehicles will be electric by 2040 and new road freight transport vehicles will be

Table 6 Classification of vehicles by segment

Segments	Category	Representative vehicles
Passenger transport	PV-1	Taxis, online ride-hailing vehicles and time-sharing rental vehicles
	PV-2	Official vehicles
	PV-3	Private vehicles
	PV-4	Buses
	PV-5	Intercity buses
	PV-6	Commuter buses
Freight road transport	TV-1	Urban light-duty logistics and postal service vehicles
	TV-2	Intercity logistics vehicles
	TV-3	Medium- and heavy-duty trucks

Source China EV100, Comprehensive Evaluation of the Timetable for China's Full Shift to EVs and Suggestions for Its Promotion

Table 7 City tiers in China

Tier	Definition	Cities and provinces
Tier 1	• Megalopolis • Development demonstration area	• Beijing, Shanghai, Shenzhen • Hainan, Xiong'an New Area
Tier 2	• Traditional cities that have vehicle purchase and traffic restrictions • Provincial capitals in key regions of the "Making our skies blue again" campaign	• Tianjin, Hangzhou, Guangzhou • Shijiazhuang, Taiyuan, Zhengzhou, Ji'nan, Xi'an, Nanjing, Hefei, Wuhan • Guiyang, Fuzhou, Xiamen, Nanchang • Chongqing, Qingdao, Chengdu, Changsha, Kunming, Ningbo, Suzhou
	• Provincial capitals and core cities in national ecological civilisation pilot zones • Leading cities in new energy vehicle promotion, core cities in industrial clusters and coastal economically developed cities	
Tier 3	• Key provinces of the "Making our skies blue again" campaign • Provinces with new energy vehicle industrial clusters • New energy vehicle promotion or low-carbon development demonstration cities and national ecological civilisation pilot zones	• Hebei, Henan, Shandong, Jiangsu, Zhejiang, Anhui, Shanxi • Guangdong, Hunan, Hubei, Jiangxi, Guizhou, Fujian, Shaanxi • Liuzhou
Tier 4	• Other provinces	• Xinjiang, Tibet, Ningxia, Gansu, Qinghai, Heilongjiang, Liaoning, Jilin, Guangxi, Yunnan, Sichuan, Inner Mongolia

Source China EV100, Comprehensive Evaluation of the Timetable for China's Full Shift to EVs and Suggestions for Its Promotion

Table 8 Four categories of small and medium-sized cities and rural areas for electric vehicle development

	Category 1	Category 2	Category 3	Category 4
Economy	Not obviously economically undeveloped	Not obviously economically undeveloped	Good or relatively good	Poor
EV-related industries	Good			Poor
Acceptance of EVs (including low-speed EVs)	Good	Relatively good	Average	Poor
Government involvement	Strong	Relatively good	Average	Poor
Other conditions	Good or relatively good EV circulation, aftermarket services, electricity capacity and other conditions			Relatively poor
Typical areas	Liuzhou, Wuhu and Baoding	Provinces where there is a consumer base for low-speed EVs, such as Henan, Shandong and Hebei; provinces where the EV uptake has made progress, but the industry base is relatively poor	Eastern coastal areas and relatively economically developed areas where there is a good consumer base for vehicles	Economically underdeveloped areas in western China

Note (1) Rural areas are included. However, the rural areas in economically developed cities—such as Beijing, Shanghai, Guangzhou and Shenzhen—have unique economic and development policies, which are not included. (2) Government involvement includes willingness to develop electric vehicle uptake, as well as the presence of environmental governance, electric vehicle targets and other policies. (3) This is a preliminary assessment according to current conditions and is for guidance only. Each province needs to promote electric vehicle uptake in its own way and in accordance with its own needs and resources
Source Research by the project team

Table 9 Full electrification of vehicles by city tier and transport segment

City	Transport segment	Full electrification of new vehicles
Tier 1	Passenger road transport	2030
	Freight road transport	2035
Tier 2	Passenger road transport	2035
	Freight road transport	2040
Tier 3	Passenger road transport	2040
	Freight road transport	2040
Tier 4	Passenger road transport	2040
	Freight road transport	2045

Source Research by the project team

Table 10 Electrification pathways in the Carbon-neutral Scenario

Vehicle category	2020	2025	2030	2035	2040	2045	2050
PV-1	I, II	III	IV				
PV-2		I, II	III	IV			
PV-3			I	II	III, IV		
PV-4	I, II	III	IV				
PV-5			I	II	III, IV		
PV-6		I, II	III	IV			
TV-1	I	II	III	IV			
TV-2			I, II		III	IV	
TV-3				I	II, III	IV	

Note The information in the table refers to zero emissions from vehicles. As cities in China vary greatly in their development, each city should carry out its own assessment based on its specific circumstances
The Roman numerals refer to the four city tiers
PV = passenger vehicles; TV = transport vehicles
Source Research by the project team

electric by around 2045, as shown in Table 9. Around 2050, all vehicles in China are expected to be electric, except for in a few special regions, as shown in Table 10.

6.4.2 Pathway Options for Less Developed Areas

Stage 1 (to be achieved by 2021–25, in which electric vehicles become more competitive).

In Category 1 areas, government should guide the shift to all types of electric vehicle (high-, medium, and low-end) in the public sector. It should open the market to non-local products, strengthen the aftermarket to make electric passenger vehicles attractive for consumers, and introduce incentive policies and support infrastructure in towns and rural areas.

Category 2 areas should improve non-financial policies for charging, driving and parking to make electric vehicles more popular. They should improve the support system to encourage users to switch from low-speed electric vehicles to mid-range vehicles in towns and rural areas to bring electric vehicle use closer to that of Category 1.

Category 3 areas should clearly define their electric vehicle uptake targets, improve electric vehicle use in the public sector and stimulate demand in the private sector.

Category 4 areas should improve electric vehicle demonstration projects in the public sector, construct the necessary support infrastructure including chargers, and gradually encourage the purchase of mid-range electric vehicles in the private sector.

Stage 2 (2026–30, in which electric vehicles become competitive).

Category 1, 2 and 3 areas should improve the user experience and aftersales market (such as residual value) for electric vehicle owners. They should strengthen the service system in rural areas and initiate the wider uptake of multi-purpose electric vehicles. Category 4 areas should start to improve the development of their electric vehicle aftersales market.

Stage 3 (2031 and beyond).

Cities in the four categories are expected to achieve an electric vehicle market that operates according to market criteria. Electric vehicles will be increasingly present at scale in rural areas (Fig. 43).

7 Policy Recommendations for Transport Decarbonisation

The development of automotive technology should be linked with the emissions trading scheme and the carbon emission abatement

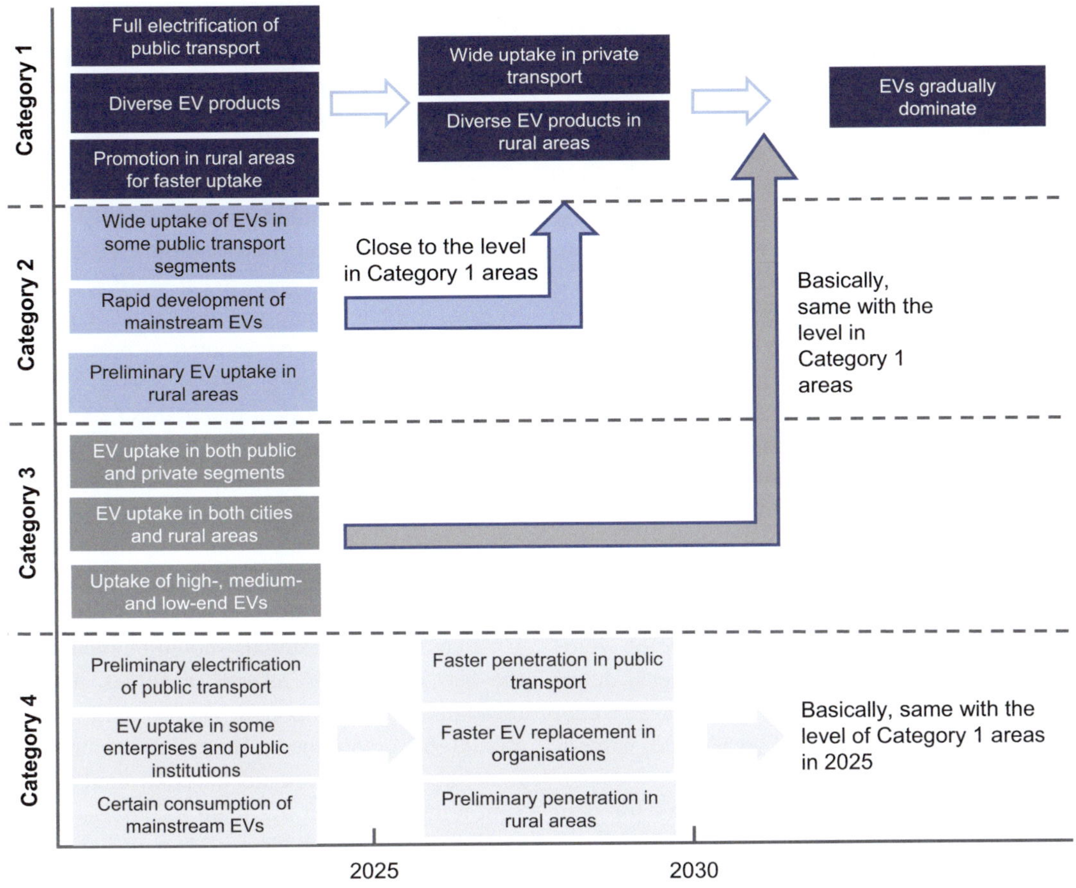

Fig. 43 Electrification pathways for small and medium-sized cities and rural areas. *Source* Research by the project team

benefits of new energy vehicles. A roadmap and timetable for the automotive industry to reach peak carbon emissions and carbon neutrality should be made, including a timeframe for carbon emission targets for cars and for a ban on the sale of petrol and diesel vehicles. Support should be given to those regions with the potential to reach peak carbon emissions in the road transport sector ahead of schedule. In the medium and long terms, a carbon reduction scheme for the automobile sector should be established through the use of carbon taxes, carbon incentives, carbon offsets or carbon emissions trading.

The supply-side management system should be further improved and a good business development environment should be created. A passenger car points pool should be set up to stabilise the pricing of points. Commercial vehicle points policies for different vehicle types, regions and applications should be designed. Automobile production capacity management should be strengthened, and a good investment environment for enterprises with capabilities and technologies should be provided.

Fiscal, tax and preferential policies should be refined, and their continuity and stability maintained. Well-designed support policies for a post-subsidy era should be provided, and the purchase tax reduction and exemption policies for new energy vehicles should be extended. Local government should be encouraged to introduce incentive policies for new energy vehicles with regard to parking, charging and the use of infrastructure like roads and bridges. Incentives for replacing old vehicles with new energy vehicles should be assessed. Policies to encourage the shift to electric in public transport segments such as taxis, online car hailing, rental cars and leasing cars should be introduced.

New energy vehicles should be promoted in small and medium-sized cities and rural areas. Rural consumers should be given extra purchase and replacement subsidies in addition to the national incentives. Small and medium-sized cities and rural areas that have sufficient potential should be given support to develop zero-emission transport demonstration projects. Innovative products that meet local rather than national requirements should be allowed, without compromising vehicle safety.

Recharging and refuelling facilities for new energy vehicles should be constructed at a faster pace. The development of charge points, battery-swapping stations and hydrogen refuelling stations should be included in urban development plans. Greater support from green finance should be given. Consistent technical standards for battery swapping should be developed. Subsidies for hydrogen refuelling stations should be increased and the approval procedures required by the relevant authorities should be clearly defined.

The development and improvement of the aftermarket should be accelerated. A valuation and certification system for used new energy vehicles should be established. Efforts should be made to devise a system for the annual inspection and recycling of key vehicle components like the power battery, motor and electronic controls. The power battery recycling system should be further improved and regulations to make battery recycling mandatory should be passed.

Business model innovation should be encouraged. Pilot projects in car sharing, vehicle-to-grid, mobile charging and other emerging technologies should be initiated to provide consumers with low-cost access to electric vehicles without upfront payment or purchase.

Supply chain security for large-scale electric vehicle development should be ensured. Businesses should be guided to secure the automotive chip supply chain and build local supply capability. Chinese companies should be encouraged to lock overseas mineral resource reserves. Upstream critical materials should be included in the lists of strategic reserve resources, and any dependence on imported supply and storage should be determined. The price of upstream material resources should be controlled to prevent a monopoly on upstream resources.

Co-operation with the outside world should be enhanced and Chinese new energy vehicles should be encouraged to "go global". Opportunities brought by the Belt and Road Initiative and international production capacity co-operation should be seized, and companies should be encouraged to build R&D centres, manufacturing bases and marketing networks globally. Under the framework of the United Nations and the International Organization for Standardization, China should participate in research on international standards for electric vehicle safety, power batteries, battery charging and battery swapping.

Chapter 5: Transitioning the Electricity System Towards Carbon Neutrality

Yanan Zheng, Bian Fu, Tianzi Wang and Marcello Espinoza

1 Introduction

China's electricity demand is set to increase steadily, but at a slowing rate. The country's total electricity consumption is projected to grow annually every five years from 2026 to 2060 by 3.0%, 2.1%, 1.4%, 0.9%, 0.6%, 0.4% and 0.3% respectively.

In the electric power system, renewables will gradually predominate. By 2025, total installed power capacity in China is expected to reach 3,000 gigawatts (GW), with the installed capacity and power output of renewables (including hydropower) accounting for 51% and 36% respectively. By 2030, total installed power capacity and output are forecast to increase to 3,700 GW and 11 trillion kilowatt-hours (kWh), with renewable energy representing 60% and 40% respectively. To build a carbon-neutral China by 2060, total installed power capacity and power output will have to reach 8,500 GW and 17.8 trillion kWh, with renewable energy making up more than 90% and 80% respectively. The combined installed capacity of wind and solar power will be about 6,500 GW. At that time, the energy system will be clean and low-carbon, secure, efficient and sustainable.

In June 2014, General Secretary Xi Jinping declared China's new energy security strategy of "Four Revolutions and One Co-operation" in energy consumption, energy supply, energy technology, the energy system and international co-operation,[1] indicating the direction of China's energy development in the new era. In September 2020, Xi Jinping announced at the General Debate of the 75th Session of the United Nations General Assembly that China would strive to achieve peak CO_2 emissions before 2030 and carbon neutrality before 2060,[2] further clarifying the direction that China's energy industry will take. The Chinese government then defined the task of building a new electric power system and clearly stated that the combined installed capacity of wind and solar power will reach more than 1,200 GW by 2030, giving a clear timetable for the development of China's energy system of the future. To meet the dual-carbon goal, electricity will be at the core of China's energy transition. Research on pathways for China's low-carbon power supply will therefore provide support for China's medium- and long-term energy policies and plans.

Y. N. Zheng (✉) · B. A. Fu · T. Z. Wang
Energy Research Institute of the China Academy of Macroeconomic Research, Beijing, China

M. Espinoza
Shell International B.V., The Hague, The Netherlands

[1] The 6th meeting of the Central Leading Group for Financial and Economic Affairs in June 2014.

[2] Xi Jinping's Speech at the General Debate of the 75th Session of the United Nations General Assembly.

Shell International B.V. and Development Research Centre of the State Council (DRC) of People's Republic of China, *Embracing the Future, Powering Growth: An Energy System Renewed for China*, Advances in Oil and Gas Exploration & Production,
https://doi.org/10.1007/978-3-031-73075-7_6

2 The International Energy Landscape

Renewable energy is the foundation of the global energy transition and an important energy source for tackling climate change. To address the increasingly severe challenges of energy supply security, environmental pollution, greenhouse gas emissions and climate change, many countries are orienting their energy development strategies towards a low-carbon and efficient energy system. Green, low-carbon and clean have become the key words in the global energy industry.

Renewables are an important part of the global energy supply system and a main component of the energy transition and the fight against climate change. The share of renewables in global power output is forecast to surge from 25% today to 86% by 2050.[3] Almost all developed and developing countries have introduced policies to support the development of renewable energy and reduce their dependence on fossil fuels as much as possible. By the end of 2017, 179 countries had announced their renewable energy goals and formulated targets, policies, legislation and financial incentives to support renewable energy.

In recent years, technological innovations have helped drive down the cost of renewable power generation. Falling equipment costs and improved efficiency have made renewable energy increasingly competitive. In many parts of the world, the levelised cost of electricity from solar power and offshore wind without government subsidies is on parity with or lower than most other power generation technologies. According to the International Renewable Energy Agency, the average generating cost of hydropower projects was $0.05 per kWh in 2017, while that of onshore wind was $0.06 per kWh and biomass and geothermal power were $0.07 per kWh. The cost of large-scale solar photovoltaic power projects also decreased sharply, from $0.36 per kWh in 2010 to $0.02 per kWh in 2017. The continuous decline in the cost of renewable energy generation across the world is making grid parity inevitable.

International co-operation on energy and electricity continues to deepen, and the importance of soft power has become increasingly important. Countries and international organisations are now working together to define bilateral and multilateral mechanisms to increase the scope of international co-operation. Global co-operation focuses on renewable energy mechanism creation, project development, capacity building and the research and development of equipment and materials. In addition, countries are expanding their bilateral and multilateral co-operation mechanisms in many fields, such as policy and market research, energy development and the planning and evaluation of projects. Looking ahead, global co-operation on renewable energy will deepen to facilitate continuous improvement in technology, equipment, services and standards.

3 China's Electricity Supply and Demand Landscape

3.1 Electricity Use

In 2020, total electricity use in China was 7,511 billion kWh (see Fig. 1), up 3.1% year-on-year, although 2.2 percentage points lower than that of the previous year. The electricity used by the primary, secondary and tertiary sectors and by urban and rural households increased by 10.2%, 2.5%, 1.9% and 6.9% respectively. In the secondary sector, growth in electricity consumption was 1.8 percentage points lower than in 2019, representing 68.3% of the country's total electricity use. This indicates improved optimisation of electricity demand. The tertiary sector and urban and rural households consumed 41% of China's electricity, driving up their combined share of total electricity use to 30.7%.

Electricity consumption in the north-western and southern power grids reported the highest growth, while that in the northern, central and south-western grids slowed down significantly.

[3] International Energy Agency, Net Zero by 2050: A Roadmap for the Global Energy Sector, 2019.

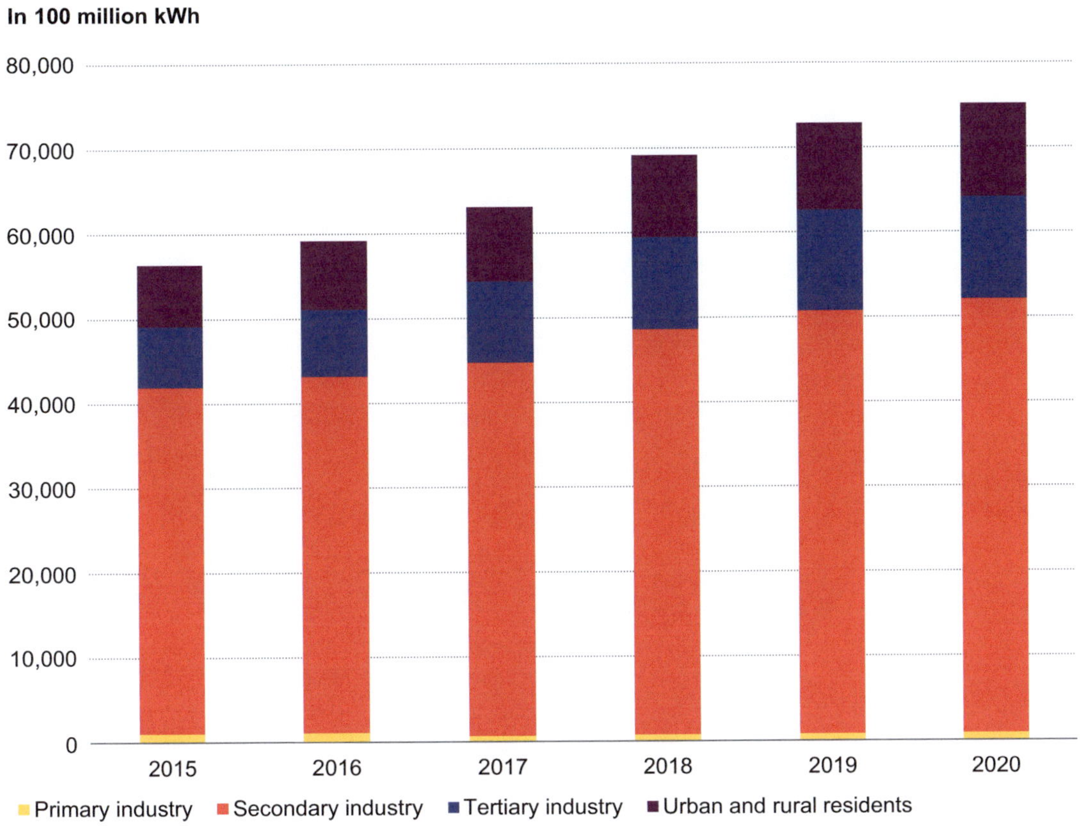

Fig. 1 China's electricity use by sectors. *Source* China Electricity Council

In 2020, electricity consumption in North China (including western Inner Mongolia), East China, Central China, north-east China (including eastern Inner Mongolia), north-west China, south-west China and South China was 1,856.3 billion kWh, 1,769.4 billion kWh, 909.5 billion kWh, 487.6 billion kWh, 717.1 billion kWh, 398 billion kWh and 1,373.1 billion kWh, an increase of 6.4%, 2.6%, 1.3%, 1.8%, 3.8%, 7.4% and 11.5% respectively compared with 2019.

3.2 Electricity Supply

In 2020, new power generating capacity of 190 GW was installed, an increase of 9.5% year-on-year. China's total installed power capacity reached about 2,200 GW, including hydropower of 370 GW, thermal power of 1,250 GW, nuclear power of 49.89 GW, wind power of 280 GW and solar power of 250 GW. The installed capacity of renewable energy accounted for 41.1%, 2.7 percentage points higher than in 2019, indicating optimisation of the power supply system (see Fig. 2).

Non-fossil power generation accounts for a growing proportion of the power generated. In 2020, China's total power output was 7,623.6 billion kWh, up 4.0% year-on-year, 0.7 percentage points lower than in 2019. Thermal power output was 5,174.3 billion kWh (including coal at 4,631.6 billion kWh), which was 67.9% of the country's total generation, 1.0 percentage points lower than the previous year. Non-fossil power output was 2.4 trillion kWh, 32.1% of China's total and 1.0 percentage point higher than in 2019. Thanks to improvements in renewable energy accommodation capacity, the wind and solar curtailment rate decreased to 3.5% and 2.0% respectively.

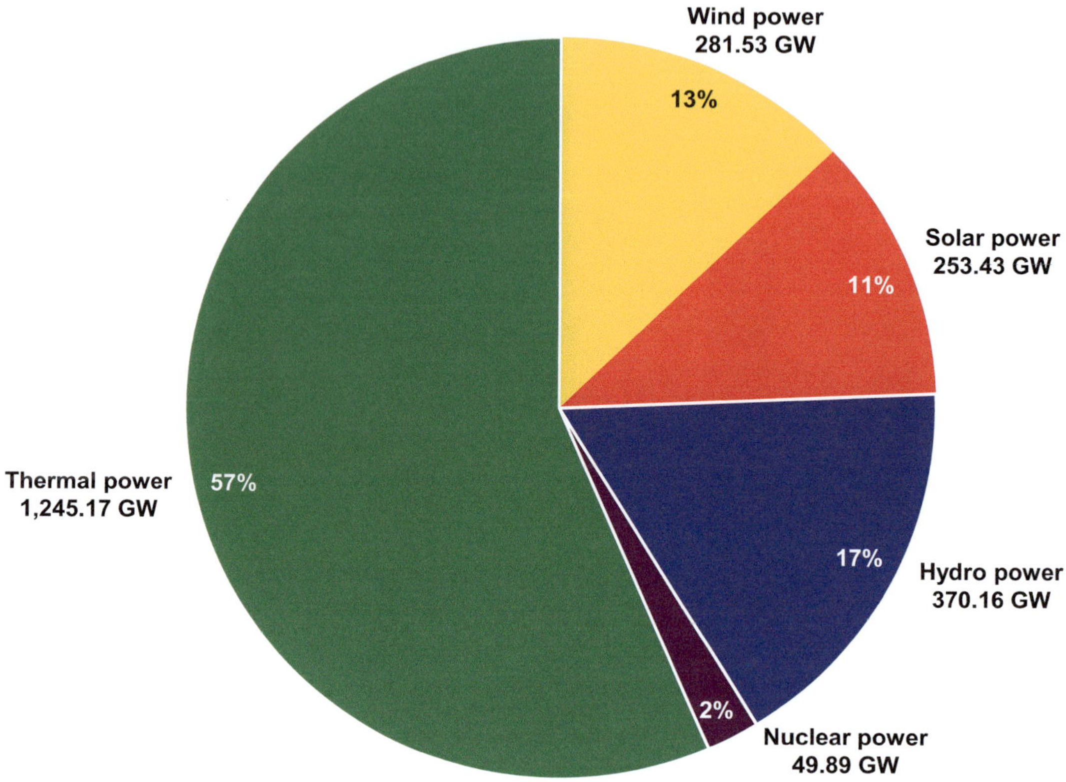

Fig. 2 China's power generation mix in 2020. *Source* China Electricity Council

Inter-regional power flows continued to increase. Electric power in China flows through three corridors: the northern corridor, which transfers mainly coal power and wind power from Shanxi, Shaanxi, Inner Mongolia, Ningxia, Xinjiang and Gansu to Beijing, Tianjin, Hebei and Shandong in North China, as well as Liaoning, Henan, Jiangsu and Hunan; the central corridor, which delivers primarily hydropower from Sichuan and the Three Gorges to Chongqing, the provinces along the middle and lower reaches of the Yangtze River, and Guangdong, and transfers coal power from northern and southern Anhui to the provinces in the Yangtze River Delta; and the southern corridor, which transports mainly hydropower from Yunnan and coal power from Guizhou to Guangdong and Guangxi.

3.3 Electricity Supply and Demand Balance

Generally, China balances electricity supply with demand. In the 13th Five-year period (2016–20), East, Central and South China achieved a power supply-demand balance; north-east and north-west China generated a power surplus; and North China achieved a power supply-demand balance for the most part, but with occasional imbalances.

3.4 Challenges

On the path towards the dual-carbon goal and a new electric power system, China's power supply still faces challenges, especially in technology and market mechanisms.

To achieve high-quality economic and social development, faster transformation of the electricity sector is imperative. China's economy has shifted from high-speed growth to high-quality development. Electricity consumption, as a measure of the nation's economy, has slowed down. The country's power supply system is relatively balanced, but some areas generate a surplus. The problems that were previously hard to notice or ignored in the high-speed stage of growth are now visible. First, the risk of coal power overcapacity and limited consumption of non-fossil energy can slow down optimisation of the power supply mix, which in turn constrains the shift to a high-quality economy. Second, the low efficiency of power equipment and the high cost of electricity use, coupled with a lack of well-designed electricity market mechanisms, hinder the efficiency transition. Third, insufficient understanding of the environmental costs of coal power, constraints on the development of the clean energy industry, and the absence of a fair competitive environment are barriers to change.

The electric power system must adapt to the predominance of new energy. Thanks to abundant wind and solar resources and the rapid decline of generating costs, the comparative advantages of new energy over conventional power sources like coal are increasingly evident. Given this, China's wind and solar power generation will maintain rapid growth for some time. However, the inherent intermittency and volatility of wind and solar power pose increasingly demanding requirements on balancing grid supply with demand. At the same time, the rapid development of wind and solar has pushed back coal and other fossil-fired power generation, which traditionally provide base load and grid balancing. Since 2016, the operating hours of coal power plants have declined, approaching their financial break-even point of 4,000 hours at the end of 2019. This forced the operators of coal power and other fossil energy plants to seek new ways to make a profit.

The development of the electricity market lags behind the pace of its transformation and does not reflect the growth of non-fossil energy. The accuracy of non-fossil energy forecasts gradually improves as the time for trading draws near, but the trading windows are rigid. The day-ahead trading window is often closed hours before the trading day, and although the intra-day trading mechanism is used in some areas, liquidity is low. There are no incentive mechanisms for timely disclosure of forecasts and the price signals are inefficient. Inadequate flexibility in resource allocation also presents a barrier. Under the existing electricity market mechanism, the uncertainty of non-fossil power generation can lead to power companies failing to submit accurate output data, subjecting them to penalties. The uncertainty of wind and solar output also hinders strategic bidding in a multi-time scale market. In addition, electricity price volatility increases with the inclusion of large-scale renewable energy generation, which raises financial risk in the electricity market.

The electricity sector is not aligned with the requirements of urban and rural development and environmental protection. Wind and solar power generation are abundant, but there is no effective planning alignment between the non-fossil energy and electricity sectors and other related land, environment and forestry plans, which creates problems in implementation. Unaligned targets, policies and actions between the national, local and sectoral levels create friction. There is an urgent need to align all related plans and legislation to determine the direction of development and the targets and actions needed.

4 China's Electricity Demand Forecasts

4.1 Electricity Demand Forecasts

China's electricity demand has great potential for growth. The country's economy has shifted from high-speed to moderate growth, which will have a direct effect on demand for electricity. As economic restructuring and industrial transformation bear fruit and electricity consumption for manufacturing decreases, total electricity demand will decline. Improved energy efficiency and lower power line losses will also help reduce electricity demand. On the other hand, China is expecting to see the end of rapid urbanisation and achieve an urbanisation rate of about 80% of its population by 2060, which will increase electricity demand. To control air pollution and tackle climate change, China will continue to optimise its energy demand mix. More primary energy will be converted into electricity, which will represent a growing share of total final energy use.

China's electricity demand is expected to grow steadily, but at a slower pace. Between 2021 and 2025, total electricity consumption will grow annually by about 4.5% on average and exceed 9.3 trillion kWh in 2025. The average annual growth rate of China's total electricity consumption in five-year periods from 2026 to 2060, will be 3.0%, 2.1%, 1.4%, 0.9%, 0.6%, 0.4% and 0.3% respectively. In 2060, total electricity use is estimated to be more than 14.0 trillion kWh, equivalent to a per capita electricity consumption of about 11,000 kWh and an electrification rate of more than 65%.

Resource allocation and grid balancing will be the two main functions of inter-regional power grid links in the future. The long distances that separate energy resources from demand centres in China create the need for long-distance ultrahigh voltage transmission links to optimise resource allocation across the country. If the current project schedules are followed, by 2025 the northern grid, western Inner Mongolia grid, north-eastern grid, central grid, south-western grid and north-western grid are expected to provide power transmission capacity of 21.1 GW, 64 GW, 18.6 GW, 26.5 GW, 47.4 GW and 108 GW respectively. Inter-regional power transmission links will also be enhanced. In 2030, regional power flows will increase, mainly with clean energy, in accordance with the "West-to-East power transmission" strategy. Coal and wind power from West China will be transferred to North and East China. Hydropower from the south-west will flow to Central and East China, Guangdong and Guangxi to meet growing demand in the eastern and central regions and dampen the building of new generating units, especially coal power plants. The new electric power system will include more intermittent renewables, which increase the need for system flexibility. Inter-regional grid connections are required to operate in a more flexible and optimised way, and schedule power across the country to achieve a dynamic balance between supply and demand, thereby effectively increasing the use of new energy. According to the State Grid Corporation of China, the capacity of inter-regional transmission links in China will increase to about 400 GW by 2035 and 500 GW by 2050, a tripling and quadrupling of their current levels respectively.

4.2 Ideas to Improve Electricity Supply

China's electricity supply will closely follow the energy revolution strategy of "Four Revolutions and One Co-operation" and will be aligned with the goals of peak carbon emissions and carbon neutrality. The approach is innovation-driven and market-oriented. An energy supply mix with a high proportion of new energy will be forged to increase the use of new energy. The new energy system will also be based on grid-to-grid transfer and supply-demand balancing. A market-oriented approach should foster competition in power generation in multiple ways and form an efficient multi-level market system. The principles that should guide China's electricity supply are as follows.

1. The role of the market is to allocate resources. Competitive allocation of resources should be encouraged to speed up cost reductions and phase out subsidies for new energy generation. Efforts should be made to build a renewable energy development system characterised by effective market mechanisms, vibrant micro-entities and macro-regulation.
2. Innovation and the wide deployment of new technologies should be encouraged. Innovation is key to developing new technologies and equipment. Applied basic research should be strengthened and the innovation capacity of the new energy industry should be developed to optimise and renew the entire new energy value chain.

3. The principle of adapting to local conditions should be followed to ensure parallel development of centralised and distributed new energy systems. Equal focus should be placed on the transmission and local consumption of new energy and the development of centralised and distributed new energy systems. In this way, a landscape of large-scale centralised new energy use, combined with distributed production and local consumption, will take shape.
4. An efficient integrated energy supply system will consist of different energy sources. By integrating wind, solar, hydro, thermal and energy storage, the complementary benefits of each power source can be harnessed. Different market development strategies should be adopted to form a multi-energy and efficient electric power system that has energy storage at its core to balance capacity.

5 A Roadmap for China's Low-carbon Electricity Supply

China's economic and social development goal is to achieve socialist modernisation by 2035 and build China into a great modern socialist country by 2050. China also aims to reach peak CO_2 emissions before 2030 and carbon neutrality before 2060. Taking these goals into account, this section analyses China's electricity demand based on supply constraint scenarios. It also uses China's integrated power resource planning model to analyse the power supply roadmap for 2020–60 and the impacts of carbon emissions from electricity consumption.

5.1 Scenario Setting

Renewable energy will be competitive by 2020–25, with all new projects expected to achieve grid parity or better. Renewable energy will be the main contributor to new energy supply in most provinces. Technologies will progress and new business models will take shape. Electricity, heat and gas pipeline networks will be transformed into open service platforms. Management, policy and market systems based on a market-oriented approach to renewable energy will be established. The international competitiveness of the new energy industry will be sharpened. Between 2026 and 2035, the renewable energy sector will be fully market-driven and include market trading and competition in most provinces. Business models will become increasingly mature, allowing China's renewable energy sector to thrive. By 2060, a clean and low-carbon, secure and efficient energy system dominated by renewable energy will exist and China's renewable energy sector will lead the world.

5.2 A Roadmap for Power Supply

Renewable energy will transition from being the main contributor of additional energy supply to being the major supply source in the energy mix. Coal power will transform gradually. Between 2021 and 2025, wind and solar power will reach grid parity and be market-driven; they will be the main contributor of additional electric power in most provinces. China's total installed power capacity is expected to reach 3,000 GW by 2025, when renewables (including hydropower) will account for 51% and 36% of the country's total installed power capacity and total electricity output respectively. By 2030, the country's total installed power capacity and electricity output are expected to reach 3,700 GW and 11 trillion kWh, with renewables accounting for 60% and 45% respectively. To build a carbon-neutral country, China's total installed power capacity and electricity output will need to be 8,500 GW and 17.8 trillion kWh by 2060, with renewables accounting for more than 90% and 80% respectively. China will then have a clean and low-carbon, secure and efficient, sustainable energy system.

5.3 Power Generation Technologies

Conventional hydropower will maintain steady growth, and the development of pumped storage resources will be completed around 2050. For environmental reasons, hydropower plants will be built mainly in south-west China and the upper reaches of the Yellow River. By 2025 and 2030, China's installed capacity of conventional hydropower is expected to reach 388.61 GW and 430 GW respectively. By 2050, the development of hydropower resources in the country will be more or less completed, increasing only slightly in the following decade. By 2060, the installed capacity of conventional hydropower and pumped storage will reach 497.75 GW and 421 GW respectively, while hydropower output will reach 2.8 trillion kWh.

The installed capacity of coal power will peak around 2025 at about 1,250 GW, falling to 420 GW by 2060, reducing emissions while ensuring energy supply security. As coal power is gradually phased out after reaching its peak, gas will play a transitory role by ensuring stability of the power system during the transition. The installed capacity of gas-fired power generation will grow slightly between 2020 and 2040, then plateau and gradually decrease. According to projections, the installed capacity of gas-fired power generation will be about 160 GW, 180 GW and 190 GW by 2025, 2030 and 2040 respectively. By 2060, the installed capacity and output of gas-fired power generation are expected to reach 183.6 GW and 760 billion kWh respectively.

Nuclear power projects will be built without compromising safety. As breakthroughs in nuclear technologies are made, the installed capacity will continue to grow. China's total installed capacity and output of nuclear power are projected to reach 69.82 GW and 500 billion kWh by 2025, 100 GW and 700 billion kWh by 2030, and 200 GW and 1.4 trillion kWh by 2060 respectively.

The installed capacity of wind and solar power will continue to grow rapidly. As onshore wind power achieves grid parity and the cost of generating offshore wind power falls, installed capacity will increase. By 2025, the installed capacity of wind power in China will reach 520 GW, including 492 GW onshore (454 GW of centralised wind power and 37.54 GW of distributed wind power) and 28.22 GW of offshore wind power, generating around 1.1 trillion kWh of electricity. By 2030, China's installed wind power capacity will be about 750 GW, including 649 GW onshore (514 GW centralised and 135 GW distributed) and 100 GW offshore, generating around 1.7 trillion kWh. By 2060, the installed capacity of wind power across China will increase to 3,200 GW, including 3,000 GW onshore (2,200 GW centralised and 800 GW distributed) and 200 GW offshore.

As solar photovoltaic power has already achieved grid parity and is supported by industrial policies, the installed capacity of solar power will continue to grow and exceed that of wind power in the 14th Five-year Plan period (2021–25). By 2025, China's installed capacity of solar power is projected to reach 550 GW (including 48.9% centralised and 51.1% distributed), while solar power output is expected to be about 516 billion kWh. By 2030, the installed capacity of solar power will reach 900 GW (including 380 GW centralised and 520 GW distributed), with output about 1 trillion kWh. By 2060, the installed capacity of solar power will increase to 3,500 GW, of which 1,500 GW will be centralised and 2,000 GW distributed.

In the near and medium terms, biomass power generation will grow rapidly and is expected to plateau around 2035 when the biomass resource constraints take effect. The installed capacity of biomass power generation in China will increase from 27 GW in 2025 to 55 GW in 2030 and 65.51 GW in 2035, and then plateau. By 2060, the installed capacity of biomass power generation is projected to be about 66 GW.

5.4 Demand-response and Energy Storage

Demand-response, energy storage and other grid flexibility measures play an important role in the power system. With increasing amounts of variable wind and solar power generated, the need for grid flexibility technologies increases. Under current conditions, the need for balancing capacity is expected to reach 270 GW in 2025 and exceed 700 GW in 2030. That's where grid-to-grid transfer, demand-response, vehicle-to-grid and energy storage come in. To meet flexibility requirements and ensure supply security of a power system with a high proportion of renewables, energy storage capacity of up to 2,260 GW (including electric vehicles), pumped storage capacity of 420 GW and other resources will be required by 2060. Looking ahead, electric vehicles must become an important part of power system flexibility, with each vehicle potentially providing about 7 kilowatts of system support. If there are 200 million electric vehicles supporting the power system in 2060, they can provide as much as 1,400 GW of electricity.

5.5 Hydrogen's Pivotal Role

Hydrogen supply will gradually shift from fossil fuel-based production to renewables-based electrolysis. Thanks to rapid advances in fuel cell technology and cost reductions, China's total hydrogen demand is estimated to rise to 120 million tonnes by 2060. Industry will make up 73%, transport and buildings 17% and energy 10% of total hydrogen demand. In the hydrogen supply structure, grey hydrogen made from natural gas or methane dominates with a share of about 70%. By 2060, decarbonised (green) hydrogen made from renewable energy will account for more than 75% of demand. In the short term, decarbonised hydrogen is not sufficiently cost effective for large-scale production. Currently, hydrogen made from fossil fuels or by-products has a cost advantage. In the transition period to 2035, industrial by-product hydrogen will play an important role and account for around 30% of total hydrogen supply. However, in the longer term, the cost of production, storage, transport and refuelling technologies will decline, and decarbonised hydrogen will account for a rising share of supply and gradually replace much of grey hydrogen (see Fig. 3).

5.6 Ideas for Regional Energy Development

The distribution of new energy resources like wind and solar is a challenge in China, as they are located far from demand centres. Given the potential supply security risks of these intermittent sources, an overall plan should be developed. New energy systems should be built according to the principle of "large power generation clusters + strong networks + distributed energy systems + local power generation facilities that ensure energy supply security". In addition, the distribution of industries should also be optimised.

China has abundant wind resources. The exploitable wind power reserves are about 4,260 GW (calculated at a height of 100 metres), including exploitable wind power resources of 3,900 GW onshore and 360 GW offshore (within 50 km of the coast). However, the distribution of wind resources is unbalanced, with onshore wind resources mainly concentrated in the three provinces of north-east China, as well as Hebei, Inner Mongolia, Gansu, Qinghai, Tibet and Xinjiang. Taking the current generating capacity of wind turbines and return on investment into account, the total onshore wind resources that are cost effective to exploit amount to 1,750 GW. The top wind-rich provinces are Inner Mongolia with 272 GW, Heilongjiang with 207 GW, Jilin with 121 GW, Xinjiang with 112 GW, Gansu with 105 GW, Liaoning with 87 GW, Qinghai with 72 GW, Guangxi with 66 GW, Shandong with 66 GW and Guangdong with 58 GW.[4] For offshore wind power resources at a height of 100 metres and within 25 km and 25–50 km of shore, the exploitable resources are 190 GW and 170 GW respectively.

Solar energy resources are found mainly in West China. The installed capacity of exploitable solar energy in China is estimated to be about 15,600 GW, including 13,795 GW in West China, which is equivalent to 88.4% of the total. The top resource-rich provinces are Xinjiang, with about 4,200 GW (26.92% of the country's total); Qinghai, with 3,400 GW (21.79% of China's total); western Inner Mongolia, with 2,600 GW (16.66% of the country's total);

[4] Only the provinces with a large amount of resources are listed here.

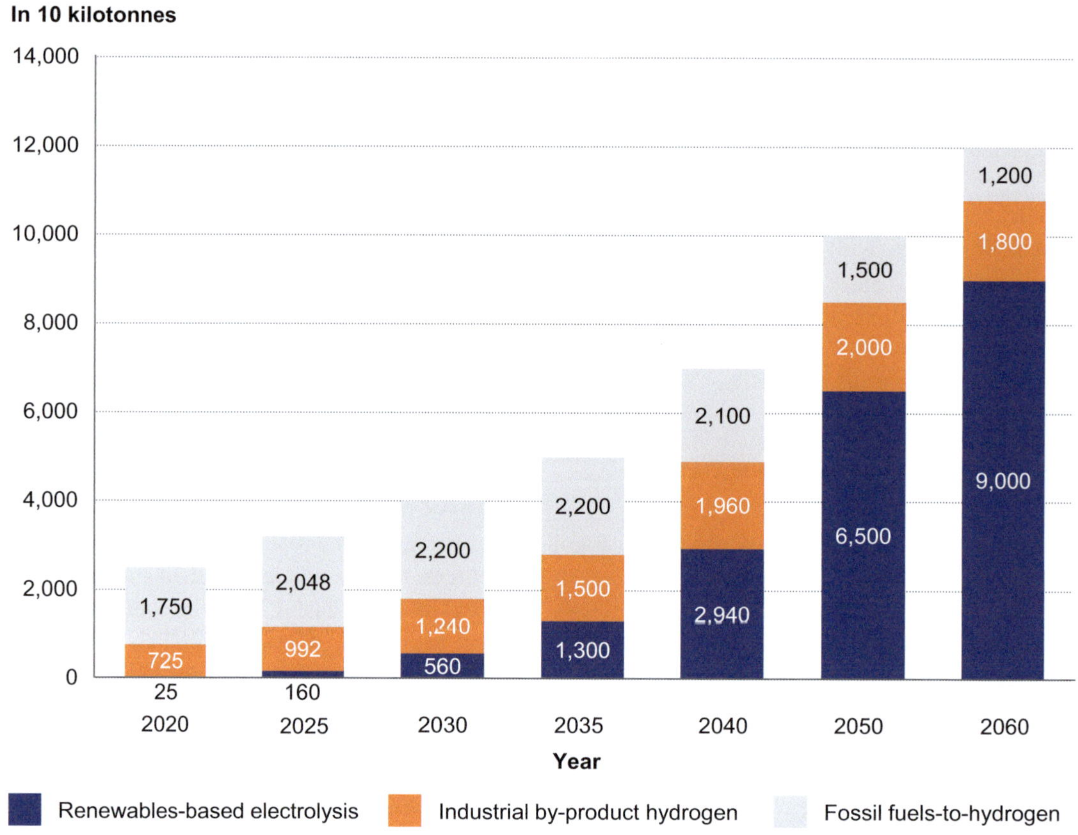

Fig. 3 Hydrogen supply structure, 2020–60. *Source* China Hydrogen Alliance

Gansu, with 2,130 GW (13.65% of the country's total); and Tibet with 700 GW (4.49% of China's total).[5]

Hydropower resources are mainly concentrated in south-west and central-south China, while potential biomass resources are found mainly in South China, north-east China and parts of East China. South-west China accounts for 67.8% of China's total exploitable hydropower resources, including 26.8% in Sichuan, 20.9% in Yunnan and 17.2% in Tibet. In terms of biomass resources, South China is suitable for planting bioenergy crops, while north-east China has a high concentration of corn, rice and wheat, which are the main source of agricultural residues.

The regional energy systems should be planned in line with the principle of "large power generation clusters + strong networks + distributed energy systems + local power generation facilities that ensure energy supply security". With a unique endowment of new energy resources, West China has great potential to develop large-scale concentrations of power generation. China will make full use of its abundant wind, solar and hydropower resources in the western region and continue to develop large wind and solar power clusters in Inner Mongolia, Gansu, Qinghai and Xinjiang. By 2060, the total installed capacity of wind and solar power in north-west China is expected to exceed 2,000 GW. China also aims to develop large hydropower clusters and pumped-storage stations in south-west China.

North-east China's future new energy investments will focus on power transmission capacity and hydrogen development. The north-east is relatively abundant in new energy resources. For

[5] Only the provinces with a large installed capacity are listed here.

example, Jilin province is located in a wind-rich belt across northern China, and is one of the country's nine largest wind power clusters. Moreover, Heilongjiang and Jilin also have abundant agricultural residues and other biomass resources. Because of limited local demand for electricity, north-east China will also become an important base of power transmission to other provinces. In view of the region's geographical advantages and industrial base, a hydrogen economy should also be developed to make good use of the diverse energy resources available.

Geographically, Central China connects North China and South China. Economically, it connects East China and West China. In terms of energy supply and demand, Central China serves as an important bridge between energy resources in West China and load centres in East China. The region needs to increase investment in energy infrastructure and play a co-ordinating role between the west and east. In view of the resource endowments of and local demand for new energy, as well as the need to address the existing structural power shortage, bioenergy and low-speed wind power are likely to become two priorities in the new energy development of Central China.

East China should develop distributed power generation and microgrids to provide local supplies of electricity. For a long time to come, the eastern region will remain the centre of China's economic development and population distribution, as well as the energy load centre. East China is not rich in new energy resources, so it makes more sense to build distributed wind and solar power systems and offshore wind power facilities within the region. However, in the medium and long terms, distributed new energy systems cannot meet the power requirements of East China. In addition to the local supply of conventional power, west-to-east power transmission is still the most effective means to narrow the power supply gap in the east. Importing or transporting hydrogen and hydrogen-based fuels from West China is also an option.

China should focus on developing inter-regional power transmission networks and on improving its electricity trading mechanisms to promote the use of new energy. Looking ahead, the development of large new energy clusters in the central and western regions will require additional long-distance power transmission links. On the one hand, China should establish more secure, reliable, intelligent and strong power grids to ensure supply security of west-to-east power transmission. On the other hand, the country should continuously refine its electricity trading mechanisms and remove interprovincial barriers that hinder the large-scale inter-regional transmission of new energy.

Coal power capacity that stabilises the power system should be retained to ensure supply security and people's well-being. If there are no other feasible alternatives that provide 100% supply security, the stabilising role of coal is essential to counteract the supply risks of geographically sensitive and time-sensitive variable new energy.

6 Policy Recommendations

Strengthen Power System Planning. Power system planning is an important measure of macro-energy management and the basis for energy development management. To achieve the dual-carbon goal, China needs to improve its power system planning. The various energy administrations must co-ordinate their planning and introduce policies to guide market development to achieve the long-term objectives of the energy strategy. Policies, legislation, standards and regulatory measures should also be introduced to meet strategic and planning requirements. The energy and electric power administrations of each province should, in accordance with the requirements of the national energy and electric power development plan, develop a provincial energy and electric power development plan based on their local resources and electricity market conditions. They should clearly define targets, plans and support measures to deliver projects as per their annual implementation plan.

Advance the Regional Energy Development Strategy. Eastern and central China should capitalise on their regional market advantages, develop new energy sources of supply and use, and build large wind and solar power clusters for local supply in the Songliao Plain, northern Hebei and the lower reaches of the Yellow River. Western China should rely on the existing and new inter-provincial and inter-regional transmission corridors; build an array of eco-friendly and cost-efficient large-scale wind and solar power clusters in the upper reaches of the Yellow River, Hexi Corridor, Yellow River Ji-shaped bend, Gobi Desert and other desert areas; and facilitate photovoltaic-based desert prevention and sand control, renewables-to-hydrogen and complementary multi-energy development projects.

Improve the Mechanisms for Integrating Conventional and New Energy. China should define the powers and obligations of government sectors, continuously optimise the market-oriented and law-based business environment for new energy, and introduce a regulatory mechanism that facilitates new energy development. A project development management mechanism based on market-oriented resource allocation should be established. The country should refine the rules for new energy participation in electricity market trading, play the guiding role for price signals in the national electricity market, remove market and administrative barriers, and create a market mechanism that reflects the environmental value of new energy and enables fair competition with conventional power sources.

Increase Energy and Electric Power Innovation. China should increase support for energy research and development, focus on technologies like renewable energy, a new electric power system, large-scale energy storage and hydrogen, and pool resources into collaborative research on core technologies and innovation for major projects. The country should improve the environment for electric power technology innovation and optimise investment and financing mechanisms. China should set up investment funds to support the development of electric power technology demonstration projects and businesses, and guide venture capital and private equity investments to support electric power technology innovation. China should encourage the formation of alliances that promote collaboration between enterprises, universities and research institutes around key and emerging energy technologies, build a scientific research management system that respects innovation and constructive competition, and improve the creation and protection mechanisms for intellectual property rights for new technologies.

Plan Power Grid Development Early. China should strengthen its grid infrastructure, coordinate the development of wind and solar power clusters and increase the amount of new energy in inter-provincial and inter-regional transmission corridors. The country should optimise the construction of new transmission and distribution networks, distribution networks, and promote the cross-provincial, cross-regional and local use of new energy, and develop an electric power system that is adapted to new energy.

Chapter 6: The Outlook for Wind and Solar Power

Dengfeng Liu, Jian Yue, Jingsheng Wang and Marcelo Espinoza

1 Introduction

China's exploitable wind energy resources will increase as wind turbine technology advances, potentially making wind a key component of the country's energy demand mix in the future. The country's exploitable onshore wind energy reserves are 1,750 GW, assuming that mature technologies are used and taking the return on investment into consideration. The data for average wind energy reserves and average annual turbine operating hours in the provinces show that the onshore wind power fleet can provide 4,897.1 billion kWh of electricity annually. Onshore wind energy resources alone could supply 58.9% of the country's total electricity demand of 8,312.8 billion kWh in 2021.

China's solar resources can help the country achieve carbon neutrality. Endowed with some of the most abundant solar energy resources in the world, China records annual sunshine of more than 2,200 hours and annual solar radiation of more than 5,000 millijoules per square metre (MJ/m^2) in more than two-thirds of its total area. The total solar radiation received by China's land area is between 3,300 and 8,400 MJ/m^2 annually, which corresponds to energy reserves of 2,400 billion tonnes of coal equivalent (tce). Preliminary analysis indicates that the installed capacity of exploitable solar energy across the country is 15,600 GW.

The Energy Internet can help solve the challenges facing the generation, grid and demand sides of the power system. The Energy Internet will stimulate change in the entire energy value chain from production and transmission to storage and use, and facilitate the formation of an energy supply system where centralised and distributed energy work together in a co-ordinated and complementary way. The Energy Internet increases the flow of information in the energy system and enables flexible management of energy flows based on that information. The Energy Internet is expected to become the main driver of the Fourth Industrial Revolution by increasing the generation of renewable energy, promoting the clean and efficient use of fossil fuels and improving the overall efficiency of energy.

Renewable energy development needs strong support from government policies. China will promote an energy supply system that is dominated by clean and low-carbon energy. Large wind and solar power clusters should be built at a faster pace, mainly in the Gobi Desert and other desert areas, and existing coal-fired power plants should be upgraded. The country should also explore introducing a co-ordination mechanism

D. F. Liu (✉) · J. Yue · J. Sh. Wang
Goldwind Science and Technology Co. Ltd., Beijing, China

M. Espinoza
Shell International B.V., The Hague, The Netherlands

Shell International B.V. and Development Research Centre of the State Council (DRC) of People's Republic of China, *Embracing the Future, Powering Growth: An Energy System Renewed for China*, Advances in Oil and Gas Exploration & Production,
https://doi.org/10.1007/978-3-031-73075-7_7

to regulate new energy-based power transmission between the sending and receiving ends of transmission links, and provide support to ensure that all new energy generation projects are developed and completed.

2 Exploitable Wind Energy Resources in China

Thanks to continual advances in wind power technologies, such as increased tower height and longer blade diameter, the amount of exploitable wind energy resources is rising. The exploitable wind power reserves across China are about 4,260 GW, calculated at a height of 100 metres. These include exploitable wind power resources of 3,900 GW onshore and 360 GW offshore (within 50 km of the coast).[1] If the onshore wind power fleet operates at full load for 2,000 hours, 7,800 billion kWh of electricity can be supplied per year. If the offshore wind power fleet operates at full load for 2,500 hours, 900 billion kWh of electricity can be produced per year. These amount to a total wind power output of around 8,700 billion kWh.

China's exploitable onshore wind energy reserves are 1,750 gigawatts (GW), assuming that mature technologies are used and taking the return on investment into consideration. The data for average wind energy reserves and average annual turbine operating hours in the provinces show that the onshore wind power fleet can provide 4,897.1 billion kWh of electricity annually. Onshore wind energy resources alone could supply 58.9% of the country's total electricity demand of 8,312.8 billion kWh in 2021.

Exploitable wind energy resources can be increased by improving tower height and blade diameter. Increasing the tower height from 100 metres to 140 metres adds another 800 GW to the exploitable wind energy reserves. Mature onshore wind turbines have a maximum height of more than 160 metres and a blade diameter of more than 190 metres. Technology improvements are also driving down turbine construction costs. Wind power will undoubtedly become an important part of the energy mix of the future.

2.1 Distribution of Wind Energy Resources in China

China's border reaches from east Eurasia to the western part of the Pacific Ocean. Due to the thermodynamic differences between sea and land, a sea-to-land monsoon blows from the south-east in the summer, and a land-to-sea monsoon from the north-west in the winter.

The typical monsoon climate in China brings abundant wind energy resources, with exploitable wind power reserves amounting to about 4,260 GW (calculated at a height of 100 metres). However, the distribution of wind resources is unbalanced. This is mainly caused by geographical location and terrain: flat terrain offers little resistance to wind and therefore has abundant wind resources; rugged terrain creates strong resistance to wind, and has relatively poor wind energy resources.

2.1.1 Onshore Wind Power

The east Qinghai–Tibet Plateau, Hengduan Mountains, Yunnan–Guizhou Plateau, south-east hilly areas, Loess Plateau and Tarim Basin have scarce wind energy resources due to their terrain. Onshore wind power resources are mainly distributed in the following regions.

1. **Inner Mongolia and North Gansu**

This region has prevailing westerlies all year and is the first to be hit by cold air in the winter. The wind energy density is 200–300 watts per square metre (W/m^2) and the effective generating time is about 70%. Wind duration with a speed equal to or greater than 3 metres per second (m/s) and 6 m/s is more than 5,000 hours and 2,000 hours per year respectively. Wind velocity gradually decreases from north to south, but at a gradient less than that of the coastal areas of the south-east.

[1] Zhu Rong, Wang Yang and Xiangyang et al., A Study on Climate Characteristics and Development Potential of Wind Energy Resources in China, Acta Energiae Solaris Sinica, Issue 6, 2021.

In Hulegaier, where wind energy resources are the most abundant, wind duration with a velocity equal to or greater than 3 m/s and 6 m/s can be as much as 7,659 hours and 4,095 hours respectively. At a height of 110 metres, the average wind duration can reach 3,200 hours. Despite a wind density less than that of the south-east coastal areas, the region's wider geographical extent makes it the most wind-abundant in China.

2. **Heilongjiang, Eastern Jilin and the Coastal Areas of Liaodong Peninsula**

The wind energy density in this region is more than 200 W/m^2, and the wind duration with a velocity equal to or greater than 3 m/s and 6 m/s can reach 5,000–7,000 hours and 3,000 hours per year respectively. At a height of 110 metres, the average generating time of wind energy can reach 3,200 hours.

3. **Qinghai–Tibet Plateau, North and North-west China and Coastal Areas**

The wind energy density in this region is 150–200 W/m^2, and the wind duration with a velocity equal to or greater than 3 m/s and 6 m/s is 4,000–5,000 hours and more than 3,000 hours per year respectively. In the Qinghai–Tibet Plateau, the wind duration with a velocity equal to or greater than 3 m/s can be up to 6,500 hours, but the wind density is relatively low due to high altitude. At an altitude of 4,000 metres, the air density in the Qinghai–Tibet Plateau is roughly 67% of that on the ground. In other words, the wind density at a wind velocity of 8 m/s is 313.6 W/m^2 in flat areas, but only 209.3 W/m^2 at an altitude of 4,000 metres.

If wind energy is calculated only by wind duration with a velocity equal to or greater than 3 m/s and 6 m/s, the Qinghai–Tibet Plateau is the most wind-abundant area in China. However, the wind energy reserves there are far less than the south-east coastal islands. Exploitable onshore wind energy resources are in the northern parts of the north-east, North China, the north-west and coastal areas.

4. **Yunnan–Guizhou–Sichuan Region, Gansu, South Shaanxi, Henan, West Hunan, the Mountainous Areas of Fujian, Guangdong, Guangxi and the Tarim Basin**

The effective wind energy density in these areas is less than 50 W/m^2. Only about 20% of their wind resources can be exploited. Wind duration with a velocity equal to or greater than 3 m/s and 6 m/s is less than 2,000 hours and no more than 150 hours per year respectively. The Sichuan Basin and Xishuangbanna have the lowest wind energy reserves in China and a year-round wind frequency of 67% in Mianyang, 60% in Bazhong, 67% in Aba, 75% in Enshi, 63% in Dege, 72% in Mengding and 79% in Jinghong. In these areas, wind duration with a velocity equal to or greater than 3 m/s and 6 m/s is only 300 hours and 20 hours per year respectively. The exploitable wind energy potential in these areas is very low, except on mountain peaks and canyon rims and on special terrain.

In practice, the development and use of wind energy should take market and policy factors into account. Assuming that well-proven wind turbines and solutions are used and taking return on investment into consideration, the exploitable onshore wind energy resources in all provinces amount to 1,750 GW, which could produce electricity of up to 4,897.1 billion kWh.

The provinces with the most exploitable wind energy resources are Inner Mongolia, Heilongjiang, Jilin, Xinjiang, Gansu and Liaoning.

2.1.2 Offshore Wind Power

China has a long coastline of more than 18,000 kilometres. Thanks to the winter and summer monsoons and little blockage from the sea surface, there are abundant wind energy resources in China's seas, which makes them ideal locations for offshore wind farms. In addition, the economically developed south-east coastal areas have very high energy demand.

The south-east coastal areas have the most offshore wind resources. The isoline of effective wind energy density equal to or greater than 200 W/m^2 runs parallel to the coastline, and the wind energy density of coastal islands is above 300 W/m^2. The percentage of effective wind time is between 80% and 90%, and the wind duration with a velocity equal to or greater than 8 m/s and 6 m/s is 7,000–8,000 hours and about 4,000 hours per year respectively. However, the hills in these areas make it difficult for cold air in winter to penetrate inland in southern areas. Typhoons in the second half of the year reduce the effective wind time to 68% in areas that are 50 km from the coast. The wind energy density decreases sharply inland, dropping to below 50 W/m^2 less than 100 kilometres from the coast. On coastal islands such as Taishan and Pingtan in Fujian and Nanji and Dachen in Zhejiang, wind energy resources are abundant. On Taishan, the wind energy density is 534.4 W/m^2; 90% of the wind time is effective; and winds with a speed equal to or greater than 3 m/s occur for 7,905 hours of the year. In other words, the daily average wind duration with a velocity equal to or greater than 3 m/s is 21.3 hours, which makes Taishan one of the richest wind energy areas in China.

The annual average wind speed at sea is significantly higher than on land. The offshore wind energy resources of the Taiwan Strait are the most abundant, with a yearly average wind speed of 7.5–9.5 m/s. The offshore waters north of the Taiwan Strait have an annual average wind speed of 6.5–8.0 m/s and those of Guangdong, Guangxi and Hainan Island 6.0–7.5 m/s. Located in the eastern South China Sea, the Taiwan Strait has a monsoon climate, where north-east winds prevail in autumn and winter and south-west winds in spring and summer. In winter especially, the strong north-east winds, which are deflected by a mountain range on the island, create a funnelling effect inside the strait. As a result, the wind speed increases, creating an area of valuable high offshore wind speeds.

China's offshore wind resources can be assessed by water depth or distance from shore. Waters with a depth of 0–5 m are an intertidal zone that are not included in the assessment. Technologically exploitable wind energy resources in waters 5–50 metres deep are estimated to be 400 GW, including 210 GW in depths of 5–25 metres and 190 GW in depths of 26–50 metres. Technologically exploitable wind energy resources in waters that are within 50 km of the coast are estimated to be 360 GW, including 190 GW within 25 km, and 170 GW within 26–50 km.

2.2 Regional Characteristics of the Wind Power Industry

China's wind farms are mainly concentrated in north-west, North and north-east China, onshore and in eastern coastal areas. The areas with the most wind energy resources include eastern Inner Mongolia, western Inner Mongolia, Hami in Xinjiang, Jiuquan in Gansu, Bashang in Hebei, western Jilin and coastal areas in Jiangsu and Shandong.

According to data released by the China Electricity Council, at the end of 2020, the top 10 provinces ranked by installed wind power capacity were Inner Mongolia with 37.86 GW, Xinjiang with 23.61 GW, Hebei with 22.74 GW, Shanxi with 19.74 GW, Shandong with 17.95 GW, Jiangsu with 15.47 GW, Henan with 15.18 GW, Ningxia with 13.77 GW, Gansu with 13.73 GW and Liaoning with 9.81 GW. By region, the combined installed wind power capacity in Central, East and South China was 38.8%, and that of north-west, North and north-east China 61.2%.

Wind power development is faster in North China than in other parts of the country. In 2020, North and north-west China accounted for 31.3% and 20.3% respectively of the country's total additional installed capacity of wind power, up 8 and 1 percentage point on the previous year. Central areas of the South, East, south-west and north-east China accounted for 20.6%, 19.9%, 4.2% and 3.7% respectively of the country's wind power fleet, down 2.4, 2.7, 2.5 and 1.4 percentage points on 2019.

2.3 Prospects for China's Wind Power Sector

Technologically exploitable wind energy reserves are related to factors such as annual average wind speed, gradient, land use and the presence or not of water, cities or nature reserves. Plateaus, mountains and hills account for 65% of China's total land area. The gradient or slope of the land determines its suitability for wind energy development, in addition to the annual average wind speed of the site. Of the land suitable for wind power development at various gradients in China's total land area, those of grades 0.8–1.0, 0.6–0.8 and 0.2–0.4 account for 26%, 32% and 39% respectively, calculated at a height of more than 100 meters. Grade 0.8–1.0 areas are mainly in western Inner Mongolia, Jiuquan and Beishan in Gansu, and Hami and Turpan in Xinjiang. Grade 0.6–0.8 areas are mainly in eastern Inner Mongolia, Buerjin in Xinjiang, the western plain of the Qaidam River basin in Qinghai, Yanchi in Ningxia, Mu Us Sandland in Yulin, Shaanxi, Zhanjiang in Guangdong, and northern coastal areas in Hainan. Grade 0.2–0.4 areas are spread all over the country. Of the land available for wind energy development in China, grade 0.6–1.0 represents 58% of the total, calculated at a height of 100 meters. Most of this land is on the flat plateau across northern China, especially in the western part where vegetation coverage is low and the land use grade is above 0.8. North-west, North and north-east China are, therefore, ideal for large-scale wind farms or wind farm clusters.

The greater the height, the larger the annual average wind speed; and the larger the land area available for wind energy development, the higher the exploitable wind energy resources. Fig. 1 shows the change in technologically exploitable wind energy resources in areas of different land availability grades at a height of 140 metres compared with that of 100 metres. The three largest increases in exploitable wind resources are: areas with land availability grade 0.6–0.8 and "very abundant" wind energy resources increase by 143%; areas with land availability grade of 0.6–0.8 and "abundant" wind energy resources increase by 88%; and areas with "very abundant" resources in land availability grade 0.8–1.0 increase by 86%.

The technologically exploitable wind resources in "very abundant" and "abundant" areas increase by 800 GW at a height of 140 metres compared with 100 metres. These areas are mainly in Inner Mongolia, Gansu and Xinjiang and can be developed at scale using large wind turbines. The technologically exploitable wind resources classed as "relatively abundant" in areas with land availability grade 0.2–0.4 increase by 150 GW, and are widely distributed in hilly areas in northern China, the southern Qinghai–Tibet Plateau, north-west Sichuan, the Yunnan–Guizhou Plateau, Zhanjiang in Guangdong and northern coastal areas of Hainan. The technologically exploitable wind resources classed as "average" in areas with land availability grade 0.2–0.4 increase by 240 GW; these are found mainly in the Sanjiang Plain, Songnen Plain, North China Plain and Jianghuai Plain. Although these areas are of flat terrain, constraints such as farmland and towns make it advisable to adopt a small-scale or distributed wind energy development approach.

As wind energy resource assessment techniques progress, we will gain an increasingly accurate understanding of wind energy reserves. Improvements in wind energy technologies will enable a growing amount of wind energy resources to be developed.

3 Exploitable Solar Energy Resources in China

China is rich in solar energy resources. Two-thirds of the country's land area enjoys annual sunshine of more than 2,200 hours and annual solar radiation of more than 5,000 MJ/m^2. The annual solar radiation received across the country is 3,300–8,400 MJ/m^2, equivalent to reserves of 2,400 billion tonnes of coal equivalent. Preliminary analysis indicates that the installed capacity of exploitable solar energy is 15,600 GW. Xinjiang has the largest installed capacity of solar energy at about 4,200 GW, followed by Qinghai

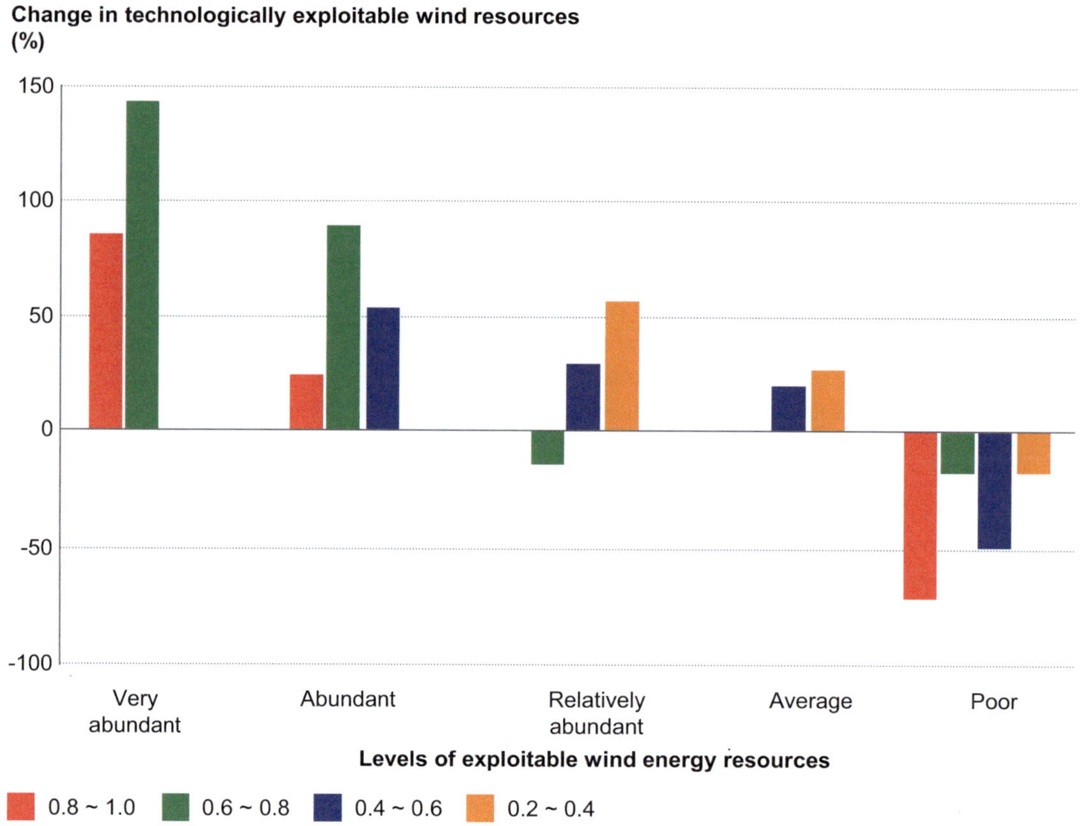

Fig. 1 Classes of exploitable wind energy resources. *Source* A Study of the Climate Characteristics and Development Potential of Wind Energy Resources in China

with 3,400 GW and Inner Mongolia with 2,615 GW. Gansu, Tibet, Ningxia and Shandong also have a considerable installed capacity. According to the Atlas of China's Solar Energy Resources by Province, published in 2006, the country's exploitable solar energy amounts to 4,017.71 trillion kWh, without taking return on investment into account (China's total electricity consumption in 2021 was 8.3128 trillion kWh).

In 2020, the average annual horizontal solar radiation on China's lands was 1,490.8 kWh/m^2, 1.04% higher than the 2010–19 average (1,475.5 kWh/m^2) and 1.38% higher than in 2019 (1,470.5 kWh/m^2).

Solar energy resources vary greatly across regions. Generally, solar energy resources are abundant in plateaus and dry areas, but poor in plains and areas with high levels of rainfall and humidity. In 2020, most parts of north-west China, south-west and central–west China, Inner Mongolia, north Shanxi, north Hebei, west and north-east Liaoning and north-east Jilin, recorded annual horizontal solar radiation of 1,400 kWh/m^2. South-west Gansu, west Inner Mongolia, west Qinghai, central–west Tibet and west Sichuan registered annual horizontal solar radiation of more than 1,750 kWh/m^2, the highest in the country. Most parts of Xinjiang and Inner Mongolia, central–east Qinghai, central Gansu, Ningxia, north Shaanxi, central–north Shanxi, east Tibet, Yunnan and west Hainan had annual horizontal solar radiation of 1,400–1,750 kWh/m^2, which is classed as "very abundant". North-east Inner Mongolia, most parts of Heilongjiang and Jilin, south Shanxi, central–south Hebei, Beijing, Tianjin, the areas between the Huanghe and Huaihe rivers and between the Yangtze and Huaihe rivers, the Changjiang–

Hanshui area, south of the Yangtze River, and most parts of South China recorded annual horizontal solar radiation of 1,050–1,400 kWh/m^2, which is classed "abundant". East Sichuan, Chongqing, central–north Guizhou, central–west Hunan and south-west Hubei had annual horizontal solar radiation of less than 1,050 kWh/m^2, which is considered "average".

China's solar energy resources have the following main characteristics:

1. Both the high-value and low-value centres of solar energy are between latitudes 22°and 35° north. The Qinghai–Tibet Plateau is the high-value centre, and the Sichuan Basin the low-value centre.
2. The total solar radiation in West China is higher than that in East China. Generally, solar radiation in South China is lower than in North China, except for Tibet and Xinjiang.
3. As most parts of South China have many cloudy and rainy days, in latitudes 30°–40° north the distribution of solar energy is contrary to the general pattern that solar energy decreases as latitude increases. In this instance, solar energy increases with a higher latitude.

According to the Wind and Solar Energy Centre of the China Meteorological Administration, there are four categories of area with solar energy resources. Categories I, II and III have annual sunshine of 2,200 hours and annual solar radiation greater than 1,388 kWh/m^2, which provide good conditions for solar energy use. These areas have abundant or relatively abundant solar energy resources and account for more than two-thirds of China's total land area. Class IV areas have relatively poor solar energy resources, but still have solar energy use value.

Category I areas have the most abundant solar energy resources in China, with annual solar radiation of 1,855–2,333 kWh/m^2 (equivalent to daily solar radiation of 5.1–6.4 kWh/m^2) and annual sunshine of 3,200–3,300 hours. These areas include north Ningxia, north Gansu, south-east Xinjiang, west Qinghai and west Tibet. West Tibet has the most abundant solar energy resources of up to 2,333 kWh/m^2 (equivalent to daily solar radiation of 6.4 kWh/m^2), second only to the Sahara Desert.

Category II areas have relatively abundant solar energy resources, with annual solar radiation of 1,625–1,855 kWh/m^2 (equivalent to daily solar radiation of 4.5–5.1 kWh/m^2) and annual sunshine of 3,000–3,200 hours. These areas include north-west Hebei, north Shanxi, south Inner Mongolia, south Ningxia, central Gansu, east Qinghai, south-east Tibet and south Xinjiang.

Category III areas have average solar energy resources, with annual solar radiation of 1,388–1,625 kWh/m^2 (equivalent to daily solar radiation of 3.8–4.5 kWh/m^2) and annual sunshine of 2,200–3,000 hours. These areas include Shandong, Henan, south-east Hebei, south Shanxi, north Xinjiang, Jilin, Liaoning, Yunnan, north Shaanxi, south-east Gansu, south Guangdong, south Fujian, north Jiangsu, north Anhui and south-west Taiwan.

Category IV areas have relatively poor solar energy resources, with annual solar radiation of 928–1,393 kWh/m^2 (equivalent to daily solar radiation of 2.5–3.8 kWh/m^2) and annual sunshine of 1,000–2,200 hours. These areas include Hunan, Hubei, Guangxi, Jiangxi, Zhejiang, north Fujian, north Guangdong, south Shaanxi, north Jiangsu, south Anhui, as well as Heilongjiang, Sichuan, Guizhou and north-east Taiwan.

3.1 Distribution of Solar Energy Resources in China

The estimated reserves of solar energy resources by province in China are given in Table 1.

1. **North-east China**

North-east China has very abundant solar energy resources and falls into Category II.

Jilin has the most abundant solar energy resources in the region, with average annual sunshine of 2,200–3,000 hours and average annual solar radiation of 1,600–1,900 kWh/m^2.

Table 1 Estimated reserves of solar energy resources by province in China

Province	Total land area (hectares)	Unused land area (hectares)	Total reserves (10^{14} kWh)	Exploitable reserves (10^{14} kWh)
Beijing	1,641,053.70	216,648.04	0.2640	0.03485
Tianjin	1,191,731.91	67,791.93	0.1620	0.00922
Hebei	18,843,061.14	4,046,516.27	2.8860	0.61976
Shanxi	15,671,124.85	5,061,140.95	2.3044	0.74423
Inner Mongolia	114,512,122.96	15,057,937.47	17.7550	2.33472
Liaoning	14,806,370.73	1,507,120.69	2.0630	0.20999
Jilin	19,112,390.97	1,126,753.84	2.4870	0.14662
Heilongjiang	45,264,501.67	4,352,413.67	5.9707	0.57411
Shanghai	823,901.21	968.47	0.0769	0.00009
Jiangsu	10,667,388.34	148,337.03	1.6540	0.02300
Zhejiang	10,539,094.51	697,753.43	1.2880	0.08527
Anhui	14,012,579.19	753,352.38	1.6950	0.09113
Fujian	12,405,796.69	957,929.16	1.5610	0.12054
Jiangxi	16,689,433.59	1,126,058.14	2.0420	0.13778
Shandong	15,705,240.90	1,654,724.55	2.1920	0.23095
Henan	16,553,641.93	1,865,865.68	2.1860	0.24640
Hubei	18,588,842.75	2,116,234.77	2.2360	0.25456
Hunan	21,185,468.75	2,035,760.11	2.4740	0.23773
Guangdong	17,975,234.85	972,591.87	2.4070	0.13024
Guangxi	23,755,809.80	5,158,298.15	2.8413	0.61696
Hainan	3,535,368.96	264,804.98	0.4759	0.03565
Sichuan and Chongqing	56,632,471.95	7,285,646.63	7.3600	0.87728
Guizhou	17,615,246.61	2,698,833.29	1.8938	0.29015
Yunnan	38,319,412.23	7,298,195.50	5.7743	1.09976
Tibet	120,207,150.77	37,049,249.94	23.6057	7.27555
Shaanxi	20,579,459.87	1,170,455.71	2.6010	0.14793
Gansu	40,409,087.32	16,114,366.21	6.7000	2.67183
Qinghai	71,748,052.29	24,841,105.45	13.6600	4.72946
Ningxia	5,195,437.51	820,985.91	1.0910	0.17240
Xinjiang	166,489,717.01	98,620,027.97	27.0600	16.02897

Source Atlas of China's Solar Energy Resources by Province, 2006

Heilongjiang and Liaoning have less solar energy resources than Jilin, with annual solar radiation of 1,222–1,500 kWh/m^2 and annual sunshine of 2,242–2,842 hours.

The technologically exploitable solar energy capacity in north-east China is about 422,170 megawatts (MW), including 34,000 MW in Jilin, 340,680 MW in Heilongjiang and 47,490 MW in Liaoning.

2. **East China**

East China has abundant solar energy resources and includes Category III and IV areas.

Shanghai has average annual solar radiation of 4,700 MJ/m^2 and annual sunshine of 2,014 hours. Fujian records annual solar radiation of 3,800–5,400 MJ/m^2; Jiangsu's annual solar radiation ranges from 4,200 MJ/m^2 to 5,400

MJ/m^2, and its sunshine duration is 1,900–2,500 hours. Zhejiang records an annual solar radiation range from 4,220 MJ/m^2 to 4,950 MJ/m^2. Anhui has an annual solar radiation range of 4,000 MJ/m^2–4,600 MJ/m^2 and sunshine duration of 1,670–2,315 hours.

The technologically exploitable solar energy capacity in East China is about 228,000 MW, including 20,000 MW in Shanghai, 46,000 MW in Fujian, 60,000 MW in Jiangsu, 42,000 MW in Zhejiang and 60,000 in Anhui.

3. **Central China**

Central China has abundant solar energy resources and falls into Category III.

Jiangxi has average annual solar radiation of 4,135.1 MJ/m^2 and Henan an annual solar radiation range of 4,300–5,000 MJ/m^2. Hubei records an annual solar radiation range of 3,450–4,800 MJ/m^2 and annual sunshine of 11,100–2,000 hours. Hunan's annual solar radiation ranges from 3,384.7 MJ/m^2 to 4,372.0 MJ/m^2 and its annual sunshine is 1,300–1,800 hours.

The technologically exploitable solar energy capacity in East China is estimated to be 228,260 MW, including 62,580 MW in Jiangxi, 63,000 MW in Henan, 69,480 MW in Hubei and 33,200 MW in Hunan.

4. **North-west China**

North-west China is very rich in solar energy resources and comprises Category I–III areas.

Qinghai has the most abundant solar energy resources, with an annual solar radiation range of 5,800–7,400 MJ/m^2 and annual sunshine of 2,300–3,300 hours. Ningxia has very abundant solar resources, with an annual solar radiation range of 5,195.3–6,344.2 MJ/m^2 and annual sunshine of 2,250–3,100 hours. Shaanxi records abundant solar energy resources, with annual solar radiation ranging from 4,100 MJ/m^2 to 5,600 MJ/m^2 and annual sunshine of 1,270–2,900 hours. Gansu has very abundant solar resources, with annual solar radiation ranging from 5,226 MJ/m^2 to 6,330 MJ/m^2 and sunshine duration of 1,912–3,316 hours. Xinjiang's solar resources are between the abundant and the most abundant levels, with an annual solar radiation range of 4,600–7,400 MJ/m^2 and sunshine duration of 2,500–3,550 hours.

The technologically exploitable solar energy capacity in north-west China is about 10,291,200 MW, including 3,400,000 MW in Qinghai, 481,700 MW in Ningxia, 80,000 MW in Shaanxi, 2,129,500 MW in Gansu and 4,200,000 MW in Xinjiang.

5. **South-west China**

South-west China has relatively abundant solar energy resources and comprises Category II and III areas.

Chongqing has average solar energy resources, with an annual solar radiation range of 3,390–4,200 MJ/m^2 and annual sunshine of 1,039.6 hours. Sichuan has very abundant solar energy resources, with annual solar radiation ranging from 3,200 MJ/m^2 to 6,900 MJ/m^2 and annual sunshine of 750–2,700 hours. Tibet holds the most abundant solar energy resources, with an annual solar radiation range of 2,000–8,200 MJ/m^2. Guizhou has average solar energy resources, with an average annual solar radiation of more than 3,400 MJ/m^2. Yunnan holds very abundant solar energy resources, with an annual solar radiation range of 3,620–6,682 MJ/m^2 and sunshine duration of 960–2,840 hours.

The technologically exploitable solar energy capacity of south-west China is estimated to be about 844,500 MW, including 26,500 MW in Chongqing, 43,000 MW in Sichuan, 700,000 MW in Tibet, 35,000 MW in Guizhou and 40,000 MW in Yunnan.

6. **North China**

North China's solar energy resources are between the abundant and most abundant levels. Western Inner Mongolia is in Category I; other areas fall into Category II or III.

Beijing's average annual solar radiation is 5,256 MJ/m^2 and its annual sunshine 2,480–2,580 hours. Tianjin has an annual radiation range of 3,780–5,240 MJ/m^2 and sunshine duration of

2,500–2,900 hours. Hebei's annual solar radiation ranges from 5,040 MJ/m^2 to 6,300 MJ/m^2 and its annual sunshine is 2,350–3,000 hours. Shandong records an annual solar radiation range of 4,824–5,292 MJ/m^2 and sunshine duration of 2,100–2,500 hours. Western Inner Mongolia has an annual solar radiation of 5,508–6,516 MJ/m^2 and sunshine duration of 2,650–3,100 hours. Eastern Inner Mongolia registers an annual solar radiation range of 5,040–7,560 MJ/m^2 and annual sunshine of 2,600–3,400 hours. Shanxi's annual solar radiation ranges from 4,770 MJ/m^2 to 5,800 MJ/m^2 and its sunshine duration is 2,200–3,000 hours.

The technologically exploitable solar energy capacity in North China is about 3,482,630 MW, including 33,130 MW in Beijing, 4,000 MW in Tianjin, 122,000 MW in Hebei, 629,000 MW in Shandong, 2,600,000 MW in western Inner Mongolia, 14,500 MW in eastern Inner Mongolia and 80,000 MW in Shanxi.

7. **South China**

In South China, Guangdong and Hainan have very abundant solar energy resources and fall into Category II areas. Guangxi has abundant solar energy resources and is in Category III.

Guangdong has an annual solar radiation range of 4,200–5,800 MJ/m^2 and average annual sunshine of about 2,000 hours. Hainan's annual solar radiation ranges from 4,600 MJ/m^2 to 5,800 MJ/m^2 and its average annual sunshine is 2,166 hours. Guangxi records an annual solar radiation range of 3,682.2–5,642.8 MJ/m^2 and average annual sunshine of more than 1,880 hours.

The technologically exploitable solar energy capacity in South China is about 129,400 MW, including 60,000 MW in Guangdong, 10,000 MW in Hainan and 59,400 MW in Guangxi (see Table 2).

3.2 Regional Statistics

Nine provinces have a technologically exploitable solar energy capacity of more than 100,000 MW, including Xinjiang, Qinghai, Inner Mongolia (west), Gansu, Tibet, Shandong, Ningxia, Heilongjiang and Hebei. Nine provinces have a technologically exploitable solar energy capacity of 50,000–100,000 MW, including Shanxi, Shaanxi, Hubei, Henan, Jiangxi, Guangdong, Jiangsu, Anhui and Guangxi. The other 14 provinces have a technologically exploitable solar energy capacity of less than 50,000 MW.

In the 14th Five-year Plan period (2021–25), China's planned solar energy capacity amounts to 357,436 MW, including 28,328 MW in the north-east, 42,650 MW in the east, 31,662 MW in the central region, 69,550 MW in the north-west, 43,476 MW in the south-west, 111,250 MW in the north and 30,520 MW in the south. North-west, North and north-east China account for a combined share of nearly 59%.

For 2021–25, 16 provinces have a planned solar energy capacity of more than 10,000 MW, including Hebei, Shanxi, Guizhou, Qinghai, Guangxi, Anhui, Gansu, Shaanxi, Inner Mongolia (west), Heilongjiang, Guangdong, Shandong, Ningxia, Xinjiang, Hunan and Sichuan. Nine provinces have a planned solar energy capacity of 5,000–10,000 MW, including Jilin, Yunnan, Jiangxi, Jiangsu, Zhejiang, Fujian, Hubei, Liaoning and Tibet. Other provinces have a planned solar energy capacity of less than 5,000 MW, including Henan, Inner Mongolia (east), Shanghai, Hainan, Beijing, Chongqing and Tianjin.

3.3 Regional Characteristics of the Solar Power Industry

China's solar power generation combines centralised and distributed approaches. Centralised solar power projects are planned and under construction in western areas where solar energy and land resources are relatively abundant, such as Xinjiang, Qinghai, Gansu, Inner Mongolia, Ningxia, Shaanxi, Tibet, Yunnan, as well as north Hebei, north Shanxi, the plateaus of Sichuan, north-west Liaoning, west Jilin, west Heilongjiang and some areas of Shandong. Distributed solar power systems are widely deployed in central and eastern coastal areas where solar energy resources are relatively abundant and economic development is strong, such as Beijing,

Table 2 Technologically exploitable solar energy resources by region

No.	Region	Province	Technologically exploitable capacity (GW)	Proportion (%)
Total			15,627	100
1	North-east China	Jilin	34	22
2		Heilongjiang	341	218
3		Liaoning	47	30
Subtotal			422	270
4	East China	Shanghai	20	13
5		Fujian	46	29
6		Jiangsu	60	38
7		Zhejiang	42	27
8		Anhui	60	38
Subtotal			228	146
9	Central China	Jiangxi	63	40
10		Henan	63	40
11		Hubei	69	44
12		Hunan	33	21
Subtotal			228	146
13	North-west China	Qinghai	3,400	2,176
14		Ningxia	482	308
15		Shaanxi	80	51
16		Gansu	2,130	1,362
17		Xinjiang	4,200	2,688
Subtotal			10,292	6,586
18	South-west China	Chongqing	27	17
19		Sichuan	43	28
20		Tibet	700	448
21		Guizhou	35	22
22		Yunnan	40	26
Subtotal			845	541
23	North China	Beijing	33	21
24		Tianjin	4	3
25		Hebei	122	78
26		Shandong	629	403
27		West Inner Mongolia[a]	2,600	1,664
28		East Inner Mongolia	15	10
29		Shanxi	80	51
Subtotal			3,483	2,229
30	South China	Guangdong	60	38
31		Hainan	10	6
32		Guangxi	59	38
Subtotal			129	83

Note [a] East Inner Mongolia and west Inner Mongolia are divisions made by power grid companies. East Inner Mongolia Electric Power Company Limited is a subsidiary of State Grid Corporation of China, while the West Inner Mongolia Power Grid is a local state-owned enterprise of the Inner Mongolia Autonomous Region

Source An Analysis of China's Solar Energy Resources and Exploitable Level, WeChat Official Account "BBS.21SPV.COM e-SPV"

Tianjin, Shanghai, Chongqing, Henan, Jiangsu, Zhejiang, Anhui, Hunan, Hubei, Jiangxi, Fujian, Guangdong, Guangxi, Guizhou and Hainan, as well as central–south Hebei, central–south Shanxi, Shandong, Sichuan and industrial parks and large industrial companies in all major cities of the north-east.

According to the China Electricity Council, at the end of 2020 the top 10 provinces ranked by installed solar power capacity included Shandong with 22.72 GW, Hebei with 21.90 GW, Jiangsu with 16.84 GW, Qinghai with 16.01 GW, Zhejiang with 15.17 GW, Anhui with 13.70 GW, Shanxi with 13.09 GW, Xinjiang with 12.66 GW, Inner Mongolia with 12.37 GW and Ningxia with 11.97 GW.

As investment growth accelerates, the installed capacity of solar power in all regions increases significantly year-on-year. South-west China had the highest year-on-year growth, while in North China capacity doubled. In 2020, the new installed capacity of solar power in East, north-west, North, central–south, south-west and north-east China was 13.44 GW (up 50.0% year-on-year), 11.35 GW (up 81.3% year-on-year), 11.14 GW (up 102.2% year-on-year), 4.86 GW (up 22.1% year-on-year), 5.83 GW (up 444.9% year-on-year) and 1.59 GW (up 55.9% year-on-year) respectively.

East China had the biggest decrease in new installed solar power capacity and south-west China the largest increase. In 2020, East, central–south and north-east China contributed 27.9%, 10.1% and 3.3% of China's total new installed capacity of solar power, down 5.5, 4.8 and 0.5 percentage points from the previous year. North-west, North and south-west China accounted for 23.5%, 23.1% and 12.1% of the country's total new installed capacity of solar power, up 0.1, 2.5 and 8.1 percentage points on the previous year.

3.4 Prospects for China's Solar Power Sector

At present, China's centralised solar photovoltaic power plants are mainly installed in the western region. Concentration in one region has made it difficult to use grid-connected solar power efficiently and has led to high transmission line losses and other challenges. As a result, the solar curtailment rate in some areas has been as high as 20%. Distributed solar energy systems are largely deployed in central and eastern China, which are also the main electricity consumers.

In the first half of 2021, China's new installed capacity of solar power was about 14 GW, an increase of 22% year-on-year. Total installed solar power capacity was 267.09 GW at the end of June 2021 (up from 253.17 GW at the end of 2020). Due to the impact of the COVID-19 pandemic in the first half of 2020, significant year-on-year growth of new installed solar power capacity occurred from February to May 2021. In 2021, there were no clear signs of project owners rushing to install new solar power before June 30 to gain higher subsidies. Rather, the months were relatively even in terms of new installed solar power capacity.

In the 14th Five-year Plan period (2021–25), China aims to implement its strategy of a transition to clean and low-carbon energy, accelerate the clean and efficient use of fossil energy and advance the development of non-fossil energy. Distributed solar power, as a green and eco-friendly power generation technology, is in line with China's energy transition strategy based on quality and efficiency. All things considered, distributed solar power technology has good prospects.

In Hebei, the average annual solar radiation is higher in the north than in the south, and it decreases from the west and east towards central areas of the province. Western and northern plateau areas have Hebei's highest average annual solar radiation of 5,800 MJ/m^2. Rongcheng and Yongqing have the lowest average annual solar radiation of less than 4,900 MJ/m^2. The average annual solar radiation in the central–south and eastern parts of the province is 5,200 MJ/m^2, while that in other areas is 5,600 MJ/m^2 (see Fig. 2).

Analysis of the unused land area available for solar power generation in the cities of Hebei province shows that Zhangjiakou and Baoding have large unused land areas, making it relatively

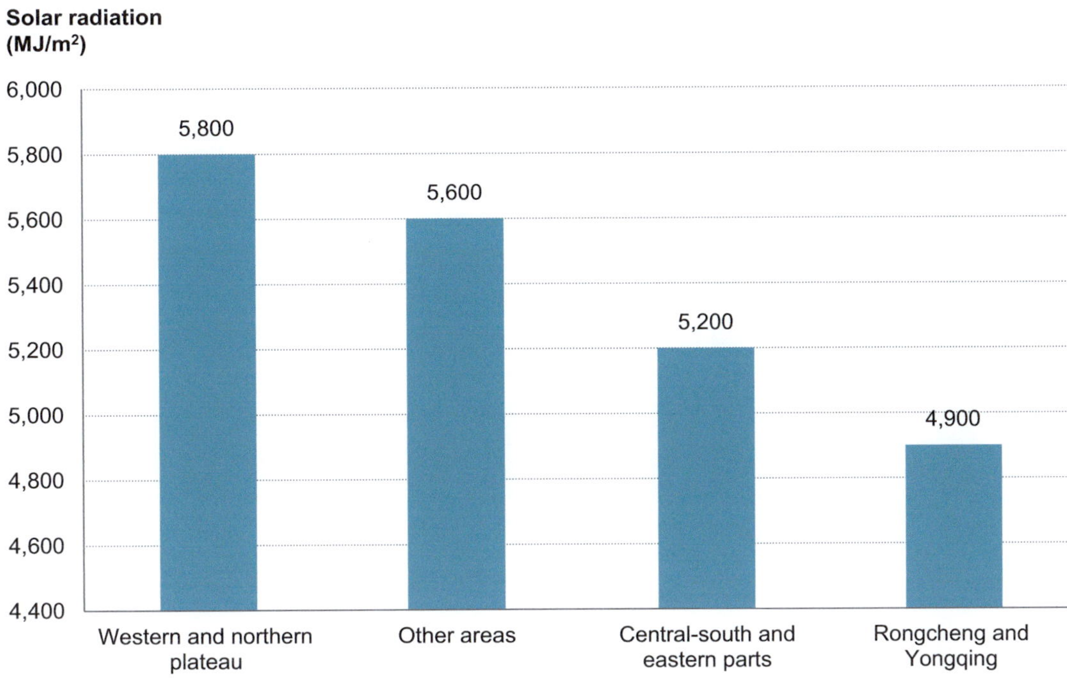

Fig. 2 Distribution of solar energy resources in Hebei. *Source* A Look at Wind and Solar Energy Plans and Current Resource Conditions in Hebei Province, Jipeng New Energy

easy for them to initiate project development opportunities. The two cities are therefore most suitable for the development of solar power projects in Hebei in 2020–30. Dingzhou and Xinji have the lowest solar energy resources in the province (see Fig. 3).

4 Energy Internet Trends

4.1 Overview of the Energy Internet

The Energy Internet integrates smart energy with the energy system. Technically, it works in harness with distributed energy systems, energy storage units, energy efficiency measures and low-carbon technologies. Operationally, it combines the Internet with innovative business models to achieve secure, efficient and low-carbon operation of the energy system. In a nutshell, the Energy Internet = smart energy + the energy system.

The smart grid, equipped with information and intelligent technologies, solves by itself to some extent, challenges that arise in the electric power system, such as improving equipment availability and use, system security and reliability, and power quality. The purpose of the Energy Internet is to connect large-scale distributed and renewable energy generation with users. Energy trading based on information is at the heart of the Energy Internet, and the integration of the cyber and physical worlds is pivotal to its operation.

The Energy Internet connects hundreds of millions of devices, machines and systems in energy production, transmission and consumption. It uses advanced sensors, control and software applications to form an internet of things. Big data analysis, machine learning and prediction provide important technological support for the Energy Internet. By integrating data on operations, the weather, the power grid and the electricity market, the Energy Internet analyses these "big" data to balance supply with demand in real time.

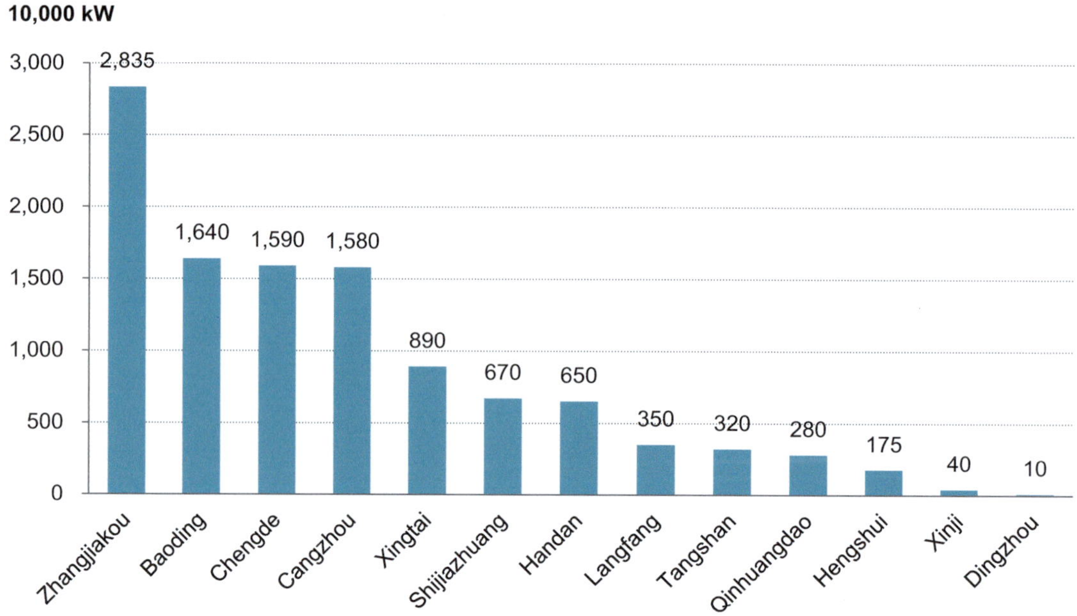

Fig. 3 Exploitable centralised solar power capacity in the cities of Hebei, 2020–30. *Source* A Look at Wind and Solar Energy Plans and Current Resource Conditions in Hebei Province, Jipeng New Energy

4.2 Prospects for the Energy Internet

Continuous technological development of the internet, computers, communications and power electronics will spark change throughout the energy value chain from generation and transmission to storage and consumption. This in turn will shape an energy supply model in which centralised and distributed energy systems are co-ordinated and complementary. Energy flows will carry information to enable power system flexibility and control. The Energy Internet is expected to become the decisive driver of the Fourth Industrial Revolution, thereby increasing the share of renewables in the energy mix and driving the energy system towards the clean and efficient use of fossil energy, energy efficiency, an open energy market, industrial modernisation, new economic growth engines and better international co-operation on energy.

The path to carbon neutrality presents varying challenges for generation, the grid and demand.

4.2.1 Power Generation

China is set to build an electric power system with a high proportion of new energy. According to estimates, new energy generation will cover about 57% of China's electricity demand in 2060. Such a high proportion can cause large fluctuations in output and uncertainty for consumers and power grids. This can be mitigated by deploying integrated new energy management platforms, trading mechanisms and other technologies. The new digital power plant technology can further optimise power generation efficiency, save energy and reduce carbon emissions.

1. **Make Energy Production Intelligent**

Internet technology should be used to enable high-quality coal processing, the use of coal according to its quality grade, the recycling of coal mine waste and the development of projects that demonstrate the clean and efficient use of coal. Efforts should be made to improve coal quality management; accelerate digital and high-quality coal processing;

research coal power technologies and equipment and their commercialisation; steadily modernise related industries; establish a mechanism for clean and efficient coal use that combines policy-guided and market-driven approaches; and build a clean, efficient, low-carbon, secure and sustainable modern coal use system.

2. **Diversify Energy Supply**

Sufficient renewable energy generation facilities should be built to gradually replace fossil fuel-fired power generation and meet growing end-use electricity demand. Fossil energy should be replaced by clean energy and a low-carbon and green development path should be followed, gradually transitioning from the dominant fossil energy + supplementary clean energy paradigm to the clean energy + supplementary fossil energy alternative.

Fossil fuels such as coal and oil in the energy demand mix should be gradually replaced by electricity, and the electricity market should be expanded to speed up electrification.

The Energy Internet, which co-ordinates and intelligently controls multiple energy sources and in which renewables play a predominant role, should be further developed to avoid under-utilisation of the renewable energy fleet. Efforts should be made to achieve technological breakthroughs in key areas such as distributed power generation, smart microgrids, active distribution networks, intelligent power system monitoring and management, internet-based two-way communication between power equipment, an open and shared energy network, and optimised energy supply mix. Different energy sources should be co-ordinated to improve planning and clean energy consumption and create a diverse energy supply system. The Energy Internet can effectively co-ordinate and support the grid connection of renewable energy, especially on the power demand side. End-user information should be timely and adequately collected and understood to make distributed renewable energy systems efficient, and fossil fuels should be replaced by clean energy.

4.2.2 Power Loads

The demand side, where energy consumption occurs, holds great potential for improvement. Because China's current industrial structure consumes large amounts of energy, it is difficult for off-grid new energy generation equipment to be used effectively, as the control capability has yet to be developed. The transformation of the energy structure therefore requires faster replacement with electricity and a reduction in direct carbon emissions by end-use sectors.

As stated in the Action Plan for Distribution Network Construction and Retrofit (2015–20) of the National Energy Administration, the collection of electricity use data needs to be improved. By 2020, the penetration of smart meters is estimated to reach 90%. Intelligent service platforms should be set up to fully support user information interaction, the connection of distributed power sources, electric vehicle charging and discharging, shore-to-ship power for ports, electric heating and other services. In addition, users should be incentivised to participate in grid load regulating, enabling co-ordinated interaction between users and the grid.

Loads on the user side are shifting to clean and intelligent technologies, and new applications such as smart cars, intelligent buildings, smart homes and photovoltaic air conditioners have been developed.

Digital technologies have a wide range of applications on the load side. The internet of things, big data, cloud computing and other technologies can be deployed to carry out comprehensive energy management, optimise energy use by businesses and transform the energy mix. Virtual power plants can be developed to integrate installations such as distributed power systems and energy storage units, and increase the use of clean energy. Enterprise carbon asset management, power trading and carbon emission trading systems should be created to enable businesses to optimise and trade their carbon assets, enhancing their motivation to reduce carbon emissions.

4.2.3 Power Transmission and Distribution

State Grid Corporation of China (SGCC) has 17 ultrahigh-voltage alternating current (UHVAC) and ultrahigh-voltage direct current (UHVDC) transmission links in operation or under construction. The links have a total length of more than 29,000 km. The development of new energy should rely on large power grids. SGCC's installed capacity of grid-connected new energy is more than 140 GW, making it the world's largest power grid in terms of grid-connected wind and solar power. China's UHVAC and UHVDC technologies have set benchmarks for the world and have been widely deployed in other parts of the globe.

Deep integration of information and automation technologies is at the core of interconnected power grids, which is designed to provide intelligent support for power grid companies and better energy services for customers. "Internet +" is a new form of energy industry development in which the internet integrates deeply with energy production, transmission, storage, consumption and the energy market, as advocated by the national government.

4.3 Drivers of New Energy Development

In 2020, thanks to the success of a feed-in-tariff policy, new wind power capacity reached a record high. In 2021, the first year of the 14th Five-year Plan (2021–25), wind power entered a new stage of development as part of the dual-carbon goal. The Notice on Matters Related to the Development of Wind and Solar Power Projects in 2021, issued by the National Energy Administration, clearly states that China will speed up the construction of wind and solar power projects. The goals of wind power development are as follows. First, China will actively advance the construction of distributed wind power projects and start the initiative of Distributed Wind Power in Towns and Villages as part of the rural revitalisation strategy. Second, the country will promote offshore wind power projects in an orderly manner. Relevant resources will be engaged to prepare or revise provincial development plans for offshore wind power, in line with the requirements in the 14th Five-year Plan. The National Energy Administration will work with natural resource authorities to formulate plans and regulations for offshore wind power development and demonstration projects in deep sea and open sea areas. Third, China will promote the orderly development of wind energy clusters. Fourth, old wind power projects will be modernised. Fifth, innovations will be made to enable more demonstration projects.

Based on the expectation that non-fossil energy will account for 20% of total primary energy demand in the 14th Five-year Plan, and in accordance with the goal of achieving peak carbon emissions before 2030, the Chinese Wind Energy Association predicts: (i) that China's new installed wind power capacity needs to be not less than 50 GW in 2021–25; and (ii) that the average annual installed wind power capacity will be close to 90 GW in 2026–35. In addition, offshore wind power projects that were approved or completed before December 31, 2021, qualified for government subsidies, starting from 2020. As a result, project owners rushed to build offshore wind power projects in 2021 and 2022. Local subsidies in coastal provinces and lower installation costs helped drive the rapid development of offshore wind power.

In October 2020, representatives of more than 400 wind energy companies from across the world adopted the Beijing Declaration on Wind Energy. The companies agreed to develop the global wind power industry and to plan China's wind power development over the next five years and beyond, with the aim of building a carbon-neutral country by 2060.

Taking resource potential, technology, grid connections and other factors into consideration, a wind power target was set to achieve average annual new installed wind power capacity of more than 50 GW.

From 2025 onwards, China's average new annual installed capacity of wind power should be no less than 60 GW and should reach at least 800 GW by 2030 and at least 3,000 GW by 2060.

In the long term, offshore wind power holds great growth potential. As wind turbine technology progresses and more ultrahigh-voltage transmission links are constructed, large wind power clusters operating at grid parity in northwest China should thrive. The central and eastern regions will focus on distributed wind power systems. The cost of offshore wind power, engineering and equipment are expected to fall in the coming years, allowing rapid development of offshore wind power.

The cost of solar photovoltaic will drop sharply. After 20 years of development, China's solar photovoltaic industry leads the world by manufacturing output, installed capacity and power output. The dual-carbon goal also brings unprecedented opportunities for the solar power sector.

Distributed solar power is a highlight of the new energy industry. The new installed capacity of solar power was 13.01 GW in the first half of 2021, a growth rate of 13.1% year-on-year. New centralised capacity decreased by 24.2%, while new distributed capacity increased by 72.7%, outpacing centralised capacity for the first time. New installed capacity of household distributed solar power exceeded 7.661 GW, up 263% year-on-year. In 2030, China's new installed capacity of solar power is expected to reach between 416 and 537 GW, with a compound annual growth rate of 24%–26%. In the coming decade, installed solar power capacity is projected to grow by a factor of 10, indicating enormous market potential.

5 Policy Recommendations

5.1 Create a Roadmap for Wind and Solar Energy Development as Soon as Possible

A mechanism to speed up the development of renewable energy should be introduced. Wind and solar energy cluster projects in the Gobi and other desert areas should be closely monitored and the relevant provinces should be urged to facilitate the completion of projects as quickly as possible. A fast track should be opened to approve the grid connection of large wind and solar power cluster projects to ensure that grid connection takes place. Without compromising environmental protection requirements or using arable land, government sectors should work together to accelerate approval of large cluster projects, the timely completion of environmental impact assessments, the introduction of financial incentives and other support policies.

5.2 Accelerate the Development and Local Use of Distributed Energy

Distributed energy does not require long-distance transmission. It therefore saves on transmission costs and can motivate energy users to take part in project execution. Local supply–demand balance can be achieved through microgrids, which reduce the need for power grid expansion and maximise new energy consumption. For this reason, China should, first, actively promote the construction of distributed ground-mounted solar power plants as well as solar power plants in villages, distributed wind power projects and energy storage systems. A second reason for building grid-connected solar power projects in villages is to alleviate poverty. Third, there is a large number of industrial parks in central and western China that can use distributed energy in large quantities.

To encourage local consumption of new energy, the following steps should be taken. First, increase the rate of grid connection. Second, adopt market practices in auxiliary services, power trading and the spot market. Third, build microgrids and connect them to the regional grid. Fourth, replace fossil fuels with new energy and clean heating. Fifth, improve energy storage technology to increase the controllability of new energy generation. Sixth, provide policy support for the local consumption of new energy, including flexible pricing mechanisms, fair and open grid connections and strong support from local governments.

5.3 Build a Diverse and Resilient Low-carbon Energy Supply System

To transition to a low-carbon energy system, China should build new energy infrastructure, accelerate energy and digital innovation, and establish energy market mechanisms and policies. The country should enhance innovation and develop hydrogen, energy storage, smart grid and carbon capture, utilisation and storage technologies to lay a solid foundation for the low-carbon energy transition. For example, hydrogen, which is widely available, is a clean, low-carbon, efficient and flexible energy form that can help decarbonise transport and industry. Hydrogen can be transported over long distances, stored at scale and support two-way hydrogen–electricity conversion. Both distributed and centralised hydrogen systems should be developed to increase use and improve the resilience and flexibility of the energy system. Hydrogen is, therefore, an important component of the low-carbon energy system and should be strategically developed.

5.4 Advance Research and Innovation in Power Generation, Energy Storage, Transmission and Demand-side Management to Create an Energy System Based on New Energy

Renewable energy output depends on the weather. This poses a challenge to power grids in terms of reliability, stability and balance. Research and pilot work are required to create a new electric power system based predominantly on new energy and greater electricity consumption.

First, increase support of new energy generation to turn new energy from an auxiliary to the dominant power source. Research should be carried out on new energy generation and large-scale grid-connected distributed power sources, the co-ordinated control of new energy generation clusters and flexible AC and DC power transmission, and the connection and control of deep-sea and open-sea offshore wind power.

Second, improve the safe and stable operation of the power system to help the sustainable and rapid development of new energy. Research should be conducted on balancing supply with demand and stabilising the new electric power system. A simulation and analysis platform that features multiple timescales, a high level of precision and intelligence should be built. Efforts should be made to develop new DC transmission and new flexible power transmission equipment.

Third, integrate generation, grid, demand and storage, and support the development and efficient use of new energy through research on integration, multi-energy conversion and use, optimised control of demand-side resource clusters, and the flexible interaction and regulation of electric vehicles and the power grid.

Fourth, research energy storage technologies and improve the flexible regulation of an electric power system based on a high proportion of new energy. Key technologies for the integration of energy storage devices and systems in different applications should be developed and large-scale energy storage systems demonstrated. Research on energy storage batteries, pumped storage, distributed energy storage and the co-ordination of multiple power sources should be carried out.

5.5 Deepen Institutional and Mechanism Reforms, Improve Energy Legislation and Create a Market System Based on Effective Competition

China should accelerate the development of a national unified electricity market system. Trading rules and technical standards for the co-ordinated operation and integrated development of power markets should either be established or improved. An electricity spot market should be created and pilot spot markets tested over the long term. The cost of auxiliary services should be transferred to users and the flexibility of supply and demand facilitated through market mechanisms. New energy should be included in market trading. China should continuously deepen reforms to streamline administration, strengthen regulations and improve services. The approval system should be reformed and the granting of business licences and permits in the energy sector separated. Enterprise-related services should be optimised and barriers removed to create a healthy environment for the development of new industries and new business models such as local trading of distributed power, microgrids and integrated energy services.

5.6 Create a Carbon Account Mechanism for Businesses and Public Institutions to Promote Green Electricity Consumption

The carbon account mechanism is a data-based governance tool that defines rights, obligations and performance criteria on carbon emissions. As such, it should become an important enabler for the green and low-carbon transition and the dual-carbon goal. Green electricity consumption is one of the most important ways for businesses and public institutions to reduce carbon emissions. The wide uptake of the carbon account mechanism and the carbon market management system will effectively address the emissions of businesses and public institutions to form a nationwide "green electricity + carbon management system". China should introduce measures to help businesses and public institutions create and manage carbon accounts and improve the carbon management systems in towns, counties, cities and provinces.

Chapter 7: The Outlook for Bioenergy and Ocean Energy

Caifu Zhong and Marcello Espinoza

1 Introduction

China's bioenergy industry is growing. At the end of 2020, the total installed capacity of bioenergy was 29.52 gigawatts (GW). Power generation using agricultural and forestry residues and municipal waste is the main commercial use of bioenergy at present. Municipal waste-to-energy power generation is growing the fastest, with a total installed capacity of 15.22 GW, compared with 12.98 GW for power generated from agricultural and forestry residues.

Bioenergy will continue to grow steadily. Installed capacity is expected to exceed 45 GW in 2025 and about 56 GW in 2030. From 2030 onwards, constraints such as the closure of some agricultural and forestry residue power projects and a fall in the amount of municipal waste available, will force the bioenergy sector into a period of slower growth. The installed capacity of bioenergy is projected to increase to about 65 GW in 2050, and then plateau.

Policy support is needed to meet bioenergy targets. In the near and medium terms, a national electricity price subsidy that follows the principle of "expenditure determined by revenue" should be used to develop bioenergy projects; the bioenergy business model should be changed to gradually reduce its dependence on subsidies; and market trading of distributed bioenergy should be introduced at county level to improve the economy of distributed combined heat and power projects. In addition to subsidies, new bioenergy projects will need other types of national support such as higher waste disposal fees, green electricity trading or carbon pricing to meet the dual-carbon goal and support the rural revitalisation strategy.

China's geothermal and ocean power generation technologies are still far from industrial scale. Due to barriers such as their small size, immature technologies and high costs, geothermal and ocean power projects are mainly at the innovation or demonstration stage, and the related equipment manufacturing sector has yet to take shape. In the coming years, China should aim to make breakthroughs in the technological maturity and cost-effectiveness of geothermal and ocean power generation, identify future areas of focus, and promote the commercial-scale development of geothermal and ocean power through support policies and demonstration projects.

In recent years, green and low-carbon development has gradually taken root, with China increasing its demands for clean energy use and environmental protection. In this new setting, there is more demand for renewable power

C. F. Zhong (✉)
Energy Research Institute, China Academy of Macroeconomic Research, Beijing, China

M. Espinoza
Shell International B.V., The Hague, The Netherlands

Shell International B.V. and Development Research Centre of the State Council (DRC) of People's Republic of China, *Embracing the Future, Powering Growth: An Energy System Renewed for China*, Advances in Oil and Gas Exploration & Production,
https://doi.org/10.1007/978-3-031-73075-7_8

generation other than wind and solar, especially for bioenergy for industry. Agricultural and forestry bioenergy and combined heat and power will play an important role in air pollution control, urbanisation and rural revitalisation. Municipal waste-to-energy power generation provides clean electricity and the harmless treatment and reuse of waste. The social and environmental benefits are far greater than the economic gains. Geothermal and ocean power still face the barriers of small scale and high cost. If they can break through those barriers, they will have growth potential in the future.

2 Bioenergy and Ocean Power Today

2.1 Bioenergy

China's bioenergy market started in 2006. Within the following 10 years, power generation from agricultural and forestry residues and municipal waste matured to become the most important use of commercial biomass. The installed capacity of power plants using agricultural and forestry residues climbed rapidly during the 11th Five-year Plan (2006–10), but tailed off in the 12th Five-year Plan (2011–15) due to less raw material collection and less advantageous pricing. In the 13th Five-year Plan (2016–20), the installed capacity increased quickly, especially for municipal waste-to-energy, as shown in Fig. 1.

In 2020, new installed capacity of bioenergy in China was 5.43 GW, taking the total installed bioenergy capacity to 29.52 GW, up 22.6% year-on-year (see Fig. 2). Shandong, Guangdong, Jiangsu, Zhejiang and Anhui were the top five provinces by total installed capacity at 36.6 GW, 28.2 GW, 24.2 GW, 24.0 GW and 21.4 GW respectively. Shandong, Henan, Zhejiang, Jiangsu and Guangdong had the most new installed capacity at 0.68 GW, 0.65 GW, 0.42 GW, 0.39 GW and 0.36 GW respectively. In 2020, bioenergy output was 132.6 billion kWh, up 19.4% year-on-year. Power output from agricultural and forestry biomass was 51.0 billion kWh, an increase of 8.7% year-on-year; power output from municipal waste was 77.8 billion kWh, up 27.7% year-on-year; and power output from biogas was 3.8 billion kWh, a rise of 12.3% year-on-year. The top five provinces for bioenergy output were Guangdong (16.6 billion kWh), Shandong (15.9 billion kWh), Jiangsu (12.6 billion kWh), Zhejiang (11.1 billion kWh) and Anhui (11.1 billion kWh), which contributed 50.8% of the country's total bioenergy output.

In 2020, the weighted average availability of bioenergy generating units in China was 5,151 hours, which was 30 hours less than the previous year. The availability of bioenergy units fuelled by agricultural and forestry residues was 4,406 hours, a decrease of 223 hours; the availability of municipal waste-to-energy and biogas power plants were 5,854 hours and 4,524 hours respectively, up 140 hours and down 193 hours respectively. As shown in Fig. 3, five provinces recorded an average availability of bioenergy plants of more than 6,000 hours in 2020, namely Shanghai, Sichuan, Beijing, Fujian and Liaoning.

2.1.1 Bioenergy Generated from Agricultural and Forestry Residues

Since China's first large-scale straw-fired power generation project started to operate in 2006, agricultural and forestry bioenergy has grown rapidly and at scale, especially in the past 10 years thanks to government support. Annual new installed capacity of bioenergy generated from agricultural and forestry residues exceeded 1 GW for the first time in 2017 and 2 GW in 2020. By the end of 2020, the installed capacity of grid-connected agricultural and forestry bioenergy reached 12.98 GW, representing about 44% of the country's total installed bioenergy capacity.

Agricultural and forestry bioenergy projects are concentrated mainly in northern, north-eastern, central and eastern China, where residues are abundant. North China has the highest installed capacity at more than 3 GW, followed by central China with 2.5 GW. The four regions account for more than 80% of the country's total

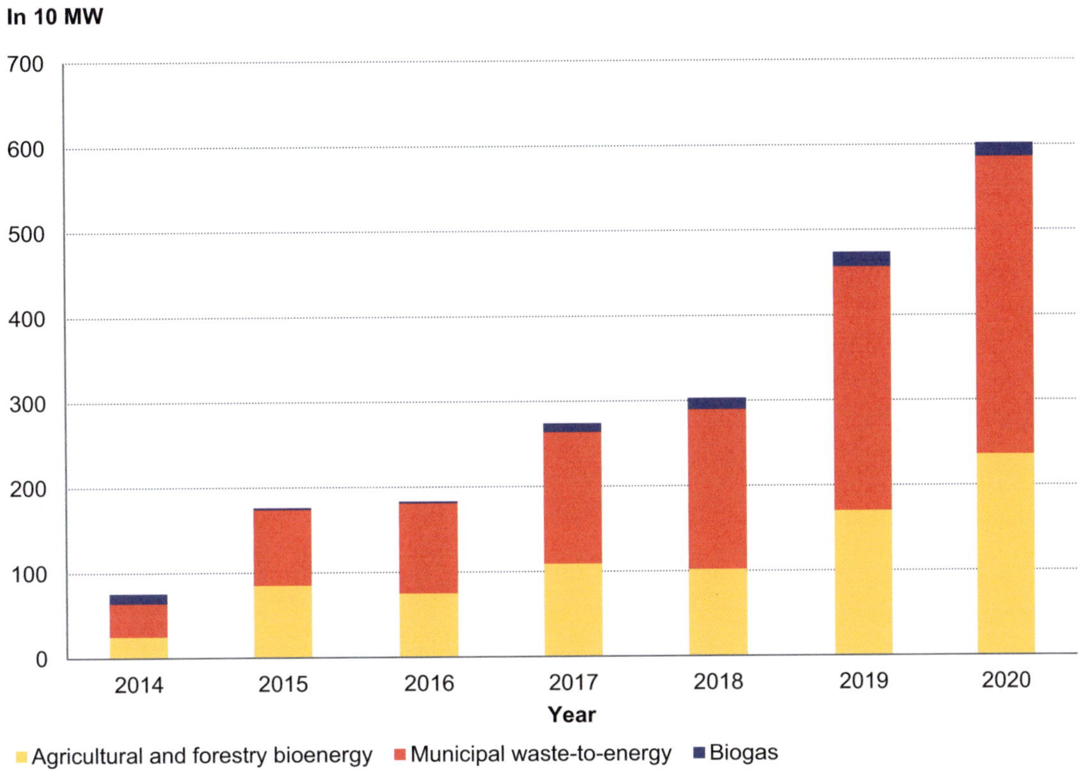

Fig. 1 New installed capacity of grid-connected bioenergy plants, 2014–20. *Source* National Energy Administration; Energy Research Institute of the National Development and Reform Commission

installed capacity of bioenergy generation. In south-west China, there are few bioenergy projects due to relatively poor agricultural residue resources, difficulties of collection and transport, and unfavourable conditions for the storage of raw materials such as a hot and humid climate. Similarly, there are few bioenergy projects in north-west China due to a shortage of straw.

As shown in Fig. 4, four provinces had a total installed capacity of agricultural and forestry bioenergy generation of more than 1 GW, namely Shandong with 1.86 GW, Guangxi with 1.72 GW, Anhui with 1.39 GW and Heilongjiang with 1.34 GW; five provinces had a total installed capacity of between 0.5 GW and 1 GW, of which Henan and Jiangsu had an installed capacity of almost 1 GW at 0.98 GW and 0.8 GW respectively. The top 10 provinces held 81% of China's total installed capacity of agricultural and forestry bioenergy. The provinces with a high installed capacity of agricultural and forestry bioenergy are the major grain producers where there are abundant resources of straw. These include the North-east China Plain, the North China Plain and the Yangtze River Delta plain.

In 2020, China's agricultural and forestry bioenergy generating units operated on average for 4,406 hours a year, a decrease of 223 hours on 2019. The decrease in availability in most provinces was due mainly to shortages of fuel, and in other provinces to the large number of new projects. Four provinces recorded average availability of more than 6,000 hours, namely Xinjiang with 7,910 hours, Sichuan with 7,203 hours, Tianjin with 7,162 hours and Fujian with 6,234 hours. These provinces had a relatively small installed capacity of agricultural and forestry bioenergy projects, but the local supply of agricultural and forestry residues was relatively stable. Most provinces that had a large installed capacity—other than Guangxi, Hunan

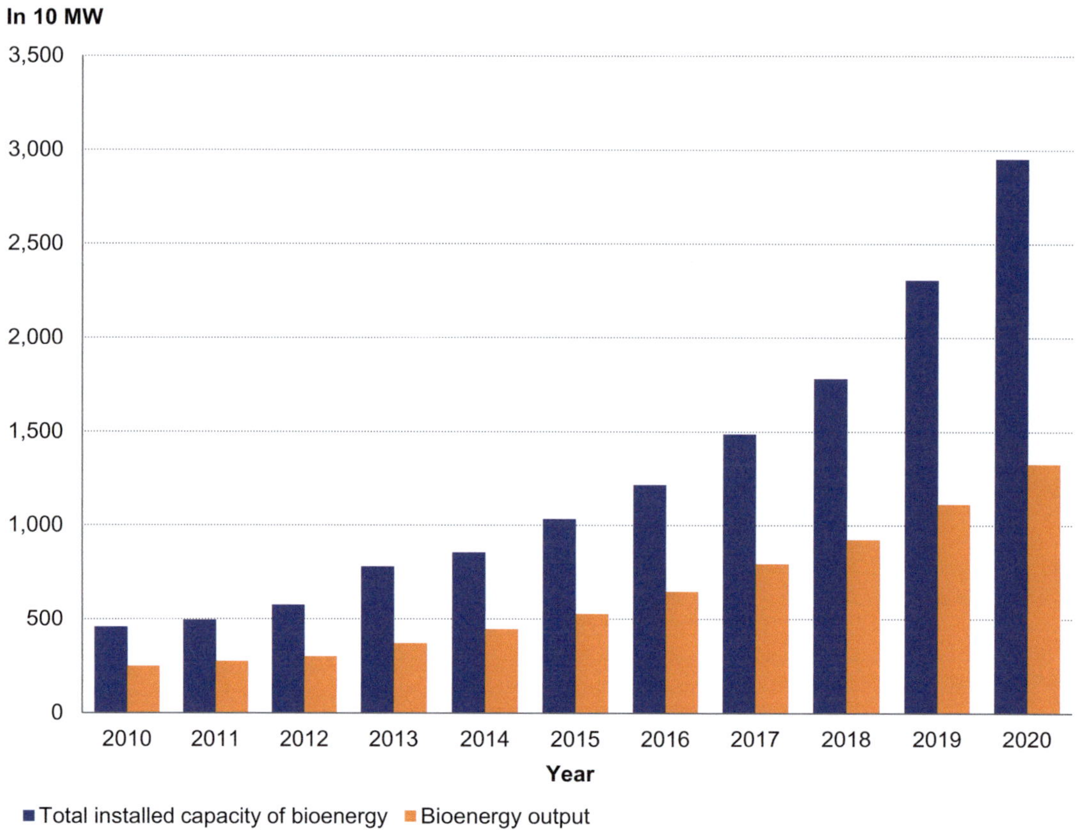

Fig. 2 China's total installed capacity and output of grid-connected bioenergy, 2011–20. *Source* National Energy Administration; Energy Research Institute of the National Development and Reform Commission

and Hubei which had availability of around 2,000 hours only—had availability of 4,500–5,500 hours.

2.1.2 Municipal Waste-to-Energy

In recent years, municipal waste-to-energy has become the main growth driver of bioenergy. From 2014 onwards, the additional installed capacity of municipal waste-to-energy has outgrown that of agricultural and forestry bioenergy. As municipal waste-to-energy does not incur peak regulating or rationing during shortages, along with the growing need to effectively manage municipal waste, bioenergy is an increasingly popular solution. In 2020, the new installed capacity of municipal waste-to-energy reached 3.08 GW. By the end of 2020, the total installed capacity of grid-connected municipal waste-to-energy in China was 15.22 GW, accounting for about 52% of the total installed capacity of grid-connected bioenergy.

Thirty provinces in China have installed capacity of municipal waste-to-energy. More than 80% of this capacity is in eastern China (see Fig. 5). East China alone accounts for nearly 50% of the total. The installed capacity of provincial municipal waste-to-energy corresponds to their economic development. Four provinces have an installed capacity of more than 1 GW, namely Guangdong with 2.24 GW, Zhejiang with 2.1 GW, Shandong with 1.71 GW and Jiangsu with 1.54 GW. As China's urbanisation increases, municipal waste-to-energy is expected to shift from large and medium-sized cities to smaller urban areas.

Thanks to increasing installed capacity, the power output from municipal waste was 77.8 billion kWh in 2020, an increase of 27.7%

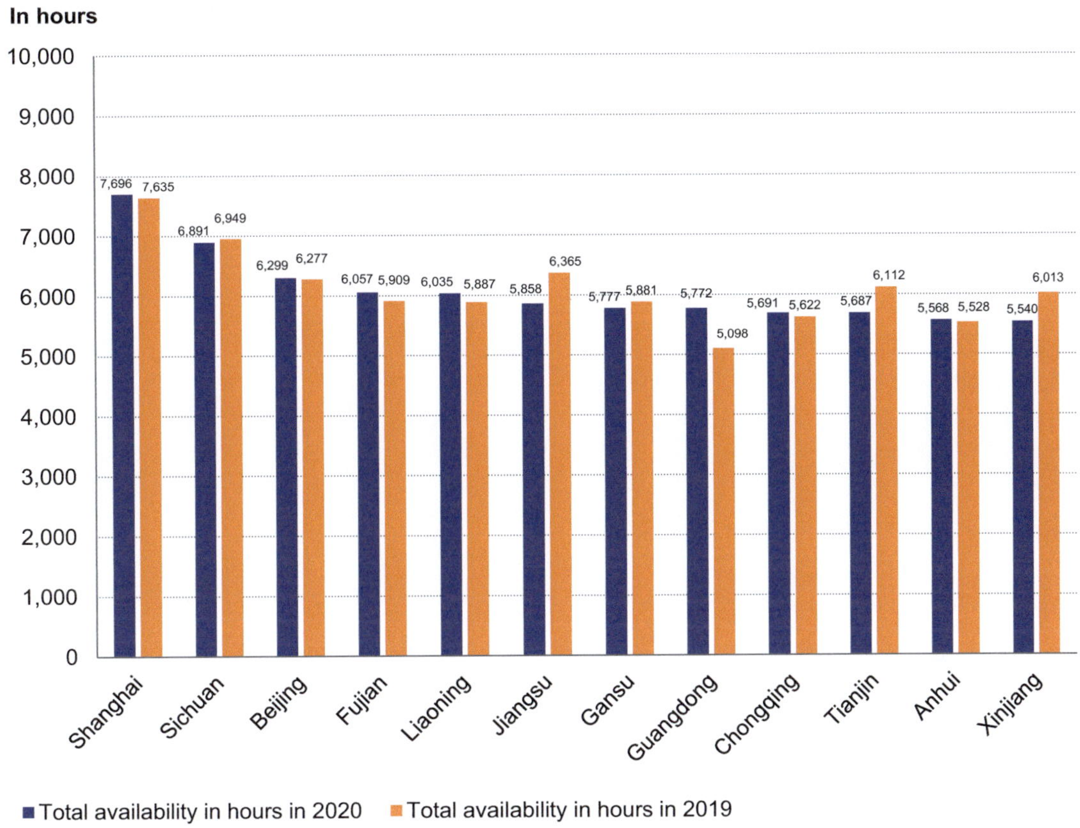

Fig. 3 Availability of bioenergy generating units in hours in major provinces in 2019 and 2020. *Source* National Energy Administration; Energy Research Institute of the National Development and Reform Commission

year-on-year. By converting municipal waste to electricity, about 140 million tonnes of municipal waste were treated in a harmless way, reducing the need for landfills. In 2020, China's average municipal waste-to-energy unit availability was 5,854 hours, an increase of 140 hours compared with 2019. Twelve provinces recorded average availability of more than 6,000 hours, headed by Shanghai with 7,761 hours, Hunan with 7,675 hours, Sichuan with 6,859 hours, Gansu with 6,851 hours and Liaoning with 6,745 hours.

2.1.3 Biogas

Biogas makes up a relatively small proportion of the biomass power fleet. The main purpose of biogas projects is to recycle organic waste, which means the demand for them is relatively stable. During the 13th Five-year Plan (2016–20), except for 2016, the new installed capacity of biogas was about 100 MW a year, indicating a relatively stable development. By 2020, the total installed capacity of biogas was 890 MW, accounting for 3% of the total installed capacity of bioenergy. Ninety per cent of biogas facilities are located in eastern regions. The five leading provinces by installed capacity are Guangdong with 159 MW, Shandong with 95 MW, Henan with 87 MW, Jiangsu with 80 MW and Zhejiang with 63 MW (see Fig. 6).

The availability of biogas units has grown in line with installed capacity. In 2020, biogas power output was 3.8 billion kWh, up 12.3% year-on-year, even though average generating unit availability was 4,524 hours, a decrease of 193 hours on 2019. Five provinces had average availability of more than 6,000 hours, including Shaanxi with 7,211 hours, Guangdong with

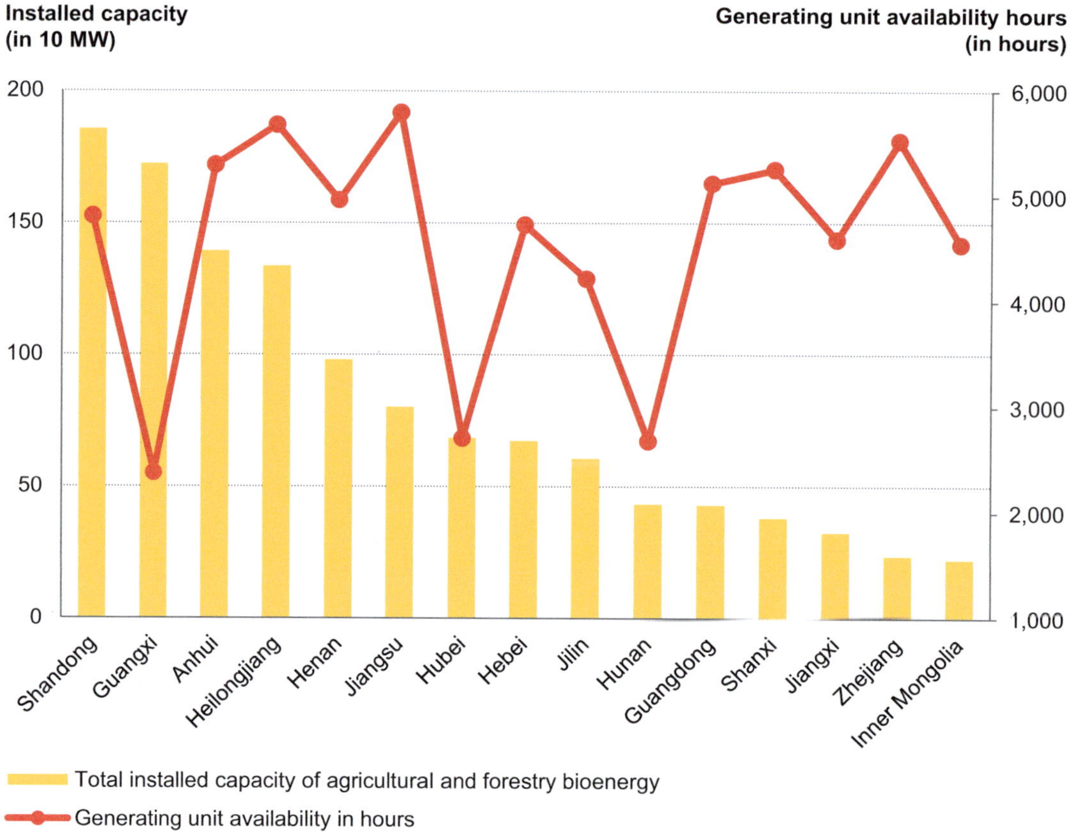

Fig. 4 Total installed capacity of agricultural and forestry bioenergy in 2020 and generating unit availability in hours in the main provinces in 2020. *Source* National Energy Administration; Energy Research Institute of the National Development and Reform Commission

6,363 hours, Shanghai with 6,296 hours, Liaoning with 6,180 hours and Sichuan with 6,117 hours (see Fig. 6).

2.1.4 The Cost of Bioenergy Technologies

The cost of bioenergy power generation depends on the fuel and technology used. Unlike solar and wind power, the cost efficiency of bioenergy is largely dependent on the availability of safe and sustainable biomass feedstocks and the conversion technology used. In addition, reserves of biomass differ between regions and countries, which also affects the cost of generating bioenergy. The weighted average cost of installed bioenergy in the world fluctuates, as shown in Fig. 7. In 2020, the cost was $2,543 per kilowatt, increasing from $2,173/kW in 2019, but flat when compared with the average over the past decade. In 2010–20, the global weighted average cost of bioenergy decreased from $0.076/kWh in 2010 to $0.066/kWh in 2019, rebounding to $0.076/kWh in 2020. The cost per kWh of bioenergy varied greatly in countries, with India low at $0.057/kWh and Europe and the USA high at $0.087/kWh and $0.097/kWh respectively.

1. **Agricultural and Forestry Bioenergy**

Agricultural and forestry bioenergy projects require large capital investment and have high operating costs. Without subsidies, these projects have a poor return on investment compared with conventional thermal power. The main reasons for this are the high cost of about RMB 9,000/kWh per unit of installed capacity and the

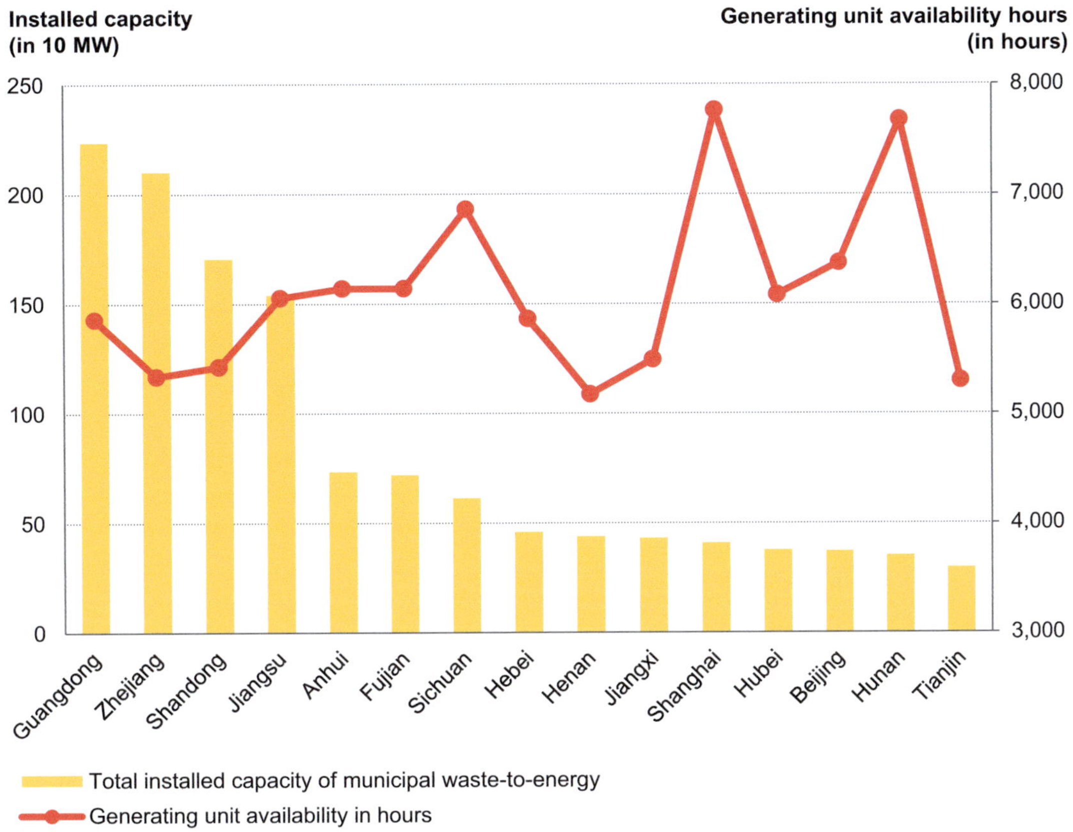

Fig. 5 Total installed capacity of municipal waste-to-energy in 2020 and generating unit availability in hours in the main provinces in 2020. *Source* National Energy Administration; Energy Research Institute of the National Development and Reform Commission

high cost of fuel of about RMB 0.4/kWh, which is much higher than those for coal power. The main costs of generating bioenergy are for installed capital, feedstock, labour and management, and operations and maintenance. The composition of installed capital cost is shown in Fig. 8.

Installed Capital Cost: In the future, bioenergy projects will be built in modules and the manufacturing technologies of biomass-fired boilers, steam turbines and other key equipment will be relatively mature, lowering the installed capital cost.

Feedstock Cost: Alongside the orderly development of the raw materials market and the improvement of feedstock collection, a gradual shift to the automated and mechanised collection of biomass and its use at scale will occur. The cost of feedstock is expected to rise during this modernisation process and then stabilise. The cost of agricultural and forestry residues when used as fuel comprises purchase, collection, transport and storage, which typically amounts to RMB 280–320 per tonne. The collection cost refers to the expenses incurred in purchasing lignocellulosic biomass from farmers and stacking or storing it temporarily, which is generally RMB 110 per tonne. Transport cost refers to the expense of transporting the raw materials, which depends on the freight transport rate, freight volume and distance from the point of origin, which is RMB 1 per tonne per kilometre on average. Storage cost refers to the maintenance, labour and other expenses incurred during storage, including fire prevention and electricity.

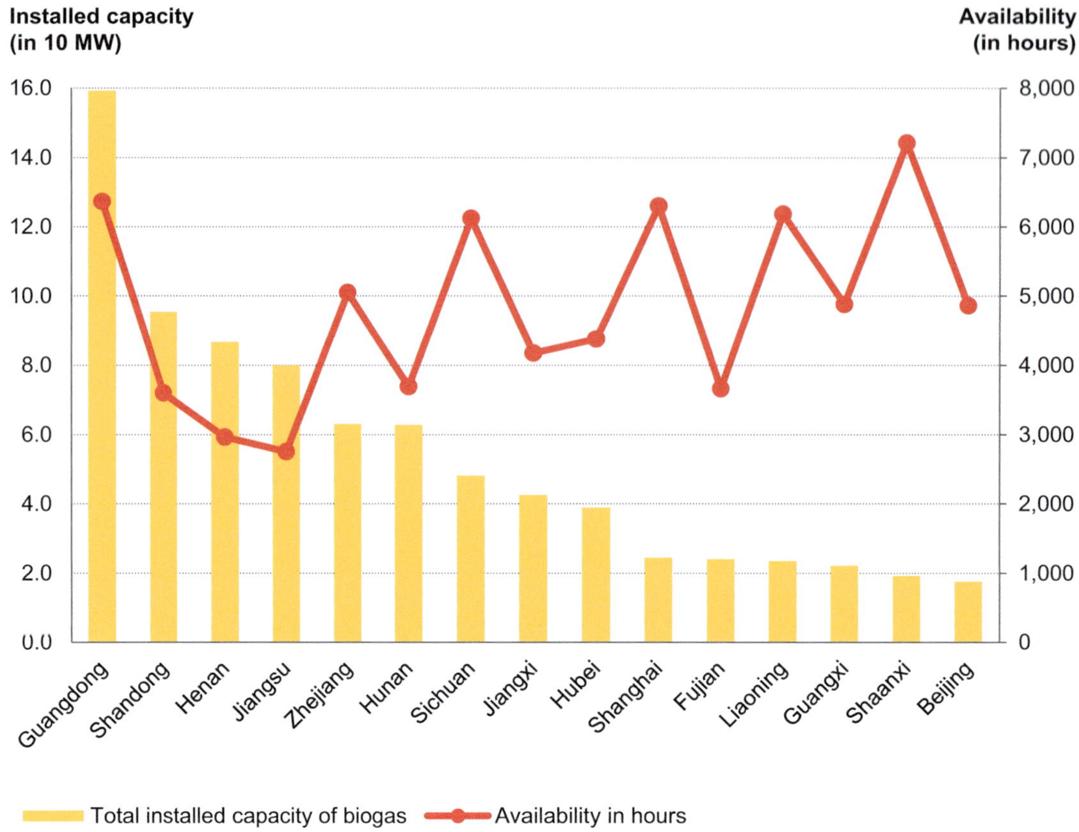

Fig. 6 Total installed capacity of biogas and generating unit availability in the main provinces in 2020. *Source* National Energy Administration; Energy Research Institute of the National Development and Reform Commission

Production technology has steadily improved. Modern manufacturing equipment and better management will increase overall system efficiency. By 2025, the thermal efficiency of biomass-fired boilers is expected to improve by 5%; steam turbine efficiency will increase by 3%; and the average efficiency of agricultural and forestry bioenergy will improve by 5%.

Management, operation and maintenance costs are on the rise. Despite improvements in automation and system integration, increased labour costs have become the main driver of substantially higher operation and maintenance costs.

2. **Municipal Waste-to-Energy**

The municipal waste-to-energy industry in China has high investment costs and stable revenues. Thanks to improvements in equipment manufacturing, construction and operations, China's advantages in this industry have become more tangible. The installed capital cost and construction time of a municipal waste-to-energy plant in China is only a third and half respectively that of a facility of the same size in developed countries. Figure 9 shows the installed capital cost composition of municipal waste-to-energy projects in China. These facilities use mainly circulating fluidised bed incinerators or grate incinerators. The unit capital cost of municipal waste-to-energy plants using circulating fluidised bed incinerators is RMB 250,000–400,000 per tonne per day, and that of facilities using grate incinerators is RMB 400,000–600,000 per tonne per day.

Power Generation Revenues: In China, the calorific value of municipal waste for

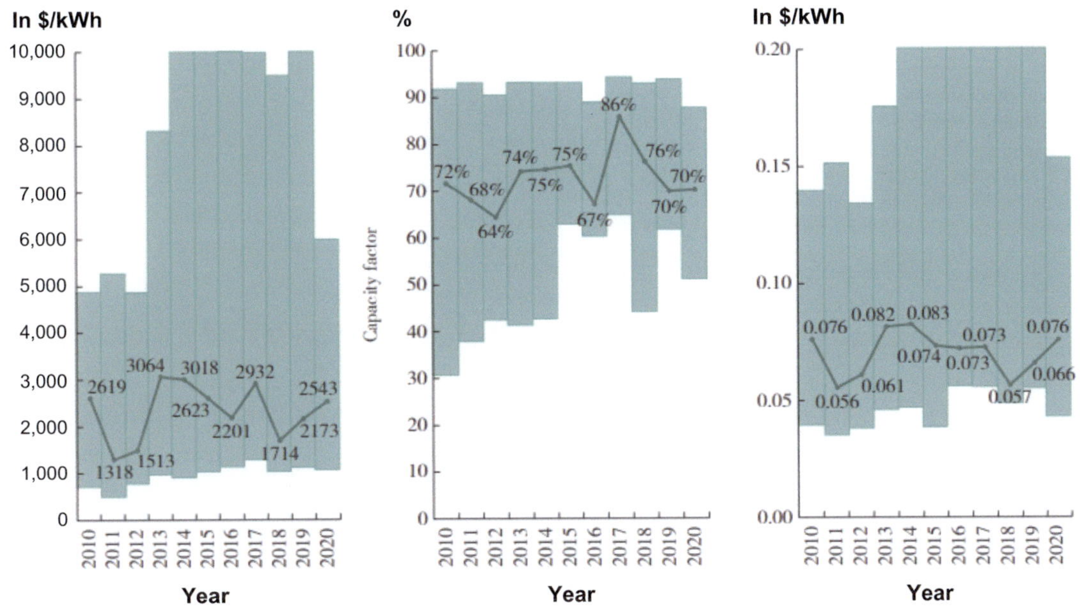

Fig. 7 The global weighted average cost of installed bioenergy capacity (left), capacity factor (middle) and levelised cost of electricity (right), 2010–20. *Source* International Renewable Energy Agency

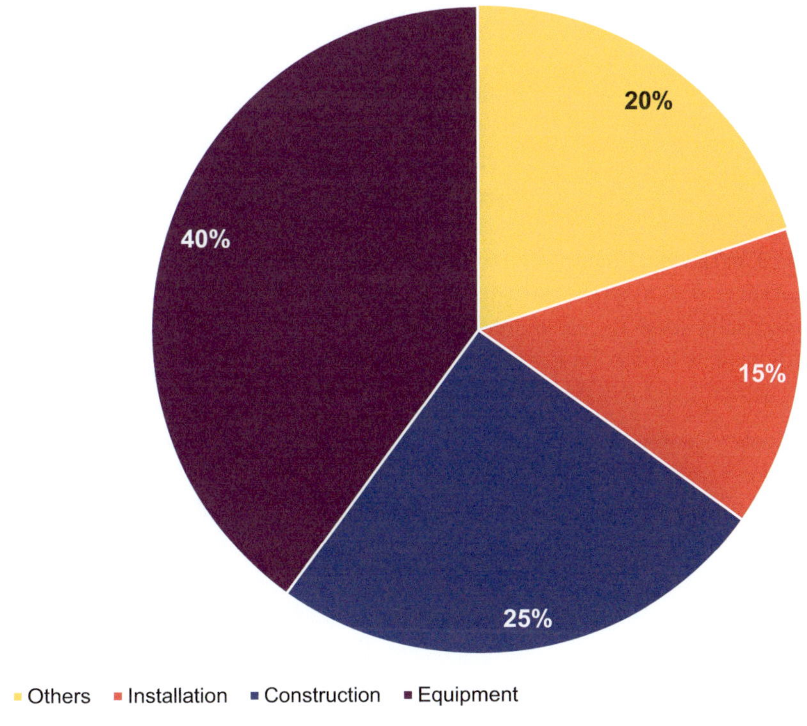

Fig. 8 Installed capital cost of agricultural and forestry bioenergy projects in China. *Source* Report of China's Biomass Power Industry Development 2019; China Renewable Energy Engineering Institute, 2020

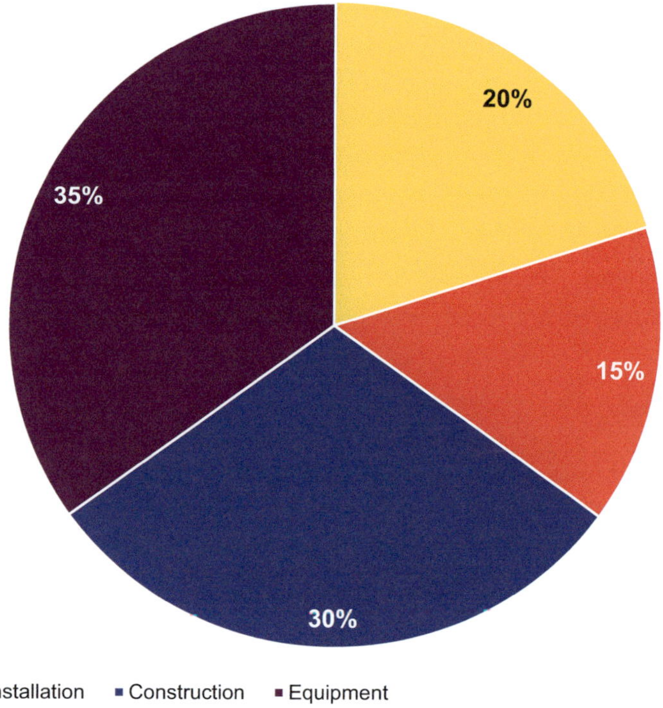

Fig. 9 Installed capital cost composition of municipal waste-to-energy projects in China. *Source* China Renewable Energy Engineering Institute, Report of China's Biomass Power Industry Development 2019

incineration is 1,500–1,800 kilo calories per kilogram (kcal/kg). Other parameters of a municipal waste-to-energy plant include a boiler thermal efficiency of 60%, steam turbine efficiency of 80% and plant electricity consumption of about 20%. Overall generation efficiency is less than 25% and the feed-in tariff is RMB 0.65/kWh. If a tonne of municipal waste can produce 350 kWh of electricity, excluding the electricity used by the plant, 280 kWh of electricity will be fed into the grid, generating revenues of about RMB 180.

Waste Incineration Costs: In recent years, China's municipal waste-to-energy plants have used locally developed and manufactured equipment. Improvements in the calorific value of municipal waste have driven up revenue expectations. For a plant with a capacity to incinerate 2,000 tonnes of waste per day, the capital and depreciation costs are about RMB 120 per tonne, and the cost of chemicals, maintenance, labour and management are around RMB 80 per tonne, amounting to about RMB 200 per tonne.

3. **Combined Heat and Power Plants**

As combined heat and power (CHP) plants use waste steam to generate heat, the overall investment cost in equipment is not significantly high and the amount of electricity needed to convert steam into heat is comparatively low. Many bioenergy plants have installed power generating units that provide such co-generation capabilities. However, due to insufficient demand for heating, the heating part of the plants is often not in operation, resulting in higher costs.

Under current electricity and heat pricing policies, 88% of the revenues of biomass-fired CHP projects come from power generation. Even if heat is produced at full capacity, power generation would still represent 82% of total revenues. If heat production is increased but the overall efficiency of the CHP plant does not improve significantly, the economics of the project worsen. The profitability of a plant contends with many factors: the price of electricity, taxes, bank loans, feedstock prices, power generation efficiency and local heating prices.

2.2 Geothermal Power

2.2.1 Policies for Geothermal Power Development

Geothermal energy attracts widespread attention, both nationally and locally.

National policies focus on basics, such as issuing guidance for the geothermal energy industry and resource evaluation, as shown in Table 1. In 2006, the Ministry of Finance issued its Interim Measures for the Management of Special Funds for Renewable Energy Development. The document clearly states that the management of funds for renewable energy development should be improved, with support given mainly to the development and use of bioethanol, biodiesel, solar energy, wind energy and geothermal energy. To promote China's development and use of geothermal energy, several government ministries jointly issued a policy document in 2013. This was followed a year later by a national conference to promote the wider uptake of geothermal energy across the country. To implement the policy, the National Energy Administration and the former Ministry of Land and Resources issued a notice requiring relevant local authorities to formulate local plans for geothermal energy development and use.

In 2017, the National Development and Reform Commission, the National Energy Administration and the former Ministry of Land and Resources jointly issued the 13th Five-year Plan for the Development and Use of Geothermal Energy. The document sets out the guidelines, objectives, key tasks and major investments needed for the development and use of geothermal energy, as well as the support measures required to implement the plan. As such, it was the reference point for geothermal energy development and use in China during the 13th Five-year Plan (2016–20). In 2021, eight central departments, including the National Development and Reform Commission, issued Several Opinions on Promoting the Development and Use of Geothermal Energy. This document makes a detailed description of the near- and medium-term development objectives, key tasks, management system and support measures needed for China's geothermal energy industry in the new era.

Local government across China also attach importance to the development of geothermal energy. Several have introduced support policies for the deployment of geothermal energy and ground-source heat pump technology, as shown in Table 2.

2.2.2 Geothermal Power Generation

In recent years, the installed capacity of geothermal energy in China has grown, albeit slowly. In the late 1970s, China started to use high-temperature geothermal resources for power generation. The Yangbajing geothermal power station in Tibet, built in 1977, has an installed capacity of 26.18 MW and has been in reliable operation for more than 40 years. At present, its annual power output is stable at about 100 million kWh, accounting for about 10% of the power in the central Tibet power grid. Over the years, it has generated more than 2.4 billion kWh of electricity. During the 13th Five-year Plan (2016–20), the new installed capacity of geothermal power in China was 18.08 MW, including the 16 MW Yangyi geothermal power station in Tibet and several small geothermal power units with a combined capacity of 2.08 MW. China has only completed 3.6% of its geothermal power development target.

Since the 1970s, China has made efforts to research and test low- and medium-temperature geothermal power generation. Currently, China has successfully deployed 300 kW units using waste heat for power generation in factories and has manufactured 1,500 kW units, but this technology has not yet seen significant market and application expansion. In addition, China's national high-tech 863 development programme aims to achieve breakthroughs in low- and medium-temperature geothermal power generation.

The development and use potential of enhanced hot dry rock geothermal energy is huge. China has begun to pay much attention to this energy source, with government departments, research institutions and some businesses exploring its development. China Geological

Table 1 National policies for the development of geothermal energy

Year	Department(s)	Related documents	Main requirements
2006	Ministry of Finance	Interim Measures for the Management of Special Funds for Renewable Energy Development	In Article 7 under "Key objects of support" in Chapter 2, it says: "For the development and use of renewable energy for heating and cooling buildings, support should be given to encourage the wider uptake of solar and geothermal energy in buildings under construction"
2013	National Energy Administration, Ministry of Finance, the former Ministry of Land and Resources and the Ministry of Housing and Urban–Rural Development	Guiding Opinions of the National Energy Administration, the Ministry of Finance, the Ministry of Land and Resources, and the Ministry of Housing and Urban–Rural Development on Promoting the Development and Use of Geothermal Energy	This document states that geothermal energy is a clean and eco-friendly type of renewable energy that is widely distributed and has large reserves and huge market potential. The development and use of geothermal energy is of practical and strategic significance now and in the long run to ease the pressure on China's energy resources, achieve the targets for non-fossil energy use, advance the revolution in energy production and consumption, and enable the formation of an ecological civilisation
2014	National Energy Administration and the former Ministry of Land and Resources	Notice of the General Administration of the National Energy Administration and the General Office of the Ministry of Land and Resources on Formulating Local Plans for Geothermal Energy Development and Use	Relevant local authorities are required to formulate local plans for geothermal energy development and use as soon as possible
2017	National Development and Reform Commission, IEA, and the former Ministry of Land and Resources	13th Five-year Plan for the Development and Use of Geothermal Energy	Guidelines and objectives, key tasks and major investments needed for the development and use of geothermal energy and the support measures required to implement the plan
2021	Eight central departments including the National Development and Reform Commission and the IEA	Several Opinions on Promoting the Development and Use of Geothermal Energy	The targets for geothermal energy development are clearly defined in this document. By 2025, all provinces should have established a sound and standard management process for geothermal energy development and use; the national geothermal energy development and use data and monitoring system should be running perfectly; geothermal energy used for heating or cooling should have increased by 50% compared with 2020; and the country's total installed capacity of geothermal power generation should have doubled relative to 2020. By 2035, the geothermal energy used for heating or cooling and the installed capacity of geothermal power should be as close as possible to double those of 2025. Key tasks, standard management processes for geothermal energy development and use, and support measures are included in the document

Source Based on publicly available information

Table 2 Local policies for the development of geothermal energy

Local government	Year	Policies
Beijing	2013	Implementation Opinions on Further Promoting Geothermal Energy Development and Heat Pump System Use in Beijing
Shenyang	2006	Development Plan of Shenyang for Promoting Ground-source Heat Pump Technology
	2007	Administrative Measures of Shenyang for Ground-source Heat Pump System Construction and Deployment
Wuhan	2006	Special Plan for Heat Pump Technology Promotion and Deployment
Ningbo	2004	Interim Measures of Ningbo for the Management of Special Funds for Energy-efficient and Cleaner Energy Production
Chongqing	2007	Opinions of Chongqing Municipal People's Government on Strengthening Geothermal Resource Management; Interim Measures of Chongqing for the Management of Special Subsidy Funds for Renewable Energy Building Application Demonstration Projects
Hebei	2006	Administrative Regulations of Hebei on Geothermal Resources
	2013	Opinions on Accelerating the Development and Use of Shallow Geothermal Energy
Henan	2013	Technical Regulations for Sewage-source Heat Pump System Applications; Technical Regulations of Henan for Testing and Acceptance of Ground-source Heat Pump Building Applications; Technical Regulations of Henan for the Testing and Acceptance of Solar Water Heating Building Applications
Tianjin	2006	Interim Provisions of Tianjin on the Management of Ground-source Heat Pump Systems
	2013	Notice of the Tianjin Land Resources and Housing Administration on Further Strengthening Geological Monitoring and Management of Shallow Geothermal Energy
	2014	Technical Requirements of Tianjin for Single- and Double-well Geothermal Resource Evaluation
Fujian	2014	Implementation Plan of Xiamen for the Green Building Initiative
Shandong	2015	Notice on Strengthening the Management of Renewable Energy Building Applications; Notice on Regulating the Management of Ground-source Heat Pump Systems for Building Applications; Notice on Regulating the Management of Ground-source Heat Pump Systems for Building Applications
Anhui	2015	Implementation Plan for Promoting the Large-scale Deployment of Shallow Geothermal Energy in Buildings

Source Based on publicly available information

Survey has prepared a National Implementation Plan for Hot Dry Rock Exploration and Development, which proposes that China should assess its hot dry rock resources and potential across the country, identify priority target areas for development, build hot dry rock exploration bases and develop technologies to facilitate exploration and commercialisation. Shandong, Qinghai, Guangdong and other provinces have carried out hot dry rock surveys to determine the lithologic structure and thermal characteristics of hot dry rock in selected areas, identify the extent and estimate the reserves of hot dry rock, and prioritise zoning for the development and use of hot dry rock resources.

2.3 Ocean Energy

Ocean energy comprises tidal energy, wave energy, tidal current energy, ocean thermal energy conversion, and salinity gradient energy. China has invested heavily in the research and development of ocean energy technologies and has achieved breakthroughs in their use. Tidal energy technology is relatively mature but ocean energy use projects are still small in scale and far from forming an industry. Related technologies are still immature, and demonstration projects are needed to create an industry value chain.

In 2016, the State Oceanic Administration issued its 13th Five-year Plan for the Development of Ocean Renewable Energy. The document states that by 2020 ocean energy development and use should be significantly improved and technological innovation capacity greatly enhanced; core technologies and equipment should be developed and commercialised; an industry supply chain and standards system should be established; national ocean energy testing sites should be set up and megawatt-level tidal energy grid connection demonstration bases and 500 kilowatt-level wave energy demonstration bases built; 10 megawatt-level tidal energy demonstration projects should be in operation and China's total installed capacity of ocean energy should exceed 50 MW; more than five multi-energy independent electric power systems integrating island ocean energy and renewable energy such as wind and solar power should be formed; and the production and use of various ocean energy types should be internationally advanced.

2.3.1 Tidal Energy

China is a global leader in tidal barrage energy. Two facilities are using this technology: the Jiangxia and Haishan tidal power stations. In 2012, supported by funds for ocean energy, China Longyuan Power Group, PowerChina Huadong Engineering, Tsinghua University and other organisations started an expansion project at Unit 1 of the Jiangxia tidal power station, increasing the single turbine capacity from 500 to 700 kW. The unit was successfully started and connected to the grid for trial operation in 2015.

In 2011–15, pre-feasibility studies for tidal power stations at Jiantiao Port, Rushankou, Baqimen and Maluan Bay were carried out. Efforts were also made to develop a new high-efficiency low-head large-flow bidirectional tubular turbine, the use of phase difference between tidal waves in and out of the bay, dynamic tidal energy technology and other new tidal power technologies.

In 2015, tidal phase difference power generation passed the acceptance tests of the State Oceanic Administration at Sansha Bay, Fujian province. The site exploration and pre-feasibility study of Bachimen tidal power station at Shacheng Port, Fujian province, undertaken by the Ocean University of China, passed the national review. By the end of the 12th Five-year Plan (2011–15), the pre-feasibility study, demonstration and marine environmental impact assessment for the Oufei tidal power station had been completed. With a total installed capacity of 451 MW, the power station comprises 41 units, each with a capacity of 11 MW and a rated head of 3.5 meters (rated head is the difference in height between low and high tides).

2.3.2 Wave Energy

In wave power generation, China has carried out R&D tests of more than 15 small power units (below 100 kW). China's 10 kW Sharp Eagle floating oscillating surge wave energy converter (WEC) completed sea tests lasting more than a year. Based on the results of these tests, the 100 kW Sharp Eagle Wanshan WEC was developed and tested in 2015. As of early 2017, it had generated 30,000 kWh of power. Multi-point absorber WEC is still under development after sea trials. The wave energy demonstration project near Daguan Island, Qingdao, Shandong province has provided residents on the island with power since 2011. As of 2017, the South China Sea Island Ocean Energy-based Independent Electric Power System Demonstration Project, developed by Guangzhou Institute of Energy Conversion and based on the Sharp Eagle WEC technology, had generated more than 2.08 million kWh of power over 36 months.

2.3.3 Tidal Current (Ocean Current) Energy

China moved early in tidal current power generation. With strong support from the 863 Programme and the National Science and Technology Support Initiative, more than 10 tidal current energy experimental units have been developed. As tidal current energy technologies advanced, some large companies also entered the industry. China's main tidal current power generation technology has reached the testing stage, the challenges facing the main technologies have been resolved and key turbine components have been developed and are manufactured locally.

Since the semi-direct-drive horizontal-axis turbine developed by Zhejiang University was tested at sea in 2014, it has generated more than 20,000 kWh of electricity. Following this success, the university designed and made two 300 kW tidal current turbine prototypes. In 2016, the Zhejiang Lindong modular tidal current turbine was installed offshore, feeding almost 10,000 kWh of commercially generated electricity into the grid. The Haineng series of 150 kW, 300 kW and 600 kW tidal current turbines, developed by Harbin Engineering University, have been undergoing long-term sea tests since 2012. Other tidal current energy demonstration projects are also under construction. In June 2015, a 20 kW horizontal-axis tidal current turbine was tested for 10 months in Zhaitang Island, Qingdao. In September 2015, a vertical-axis direct-drive tidal current turbine prototype with a capacity of 15 kW was put into trial operation in Changhai County, achieving an efficiency rate of 27%. And in December 2015 on Zhaitang Island, a tidal current multi-energy power system passed the acceptance tests of the State Oceanic Administration. This demonstration project comprised three 50 kW wind turbines, one 50 kW solar power generating unit and two tidal current turbines.

3 Resources and Their Development Potential

3.1 Biomass Resources

There are many different types of biomass. They include straw and agricultural residues, forestry cuttings, wood processing residues, livestock and poultry waste, municipal waste and domestic sewage, industrial organic waste and high-concentration organic waste water. Based on current bioenergy technologies, the total available biomass resources in China are about 460 million tonnes of coal equivalent (tce) annually.

The production of agricultural residues and the availability of biomass for energy generation are not expected to change significantly in the future. China's forest resources will continue to expand as more natural forests are protected with felling bans. The collection of commercial forestry residues is inefficient, even though demand for residues for industrial use is increasing. Demand is expected to outpace supply for some time to come, and the resources available for energy production will be limited. As living standards rise and the consumption of meat, eggs and milk increases, animal husbandry will continue to expand. At the same time, breeding costs, management technology and the requirements and standards of export markets will drive a gradual shift from small-scale to large-scale production. As a result, livestock and poultry manure will increase and the resources available for energy production will grow.

3.1.1 Agricultural Residues

According to the China Statistical Yearbook for 2017, the stock of agricultural residue resources across the country was about 820 million tonnes, mainly distributed across 13 grain-producing provinces including the North China Plain, the Middle–Lower Yangtze plains and the North-east Plain. Straw resources for fertiliser, feed and paper production amounted to about 400 million tonnes a year, and the agricultural residues available for energy production were around 420 million tonnes per year (see Fig. 10).

3.1.2 Forestry Residues

Forest cover in China amounts to around 326 million hectares. Natural forest and planted forest cover 140 million hectares and 80 million hectares respectively. The forest growing stock is 17.56 billion cubic metres, with living trees amounting to 19 billion cubic metres.

The resources available for energy production include fuelwood forest, forest harvesting residues and wood processing residues. Forestry

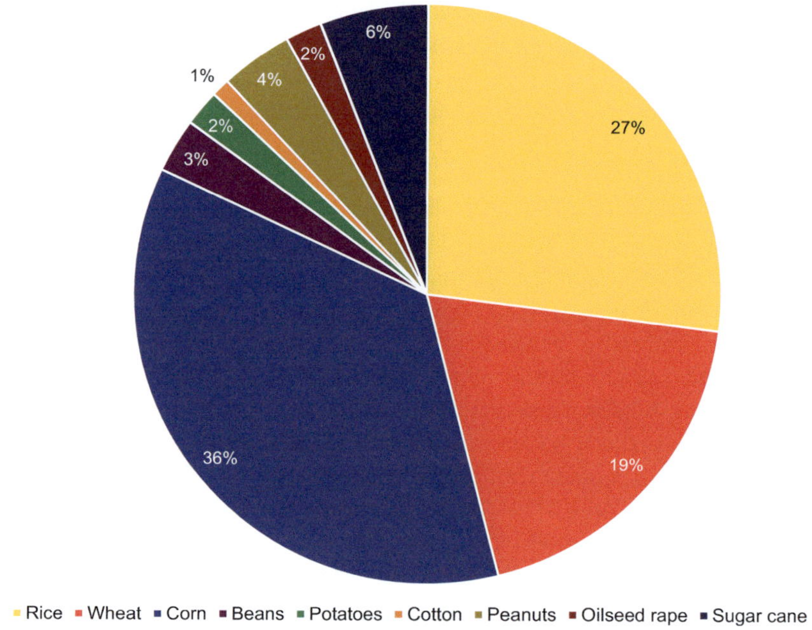

Fig. 10 Agricultural residues in China in 2016. *Source* National Bureau of Statistics

resources that could be used to generate energy amount to about 350 million tonnes annually, of which a considerable part is already used as cooking fuel by farmers or as industrial raw materials in composite wood manufacturing. If all forestry residues were used for energy, 200 million tonnes of coal equivalent fossil fuels could be replaced.

3.1.3 Municipal Waste

China's municipal waste is increasing year by year. In 2020, it was about 235 million tonnes, an increase of almost 50% over the past decade. Behind this increase lies the growing urbanisation of the population. Landfills were the traditional way to dispose of municipal waste. In 2011, 100 million tonnes of municipal waste were deposited in landfills, 80% of the total. This figure gradually climbed to 120 million tonnes in 2017. In recent years, the volume of municipal waste incinerated has increased rapidly. In 2020, the volume of municipal waste disposed of in landfills dropped to 78 million tonnes, while that incinerated rose to 146 million tonnes. The percentage of municipal waste incinerated increased from about 20% in 2011 to around 60% in 2020 (see Fig. 11).

The average daily municipal waste produced per person averages 0.66 kg per day, although the amount varies in cities and rural areas. Based on the existing electricity price subsidies and waste disposal fees, municipal waste-to-energy projects are viable only in counties or cities that collect more than 300 tonnes of waste per day.

As China's urbanisation progresses, municipal waste-to-energy projects will gradually shift from large and medium-sized cities to smaller towns and counties. According to the World Bank, China's effective use of waste in 2050 is expected to match that of the members of the Organisation for Economic Co-operation and Development (OECD) in 2025. Between 2020 and 2030, China's urbanisation rate will change from rapid to steady, and the amount of municipal waste collected will increase at an average annual rate of 3–4%, reaching about 350 million tonnes in 2030. The amount of municipal waste used for power generation is projected to be about 310 million tonnes, which is nearly 90% of the total volume collected. From 2030 to 2050, the volume of municipal waste collected is expected to grow slowly at about 2%, reaching 480 million tonnes in 2050. About 450 million

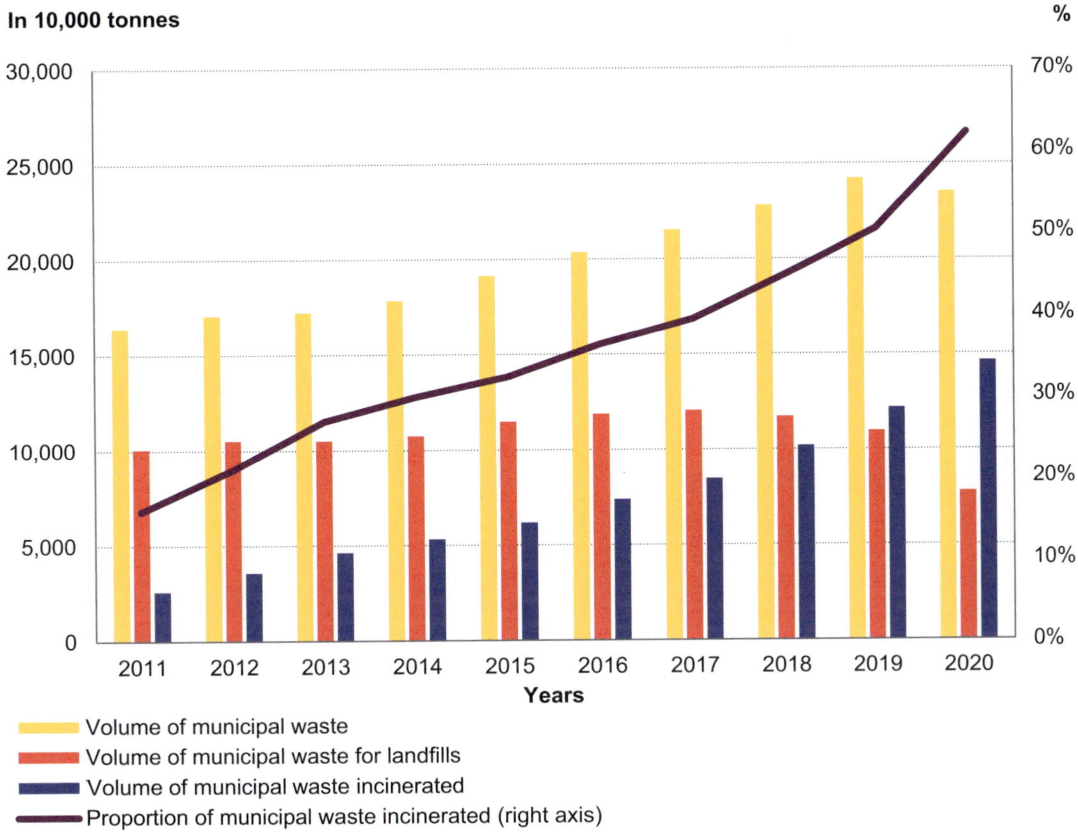

Fig. 11 Volume of municipal waste and methods of disposal in China, 2011–20. *Source* National Bureau of Statistics

tonnes of waste will be used for power generation, accounting for 94% of the total volume collected. Between 2050 and 2060, the volume of municipal waste collected and the resources available for power generation are expected to be stable.

3.1.4 Livestock and Poultry Manure

Livestock and poultry manure comes mainly from cattle, pigs and chickens raised in captivity. In 2016, China reported livestock and poultry manure of 1.34 billion tonnes. The manure from large-scale livestock and poultry farms was about 1 billion tonnes annually, which means biogas potential of about 50 billion cubic metres (about 35 million tonnes of coal equivalent), assuming 1 tonne of livestock and poultry manure can produce 50 cubic metres of biogas on average.

3.2 Geothermal Resources

China holds huge geothermal resource potential. According to the data released by the former Ministry of Land and Resources in 2011, the geothermal resources in the country's major basins were 853 billion tonnes of coal equivalent (tce), including exploitable resources of 640 million tce. The shallow geothermal resources of cities at or above the prefecture level were 9.5 billion tce, including exploitable reserves of 350 million tce.

China's conventional geothermal resources are mainly at medium and low temperatures, between 200 and 4,000 meters underground. There are two types of geothermal heat transfer: conductive and convective. Conductive geothermal resources are concentrated mainly in the central–eastern sedimentary basins, where the geothermal resources are widely distributed and the temperature increases with depth. Convective geothermal resources are found mainly in Yunnan, Sichuan, Guangdong, Fujian, Shandong and the Liaodong Peninsula, where the resources are distributed zonally along the main geological faults. China's high-temperature geothermal resources are very limited, and are only found in Tibet, Tengchong (Yunnan) and northern Taiwan, which accounts for the limited and slow development of geothermal power generation in the country.

3.3 Ocean Energy Resources

In 1958, China carried out its first national comprehensive marine survey. In 1985, the country completed a second national survey, this time of coastal tidal energy resources. In 1989, the zoning of ocean energy resources in China's coastal rural areas was completed. However, no formal survey of the salinity gradient energy resources in coastal areas or of the wave energy or ocean thermal energy conversion resources has been conducted. Some scientists have made estimates in their studies.

Total ocean energy resources are related to many complex factors, such as the characteristics of that part of the sea under investigation, the energy calculation method used, the available data and the type of energy generation to be tested. For a long time, there have been no accurate and scientific data about ocean energy reserves in China.

4 Issues and Challenges

4.1 Bioenergy

4.1.1 Subsidies

Table 3 shows the benchmark prices of bioenergy from 2006 to 2020. The disparity between commercial revenues and national renewable energy subsidies places increasing pressure on the ability of bioenergy producers to make a profit.

For agricultural and forestry bioenergy companies, the disparity has an impact on cash flow. On the one hand, power generation is more or less the companies' sole source of revenue, along with the high subsidies paid per kilowatt-hour of power generated. If the national average subsidy was RMB 0.393/kWh, this accounted for more than 50% of the companies' total income (even in Guangdong, where the subsidies were the lowest, subsidies accounted for more than 40%

Table 3 Biomass power benchmark prices, 2006–20

Type of energy	Period	Electricity price (RMB/kWh)	Remarks
Power generated from agricultural and forestry residues	January 2006 to June 2007	0.50–0.689	Varied by province + RMB 0.25/kWh from the coal power benchmark price in 2005
	July 2007 to June 2010	0.60–0.789	Varied by province + RMB 0.35/kWh from the coal power benchmark price in 2005
	July 2010 to the end of 2020	0.75	Unified electricity price across the country
Biogas and biomass gasification	January 2006 to the end of 2020	0.50–0.689	Varied by province + RMB 0.25/kWh from the coal power benchmark price in 2005
Municipal waste-to-energy	January 2006 to March 2012	0.50–0.689	Varied by province + RMB 0.25/kWh from the coal power benchmark price in 2005
	April 2012 to the end of 2020	0.65	Unified electricity price across the country, with provinces required to share an electricity price subsidy of RMB 0.1/kWh

Source Electricity pricing data from the National Development Reform Commission of income). On the other hand, the agricultural and forestry residues used to generate bioenergy were purchased from farmers, which required settlement within a short time. The payment of subsidies did not always meet, therefore, the cash flow needs of the bioenergy companies.

Municipal waste-to-energy companies still performed well, despite the effects of delayed subsidy payments. On the one hand, the waste disposal fee accounted for a relatively large proportion of their total income. Assuming a national average waste disposal fee of RMB 69 per tonne, the income from waste disposal made up about 30% of the companies' income. On the other hand, the subsidies were relatively low, especially the national subsidy which was about RMB 0.19/kWh on average and which made up only about 30% of income. The impact of delayed subsidy payments on business performance was much smaller than that on agricultural and forestry bioenergy projects.

4.1.2 New Projects Had Insufficient Cost Effectiveness After Subsidies Were Phased Out

In the near term, the efficiency of bioenergy generation is expected to improve through technological advances. However, the cost of biomass is expected to increase, which will counteract the cost reduction enabled by technology. By 2025, the cost reduction scope of bioenergy will be small. From 2030 onwards, the cost of generating bioenergy is projected to rise.

1. **Agricultural and Forestry Bioenergy**

Agricultural and forestry bioenergy plants usually have a production capacity of 25–30 MW. They use agricultural and forestry residues as their main feedstock, including corn straw, rice straw, oil-bearing crop straw, cotton straw, rice husks and tree cuttings. Biomass fuels can be blended. Ideally, an agricultural and forestry bioenergy plant runs 7,500–8,000 hours per year and generates 180–220 million kWh of electricity from 280,000 to 300,000 tonnes of agricultural and forestry residues. However, most agricultural and forestry bioenergy plants in China operate for less than 5,000 hours per year.

Estimates for the levelised cost of electricity (LCOE) of agricultural and forestry bioenergy over the coming decades are shown in Table 4. In the short term, the LCOE is forecast to remain stable at around RMB 0.6/kWh including tax, of which the cost of fuel accounts for more than 60%. Thanks to improvements in technology and management, agricultural and forestry bioenergy plants are expected to achieve higher efficiency and lower fuel consumption per kWh of electricity generated. However, as a result of rising labour costs and higher collection and storage costs per unit of fuel mass, the fuel delivery price will gradually increase for the plants. The two components counteract each other, which means the power generation cost should remain stable.

In the long term, the LCOE of agricultural and forestry bioenergy will be under pressure. The

Table 4 LCOE forecasts for agricultural and forestry bioenergy

Year	Installed capital cost (RMB/kW)	Delivery price of agricultural residues to plants (RMB/tonne)	Operating cost (RMB/kW per year)	Generation efficiency (%)	LCOE (RMB/kWh)
2020	8,700	300	435	31	0.59
2030	8,500	350	460	33	0.62
2050	8,300	400	500	36	0.64
2060	8,200	420	520	37	0.65

Note The estimates are based on the following assumptions (per year): a plant operating time of 5,500 hours, plant power consumption of 10%, calorific value of straw of 3,500 kcal/kg, plant operating lifetime of 20 years, own capital of 20%, loan interest rate of 4.9% and internal rate of return on capital of 8%

Source Calculated by the authors of this chapter

additional expenses, driven by rising labour costs and more stringent emission standards, will outweigh the cost reduction achieved by efficiency improvements and other technological advances. After 2050, when the straw delivery price reaches RMB 400 per tonne, the LCOE is expected to increase to RMB 0.65/kWh even if generation efficiency improves. This price level remains more or less flat, even with feed-in tariffs, leaving no room for profitability. If the tariffs are phased out, the projects that generate power only will be significantly in the red, making it difficult to achieve normal operation.

2. **Municipal Waste-to-Energy**

The income for waste-to-energy plants comes mainly from power generation and waste disposal fees. Before 2021, the income for delivering power to the grid from waste-to-energy plants was set nationally, while the waste disposal fees varied greatly across the country depending on the size of the project, competitive bidding and other factors. In general, operating costs will increase in the coming years, which will raise the cost of waste-to-energy, but at a slow rate, as shown in Table 5.

Waste disposal fees have the biggest impact on the profitability of municipal waste-to-energy plants. Assuming a plant operating lifetime of 15 years and no waste disposal fee, the LCOE of waste-to-energy is close to RMB 0.8/kWh, which is unprofitable when the feed-in-tariff is RMB 0.65/kWh. When the waste disposal fee is RMB 50 per tonne, the estimated LCOE is RMB 0.57/kWh; if the feed-in-tariff in this example is RMB 0.65/kWh, plant profitability depends on the size of the national subsidy. If we use the average waste disposal fee of bid winners for power generation projects in 2019—RMB 69 per tonne—the LCOE for waste-to-energy is RMB 0.44/kWh. In this case, as long as the national or local subsidies are paid, most projects can make a profit. Considering that waste disposal fees will gradually increase in the future, especially in economically developed areas and large cities, a waste disposal fee of RMB 100 per tonne and a feed-in tariff of RMB 0.37–0.41/kWh would enable plants in most parts of the country to achieve a return on investment after tax of 8% without subsidies.

In the medium and long terms, waste disposal fees are expected to increase steadily across the country as the economy grows. Assuming that the national average waste disposal fee and waste-to-energy LCOE are RMB 100 per tonne and RMB 0.39/kWh in 2030, RMB 110 per tonne and RMB 0.37/kWh in 2050, and RMB 120 per tonne and RMB 0.35/kWh in 2060, municipal waste-to-energy plants can be profitable.

If the plant operating lifetime is extended, the plant's income will improve. Assuming a plant operating lifetime of 20 years and a waste disposal fee of either RMB 50 per tonne, RMB 69 per tonne or RMB 100 per tonne, the LCOE of a waste-to-energy project in 2020 would be RMB

Table 5 LCOE forecasts for municipal waste-to-energy

Year	Installed capital cost (RMB/kW)	Operating cost (RMB/kW per year)	Generation efficiency (%)	LCOE (RMB/kWh)			
				No waste disposal fee	Waste disposal fee of RMB 50/t	Waste disposal fee of RMB 100/t	Waste disposal fee of RMB 150/t
2020	20,000	1,300	24	0.77	0.57	0.37	0.18
2030	19,000	1,400	25	0.78	0.58	0.39	0.20
2050	18,000	1,500	26	0.79	0.59	0.41	0.23
2060	17,500	1,550	26	0.80	0.60	0.42	0.24

Note The estimates are based on the following assumptions (per year): a plant operating time of 6,000 hours, plant power consumption of 20%, calorific value of municipal waste of 1,300 kcal/kg, own capital of 20%, loan interest rate of 4.9%, internal rate of return on capital of 8% and a plant operating lifetime of 15 years
Source Calculated by the authors of this chapter

0.52/kWh, RMB 0.39/kWh or RMB 0.32/kWh respectively. The LCOE is RMB 0.05/kWh lower than if the plant had an operating life of 15 years under the same conditions. That means that if the national average municipal waste disposal fee would be in the range of RMB 70–100 per tonne, waste-to-energy projects in most areas would achieve grid parity.

It can be seen from this analysis that technology and fuel determine the LCOE of bioenergy plants, with little scope for change. Fuel price is the biggest factor. In 2021, the cost of fuel for agricultural and forestry bioenergy exceeded RMB 0.4/kWh, which was around 60% of the total power generation cost. The fuel cost of waste-to-energy is negative (thanks to the waste disposal fee), while the fuel cost of biogas varies greatly depending on the organic matter used.

4.1.3 Bioenergy Project Planning Needs to Improve

Biomass supply security and industry data need to be improved. The Renewable Energy Law of the People's Republic of China defined the activities needed to survey biomass resources, but the mechanisms for doing this regularly have yet to be determined. The distribution of biomass resources across China is uneven, with large differences between provinces. Because the approval process for agricultural and forestry bioenergy plants is decentralised, local governments have not formulated a common plan for plant approval. Moreover, project developers didn't have sufficient understanding of the availability of local biomass resources. This resulted in unviable plant projects, raw material competition and insufficient supply. In addition, most bioenergy projects failed to take into account local demand for heat during their site selection.

4.1.4 The Operating Efficiency of Bioenergy Plants Needs to Improve

China is still behind Europe in terms of the operating efficiency of its bioenergy technologies. In 2020, the average full-load operating hours of bioenergy projects across China was 5,151 hours per year, 30 hours less than the previous year. In comparison European facilities achieved more than 7,000 hours. Besides the fuel supply challenges in China, the lower operating hours were also due to lower generating efficiency. It is therefore necessary to modernise the biomass power industry. For example, combined heat and power technology can be deployed to improve system efficiency and the economics of bioenergy.

4.2 Improvements Are Needed in Geothermal and Ocean Power Generation

Data on the resource reserves and the development and use of geothermal and ocean energy in China are insufficient. The standards for geothermal energy exploration do not match the pace of exploration, and the scientific assessment of geothermal resources is inadequate. Geothermal energy is not included in the national energy statistics, and local energy administrations often do not monitor geothermal heating and cooling.

The technology readiness level of geothermal and ocean energy development and use needs to improve. Technological challenges remain, including in geothermal power generation. There are also technological barriers to the development and use of ocean energy, be it tidal energy, wave energy or ocean thermal energy conversion.

The administrative system for geothermal and ocean power generation does not function well. Despite the small scale of both geothermal and ocean energy, the demarcations between public authorities are unclear and an excessive number of government agencies are involved. For example, the geothermal energy administrations include the Ministry of Natural Resources, the Ministry of Housing and Urban–Rural Development, the Ministry of Water Resources, the National Energy Administration, and the Administrative and Law Enforcement Bureau. China's maritime administration is also divided into administrative jurisdictions, so multiple agencies, cross-jurisdictions and administration overlaps easily occur.

There is still a long way to go before geothermal and ocean power generation technologies can be commercialised. The support policies are inadequate and not standardised. Despite years of research and trials, geothermal and ocean power technologies still face high capital investment costs, small scale and poor profitability, which means large-scale commercial operation is impossible at present. Moreover, the equipment manufacturing industries related to geothermal and ocean power generation have not yet taken shape. At present, there are only small ocean energy generation facilities in operation, so the mass production of facilities and equipment is not possible. Support policies are therefore required to enable demonstration and wide uptake; current incentive policies are insufficient. For instance, it is difficult to implement the fiscal and tax incentive policies related to geothermal energy, and price policies similar to those for early-stage solar and wind energy are absent.

5 Trends and Outlook

The national electricity price subsidy for agricultural and forestry bioenergy projects is to be phased out. Uncertainties remain as to whether government in those areas with relatively abundant agricultural and forestry residue resources will provide local subsidies or not. It will therefore be difficult for bioenergy projects that rely solely on agricultural and forestry residues to maintain operational viability in a period of uncertainty. Looking ahead, combined heat and power (CHP) projects could be developed in areas where conditions permit, especially distributed biomass-fired CHP facilities at the county level. Alternatively, existing bioenergy projects could be upgraded to CHP facilities to achieve continuity. Biomass-fired CHP plants have a high energy conversion efficiency. China's agricultural and forestry bioenergy industry consumes about 70 million tonnes of residues annually. Most of these facilities are designed mainly for power generation only and have an energy conversion efficiency of less than 30%, which means their financial benefits are small. Internationally, CHP plants can achieve an energy conversion efficiency of between 60% and 80%. More importantly, the inadequate cost efficiency of power generation can be offset by the value of the heat produced in CHP. Developing county level distributed CHP demonstration projects fuelled with biomass and allowing them to take part in distributed power trading is a potential future business model adapted to China's conditions. The electricity and heat generated can meet the principal user's needs, while the surplus power can be fed into the distribution network. The advantages of distributed power generation can thus outweigh the difficulties and high cost of collecting, storing and transporting biomass in decentralised systems.

In the near and medium terms, the installed capacity of agricultural and forestry bioenergy in China will increase. New bioenergy projects that generate power only will be limited. Facilities that generate power and heat (CHP) will predominate. By 2030, the installed capacity of bioenergy is expected to reach about 20 GW. From 2030 onwards, some agricultural and forestry bioenergy projects that qualify for early-stage and local subsidies will convert to CHP, while those without subsidies will be phased out. The capacity of closed facilities may exceed that of new projects, so the installed capacity of agricultural and forestry bioenergy projects will dip. It is projected that by 2050 the installed capacity of bioenergy plants will stabilise at about 18 GW.

For waste-to-energy projects, the waste disposal fee is an important component of income, which varies greatly from project to project. Electricity is essentially a by-product of waste disposal. The waste disposal fee will gradually increase as China's economy grows. This is expected to drive the transition from four revenue streams (payment from the grid, national subsidies, local subsidies and waste disposal fee) to two (grid payment and waste disposal fee). In the near and medium terms, the installed capacity of municipal waste-to-energy will continue to grow rapidly and is expected to exceed 25 GW in 2025 and 30 GW in 2030. From 2030 onwards, as

China's urbanisation slows down, the amount of municipal waste produced will also decline, as will the installed capacity of waste-to-energy. By 2050, the installed capacity of municipal waste-to-energy is projected to be close to 45 GW. After 2050, China's urbanisation and economic growth will plateau and the installed capacity of waste-to-energy will stabilise at about 45 GW due to resource constraints.

There are two main sources of biogas for power generation: landfill gas and livestock waste. Some landfill gas is free and some obtains a price. Power generation with landfill gas works in a similar way as municipal waste-to-energy, in which the fuel cost is negative. The number of revenue streams of landfill biogas projects is also expected to drop from four to two. The second type of biogas is generated from waste produced by livestock and poultry farms, wineries and food factories. Distributed biogas power systems can be built to meet the power and heat requirements of the farm or factory, or to feed into the grid and allow the facility to participate in power trading to earn additional income. Looking ahead, the installed capacity of biogas power is set to grow steadily, but will remain small. By 2030 and 2060, the total installed capacity is forecast to be about 2 GW and 3.5 GW respectively.

To sum up, the installed capacity of bioenergy in China will maintain steady growth and is expected to exceed 45 GW and 56 GW by 2025 and 2030 respectively. From 2030 onwards, due to the closure of some agricultural and forestry bioenergy projects and the decline in growth of municipal waste, the installed capacity of bioenergy will slow to about 65 GW by 2050, where it will remain (see Fig. 12).

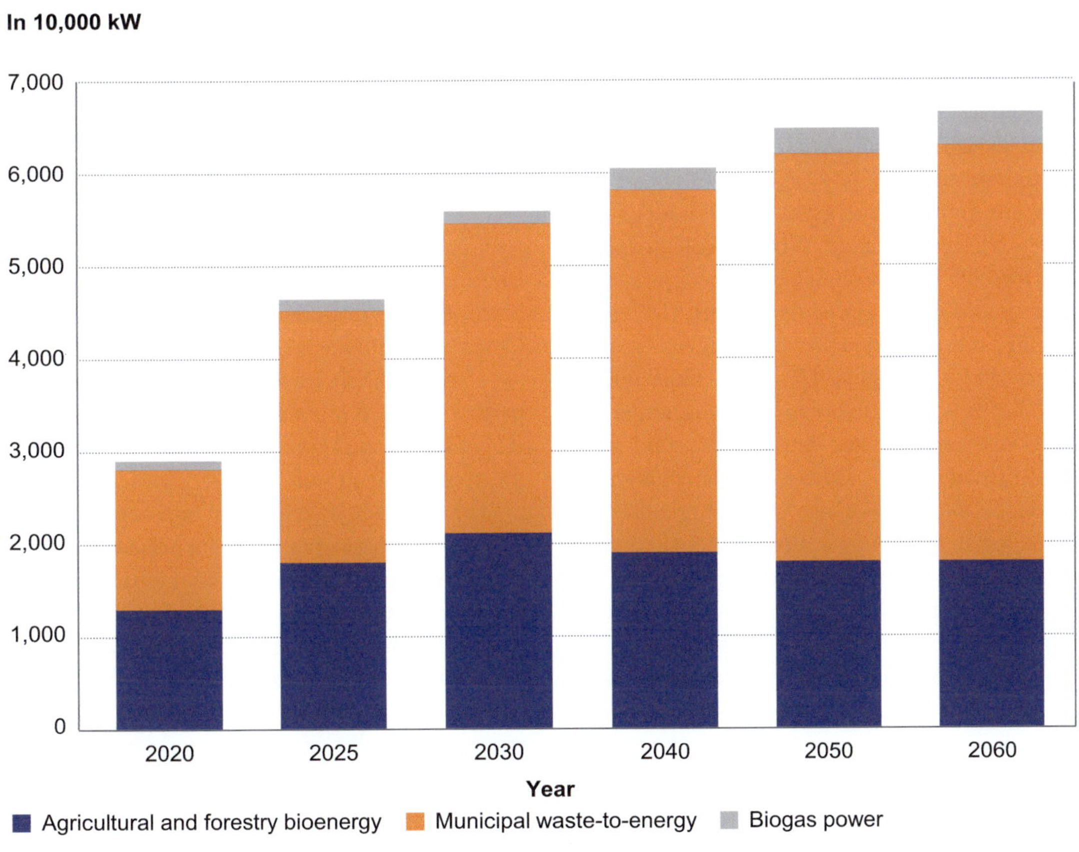

Fig. 12 Forecast of installed bioenergy capacity. *Source* Compiled by the authors

In the short run, it is difficult to develop ocean and geothermal power generation at scale. Small technology or project demonstrations will predominate for some time. In the medium and long terms, the development of large-scale projects depends on breakthroughs in ocean and geothermal power technologies and policy support. There is, therefore, uncertainty about the potential of ocean and geothermal energy.

6 Policy Recommendations

6.1 Bioenergy

In the near and medium terms, different policies should be adopted for different bioenergy technologies to ensure reasonable revenues for projects.

First, government sectors should give bioenergy a sufficient transition period in terms of the phase-out of electricity price subsidies. Projects in areas that have agricultural and forestry biomass and municipal waste resources and the appropriate conditions for bioenergy development, should be prioritised to maximise the value of support policies.

Second, the bioenergy business model should undergo gradual change. While small agricultural and forestry residue-fired power generation projects should no longer be developed, CHP plants should be encouraged for their higher cost efficiency and use of heat. Given the huge positive social effects of developing agricultural and forestry residue-to-power projects in rural areas, local governments should be encouraged to introduce support policies, including price subsidies for electricity or heating. Relevant existing plants should be upgraded to CHP, so that they can continue to operate after the subsidies expire. For municipal waste-to-energy plants, the municipal subsidy should be gradually phased out to achieve the shift from four to two revenue streams.

Third, distributed power trading at county level should be encouraged to improve the economics of distributed bioenergy CHP projects. County-level distributed bioenergy power projects, especially those using agricultural and forestry biomass and biogas CHP projects, should be developed to capitalise on the advantages of distributed power generation. Businesses should be incentivised to generate electricity to meet their own demand and to participate in power trading to improve project economics.

To achieve the dual-carbon goal and support rural revitalisation, bioenergy generation should be supported by multiple national policies. First, agricultural and forestry biomass and biogas power projects should be given national support. Renewable energy price subsidy issues are expected to be resolved around 2035, which means a national electricity price subsidy for agricultural and forestry biomass and biogas power projects could be introduced. The municipal waste disposal fee should be raised in alignment with economic growth to achieve economic and social benefits. Finally, carbon market, green electricity trading and other mechanisms should be used to reflect the environmental value of bioenergy projects, thus boosting revenues and motivating owners.

6.2 Geothermal and Ocean Power Generation

Exploration and research on geothermal and ocean energy resources should be strengthened. China should increase funding, intensify its surveys of geothermal resources, encourage the participation of private capital, and ensure sufficient geothermal resources are available for development and use.

The research, development and deployment of key technologies should be increased. For technologies critical to geothermal and ocean power generation and use, China should create roadmaps and milestones in line with the goals of geothermal and ocean energy, carry out relevant research, and develop and implement support measures to drive innovation and breakthroughs in key technologies.

The administrative system for the development and use of geothermal and ocean power generation should be streamlined. China should clarify the management of geothermal and ocean energy resources and exploration and mining rights, strengthen guidance for the development and use of geothermal and ocean energy resources, and intensify the supervision and management of geothermal and ocean energy development projects. The country should promote co-operation between the authorities involved, create favourable conditions for market players to participate in and play an important role in the development of geothermal and ocean energy, and motivate local government.

Operational support policies should be introduced. Actions should include formulating policies and subsidies to support geothermal and ocean energy as soon as possible. Given the low technology readiness level and insufficient commercialisation, focus should be placed on developing projects that supply power to islands or offshore platforms.

Chapter 8: Non-electrical Pathways to the Energy Transition

Lei Tian, Huawen Xiong and Joep Huijsmans

1 Introduction

China is accelerating its energy transition to meet its dual-carbon goal of peak carbon emissions by 2030 and carbon neutrality by 2060. As a result, its energy system will undergo disruptive change. On the supply side, dominance will shift from fossil fuels to low-carbon or zero-carbon non-fossil energy; on the consumption side, an energy mix based on electricity will gradually take shape. To mitigate the risk to energy security during the transition, the support role of low-carbon non-electrical energy—such as hydrogen, natural gas and biomass—will be crucial to building a new electric power system and decarbonising sectors like industry, buildings and transport. Hydrogen, will be an important part of the zero-carbon energy system of the future; it will play a leading role in hard-to-abate segments such as steel, chemical feedstocks, heavy-duty trucks and high-grade heat. Natural gas will serve as a key transitional energy carrier, thanks to its large-scale availability, cost-effectiveness and clean and relatively low-carbon characteristics. Biofuels will be indispensable in some sectors, such as aviation.

L. Tian (✉) · H. W. Xiong
Institute of Energy Research, China Academy of Macroeconomic Research, Beijing, China

J. Huijsmans
Shell Global Solutions International B.V., The Hague, The Netherlands

2 Low-carbon, Non-electrical Energy Supports China's Dual-carbon Goal

2.1 Primary Energy Supply

Fossil fuels will gradually be replaced by non-fossil energy and their role will change from a principal to a supplementary energy source. During peak carbon emissions, total fossil energy demand will remain stable: coal's share of the fossil energy mix will decrease while that of natural gas will rise. In the approach to carbon neutrality, demand for fossil energy will decline slowly then rapidly, and the main use of fossil energy will shift from fuel to raw material.

As the pace of electrification in China intensifies, non-fossil energy will gradually dominate electricity supply and demand. Electricity will become the fastest growing energy carrier, and the electrification rate of end-use sectors will increase significantly. An electricity-centred energy mix on the demand side will gradually take shape, as will a supply system dominated by low-carbon and zero-carbon energy. The transition of the power system to low carbon is essential to achieve the dual-carbon goal and the deep decarbonisation of end-use sectors like industry, buildings and transport.

Shell International B.V. and Development Research Centre of the State Council (DRC) of People's Republic of China, *Embracing the Future, Powering Growth: An Energy System Renewed for China*, Advances in Oil and Gas Exploration & Production,
https://doi.org/10.1007/978-3-031-73075-7_9

2.2 The Energy Transition

The transition from the old to the new energy system will not be a smooth one. It will incur noticeable energy security risks. On the one hand, the rapidly increasing proportion of renewables in the energy mix undermines the secure and stable operation of the power system. On the other hand, carbon abatement constraints erode the confidence of the market to invest in conventional fossil energy. Since 2021, prices have risen and supply tightened in global energy markets. This further highlights the urgency of accelerating the energy transition and how formidable and complex the transition will be.

Low-carbon, non-electrical energy will be essential to the development of the new power system and the decarbonisation of hard-to-abate sectors. Natural gas power generation provides flexibility, and electrolysis-based hydrogen production enables energy storage. Integrating natural gas and hydrogen is therefore a promising approach to a new electric power system. Biomass and hydrogen also have potential for low-carbon applications in buildings, transport and industry. Efforts should be made to: (i) explore the negative emissions role of bioenergy with carbon capture and storage (BECCS); (ii) develop technologies such as decarbonised hydrogen production for long-distance road transport and other end uses; (iii) promote electrolysis-based hydrogen production in western China; and (iv) build integrated electricity and hydrogen infrastructure to replace fossil fuels with decarbonised hydrogen in hard-to-abate sectors.

2.3 Trends and Focus Areas of Low-carbon, Non-electrical Energy in China

In line with the requirements of a “clean and low-carbon and secure and efficient” energy system, non-electrical low-carbon energy should generate economic and social development, control the cost of the shift to a low-carbon energy system and foster new growth in the energy economy. Based on current developments and future projections, the three energy carriers of focus are hydrogen, natural gas and biomass.

2.3.1 Hydrogen and the Zero-carbon Energy System of the Future

“Renewable energy + electrification” is not a panacea for all carbon emissions. In hard-to-abate sectors such as steel, chemical feedstocks, heavy-duty road transport and high-grade heat, hydrogen is expected to provide the deep decarbonisation needed to achieve carbon neutrality.

Hydrogen is highly efficient, low carbon and flexible. The International Energy Agency, the International Renewable Energy Agency and other organisations are optimistic that hydrogen will become a pillar of the future energy system. Promoting the uptake of hydrogen in transport, industry and buildings will enable renewable power to achieve deep decarbonisation in end-use sectors. Hydrogen can also serve as a medium for energy interconnections and energy storage. Coupled with power grids and natural gas pipelines, it can provide large-scale energy storage and peak load balancing capacity, thereby playing a pivotal role in a modern Energy Internet. The co-ordinated use of hydrogen, electricity and heat produces a diverse and complementary energy supply system and the smooth realisation of China’s dual-carbon goal.

2.3.2 Natural Gas as a Transitional Type of Energy

Natural gas will be one of the main support energy sources in the transition to a low-carbon energy system. Thanks to its large-scale availability and affordability, natural gas together with renewable energy, will have the capacity both to cover rising additional energy demand and replace coal and oil. In terms of end uses, the use of natural gas in the urban residential market will increase; gas will replace the declining use of coal and support the development of new energy; and the switch from coal to gas in industry will advance steadily and its use in chemical production remain stable.

Natural gas will go hand in hand with renewable energy to meet the fuel demand of industry, urban homes and power generation. In transport, natural gas demand will increase before it declines. Natural gas is likely to be the only fossil fuel used at scale in the future. The scale and extent of this use will depend on its co-ordinated development with renewable energy and hydrogen, and the advance of carbon capture and storage. During this window of success, the natural gas industry should make every effort to address challenges such as supply security and cost efficiency in its value chain. However, carbon neutrality places constraints on fossil energy consumption. As a result, natural gas demand is set to decrease gradually over time.

2.3.3 Bioenergy's Multiple Roles

Biomass is an important renewable energy carrier. It is used mainly for power generation and for liquid biofuels, biogas and biomass briquette fuels. Liquid biofuels are used widely as a substitute for fossil fuels in road transport, mainly as bioethanol and biodiesel. Liquid biofuels play an important role in energy supply security, tackling climate change and promoting social and economic development. They also hold great potential in hard-to-electrify sectors such as heavy-duty road transport, shipping and aviation.

3 Outlook for Low-carbon, Non-electrical Energy Supply in China

3.1 Hydrogen

Hydrogen is an important component of the future zero-carbon energy system. It will play an essential role in hard-to-abate sectors such as steel, chemical feedstocks, heavy-duty road transport and high-grade heat. Coal-to-hydrogen is currently the main method for the large-scale, stable and inexpensive production of industrial hydrogen in China. Decarbonised hydrogen made from renewables will gradually improve its competitiveness as the cost of renewable energy falls and carbon market incentives take effect. By 2050, the production of decarbonised hydrogen is expected to account for more than 80% of hydrogen production.

3.1.1 Hydrogen Production

Hydrogen is mainly produced using fossil fuels, especially coal and natural gas. Large-scale, low-cost industrial hydrogen is produced mainly from coal in China. Conventional coal-to-hydrogen processes include fixed bed, fluidised bed, and entrained-flow bed, which have high carbon emissions and contain corrosive gases like sulphides. In recent years, new coal gasification hydrogen technologies have advanced. For example, supercritical water gasification offers high gasification efficiency, high hydrogen yield and less pollution, but it needs further commercialisation. Hydrogen produced from natural gas through steam methane reforming is relatively mature and widely deployed. The cost of hydrogen production from natural gas varies greatly, as the price represents 70–90% of the cost of production.

Industrial by-product hydrogen is produced from industrial tail gas. Industrial tail gases that contain hydrogen include coke oven gas, chlor-alkali by-product hydrogen, refinery dry gas, synthetic methanol and purge gas from ammonia synthesis. Despite the relatively complex purification process of industrial by-product hydrogen, its technological maturity, low cost and environmental friendliness make it a potentially important source of high-purity hydrogen in the near future. The supply potential and cost of industrial by-product hydrogen depend on the product and revenue mix of the producers.

Producing hydrogen from water electrolysis will be the main route forward in the future. Water electrolysis technologies include alkaline water electrolysis (AWE), proton exchange membrane (PEM) electrolysis, and solid oxide electrolysis (SOE). AWE technology has been used in large-scale commercial production, giving a large hydrogen yield from a single electrolytic cell. The main parameters of locally manufactured AWE equipment in China have been close to or reached internationally advanced levels. SOE technology provides high operational flexibility and reaction efficiency and can operate in standby mode at low power. This technology makes a good match with solar photovoltaic systems that have a small installed capacity and large volatility. However, it is still behind internationally in terms of technological maturity, installed capacity, lifetime and cost effectiveness. SOE technology consumes less electricity than AWE and PEM, but has not yet been widely commercialised.

3.1.2 Hydrogen Storage and Transport

Hydrogen storage refers mainly to compressed hydrogen storage, cryogenic liquid hydrogen storage, solid-state hydrogen storage and hydrogen storage in liquid organic hydrogen carriers. The latter two storage technologies allow hydrogen to be stored at normal temperature and atmospheric pressure with a high level of safety. However, solid-state hydrogen has a low energy density, which can result in high energy losses when stored. Solid-state hydrogen storage is still at the R&D and demonstration stage, with few commercial projects in operation. Compressed hydrogen storage and cryogenic liquid hydrogen storage have a relatively high technology readiness level.

Compressed hydrogen storage technology is relatively mature and low cost, and it can charge and discharge hydrogen efficiently. However, compressed hydrogen has a relatively low density, which means safety is a risk. The transport cost of compressed hydrogen depends on the distance transported.

Cryogenic liquid hydrogen storage technology stores the liquefied hydrogen, which is cooled to minus 253 °C in a cryogenic adiabatic liquid hydrogen tank. Liquefied hydrogen has a volume density that is 845 times higher than gaseous hydrogen, which means better hydrogen storage and transport efficiency. However, hydrogen liquefaction consumes lots of energy, accounting for about 40% of hydrogen's energy content. The transport cost of liquid hydrogen tankers mainly comprises electricity, vehicle depreciation and labour costs, and has a low correlation with the transport distance.

Pipelines can transfer large volumes of hydrogen and are suitable for large-scale, centralised, point-to-point transport. However, pipelines are prone to hydrogen embrittlement (metal reacts with hydrogen) causing hydrogen to escape. In addition, hydrogen has a small volumetric energy density—the energy density of hydrogen is only a third that of the same volume of natural gas. That means that the pump station compressors use more power to transport hydrogen through the pipeline than for natural gas, which means the variable costs for piping hydrogen are higher.

Hydrogen can be synthesised into ammonia and methanol, which have an energy density far higher than that of high-pressure gaseous hydrogen and liquid hydrogen. Ammonia and methanol can also be stored and transported at nearly normal temperature and pressure. In this way, hydrogen can be stored for a long time and transported safely, which reduces the high cost and safety risk of hydrogen transport.

3.1.3 Hydrogen Supply Prospects

The cost of all types of hydrogen production, storage, transport and refuelling will gradually decline over time and become closer to or lower than the acceptable cost boundaries of applications. The cost acceptability of hydrogen for stationary applications will remain stable, while that of hydrogen for vehicles will vary over time as the cost effectiveness of fuel cell technology improves. The combination of low-cost renewables-based electrolysis and centralised storage and transport is expected to enable decarbonised hydrogen to replace fossil fuels in steel, synthetic ammonia and synthetic methanol by 2050.

Thanks to the lower cost of renewable power and carbon market incentives, decarbonised hydrogen will gradually become more cost competitive. By 2050, decarbonised hydrogen is expected to make up more than 80% of total hydrogen production. Due to the rising cost of fossil energy and intensified emission constraints, hydrogen production from fossil energy will cease by 2060, while decarbonised hydrogen will continue to grow.

3.2 Natural Gas

3.2.1 Natural Gas Supply in China

Domestic natural gas production has maintained strong momentum in China in recent years. In 2021, the country's natural gas output was 205.3 billion cubic metres (m^3), an increase of 8.2% year-on-year. Natural gas production is expected to maintain rapid growth and reach 270 billion m^3 by 2035. In the long run, China's natural gas supply will still need to be supplemented by imported resources.

3.2.2 Measures to Enhance Supply Capacity

China will raise domestic natural gas supply capacity through multiple sources. Increasing upstream exploration and development is the main priority. Insufficient effort has been made to explore resources in most natural gas basins in China. Proven conventional gas resources account for less than 20% of estimated reserves, and the exploration and development of shale gas and coalbed methane have just started. This suggests there is great potential to explore and develop natural gas resources. In addition, China aims to; (i) unplug the bottlenecks that hinder the development and use of coal-to-gas, biogas and natural gas hydrates; (ii) increase its efforts to develop pilot and demonstration projects; (iii) achieve economies of scale; and (iv) expand natural gas supply channels. In addition, China will speed up coal-to-gas projects and technologies without compromising the environment and affordability.

China will take various steps to improve natural gas import capacity. Currently, a strategic mix of imports exists, with pipeline natural gas coming mainly from Central Asia and Russia, and liquefied natural gas (LNG) from Australia, Qatar, Malaysia and other countries. Since 2017, LNG imports have exceeded pipeline imports. Looking ahead, it is necessary to improve pipeline import capacity and build more LNG terminals in an orderly manner. The scale of imported natural gas depends largely on market demand.

3.3 Liquid Biofuels

The supply of raw materials for bioethanol depends mainly on land resources. The raw materials for making cellulosic ethanol are agricultural and forestry residues. If non-grain crops such as cassava or sorghum are used, large-scale cultivation will be necessary to ensure their supply.

In China, the area of land (planted and reserved) for bioethanol raw materials is 5.5 million hectares and 6.741 million hectares respectively. The proportion of reserved land developed is forecast to be 10% in 2025, 15% in 2030, 25% in 2040 and 35% in 2050. The crops grown on this land will primarily be cassava, sorghum and sweet potatoes.

According to the national agricultural zoning system, there are eight farming zones in mainland China (see Table 1). The factors that influence the zoning of an area of land are environmental protection; maintaining a dynamic balance of farmland, water resource availability, urbanisation and reclamation; the suitability of the land for energy crops; and the expected yield of the crops. Table 2 shows the land resources developed for energy crops in 2020–50.

In the near term (before 2025), China should: (i) develop bioethanol fuel using starchy non-grain crops (such as cassava and sweet potatoes) and enhance the cost effectiveness of bioethanol production by improving production technologies; (ii) carry out commercial demonstration projects and encourage the uptake of sorghum-based bioethanol fuel, increase R&D efforts in cellulosic bioethanol, achieve breakthroughs in cellulose enzymolysis technology and build a complete cellulosic ethanol production chain; (iii) plant stress-resistant energy crops; and (iv) improve infrastructure and lay a solid foundation for the large-scale development of transport biofuels.

By 2025, a biomass resources database will be established and land reserved for biomass cultivation across China will be developed. Issues related to the collection, storage and transport of non-grain crops will be largely solved.

Table 1 China's eight farming zones

Region	Provinces concerned	Arable land reserves (In hectares)
Northern China	Beijing, Tianjin, Hebei, Shandong, Henan	571,000
Middle and lower reaches of the Yangtze River	Shanghai, Jiangsu, Zhejiang, Anhui, Jiangxi, Hubei, Hunan	697,000
North-east China	Heilongjiang, Jilin, Liaoning	453,000
South-west China	Chongqing, Sichuan, Guizhou, Yunnan	321,000
Inner Mongolia–Xinjiang region	Inner Mongolia, Ningxia, Xinjiang	3,696,000
Southern China	Fujian, Guangdong, Guangxi, Hainan	124,000
Loess Plateau region	Shanxi, Shaanxi, Gansu	879,000
Others	Qinghai, Tibet	–

Source Yan Jinyue, Zhao Lixin, Research on a Sustainable Development Strategy for Energy Crops in China, China Agriculture Press, 2009; Energy Research Institute under the National Development and Reform Commission

Table 2 Land under development for energy crops, 2020–50

Raw materials	Land development status	2020	2025	2030	2040	2050
Cassava	Planted area (hectares)	265,800				
	Reserved land (hectares)	124,000				
	Estimated proportion (%) under development	5	10	15	25	35
	Estimated area of land under development (hectares)	10,000	10,000	20,000	30,000	40,000
Sorghum	Planted area (hectares)	570,000				
	Area of replaced land development (hectares)	28,500	57,000	85,500	142,500	199,500
	Reserved land (hectares)	5,599,000				
	Estimated area of land under development (hectares)	280,000	560,000	840,000	1,400,000	1,960,000
Sweet potatoes	Planted area (hectares)	4,650,000				
	Reserved land (hectares)	1,018,000				
	Estimated area of land under development (hectares)	50,900	101,800	152,700	254,500	356,300
Total	Planted area (hectares)	5,485,800				
	Reserved land (hectares)	6,741,000				
	Estimated area of land under development (hectares)	337,050	674,100	1,011,150	1,685,250	2,359,350

Source Energy Research Institute under the National Development and Reform Commission

Cultivation will be significantly improved and the production technologies used will be mature. Waste-water treatment capacity will match production capacity, and energy consumption in production will be greatly reduced. Modern management and business operational models will be introduced. The main technological barriers restricting the production of cellulosic ethanol will be addressed and the manufacturing cost of cellulase will decrease from around 70%

Table 3 Bioethanol fuel supply capacity in the future

Raw materials	2022	2025	2030	2040	2050
Cassava	30,000	60,000	110,000	210,000	390,000
Sorghum	1,160,000	2,310,000	3,760,000	7,200,000	11,520,000
Sweet potato	437,500	925,000	1,500,000	3,214,000	5,000,000
Cellulose	833,300	2,000,000	6,000,000	15,000,000	25,000,000
Aged grain	3,000,000	6,000,000	6,000,000	6,000,000	6,000,000
Industrial exhaust gas fermentation	90,000	300,000	1,000,000	3,000,000	6,000,000
Total (tonnes)	5,550,000	11,600,000	18,370,000	34,620,000	53,910,000

Source Energy Research Institute under the National Development and Reform Commission

today to less than 30% of the total cost. Large-scale cellulosic ethanol demonstration plants will be built. Policies on biofuels will be optimised, a regulated market access mechanism will take shape and a nationwide biofuel market trading system will be established.

In the medium term (2025–30), the large-scale commercial development of bioethanol from sugar crops like sorghum will be possible with advanced microbial fermentation technology. The cost competitiveness of bioethanol with petroleum will be continually enhanced. Breakthroughs in the co-fermentation of pentose and hexose for bioethanol production will be achieved and the production of cellulosic ethanol will take off. Energy crops will be planted at scale in marginal saline–alkali land and sandy wasteland. Methods to increase the starch content in starchy crops and sugar content in sugar crops will be developed. Shrubs and grasses that provide high yields and resist wind and drought will be cultivated. And automotive technologies that allow bioethanol to play an important role in the transport fuel mix will take shape.

By 2030, a complete database of biomass resources will exist, available land resources will be cultivated and the use of agricultural and forestry residues will be intensified. The technologies used to produce biofuels will be comparable to those in developed economies, and the large-scale production of second-generation bioethanol will be the focus of the biofuel industry. Biofuels will serve as the main alternative to fossil fuels (see Table 3).

By 2025, 3.295 million tonnes of bioethanol fuel will be produced from energy crops such as sorghum, cassava and sweet potato on 10% of China's marginal lands. The yield per unit of land will increase for cassava and sweet potato, and the starch saccharification process and fermentation level improve. The crop straw used to produce cellulosic ethanol will consume 1% of total straw resources. Thanks to a significant improvement in enzymatic activity and the maturity of the production technology, the amount of straw required to produce 1 tonne of bioethanol will be reduced from 7 tonnes to 5 tonnes. As a result, the output of cellulosic ethanol will reach 2 million tonnes and that of bioethanol about 11.6 million tonnes per year.

By 2030, the marginal land used for cultivating energy crops such as sorghum, cassava and sweet potato will make up 25% of the total used for biofuels, amounting to 1.01 million hectares. Bioethanol produced from non-grain crops will be about 5.37 million tonnes. About 30 million tonnes of agricultural and forestry residues will be used to produce cellulosic ethanol, yielding about 6 million tonnes of fuel. Bioethanol production from industrial exhaust gas fermentation will be about 1 million tonnes. Total supply of bioethanol is estimated to reach 18.37 million tonnes.

By 2040, the technologies for producing bioethanol from non-grains such as sorghum, cassava and sweet potato will be mature and commercialised. The focus will shift to how to develop marginal land effectively and achieve better industrial planning and plant design. Thanks to achieving a balance between energy supply, environmental protection and energy conversion efficiency, the production capacity of bioethanol will increase to more than 23 million tonnes a year.

By 2050, the technology for producing bioethanol from different raw materials will be mature and bioethanol will be the main alternative to fossil fuels. Maximising the energy and social and environmental benefits will be the main driver in the bioethanol industry. Bioethanol fuel supply capacity will exceed 53 million tonnes a year.

4 Roadmap for Low-carbon, Non-electrical Energy Supply

As China approaches its dual-carbon goal, demand for fossil energy will reach saturation point and the role of fossil fuels in the energy system will change. In the shift to electrification, non-fossil energy will gradually dominate both electricity supply and demand. Low-carbon, non-electrical energy—such as hydrogen, natural gas and biomass—will play a crucial support role in the new electric power system and in decarbonising sectors like industry, buildings and transport (see Fig. 1). In the medium and long terms, low-carbon energy supply will develop in three stages.

Between 2021 and 2030, natural gas supply capacity will grow rapidly; grey hydrogen will continue to dominate hydrogen production and liquid biofuels will develop at a faster pace. Between 2031 and 2040, green (decarbonised) hydrogen will achieve strong momentum, the supply potential of liquid biofuels will double and domestic natural gas production will gradually plateau. In 2041–60, as the goal of carbon neutrality approaches, decarbonised hydrogen will account for more than 80% of total hydrogen production; the supply capacity of liquid biofuels will increase significantly; and the supply capacity of natural gas will dip as demand declines.

5 Support Measures to Increase Low-carbon, Non-electrical Energy Supply Capacity

Accelerating the development of low-carbon, non-electrical energy will support the dual-carbon goal and the energy transition. First, efforts should be made to develop decarbonised hydrogen production technology. Second, multiple measures should be taken to boost domestic natural gas supply capacity and to optimise the imported natural gas portfolio. Third, a liquid biofuel supply system should be built at faster pace.

5.1 Develop Decarbonised Hydrogen Production Technology

The cost of producing hydrogen from fossil fuels is low. However, when carbon emissions and decarbonisation are taken into account, the economic advantages of hydrogen produced from fossil fuels are limited. Low-cost and low-emission hydrogen made from industrial by-products will serve as an important intermediate source going forward. There are various hydrogen production technologies in use that are based on electrolysis, although their costs vary. There is a large scope for reducing the cost of hydrogen made from electrolysis in the coming years. Natural gas as a raw material for hydrogen will continue to be important.

5.2 Increase Natural Gas Supply Capacity

First, natural gas production should be positioned as a foundation stone of national energy security. Increasing domestic oil and gas production will be the most important way to ensure supply security. Second, the imported natural gas portfolio should be optimised. Lessons should be learned from the large international oil companies and the business advantages of Chinese oil companies given free rein to enhance their resources.

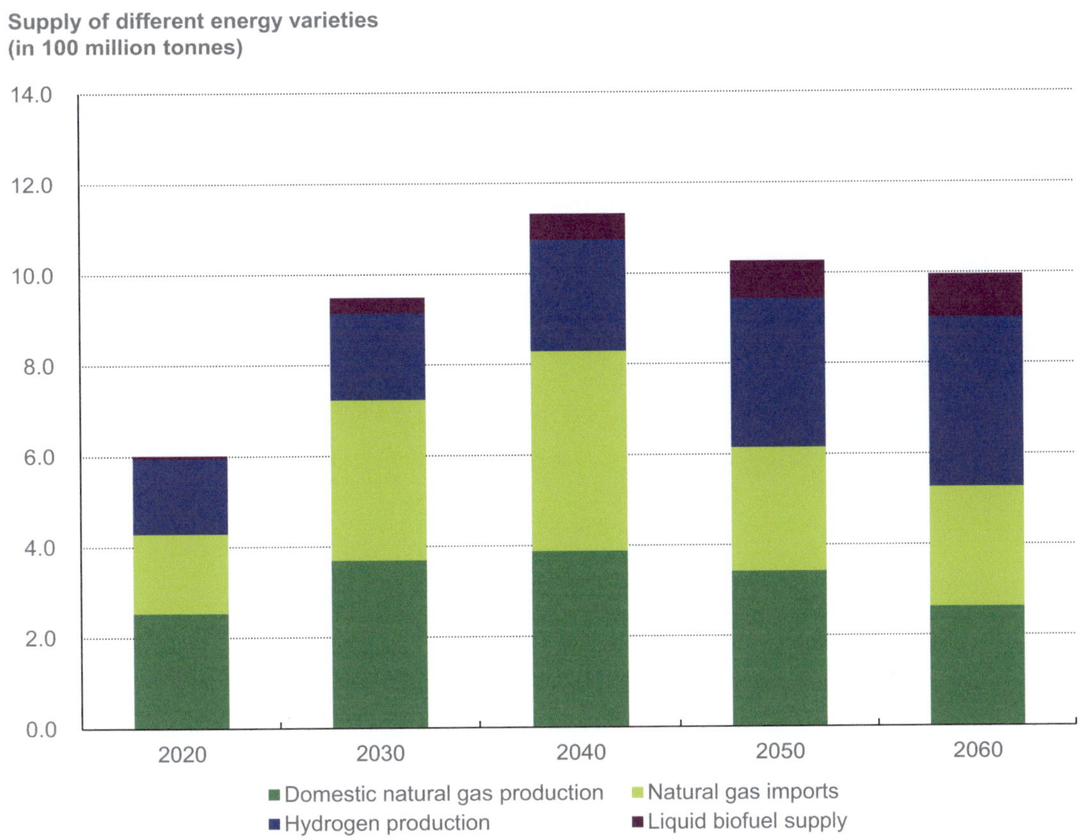

Fig. 1 Supply sources of China's low-carbon non-electrical energy. *Source* Prepared by the authors

5.3 Build a Liquid Biofuel Supply Support System Faster

First, an upstream raw material supply system should be established. The future development potential of biofuels will depend on the supply of raw materials. China should strengthen efforts to increase the supply of raw materials for biofuels and create a supply system for the cultivation, collection, transport and storage of energy crops.

Second, a biofuel standard and quality certification system should be created. Efforts should be made to study quality certification standards for bioethanol, establish a traceability system for biofuel raw materials and a carbon emission certification system, and issue clean fuel quality certification for bioethanol fuel products.

Third, a bioethanol trading market should be introduced. The biofuel green certification and trading system should create and stabilise market demand for biofuels. Fossil fuels should gradually be replaced by clean biofuels in an orderly manner.

Chapter 9: Biofuels

Kejun Dou and Joep Huijsmans

1 Introduction

In the near and medium terms, liquid biofuels will serve as a key strategic energy carrier to replace fossil fuels with cleaner energy. They will provide an effective way to flexibly decarbonise road transport.

The supply capacity of liquid biofuels is currently about 7 million tonnes of coal equivalent (tce) per year. By 2060, total supply of liquid biofuels is expected to exceed 100 million tce annually. Under current technology and economic conditions, the liquid biofuel industry still requires support from green certification, tax breaks and subsidy incentives.

There is growing global recognition of the crucial role that bioenergy can play in the energy transition. Bioenergy is an important alternative to fossil fuels and chemical products; it is an integral part of the carbon neutrality strategy of many countries around the world. As liquid biofuels demonstrate their potential and advantages in hard-to-electrify segments such as heavy-duty road transport, shipping and aviation, countries are intensifying their R&D efforts in commercial applications. Uncertainties persist around the future of liquid biofuels in factors such as resource availability, marginal land development, feedstock prices, production technologies and market demand. In response to these uncertainties, China should build a cost-effective and large-scale feedstock supply system by optimising project planning; co-ordinating feedstock collection, storage and transport; cultivating energy crops in an orderly manner; increasing investment in scientific research; and striving to achieve breakthroughs in the cost competitiveness of liquid biofuel conversion technologies. The country should create a market environment for fair and preferential use of liquid biofuels by creating credit and green certification systems and unlocking the potential for liquid biofuels in the carbon trading market.

2 The Significance of Liquid Biofuels for the Dual-carbon Goal

Liquid biofuels play a significant role in ensuring energy security, mitigating climate change and promoting high-quality social and economic development. They are a promising solution for hard-to-electrify sectors such as heavy-duty road transport, shipping and aviation. All countries are intensifying their R&D efforts in commercial applications and consider liquid biofuels a key strategic energy carrier to decarbonise transport.

K. J. Dou (✉)
Deputy Secretary-General, Biomass Energy Industry Promotion Association, Beijing, China

J. Huijsmans
Shell Global Solutions International B.V., The Hague, The Netherlands

Shell International B.V. and Development Research Centre of the State Council (DRC) of People's Republic of China, *Embracing the Future, Powering Growth: An Energy System Renewed for China*, Advances in Oil and Gas Exploration & Production,
https://doi.org/10.1007/978-3-031-73075-7_10

In 2021, IEA Bioenergy published the Role of Renewable Fuels in Decarbonizing Road Transport, which concluded that biofuels will remain the largest contributor to decarbonisation in the short and medium terms. Depending on the availability of local feedstocks, biofuels can reduce tank-to-wheel carbon dioxide emissions to near zero by 2050. By 2060, biofuels produced from sustainably available feedstocks will meet 30% of total transport fuel demand.

Of the existing low-carbon transport fuels, and before electric vehicles effectively remove carbon emissions, biofuels can reduce exhaust emissions without much need to change existing infrastructure. Biofuels are an effective way to flexibly decarbonise road transport at low cost.

3 Overview of Biofuels Worldwide

3.1 Advances in Biofuel Technologies

3.1.1 Bioethanol

There are three generations of bioethanol production, each of which relies on different feedstocks and technologies. First-generation biofuels are produced from starchy crops such as corn and sorghum, sugar crops like sugar cane and beet, oil-bearing crops (soybeans and canola, etc.) and animal fats. Sugar and starchy crops are converted into bio-alcohols including ethanol, butanol and propyl alcohol through a fermentation process. Second-generation biofuels, also known as cellulosic biofuels, are made from cellulose in non-grain crops and waste biomass, including corn straw, corncobs, rice straw, wood and wood by-products. Third-generation biofuels use algae as feedstock.

At present, bioethanol produced from sugar or starchy crops such as sugar cane, corn and potatoes predominates. Biofuels derived from non-grain feedstocks are called advanced biofuels in that the raw materials are mainly agricultural and forestry residues, organic waste and other renewable biomass resources, which have few effects on land-use change, greenhouse gas emissions and other environmental factors. Advanced biofuels have huge potential for development.

The commercialisation of cellulosic ethanol lags behind that of biodiesel. The main barriers to investment in cellulosic biofuels include high research and production costs and regulatory uncertainty.

The Argonne National Laboratory (ANL) in the USA used the GREET[1] model to calculate the life-cycle greenhouse gas emissions of hydrocarbon fuels produced from different feedstocks and using different conversion methods. The results show that fuel produced from corn grains, sugar-cane juice and cellulosic biomass like sugar cane and corn straw reduced greenhouse gas emissions by 40%, 70% and 70%–96% respectively.

3.1.2 Biodiesel

The European Union leads the global biodiesel industry. During the COVID-19 pandemic in 2020, the use of biodiesel and hydrogenated renewable diesel in the EU fell by nearly 6% (860,000 tonnes).

Hydrogenation-derived renewable diesel is a high-quality biodiesel produced using mainly animal fats and plant oils in chemical reactions and hydro-treating processes. Thanks to its high quality, it can replace diesel without the need to modify vehicle engines. Modified hydrogenation-derived renewable diesel can be used as aviation fuel, and many airlines have used this biodiesel as an alternative to kerosene. This represents the most mature advanced biofuel technology and it has the largest production capacity. The EU began to produce hydrogenation-derived renewable diesel in 2012, and produced about 2.65 million tonnes in 2020.

3.1.3 Advanced Biofuels

Advanced biofuels are mainly produced using one of three processes: biochemical, thermochemical or chemical.

The biochemical process is used mainly by second-generation bioethanol technologies such as cellulosic ethanol.

[1] GREET stands for greenhouse gases, regulated emissions and energy use in transport.

The thermochemical process typically refers to one in which biomass is decomposed into simple gas molecules, such as hydrogen, oxy-carbide, water and methane. The processes used are gasification, pyrolysis, drying and others, after which liquid or gaseous biofuels are generated via catalytic conversion. Wood chips and forestry wood residues, municipal solid waste and sulphite waste liquor are the main feedstocks used to produce Fischer–Tropsch synthetic fuels, synthetic natural gas and oxygenated liquid fuels such as dimethyl ether and alcohol fuels.

The chemical process uses mainly animal fats and plant oils to produce high-quality biodiesel through chemical reactions and hydro-treating. It generates high-quality renewable diesel, which can be used as a flexible fuel for road transport and shipping. The process can also produce renewable naphtha, propane and alkanes.

3.1.4 Aviation Fuel

Biofuels can significantly reduce greenhouse gas emissions from the aviation sector and have a positive impact on climate change. The sustainable aviation fuel market is expected to grow from $66 million in 2020 to $15.307 billion in 2030. With commercial use of alternative aviation fuels increasing, biofuel production capacity is expected to expand quickly this decade. According to the U.S. Energy Information Administration, the global commercial aviation fuel market is expected to grow from 106 billion gallons (400 billion litres) to 230 billion gallons (870 billion litres) by 2050. Cost-competitive and environmentally sustainable aviation fuels are seen as the key to decoupling market growth from increased carbon emissions.

Hydro-processed ester and fatty acid synthetic paraffin kerosene (HEFA-SPK) is the most commercially viable sustainable aviation fuel. The development of sustainable aviation fuels has evolved from one-way demonstration flights by airlines or equipment manufacturers to supply chain initiatives involving multiple stakeholders, such as equipment manufacturers, airlines, fuel manufacturers and airports.

Between 30% and 50% of the sustainable aviation fuel market will maintain a high compound annual growth rate this decade. Moderate blending requirements, refuelling facilities in existing fuel systems, supply logistics and a fleet approach will drive costs down to the minimum level and meet the volume requirements of commercial and military aviation.

A world-leading sustainable aviation fuel market has been built in Europe, where there are key players in biofuel production technologies and commercial-scale plants in operation. Moreover, Europe is actively promoting co-operation between the aviation industry and research institutions to progress the European market. The EU's RED II renewable energy directive limits the amount of food- and feed-based crops that member states can use in biofuels to 7%, with an additional 1% for flexibility. RED II also sets a minimum target for advanced biofuels of at least 3.5% by 2030. Additionally, the EU introduced sustainability criteria for biomass in RED II and expanded the sustainability criteria for biofuels. The EU is also planning to gradually increase uptake by requiring each EU airport to include at least 2% sustainable aviation fuel in the fuel it supplies in 2025 and 63% in 2050.

In 2021, sustainable aviation fuels achieved a milestone. In May, Air France–KLM, TotalEnergies, Groupe ADP and Airbus joined forces to carry out the first long-haul flight powered by sustainable aviation fuel produced in France. The biofuel used for the flight was made from waste and residues. French legislation requires aircraft to use at least 1% sustainable aviation fuel by 2022 for all flights departing from France.

3.2 The International Biofuel Market

Liquid biofuels, especially bioethanol and biodiesel, have been widely used around the world as alternatives to fossil fuels in road transport. As biofuel technologies advance, biofuels will have great potential in aviation in the future.

Bioethanol is produced mainly from sugar and cereal crops, with bioethanol fuel production from sugar and cereals accounting for about 60% and 40% respectively. Traditional crops and growing conditions determine the main feedstock used to produce biofuel in each country, which is corn in the USA, sugar cane in Brazil and aged grain in China. Global bioethanol production increased from 16.19 million tonnes in 2002 to a record 85.74 million tonnes in 2018. Due to the COVID-19 pandemic, global bioethanol production decreased slightly to 77.47 million tonnes in 2020 (see Fig. 1).

In recent years, the shift to electric and smart vehicles has posed challenges to the development of conventional transport fuels, including biofuels. In view of the important role that biofuels play in ensuring energy security, combatting climate change and generating social and economic development, there is still much scope for growth in hard-to-electrify segments such as heavy-duty road transport, shipping and aviation. Developed countries have provided financial support for the research and development of technologies and commercial applications for liquid biofuels for a long time. In the near and medium terms, biofuels will serve as an important strategic fuel in the transition to cleaner energy.

3.3 Bioethanol Developments in the USA

A fully-fledged bioethanol market has been built up in the USA over the past 50 years. Both the value chain and government administration have contributed hugely to the sustainable development of bioethanol in this country. The bioethanol industry in the USA was hit heavily by the COVID-19 pandemic. According to year-end data released by the U.S. Energy Information Administration, the country's bioethanol production was about 41.62 million tonnes in 2020, a decrease of 5.52 million tonnes from 2019 (11.7% on a year-on-year basis). It was a record low since 2014 (see Fig. 2). Even so, the USA remained the world's

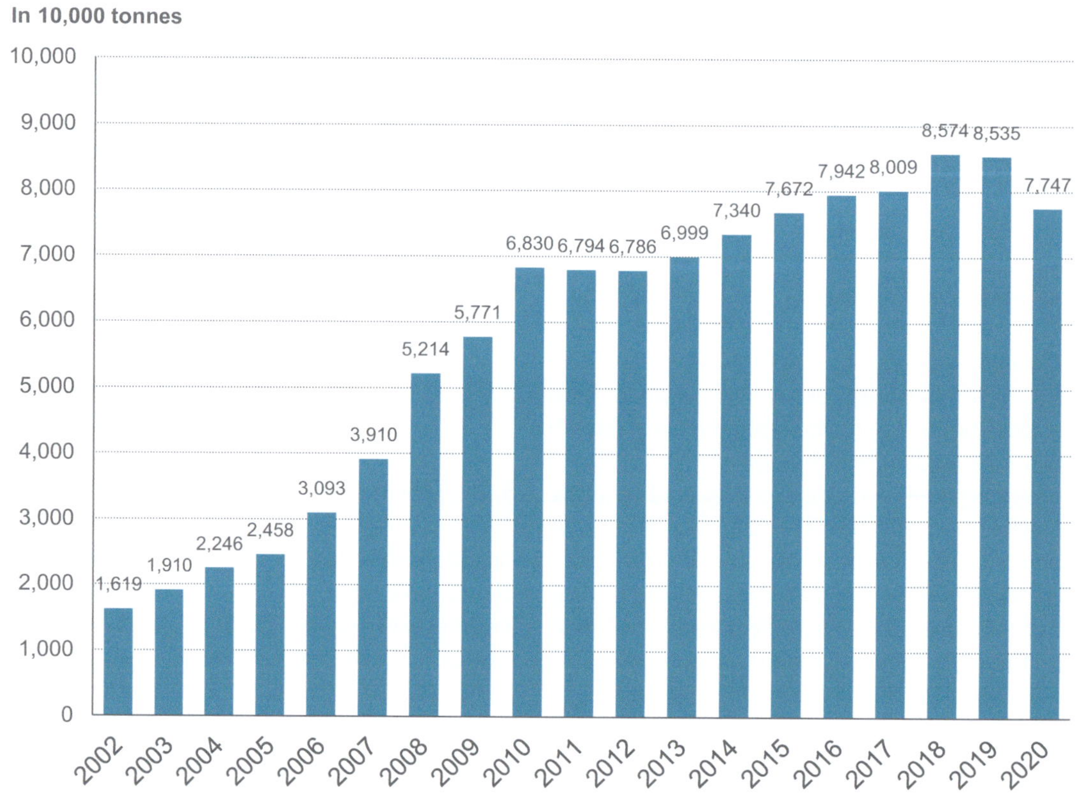

Fig. 1 Global bioethanol production, 2002–20. *Source* Renewable Fuels Association

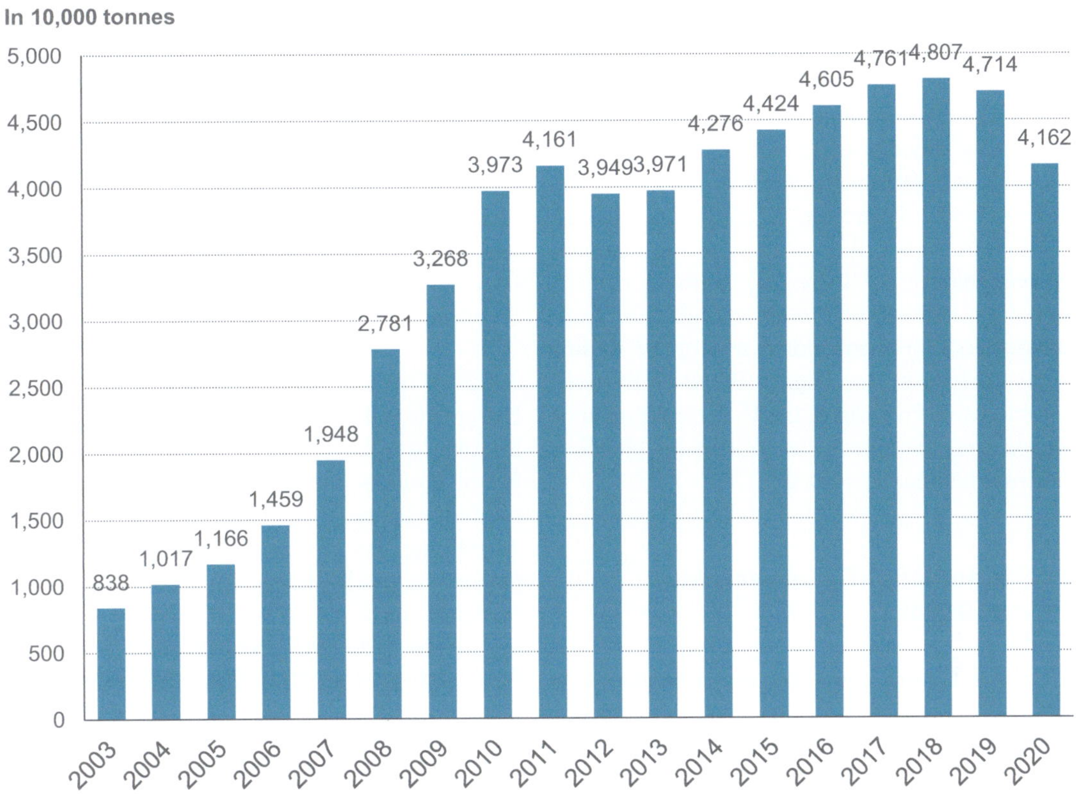

Fig. 2 Bioethanol production in the USA, 2003–20. *Source* U.S. Energy Information Administration

largest producer of bioethanol, contributing about 54% of global bioethanol fuel production. In 2020, domestic bioethanol consumption in the USA was 37.73 million tonnes, down 13.2% on 2019 and the lowest since 2009. In the same year, domestic petrol consumption was about 350 million tonnes, a decrease of 13.5% year-on-year. Despite a drop in bioethanol production and consumption of blended fuels in 2020, the proportion of bioethanol in petrol increased, with the average annual blending rate reaching a record high of 10.23%.

3.3.1 Bioethanol Blending Methods

There are three main ethanol–petrol blends in the USA, namely E10, E15 and E85. E10 is a blend in which 10% (volume ratio) ethanol is added to petrol; E15 is a blend that contains 15% (volume ratio) ethanol; and E85 is an ethanol-petrol blend containing 85% ethanol. Most petrol fuels in the USA have ethanol content of no more than 10%. E15 is sold mainly in the mid-west, where ethanol production is concentrated. Flexible fuel vehicles (FFVs) can use any ethanol–petrol blends below E85. Service stations usually provide clear labels on the ethanol content in petrol products.

3.3.2 The Bioethanol Market

Economic disruptions caused by the COVID-19 pandemic hit agriculture in the USA hard. The bioethanol industry is a major driver of rural economic growth. As the world's lowest-cost bioethanol producer, the USA has stepped up efforts to expand in international markets. Its low-cost and high-octane bioethanol fuel has attracted users in the global transport market.

End-use market conditions have driven growth in bioethanol demand. In recent years, automakers have preferred turbocharged and high-compression-ratio engines using high-octane petrol, driving up demand for bioethanol as a clean, low-cost and high-octane fuel. While the price of regular unleaded petrol dipped, the price of premium petrol remained high. The price difference between premium and regular petrol climbed to a record high.

Policies aimed at increasing E15 uptake have enlarged the bioethanol market. In 2019, the Environmental Protection Agency allowed E15 to be sold across the country all year round. Encouraged by the increasing number of E15 refuelling stations and wide recognition from automakers, the bioethanol market grew significantly. Driven by the Higher Blends Infrastructure Incentive Program of the U.S. Department of Agriculture, the amount of bioethanol blending infrastructure has increased.

3.4 Bioethanol Developments in Brazil

Brazil was the first country to use bioethanol as a road transport fuel. In 2020, total bioethanol production was about 31.35 billion litres, down 16% from 2019, due mainly to more sugar cane being used to produce sugar. Currently, there are more than 360 sugar-cane bioethanol plants in Brazil. According to data released by the Brazilian Sugarcane and Bioenergy Industry Association (UNICA), bioethanol production in the country has tripled in the past decade. In 2018, the number of newly registered vehicles in Brazil was 2.25 million, including 2.17 million vehicles using ethanol fuel (96.4%), 4,034 electric vehicles and 81,934 petrol vehicles.

3.4.1 Bioethanol Policies

Brazil has long been one of the world's largest bioethanol producers. In 1975, the Brazilian government launched its ambitious Proálcool programme authorising the national oil company Petrobras to blend ethanol with petrol at a given ratio. The government also encouraged the large-scale cultivation of sugar cane as a feedstock and the development of flexible fuel vehicles through subsidies, industrial tax and value-added tax breaks, and policies. As of 2015, the ethanol blending ratio in petrol was increased from 25% to 27%. The government is also encouraging ethanol to be made from agricultural and forestry residues to further increase production. In 2019, the national biofuels policy RenovaBio was launched, aiming to support Brazil's nationally determined contribution commitment made at COP21 in Paris in 2015. Brazil pledged to reduce greenhouse gas emissions by 37% below 2005 levels by 2025 and by 43% by 2030. In 2020, the Brazilian Stock Exchange started to trade decarbonisation credits (CBIOs).

3.4.2 The Bioethanol Market

By creating the bioethanol market for CBIOs, RenovaBio recognised the environmental benefits of biofuels and provided financial compensation for the bioethanol sector's role in reducing greenhouse gas emissions. RenovaBio set compulsory targets to reduce the carbon intensity of transport fuels from the 73.6 g of CO_2 per megajoule (CO_2/MJ) benchmark of 2018 to 66.1 g CO_2/MJ by 2029 (down about 10.2%).

3.5 Biofuels in Europe

In 2018, ethanol fuel production reached 4.27 million tonnes, up about 1% compared with the previous year. France is the largest producer of ethanol in Europe, and Germany is the largest consumer of liquid biofuels among EU member states, with liquid biofuels representing 5.7% of its road transport fuel mix. According to the German Environment Agency, the bioethanol industry created 23,000 direct jobs in Germany. Hungary and the UK also increased ethanol production in 2018 by 38% and 23% respectively.

Due to the COVID-19 pandemic, bioethanol consumption in the EU decreased by around 10% (760 million litres), falling to 2013–16 levels. The use of biodiesel and hydrogenation-derived renewable diesel fell by nearly 6% to 1.1 billion litres, the same as in 2018, which reflected the overall fall in fuel consumption.

The proportion of biofuels in transport fuel use is expected to rise steadily. In 2020, the blending rate of biofuels in fossil fuels was 8.1%, which is only a small increase on the 7.6% of 2019. The blending rate in 2020 is estimated to be 6.2%, against 5.9% in 2019. Bioethanol accounted for 3.8% of the biofuels blended (the same as in 2019), while biodiesel and hydrogenation-derived renewable diesel made up 7.1%, compared with 6.7% in 2019.

3.6 Overview of International Biofuel Policies

The global liquid biofuel industry fluctuates in line with national policy changes. The EU has introduced policies on the sustainability and emission reduction effects of feedstocks for the production of liquid biofuels, and the USA has repeatedly extended subsidies for liquid biofuels and adjusted its targets for renewable fuels. Other countries are also working to introduce blending policies or increase the blending ratio of liquid biofuels. So far, nearly 50 countries have introduced policies to support the use of renewable fuels in transport.

3.6.1 Blending Policies

About 33 countries have mandatory policies on liquid biofuel blending. In Argentina, the ethanol blending ratio has been increased from 7% (E7) to 10% (E10), and the biodiesel blending ratio has also been raised to 10% (B10). In Brazil, the blending rate of biodiesel has been increased from 5% (B5) to 7% (B7), and that of ethanol from 25% (E25) to the upper limit of 27.5% (E27.5). The biodiesel blending ratio in Malaysia has been increased from 5% (B5) to 10% (B10).

The USA started its bioethanol development programme after the global oil shocks of the 1970s. Advanced corn cultivation, efficient ethanol production, stable and well-designed policies and market-oriented operation mechanisms provided effective support for the bioethanol industry. Initially, bioethanol was considered an important part of the national energy strategy. To this end, targeted incentives were studied and formulated, bioethanol was strategically positioned as important to energy security and its positive impacts on the environment, rural development and farmers' incomes were highlighted.

3.6.2 Tax Policies

In the USA, tax breaks and tax protection policies have advanced the biofuel industry. In 1978, the US government passed the National Energy Act, which reduced income tax for users of new energy types including bioethanol. To protect the domestic ethanol industry, the government imposed tariffs on ethanol imported from Brazil from 1980 to 2011. Bioethanol production in the USA increased exponentially in 2004, spurred by a government subsidy of more than 45 cents per gallon to bioethanol producers.

Tax incentives provide important support for the sustainable development of the biofuel industry in the USA. Among the long-term incentives for advanced biofuels, the most important include the $1.01 per gallon tax break for cellulosic fuel products and the $1 per gallon federal tax break for biodiesel. The cellulosic fuel tax break has been extended to include biofuels produced from algae, from which both producers selling biofuels to end users and producers selling biofuels to refineries benefit. The enlarged scope of the tax incentive will help promote the commercial development of algae-based biofuels.

3.6.3 Legislation

Legislation has helped develop the US bioethanol market. The Renewable Fuel Standard (RFS), created under the Energy Policy Act of 2005, requires renewable fuels like ethanol to be used in transport fuels. The Energy Independence and Security Act (EISA) of 2007 updated the RFS, requiring fuel producers to use at least 36 billion gallons of renewable fuels in transport fuels by 2022. EISA requires biorefineries to reduce life-cycle greenhouse gas emissions from corn-derived ethanol by at least 20%, from biodiesel and advanced biofuels by 50% and from cellulosic biofuels by 60%, compared with the life-cycle emissions of fossil fuels. The act also provides cash incentives, grants, subsidies and loan guarantees for biorefineries and the commercial development of advanced biofuels.

3.6.4 Industry Management

To support energy supply security and tackle climate change, the European Commission passed the Renewable Energy Directive (RED) in 2009, requiring that at least 20% of the EU's energy use should come from renewable sources by 2020. The directive set a mandatory target for the use of biofuels: at least 10% of the transport energy use of each member state should be renewable fuel by 2020. Each EU member state was required to develop its own biofuel-related regulations within 18 months and set an annual target for the increased use of biofuels, which would in turn contribute to the EU's 2020 renewable fuel consumption target.

Globally, the commercial production of cellulosic ethanol has ebbed. International crude oil prices remain low, which erodes the cost competitiveness of second-generation ethanol technologies. The world's first commercial cellulosic ethanol project in Italy has shut down. DuPont has sold its cellulosic ethanol project in the USA. Several European companies with cellulosic ethanol production technologies and capacity are struggling to find partners for commercial production. In addition, the development of electric vehicles and new fuel vehicles is bound to squeeze the growth prospects of biofuels. In the past decade, the EU has started to gradually phase out support policies for conventional biofuels. In contrast, China may become a leader in commercialising second-generation ethanol technology and stimulating the development of the global bioethanol industry, driven by favourable policies. Going forward, liquid biofuels will shift to high-value and diverse bio-refined green products, which will serve as alternatives to conventional fossil-based products and aviation fuels. Where the biofuel industry goes from here depends mainly on the type of policy support it receives from countries.

4 Current Developments in Liquid Biofuels in China

4.1 Grain-based Bioethanol

Ethanol production in China uses mainly starchy grains like corn and wheat as feedstock. The traditional production method is hydrolysis and fermentation. Biomass is converted to pentose or hexose via hydrolysis reaction, which are then converted to ethanol through fermentation. The crude ethanol is then refined into fuel through distillation and rectification. This method typically has five stages: liquefaction, saccharification, fermentation, distillation and dehydration. Grain-derived ethanol is first-generation technology, which is mature in China. In the 10th Five-year Plan (2001–05), four fuel ethanol pilot companies—Jilin Fuel Ethanol, Henan Tianguan Enterprise Group, Anhui BBCA Biochemical and Heilongjiang Huarun Ethanol—were approved to produce ethanol using mainly aged grain. In recent years, ethanol-blended fuels have been promoted in several provinces and cities across the country, which have yielded good results. The number of pilot provinces for petrol containing 10% ethanol has increased from four to nine. As a result, the sales volume climbed rapidly. In 2011, the production and sales volume of ethanol was 1.9376 million tonnes, making China the world's third-largest bioethanol producer.

The main factor influencing the cost of producing bioethanol is the price of feedstock, which accounts for 60–80% of the total. The cost of processing equipment, building materials, feedstock and transport vary across regions and between processing methods. Generally, the investment cost for ethanol plants of different capacities that use starchy feedstocks is in the range of RMB 5,800–7,000 per tonne of ethanol produced. As the cost of aged grains and corn rises, first-generation ethanol producers using grain as feedstock are mostly making a loss, even though the government has introduced tax incentives. Their gross margin ranges from -20% to -10%, which can reach 10–20% with flexible financial subsidies from the government. The consolidated net profit margin of those producers is about 12%.

Table 1 shows the cost-effectiveness of first-generation corn-based ethanol technology.[2]

[2] Assuming 2% annual increases in the cost of fixed investment and management; and an annual increase in the corn price of 3% in 2017–20, 2% in 2021–30, 1% in 2031–40, 0.7% in 2041–50 and 0.5% in 2051–60.

Table 1 Cost effectiveness of first-generation corn-based ethanol production

	2020	2030	2040	2050	2060
Corn price (RMB/tonne)	1,967	2,398	2,648	2,840	2,985
Feedstock consumption (RMB/tonne of ethanol)	3.0	2.8	2.8	2.7	2.7
Feedstock cost (RMB/tonne of ethanol)	5,901	6,713	7,415	7,667	8,059
Plant capacity (tonne/year)	300,000				
Fixed investment (RMB million)	1,061.20	1,293.60	1,576.90	1,922.20	2,343.20
Auxiliary materials consumption (RMB/tonne of product)	191	233	284	346	422
Heat consumption (RMB/tonne of product)	200	200	200	200	200
Electricity consumption (RMB/tonne of product)	100	100	100	100	100
Water consumption (RMB/tonne of product)	10	10	10	10	10
Labour and management (RMB/tonne of product)	424	517	631	769	937
Total cost of ethanol (RMB/tonne)	6,826	7,774	8,640	9,092	9,728

Source Data collected and analysed by the authors

In 2020, the purchase price of corn was about RMB 2,000 per tonne, and the production cost of corn-derived ethanol was about RMB 6,800 per tonne. Corn-to-ethanol technology is relatively mature, and improvements in conversion efficiency are limited. As the price of corn rises, the production cost of corn-to-ethanol will also rise: it is estimated to reach about RMB 7,800 per tonne by 2030 and RMB 10,000 per tonne by 2060.

4.2 Non-grain-derived Bioethanol

As the global bioethanol industry advances rapidly, both the use and price of starchy feedstocks like corn are increasing, which makes grain feedstock a barrier to the further development of the ethanol industry. As a result, bioethanol produced from non-grain feedstocks is the way forward. At present, the main non-grain feedstocks for ethanol production in China are cassava and sweet sorghum.

Ethanol made from non-grain feedstocks has cost advantages over grain-based ethanol. However, the price of cassava is also rising, which means the cost of ethanol made from non-grain feedstock is rising too. Although the cassava cultivation area is more than 7 million mu (466,690 hectares) in China, it is insufficient to meet the requirements of local bioethanol production, and imports from Thailand, Vietnam and other countries are needed. Given that it is difficult for global cassava production to increase significantly in the short term, and given the significant growth in demand from the ethanol industry, the price of imported cassava will rise, leading to gradually higher production costs for non-grain-based ethanol. This will affect the profitability of ethanol producers. In the past decade, the number of new cassava-based ethanol projects in China rose slowly, and the cassava cost per tonne of ethanol produced was close to RMB 5,000. Tables 2 and 3 provide a cost analysis of ethanol made from cassava and sweet sorghum respectively.

4.3 Cellulosic Ethanol

Cellulosic ethanol is a second-generation biofuel that has three main stages of processing: feedstock pre-treatment, hydrolysis and fermentation. There are five ways to pre-treat feedstock: physical, chemical, physical–chemical, biological, and various combinations of the four. Dilute acid or steam explosion is the most widely used. Cellulose hydrolysis is performed with dilute acid or enzymes. Due to the high levels of equipment robustness needed for cellulose hydrolysis under high temperatures and acidic conditions, as well as other inhibiting factors like waste-water

Table 2 Cost effectiveness of cassava-based ethanol

	2020	2030	2040	2050	2060
Fresh sweet potato price (RMB/tonne)	765	932	1,030	1,104	1,161
Feedstock consumption (RMB/tonne of ethanol)	6.3	6.0	5.8	5.5	5.3
Feedstock cost (RMB/tonne of ethanol)	4,819	5,594	5,974	6,074	6,152
Plant capacity (tonne/year)	200,000				
Fixed investment (RMB million)	1,059.50	1,385.50	2,037.50	4,075.00	4,075.00
Auxiliary materials consumption (RMB/tonne of product)	244	298	363	442	539
Heat consumption (RMB/tonne of product)	200	200	200	200	200
Electricity consumption (RMB/tonne of product)	317	321	374	474	474
Water consumption (RMB/tonne of product)	100	100	100	100	100
Labour and management (RMB/tonne of product)	424	517	631	769	937
Total cost of ethanol (RMB/tonne)	6,104	7,030	7,641	8,059	8,402

Source Data collected and analysed by the authors

Table 3 Cost effectiveness of ethanol derived from sweet sorghum[3]

	2020	2030	2040	2050	2060
Sweet sorghum stalk price (RMB/tonne)	284	346	383	410	431
Feedstock consumption (RMB/tonne of ethanol)	16	16	15	15	15
Feedstock cost (RMB/tonne of ethanol)	4,546	5,541	5,738	6,152	6,467
Plant capacity (tonne/year)	100,000				
Fixed investment (RMB million)	31,836	38,808	47,307	57,666	70,296
Auxiliary materials consumption (RMB/tonne of product)	424	517	631	769	937
Heat consumption (RMB/tonne of product)	200	200	200	200	200
Electricity consumption (RMB/tonne of product)	300	300	300	300	300
Water consumption (RMB/tonne of product)	35	35	35	35	35
Labour and management (RMB/tonne of product)	424	517	631	769	937
Total cost of ethanol (RMB/tonne)	5,730	6,911	7,335	8,025	8,676

Source Data collected and analysed by the author

management, producers are gradually switching to enzymatic hydrolysis, which is more eco-friendly. The main fermentation methods include enzymatic simultaneous saccharification and (co-)fermentation (SSF or SSCF, cellulose hydrolysis and ethanol fermentation take place in the same process), and enzymatic separate hydrolysis and fermentation (SHF, cellulose hydrolysis and separation produce a fermentable sugar solution before fermentation). COFCO Biochemical Energy (Zhaodong), Anhui BBCA Group and Shandong Longlive Bio-technology have built kilotonne-class pilot plants to produce cellulosic ethanol. Although the cellulosic ethanol process is making progress, challenges remain in terms of feedstock collection, pretreatment, saccharification, fermentation and rectification.

In the 12th Five-year Plan (2011–15), the development of cellulosic ethanol hit a barrier,

[3] Assuming 2% annual increases in the cost of fixed investment and management, and an annual increase in the price of sweet sorghum of 3% in 2017–20, 2% in 2021–30, 1% in 2031–2040, 0.7% in 2041–50 and 0.5% in 2051–60.

Table 4 Cost effectiveness of cellulosic ethanol

	2020	2030	2040	2050	2060
Straw price (RMB/tonne)	328	400	441	473	497
Feedstock consumption (RMB/tonne of ethanol)	5	5	4	4	4
Feedstock cost (RMB/tonne of ethanol)	1,639	1,998	1,766	1,893	1,990
Plant capacity (tonne/year)	100,000				
Fixed investment (RMB million)	424.48	517.44	630.76	768.88	937.28
Auxiliary materials consumption (RMB/tonne of product)	4,000	3,268	2,670	2,182	1,782
Heat consumption (RMB/tonne of product)	300	300	300	300	300
Electricity consumption (RMB/tonne of product)	300	300	300	300	300
Water consumption (RMB/tonne of product)	120	120	120	120	120
Labour and management (RMB/tonne of product)	424	517	631	769	937
Total cost of ethanol (RMB/tonne)	6,784	6,503	5,786	5,564	5,429

Source Produced by the authors

caused by unviable commercial projects in overseas markets and slow progress in 10,000-tonne commercial projects in China. Table 4 estimates the cost of producing cellulosic ethanol over the next four decades. The variable cost consists mainly of feedstock price and management costs, which rise annually; the amount of feedstock and auxiliary materials used gradually decreases as the processing technology improves.[4] In 2020, feedstock (straw) consumption was 5 tonnes per tonne of ethanol produced. If the price of straw is RMB 328 per tonne, the cost of producing ethanol is estimated to be about RMB 6,800 per tonne. By 2030, the amount of feedstock needed to produce cellulosic ethanol is expected to stabilise at 5 tonnes of straw per tonne of ethanol produced, and the production cost is forecast to decline to about RMB 6,500 per tonne. After 2040, the amount of feedstock needed is expected to decrease to 4 tonnes of straw per tonne of ethanol produced. By 2060, the cost of producing cellulosic ethanol is estimated to be about RMB 5,400 per tonne.

[4] Assuming 2% annual increases in the cost of fixed investment and management, and an annual increase in the price of straw of 3% in 2017–20, 2% in 2021–30, 1% in 2031–2040, 0.7% in 2041–50 and 0.5% in 2051–60.

4.4 The Bioethanol Market

In China, the production and sale of ethanol follows the principles of "designated production, designated circulation and restricted operation". To this end, China has designated 10 provinces for the mandatory uptake of ethanol-blended petrol, namely Heilongjiang, Jilin, Liaoning, Hebei, Shandong, Anhui, Henan, Jiangsu, Hubei and Guangxi. After years of pilot operation and promotion efforts in the regions concerned, the pilot areas achieved an average ethanol-blended petrol coverage of more than 90%, and the use of ethanol-blended petrol accounted for more than 20% of the country's total petrol consumption. A complete value chain of ethanol-blended petrol —from production, blending, storage and transport to sale—has taken shape. Table 5 lists seven designated ethanol sellers.

Figure 3 gives bioethanol output in China from 2011 to 2020. During the 12th Five-year Plan (2011–15), China's bioethanol production grew steadily, exceeding 2 million tonnes for the first time in 2012. Driven by the Implementation Plan on Expanding Bioethanol Production and Promoting the Use of Ethanol-blended Petrol for Vehicles, demand for ethanol grew and production capacity in China rapidly increased. In 2018, output of bioethanol went beyond 3 million tonnes, an increase of about 35% on the previous year.

Table 5 Designated ethanol fuel sellers in China

Producer	Place of production	Main feedstocks	Sources of supply
COFCO Biochemical Energy (Zhaodong)	Heilongjiang (Zhaodong)	Corn	Heilongjiang
Jilin Fuel Ethanol	Jilin (Jilin)	Corn	Jilin
			Liaoning
Henan Tianguan Enterprise Group	Henan (Nanyang)	Wheat	Henan
			Hubei (9 cities)
			Hebei (4 cities)
COFCO Biochemical (Anhui)	Anhui (Bengbu)	Corn	Anhui
			Shandong (7 cities)
			Jiangsu (5 cities)
			Hebei (2 cities)
COFCO Biomass Energy (Guangxi)	Guangxi (Beihai)	Cassava	Guangxi
Shandong Longlive Bio-technology	Shandong (Dezhou)	Corncobs, corn straw	Shandong
Zhongxing Energy	Inner Mongolia (Bayannur)	Sweet sorghum	Inner Mongolia

Source Produced by the authors

According to the data (which are incomplete), at the end of 2019, the potential new production capacity of ethanol exceeded 9 million tonnes. This included about 5.6 million tonnes of bioethanol and about 3.45 million tonnes of fossil-based ethanol. Among the new ethanol production facilities, only two projects were in operation. About 45% of the new ethanol plants use industrial exhaust gas and a small amount of coal-based feedstock.

During the COVID-19 pandemic, petrol consumption dropped significantly in 2020, and the use of ethanol shifted from fuel to medical ethanol, further decreasing the ethanol blending rate in petrol. In the first quarter of 2020, petrol production and consumption in China fell by about 30% year-on-year. Record low international oil prices further dampened the willingness of oil companies to blend bioethanol in petrol, and domestic petrol prices dipped below that of ethanol from March 2020. In addition, the rising corn price pushed up the production cost of bioethanol, forcing the provinces that planned to implement E10 policies in 2020 to scale back. As a result, bioethanol producers in China reported high levels of inventory and redirected some of their production capacity towards medical ethanol to combat the virus.

4.5 Biodiesel

Biodiesel is an important component of liquid biofuels in China. Biodiesel is made from waste oils and fats, in line with the approach of the circular economy. In 2019, biodiesel production capacity was more than 2 million tonnes per year, and biodiesel yield was about 1.2 million tonnes. Europe is the main biodiesel product market. In China, only Shanghai has developed biodiesel demonstration projects. Since the launch of B5 biodiesel at service stations run by Sinopec Shanghai Petrochemical Company Limited (SPC) in 2017, an average of about 15,800 vehicles refuel with biodiesel daily. This pilot project has helped process nearly 30,000 tonnes of used cooking oil. SPC has built one B5 biodiesel blending facility, which has an annual delivery capacity of between 400,000 and 600,000 tonnes. More than 240 SPC service

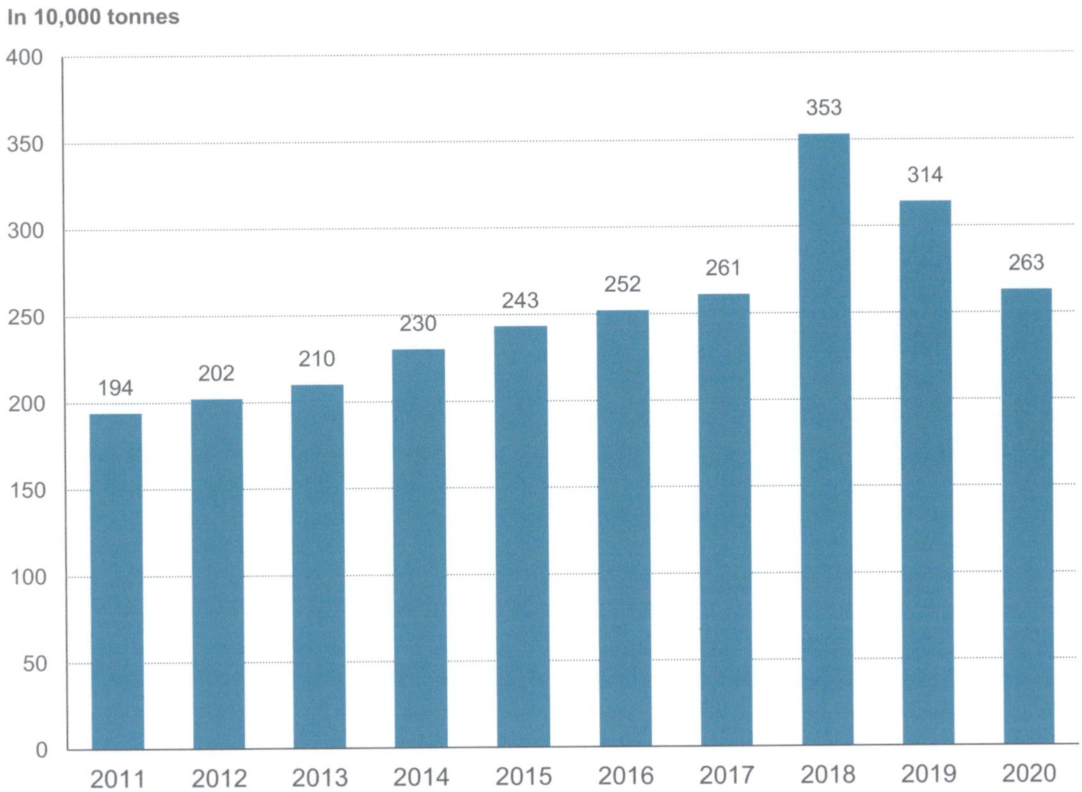

Fig. 3 China's bioethanol production, 2011–20. *Source* Produced by the authors

stations in 13 administrative districts of Shanghai supply B5 biodiesel, accounting for 41% of the total number of operating outlets.[5]

4.6 Challenges Facing the Liquid Biofuel Industry

4.6.1 It Is Difficult for Feedstock Supply to Meet Market Demand

In China, it is difficult for bio-based feedstocks to meet market demand in the short term under present ethanol policies. Achieving better control of the sources of ethanol feedstocks will influence the direction and sustainability of the industry. The challenge for the sustainable development of ethanol-blended petrol in China is to clearly define production from non-grain feedstocks and strictly control the production of ethanol from fossil-based raw materials. Corn stock has been gradually controlled to a reasonable level, as a result of the deregulation of the corn market. Existing and new production capacity of ethanol made from corn feedstock suggests it is unlikely to change in the near and medium terms. The planting area and supply capacity of domestic cassava cannot support a large increase in production in the short term, and dependence on imported cassava will not ensure the cost effectiveness and stability of feedstock supply. Sweet sorghum cultivation is risky, and sweet sorghum-derived ethanol projects have yet to reach scale.

[5] A total of 10 million vehicles had refuelled with biodiesel at SPC service stations by June 2020.

4.6.2 Inadequate Quality Testing Standards and Certification

The liquid biofuel industry has weak supervision and uneven product quality. Scientific methods for calculating life-cycle greenhouse gas emissions for bioethanol and biodiesel produced from different feedstocks are not in place, and a method for calculating carbon emission reductions from blended bioethanol and biodiesel does not exist. In addition, monitoring and evaluating emission targets have been inadequate. The supervision of clean fuel product standards and production indicators is inadequate and the country lacks certification organisations eligible to test bioethanol products. The quality certification and regulatory procedures for liquid biofuels, from product delivery to blending in petrol, need to be improved urgently.

4.6.3 Bioethanol Market Mechanisms Are Inflexible

For the first 20 years or so of bioethanol demonstration and promotion in China, closed mandatory blending in designated pilot provinces and cities was the norm and the low-carbon characteristics of bioethanol were excluded from the market trading system. Distributors and consumers have passively participated in this system for a long time and the mandatory blending approach has fallen short of market requirements. The abatement qualities of bioethanol have not been linked with the decarbonisation of transport. When fuel ethanol production dipped, some provinces and cities discontinued the sale of ethanol-blended petrol and did not implement regulatory and response measures. Bioethanol is priced against petrol of a corresponding grade multiplied by a fixed coefficient, without considering the emission characteristics of and differences between bioethanol produced from different feedstocks. As a result, the market price of bioethanol does not fully reflect its emission reduction effects and social and environmental value.

4.6.4 The Biodiesel Market in China Is Progressing Slowly

Biodiesel companies in China are mainly privately owned. The market is constrained by an imperfect feedstock collection system, no feedstock quality standards and restricted access for biodiesel products in the fossil fuel trading market. Biodiesel production capacity is limited and strong support policies for blending biodiesel with diesel are not in place. In the international market, the demand for biodiesel in Europe is growing, especially biodiesel made from used edible oils, which has good life-cycle carbon reduction benefits. China's biodiesel products are made mainly from used edible oils and waste animal and plant oils, and have life-cycle abatement effects that are actively advocated by the EU. As a result, exports of China's biodiesel feedstock and products to Europe have grown rapidly in recent years. This encourages the development of the domestic biodiesel market and the decarbonisation of transport.

4.7 China's Bioethanol Policies

Between 2001 and 2010, China implemented support policies for ethanol made from aged grain, which encouraged the production of liquid biofuels and the use of ethanol-blended petrol in motor vehicles. These policies included production tax incentives and product subsidies, government-guided prices, subsidies for feedstock production, and others.

4.7.1 Production and Use Policies

In February 2004, eight central departments including the National Development and Reform Commission (NDRC) jointly issued the Notice on Issuing the Expanded Pilot Program for Ethanol-blended Gasoline for Motor Vehicles and the Implementation Rules for the Expanded Pilot Program for Ethanol-blended Gasoline for Motor Vehicles. According to this document, China National Petroleum Corporation and Sinopec are required to buy fuel ethanol and blend petrol containing ethanol. The programme was implemented in five provinces (Heilongjiang, Jilin, Liaoning, Henan and Anhui) and 27 prefecture-level cities in Hebei, Shandong, Jiangsu and Hubei. Its purpose was to promote the uptake of petrol containing 10% ethanol.

4.7.2 Purchase and Retail Prices

The programme also states that the settlement price of denatured ethanol should be the same as that for 90 octane petrol (for military and national reserves) multiplied by a price conversion coefficient of 0.9111. In 2011, the settlement price of denatured ethanol was adjusted using the supply price of 93 octane petrol for the military and national reserves announced by the NDRC, multiplied by the same price conversion coefficient of 0.9111. This settlement price took effect in March 2011, and the retail price of ethanol-blended petrol became the intermediate benchmark retail price of regular petrol of the same grade.

4.7.3 Tax Incentives

In accordance with the Notice of the Ministry of Finance and the State Taxation Administration on the Tax Policy Issues Related to the Designated Producers of Denatured Ethanol, ethanol producers are exempt from the 5% consumption tax; in addition, the aged grain used qualifies for the aged grain subsidy. As the development direction of non-grain ethanol is clearly defined, a plan for phasing out related tax incentives has also been formulated.

In the 11th Five-year Plan (2006–10), China's ethanol producers relied mainly on government subsidies, with the annual subsidy for the four ethanol producers exceeding RMB 2 billion. In 2007, the NDRC made it clear that grain should no longer be used as feedstock for bioenergy production, and that non-grain crops should be used instead. As stated in the Interim Measures for the Management of Special Funds for the Renewable Energy Development issued by the Ministry of Finance, the development and use of renewable energy as an alternative to oil should support the development of bioethanol and biodiesel. Bioethanol refers to ethanol produced from sugar cane, cassava, sweet sorghum and other crops.

Due to constraints on resource availability, feedstock prices and environmental requirements, non-grain-derived ethanol projects have not progressed. With the rise in international grain prices, ethanol producers using grain as feedstock faced mounting pressure. To avoid "competing for grain with the populace and competing for land with grain crops", China supported the development of biofuels made from non-grains with appropriate national polices during the 12th Five-year Plan (2011–15).

In the past decade, China has introduced a series of policies to encourage the production of liquid biofuels and the uptake of ethanol-blended petrol. These policies comprise production tax incentives and subsidies, government-guided retail prices and subsidies for feedstock production. In recent years, growth in ethanol production has been driven mainly by capacity expansion by grain-to-ethanol producers. To control the output of grain-to-ethanol, China has gradually reduced its subsidies for ethanol. Fiscal and tax incentives will also be phased out. Depending on how the industry develops, the incentive policies will be adjusted to control trends and guide the industry towards non-grain-derived ethanol to ensure the healthy and sustainable development of the ethanol industry.

4.7.4 National Roll-out Policies

In September 2017, 15 central departments including the NDRC and the National Energy Administration jointly issued the Implementation Plan on Expanding Bioethanol Production and Promoting the Use of Ethanol-blended Gasoline. The plan required China to develop ethanol from grain in a scientific and controlled manner and to develop advanced liquid biofuels, including cellulosic ethanol, to meet growing market demand. Its areas of focus included access to the ethanol-blended petrol market, the sustainability of ethanol feedstocks and market prospects. The executive meeting of the State Council in August 2018 defined the measures needed to develop the natural gas sector and the bioethanol industry. According to the decisions taken at that meeting, the uptake of ethanol-blended petrol was to be expanded in an orderly manner to an additional 15 provinces including Beijing, Tianjin and Hebei. The principles of total control of production, a limited number of designated producers and fair market access were to be followed, and idle alcohol production facilities were to be used. Investments in grain-to-ethanol production were to be made and cassava-based ethanol projects were to be constructed at a faster pace. Commercial demonstration projects of ethanol produced from straw and exhaust gas from the iron and steel sector were to be developed. In 2018, the main regions of air pollution prevention and control—such as the Beijing–Tianjin–Hebei region and surrounding areas, and the Yangtze River and Pearl River deltas—were to make ethanol-blended petrol available market-wide by 2019. By the following year, ethanol-blended petrol was to be available nationwide. Exceptions were the military, national and special reserves, and industrial oil applications.

5 Roadmap for Liquid Biofuels

5.1 Liquid Biofuels

Using the current supply capacity of liquid biofuels in China, this section forecasts the supply capacity of liquid biofuels over time taking into account existing available resources, marginal land development, feedstock prices, advances in production technologies and market demand.

The current supply capacity of liquid biofuels is around 7 million tonnes of coal equivalent (tce). By 2030, the supply capacity of bioethanol, biodiesel and sustainable aviation fuel is expected to increase significantly and their blending ratios are forecast to be 15%, 10% and 5% respectively. By 2040, bioethanol will contribute the most additional supply capacity to liquid biofuels, with the supply capacity of sustainable aviation fuel exceeding 10 million tce and accounting for 20% of the air transport fuel mix. By 2050, total supply of liquid biofuels is projected to exceed 90 million tce; liquid biofuels are expected to become the main clean alternative fuels in the transport sector; and the blending ratio of bioethanol, biodiesel and sustainable aviation fuel is set to be 85%, 50% and 50% respectively. By 2060, liquid biofuels will cover the demand of the non-electrified aviation segment, and the total supply of bioethanol, biodiesel and sustainable aviation fuel will exceed 100 million tce (Table 6).

5.2 Bioethanol

The raw materials of cellulosic bioethanol are mainly agricultural and forestry residues, which require limited amounts of land. If non-grain crops such as cassava and sorghum are used as raw materials for bioethanol fuel, reserved land will have to be used and large-scale cultivation adopted to ensure raw material supplies. Bioethanol production in the future will depend mainly on the effective development of marginal lands.

In China, planted land and reserved land for growing biofuel feedstock crops (mainly cassava, sorghum and sweet potato) is currently around 5.5 million hectares and 6.741 million hectares respectively. The percentage of reserved land developed is expected to be 10% in 2025, 15% in 2030, 25% in 2040 and 35% in 2050.

Near term (before 2025): By 2025, a biomass resources database will be established and reserved land across China will be developed efficiently. The problems related to the collection, storage and transport of the raw materials used to produce bioethanol from non-grain crops will be solved. The amount of land cultivated will be significantly increased and the production technologies used will be mature. Waste-water treatment capacity will equal production capacity, and energy consumption in production will be greatly reduced. Modern management and operational models will be implemented. The main technological bottlenecks of cellulosic ethanol will be addressed, the manufacturing cost of cellulose will be less than 30% of the cost of bioethanol fuel (currently 70%), and large-scale cellulosic ethanol demonstration plants will be built. Policies related to biofuels will be optimised, a regulated market access mechanism will take shape, and a nationwide biofuel market trading system will be established.

Medium term (2025–30): The large-scale commercial development of bioethanol from sugar crops like sorghum will be possible.

Table 6 Liquid biofuel supply over the coming decades (in million tce)

	2030	2040	2050	2060
Bioethanol	17.05	32.14	43.16	50.06
Biodiesel	17.80	19.68	22.81	24.77
Sustainable aviation fuel	3.24	11.72	25.16	30.11
Total	38.09	63.54	91.13	104.94

Source Estimated by the authors

Microbial fermentation technology that provides resistance to high temperature, high ethanol concentration and high permeability will be deployed, and ethanol separation technology that eliminates phase change will be adopted. The cost competitiveness of bioethanol with petroleum will increase. Technology breakthroughs in pentose–hexose co-fermentation will be achieved, and cellulosic ethanol will enter production. Energy crops that grow on barren land will be planted at scale in saline–alkali soils and sandy wasteland. The technologies for increasing the starch in starchy crops and sugar in sugar crops will be developed, as will the strains of shrubs and grasses that provide high yields and withstand wind, sand and drought. The required vehicle technologies will be launched, allowing bioethanol to play an important role in the transport fuel mix.

By 2030, a complete database of biomass resources will exist, and available land resources will be developed and the use of agricultural and forestry residues will be intensified. Biofuel production technologies will be comparable to those of developed economies, and the large-scale production of second-generation bioethanol will be the focus of the biofuel industry. Biofuels will be the main alternatives to fossil fuels. Table 7 forecasts bioethanol supply capacity over the coming decades.

5.3 Biodiesel

Biodiesel in China comes mainly from feedstocks such as used cooking oil and animal and plant oils and fats. At present, used cooking oil in China can produce about 5 million tonnes of biodiesel annually. As China's biodiesel market evolves, biodiesel could be produced using oil processing waste and energy crops as feedstocks.

Under the dual-carbon goal, there will be limited room for growth in waste oils and fats. With the increasing pace of urbanisation and improvement in living standards, the amount of used cooking oil is expected to increase by 2% per year by 2040, after which it will gradually decrease by 1% annually. In 2020, the total supply of animal oils and fats and oil processing waste could produce about 2 million tonnes of biodiesel, which is expected to grow annually by 3% over the coming years. Table 8 forecasts biodiesel supply capacity over the coming decades.

After 2030, the main increase in biodiesel feedstocks is expected to come from oil-yielding energy forests. After 2040, energy crops will grow into one of the main feedstock sources for large-scale biodiesel production. According to the statistics of the National Forestry and Grassland Administration, China has more than 10 million hectares of barren mountains and wastelands, as well as land affected by desertification and saline–alkali lands, which could be used for the development of forestry-derived bioenergy. At present, the technology for converting biodiesel from the seeds of jatropha, Chinese pistache and *Cornus wilsoniana* is relatively mature. Among the existing oil plants in China, there are more than 150 varieties of plant that have a seed oil content of more than 40%. Arbor and shrub species that can be cultivated at scale include oil palm, soapberry, jatropha, *Cornus wilsoniana* and *Xanthoceras*. The area in which oil palm and soapberry are distributed in relatively high concentrations exceeds 1 million

Table 7 Bioethanol supply capacity over the coming decades (in tonnes)

Bioethanol feedstocks	2022	2025	2030	2040	2050	2060
Cassava	30,000	60,000	110,000	210,000	340,000	410,000
Sorghum	1,160,000	2,310,000	3,760,000	7,200,000	9,870,000	11,520,000
Sweet potato	437,500	925,000	1,500,000	3,214,000	4,286,000	5,000,000
Cellulose	833,300	2,000,000	6,000,000	15,000,000	20,000,000	25,000,000
Aged grain	3,000,000	6,000,000	6,000,000	6,000,000	6,000,000	6,000,000
Industrial exhaust gas fermentation	90,000	300,000	1,000,000	3,000,000	6,000,000	6,000,000

Source Estimated by the authors

Table 8 Biodiesel supply capacity over the coming decades (in million tonnes)

Biodiesel feedstocks	2020	2025	2030	2040	2050	2060
Used cooking oil	5.00	5.52	6.09	7.43	6.72	6.08
Animal oils and fats	1.00	1.16	1.34	1.81	2.43	3.26
Oil processing waste	1.00	1.16	1.34	1.81	2.43	3.26
Energy crops	0	0.20	1.00	2.00	4.00	6.00
Total	7.00	8.04	9.77	13.05	15.58	18.60

Source Estimated by the authors

hectares. This provides an annual fruit yield of more than 1 million tonnes, which is enough feedstock to make 400,000 tonnes of biodiesel.

5.4 Sustainable Aviation Fuel

Sustainable aviation fuel can be made from a variety of feedstocks—such as lignocellulose, oil-bearing crops and waste oils and fats—and using various technologies. At present, research on sustainable aviation fuel is focused on oil-bearing crops such as lignocellulose and algae. Feedstock selection in the future will depend mainly on the cost competitiveness of conversion technologies. China's sustainable aviation fuel technologies are still at the R&D stage. They are not expected to reach large-scale production until 2030 when supply capacity will be more than 2 million tonnes, which is 5% of all aviation fuels. By 2040, the supply capacity of sustainable aviation fuel will exceed 8 million tonnes or 20% of total aviation fuel demand. By 2050, this will reach 17 million tonnes or half of total aviation fuel demand. And by 2060, the supply of sustainable aviation fuel is projected to exceed 20 million tonnes or 100% of aviation fuel demand. Table 9 forecasts sustainable aviation fuel supply capacity over the coming decades.

6 Policy Recommendations for the Liquid Biofuel Industry

6.1 Build an Upstream Feedstock Supply System in the Near Term

The future development of biofuels depends on feedstock supply capacity. Efforts should be made to increase feedstock supply capacity and establish a logistics system for the cultivation, planting, collection, transport and storage of energy crops. China has abundant agricultural and forestry residue resources, although the existing feedstock supply system and feedstock market prices do not favour the sustainable development of a biofuel industry. China should create a system for the large-scale cultivation and collection of energy crops and intensively

Table 9 Sustainable aviation fuel supply capacity over time (in million tonnes)

	2025	2030	2040	2050	2060
Sustainable aviation fuel	0.06	2.22	8.04	17.26	20.66

Source Estimated by the authors

develop marginal lands. This would support the sustainable development of a large-scale biofuel industry.

6.2 Biofuel Standards and Quality Certification Are Urgently Needed

Research on bioethanol quality certification standards should be carried out. A biofuel feedstock traceability system and carbon emission certification system should be established to determine the carbon emission coefficient of biofuels. The carbon intensity of fuel containing different percentages of bioethanol should be calculated, based on which each grade of fuel should be given a clean fuel quality rating.

6.3 Gradually Develop a Bioethanol Trading Market

Emission reduction levels for green certification should be set and fossil fuel producers and distributors given carbon abatement targets. Fossil fuel producers and distributors should be allowed to fulfil their annual abatement targets by purchasing bioethanol green certificates. Stable market demand for biofuels should be built up through the biofuel green certification and trading system, enabling cleaner biofuels to gradually replace fossil fuels in an orderly manner.

6.4 Develop a Medium- and Long-term Strategy for Liquid Biofuels

Liquid biofuels play an important role in ensuring energy security, combatting climate change, decarbonising transport and supporting high-quality social and economic development. Because feedstocks for liquid biofuels come mainly from rural areas, the development of a liquid biofuel industry is in line with the government's rural revitalisation strategy. To this end, a medium- and long-term development strategy for liquid biofuels made from rural agricultural and forestry residues should be formulated. In addition, the strategic positioning, development targets, application markets and policy support needed for liquid biofuels should be defined. This will help revitalise rural areas and ecologies and create a sustainable transport sector.

Chapter 10: Hydrogen

Miao Li, Yongliang Li, Rui Rui Liang, Mei Dong and Joep Huijsmans

1 Introduction

Hydrogen will be an important part of the diverse low-emission energy supply system of the future. It will make a big difference in the fight against climate change and environmental pollution.

This chapter describes the hydrogen industry from upstream supply, midstream storage and transport to downstream demand in China and across the world. It summarises the main uses of hydrogen supply and demand; explains the hydrogen production, storage, transport and marketing system; and analyses the cost structure and technologies used in different parts of the hydrogen value chain. The chapter also examines the improvements needed to enable the large-scale development of hydrogen—in policies, quality standards, research and development and cost effectiveness—and it forecasts supply and demand trajectories in both international and domestic markets. The chapter offers an action plan to make blue hydrogen (produced from natural gas) a breakthrough for the development of the hydrogen industry. It discusses the current distribution and use of blue hydrogen resources in China and explores possible routes of industrial by-product hydrogen production in the future. The chapter concludes with policy recommendations for the development of the hydrogen industry around the three strategic goals of "energy transition, deep decarbonisation, and high-quality development".

As climate change and environmental pollution increase, the research, development and use of renewable energy becomes increasingly important. Hydrogen is a potentially low-emission and efficient secondary energy source that can help decarbonise the conventional energy system. Many countries regard hydrogen as an important part of their efforts to develop a cost-efficient, low-carbon and sustainable energy system. Countries such as Japan, South Korea and Germany have incorporated hydrogen into their national energy strategies. In 2014, the Japanese government pledged to create a "Hydrogen Society". In 2017, the country took a further step with its "Basic Hydrogen Strategy" to help revitalise the Japanese economy and enhance industrial competitiveness. These strategies were followed by a series of policies to achieve those goals. In 2018, South Korea announced its "Hydrogen Economy", one of three strategic areas for investment and innovation. A year later, the country released its "Hydrogen Economy Roadmap" to guide its transition from a carbon economy to a hydrogen economy. In 2022, the government announced a hydrogen development strategy and targets to be

M. Li (✉) · Y. L. Li · R. R. Liang · M. Dong
China Petroleum and Chemical Industry Federation, Beijing, China

J. Huijsmans
Shell Global Solutions International B.V., The Hague, The Netherlands

Shell International B.V. and Development Research Centre of the State Council (DRC) of People's Republic of China, *Embracing the Future, Powering Growth: An Energy System Renewed for China*, Advances in Oil and Gas Exploration & Production,
https://doi.org/10.1007/978-3-031-73075-7_11

achieved by 2040. Many international organisations recognise the potential of hydrogen. For example, the Hydrogen Council and the International Energy Agency are promoting the commercialisation of hydrogen. According to forecasts by the Hydrogen Council, hydrogen will meet about 12% of global primary energy demand and the hydrogen market will be worth more than $2.5 trillion by 2050.

Thanks to policy support and clear strategic plans, China's hydrogen industry gained momentum from 2018. Hydrogen was mentioned by the government for the first time in 2019 and is now included in national energy data, signalling the growing importance of hydrogen in national strategy. Local government has also formulated and implemented policies and plans for the hydrogen industry. According to the Research Report on Hydrogen Development Strategy in the Petrochemical and Chemical Industries published by the Special Committee on Hydrogen of the China Petroleum and Chemical Industry Federation, by the end of 2020 nearly 50 local governments had issued plans to develop the hydrogen industry in China; investments in hydrogen-centred industrial parks and towns exceeded RMB 100 billion; and the sale of more than 100,000 hydrogen fuel cell electric vehicles and the construction of more than 1,000 hydrogen refuelling stations were planned. A hydrogen industrial ecosystem in the Yangtze River Economic Belt continues to expand and the development of a hydrogen industry in the Guangdong–Hong Kong–Macao Greater Bay Area has gradually taken shape. A hydrogen industry boom is evident in almost all provinces—each province strives to build up technological competitiveness and promote its advantages. The industrial clusters centred on Beijing, Shanghai and Guangdong are gradually expanding to include the hydrogen value chain, and new hydrogen industrial clusters are growing in Shandong, Shanxi, Shaanxi, Sichuan and Wuhan.

In March 2022, the National Development and Reform Commission (NDRC) and the National Energy Administration (NEA) jointly issued the Medium- and Long-term Plan for the Development of the Hydrogen Industry for 2021–35. The plan states that hydrogen is part of China's energy system of the future and that clean and low-carbon hydrogen are integral to the low-carbon transition of end-use sectors such as transport and industry and energy-intensive and high-emission industries. The report goes on to say that hydrogen is key to emerging industries of strategic importance and is a new growth driver for a green and low-carbon industrial system and for implementing industrial transformation and modernisation.

Thanks to its high energy density, potential for zero-carbon emissions, chemical properties and suitability for different applications, hydrogen not only brings disruptive potential to transport, it can also enable the deep decarbonisation of traditional industrial sectors such as petrochemicals, chemicals and iron and steel, as well as power supply and heating. In industry, hydrogen can help deep decarbonisation both as energy and feedstock: it can replace coke in hydrogen-based direct reduced iron (H_2–DRI), it can be substituted for coal and natural gas in hydrogen-based direct synthesis of ammonia, and it can provide high-quality heat when hydrogen fuel cells are used in combined heat and power applications. In transport, hydrogen fuel cell electric vehicles have a long range, short refuelling time and tolerate low temperatures, which gives them a potentially large market in China. In buildings, fuel cells for home use are energy efficient and low carbon. For instance, the fuel cell for home use in Japan's "ENE-FARM" project provides up to 90% energy efficiency in a combined heat and power application, the cost of which is rapidly declining, making it increasingly market competitive.

Hydrogen fuel cells can also be used as a standby emergency power source. The roll-out of 5G base stations and large data centres will increase the use of hydrogen as a standby emergency power source. Hydrogen is also a flexible energy carrier; it can improve renewable energy use, provide long-term energy storage and optimise regional energy flows, helping to create a new energy development model in which multiple energy forms interplay and complement

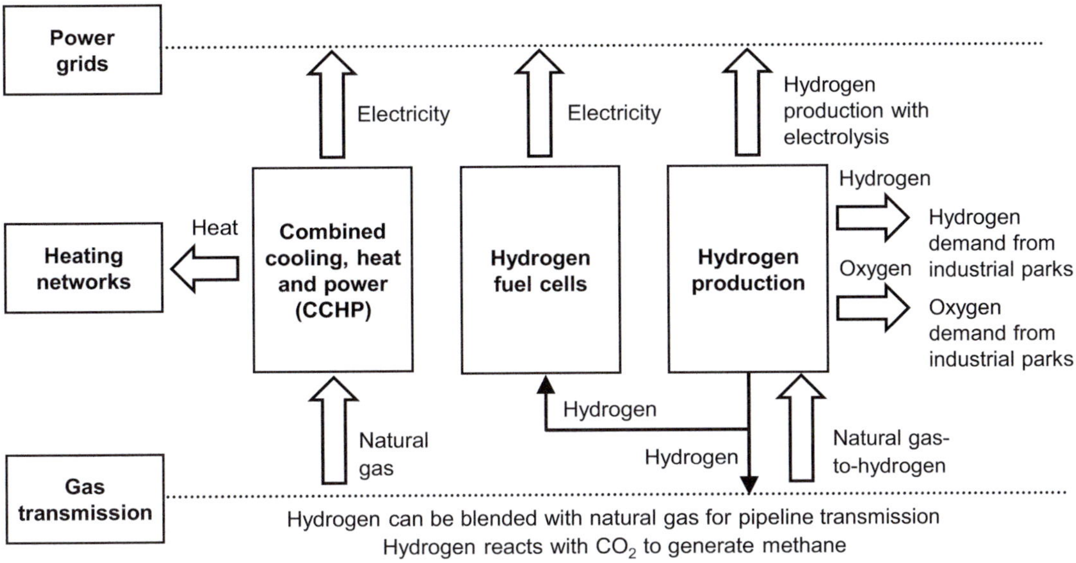

Fig. 1 Schematic diagram of a hydrogen-centred multi-energy complementary model. *Source* Produced by the authors with reference to the Positioning and Role of Hydrogen in China's Energy Transition

each other. The hydrogen-centred multi-energy complementary model is shown in Fig. 1. Hydrogen technologies such as electrolysis and fuel cells enable power grids, district heating networks and gas transmission systems to be integrated and operated in a dynamic way. Hydrogen produced with electrolysis can provide rapid response to power grid needs within minutes, helping the power system to implement peak load shaving and increase local consumption capacity for renewable energy. In short, the development of hydrogen will make a great difference to China's efforts to optimise its energy mix and achieve its dual-carbon goal.

2 The Hydrogen Industry in China and the World

This section looks at the hydrogen industry from upstream supply, midstream storage and transport, and downstream demand in China and across the world. On the demand side, it discusses the main applications of, and demand for, hydrogen as both feedstock and fuel. On the supply side, it describes the main hydrogen production technologies and their use in China and compares the cost of the various production processes. The main hydrogen storage and transport methods are explained and their features and deployment assessed.

2.1 Hydrogen Use

Hydrogen will be an important energy carrier in the diverse energy supply system of the future, which will be dominated by clean energy. This section discusses the main uses of hydrogen today and analyses the trends of hydrogen use tomorrow.

Hydrogen is currently used as an industrial feedstock and reducing agent. These two uses account for around 99% of total hydrogen use globally. By industry, 66% of hydrogen is consumed by the chemical industry, 26% by oil refining and 7% by metals and glass processing. Only 1% of hydrogen production is used by buildings, transport and other sectors. Europe, the Middle East, the USA and China are the main hydrogen consumers, with the first three accounting for 75% and China 25% of global hydrogen consumption. Between 2009 and 2019, global hydrogen consumption steadily increased, with hydrogen demand exceeding 80 million tonnes per year. Demand dipped in 2020 due to COVID-19 travel restrictions and a decline in the production of ammonia fertilisers from hydrogen. After the pandemic, the global hydrogen demand market recovered quickly.

According to projections by the Hydrogen Council, by 2050, global hydrogen demand will reach 530 million tonnes per year; the share of hydrogen in the global energy demand mix will rise to 18%; and the hydrogen market will be worth $2.5 trillion. The Research Institute of the China Hydrogen Alliance has forecast that hydrogen will represent 5% of China's total final energy consumption by 2030 and 10% by 2050. More than 10,000 hydrogen refuelling stations will be built, and a broad uptake of hydrogen will be possible in transport, industry and other sectors.

Hydrogen, which potentially is the lowest-carbon energy carrier, holds promise for China. China's hydrogen use is mainly concentrated in the chemical industry. Going forward, as technologies advance, hydrogen demand in transport, buildings and other sectors will increase.

2.2 Hydrogen Supply

Globally, hydrogen is produced mostly from fossil fuels. Hydrogen made from natural gas accounts for 48% of global hydrogen production, thanks to its lower-carbon properties, high efficiency and relatively low cost (see Fig. 2). In China—where coal dominates and natural gas has a small share of the energy mix, coupled with resource constraints and cost—64% of hydrogen is made from coal and 21% is an industrial by-product (see Fig. 3). According to data from the Research Institute of the China Hydrogen Alliance and the China National Petroleum & Chemical Planning Institute, in 2019, China's hydrogen production capacity was 41 million tonnes and its hydrogen output 33.42 million tonnes.

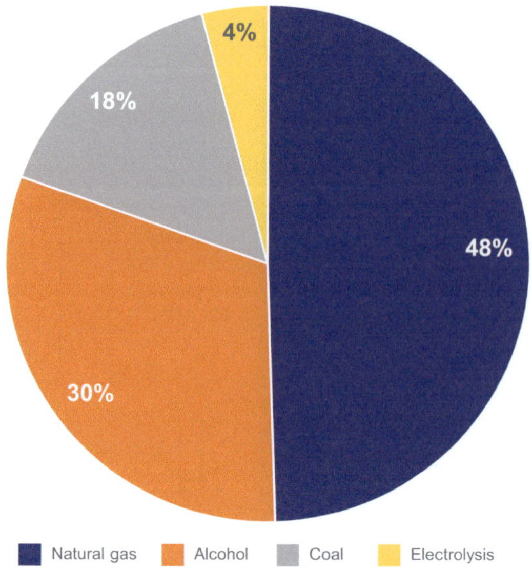

Fig. 2 Global hydrogen source mix. *Source* Produced by the authors from various sources

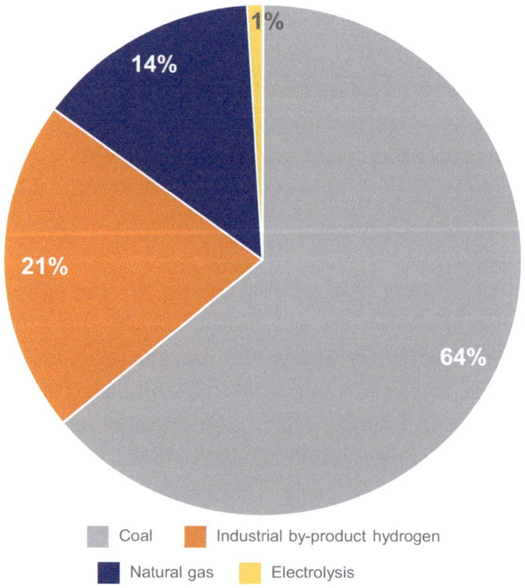

Fig. 3 China's hydrogen source mix. *Source* Produced by the authors from various sources

2.2.1 Hydrogen Made from Coal

Hydrogen made from coal has the largest share and lowest production cost of the hydrogen market in China. Coal and a gasification agent are mixed under high temperature and high pressure to produce a gas, from which impurities are removed to result in high-purity hydrogen. Data from the Research Institute of the China Hydrogen Alliance show that output of hydrogen made from coal was 21.24 million tonnes in China in 2019, representing 64% of the country's total hydrogen production. This mature and efficient technology can produce hydrogen stably, but the process is complex: it requires much investment capital and its carbon emissions are high.

2.2.2 Hydrogen Made from Natural Gas

Making hydrogen from natural gas is one of the best ways to produce hydrogen from fossil fuels. Pre-treated natural gas and water vapour reform into a synthesis gas at high temperature, which is then converted into hydrogen and carbon dioxide at medium temperature. Product-grade hydrogen is obtained after condensation and pressure swing adsorption. Natural gas has the largest proportion of hydrogen atoms at 25%. Using natural gas as feedstock offers many benefits, such as small water consumption, low CO_2 emissions, high hydrogen yield and a relatively small environmental footprint, making it the ideal feedstock for producing hydrogen from fossil fuels. According to the Research Institute of the China Hydrogen Alliance, China produced 4.6 million tonnes of hydrogen from natural gas in 2019, 14% of the country's total hydrogen output.

2.2.3 Hydrogen Made from Methanol

Hydrogen made from methanol is easy to transport and ready to use, but it is high cost. A catalytic reaction between methanol and water vapour occurs at 200 °C to produce a mixture of hydrogen and carbon dioxide, which results in high-purity hydrogen after pressure swing adsorption. The technology has a low investment cost and produces relatively little pollution. Methanol is liquid at room temperature, which is easy to store and transport, and the hydrogen can be used immediately after production. However, constrained by the high total cost of production, methanol-to-hydrogen is only suitable for small-scale production.

2.2.4 Hydrogen Made from Industrial By-products

Coke Oven Gas

Hydrogen made from coke oven gas has the largest share of the by-product market. Coke oven gas comprises mainly hydrogen and methane, which can generate hydrogen after compression, pretreatment, pressure swing adsorption and purification. To ensure that waste water discharged by the process meets environmental requirements, a waste-water treatment unit is typically installed at the plant.

In 2020, China's coke production was 471 million tonnes. Assuming that 1 tonne of coke can produce 400 cubic metres (m^3) of coke oven gas as a by-product, half of which is returned to the furnace, the country's coke oven gas output is estimated to be 94.2 billion m^3. If coke oven gas has a hydrogen concentration of about 55% and the hydrogen recovery rate in pressure swing adsorption is 92%, 4.285 million tonnes of by-product hydrogen can be produced from coke oven gas in China. Considering that in the 14th Five-year Plan period (2021–25), China's coking industry is tasked to resolve its excess capacity, an increase in by-product hydrogen from coke oven gas is unlikely.[1]

Chlor-Alkali

The chlor-alkali industry uses brine as feedstock to produce caustic soda, polyvinyl chloride (PVC), chlorine and hydrogen through membrane electrolysis or ion exchange. Hydrogen made from by-product chlor-alkali has several advantages, such as high hydrogen purity (up to about 99% before purification), low energy consumption and a high level of automation. Its production process emits no carbon dioxide, which means it is relatively pollution-free.

In 2020, China produced 36.43 million tonnes of caustic soda. Assuming 1 tonne of caustic soda generates 280 m^3 of by-product hydrogen, annual output of by-product hydrogen could reach 910,000 tonnes. Of this, 70% will be used by the plant to produce PVC and hydrochloric

[1] Research Report on Hydrogen Development Strategy in the Petrochemical and Chemical Industry.

acid, which could produce about 270,000 tonnes of hydrogen products. In the future, China expects there to be few new chlor-alkali plants, which implies the potential for increased by-product hydrogen supply is limited.

Propane Dehydrogenation

Propane dehydrogenation is an important way to produce propylene. It accounted for 17% of total propylene production capacity in 2020. Hydrogen is a by-product of propane dehydrogenation and can be sold to enhance the overall profitability of the plant. In 2020, propane dehydrogenation plants in China had a total production capacity of 7.76 million tonnes per year. Assuming an average plant operating rate of 80% and a production level of 38 kilograms of high-purity by-product hydrogen from 1 tonne of dehydrogenated propane, the total output of by-product hydrogen is estimated to be 236,000 tonnes. In coming years, the expansion capacity of propane dehydrogenation plants in China is expected to exceed 30 million tonnes per year. This translates into a potential increase of by-product hydrogen of more than 900,000 tonnes per year.

Ethane Cracking

The technology for cracking ethane to produce ethylene is relatively mature and has been successfully deployed for decades. The by-product hydrogen produced in the process has an impurity content lower than that made from coke oven gas. Because of its low investment cost, low feedstock cost, high ethylene yield and high ethylene purity, cracking ethane to produce ethylene has attracted much attention from refining and chemical companies in China. Using Satellite Chemical's 2.5 million tonnes per year and CNPC's 1.4 million tonnes per year ethane cracking capacity as a measure, the amount of by-product hydrogen produced from cracking ethane is estimated to be about 220,000 tonnes per year.

2.2.5 Hydrogen Produced Through Electrolysis

To produce hydrogen through electrolysis, direct current is fed into an electrolyser filled with electrolyte. This causes the water molecules to undergo an electrochemical reaction and split into hydrogen and oxygen. There are three main electrolysis technologies: alkaline water electrolysis (AWE), proton exchange membrane electrolysis (PEME) and solid oxide electrolysis (SOE). AWE is relatively mature and is used for industrial-scale hydrogen production, while PEME and SOE are still in the early laboratory development stage. AWE will therefore be the main electrolysis technology for producing hydrogen for some time. Compared with other technologies, AWE has low support costs and a high technology readiness level. On the minus side, it consumes large amounts of electricity. PEME produces high-purity hydrogen from low electricity consumption. However, the core components, including the proton exchange membrane, are imported, and the electrolyser is expensive. Currently, it costs 40% more to produce hydrogen with PEME. As locally developed components mature and the technology progresses, the PEME market share is expected to reach 10% by 2030, according to China EV100.

Hydrogen produced from fossil fuels continues to dominate thanks to its cost advantages. In the long term, however, the huge CO_2 emissions associated with these feedstocks run counter to the dual-carbon goal. Hydrogen produced through electrolysis, which is environmentally friendly and delivers a high-purity product, represents the future.

2.3 Creating a Hydrogen Production, Storage, Transport and Marketing System

The hydrogen value chain is huge, spanning production, storage, transport, refuelling and end uses. Hydrogen has long been an important chemical feedstock used widely in petroleum refining and ammonia synthesis. In recent years, measures have increasingly been taken to combat climate change globally, and the transition to low-carbon energy has accelerated. In this setting, hydrogen's potential as a low-carbon energy carrier has gradually come into focus in the energy market and related industries at home and abroad. This section explores how to build a hydrogen value chain in three segments: upstream hydrogen production, midstream storage and transport, and downstream applications. The section also looks at possible future trends and the deployment of related technologies.

2.3.1 Upstream Hydrogen Production

Hydrogen production is a process of converting hydrogen, an element present in natural substances or compounds, into hydrogen gas through chemical reactions. The methods used to produce hydrogen can be divided into non-renewable and renewable production, with the former using fossil fuels as feedstock and the latter using water or renewable substances as feedstock. The main production methods include: making hydrogen from fossil fuels, industrial by-products and purification, electrolysis and others, as shown in Fig. 4. The main challenge in hydrogen production is how to control carbon emissions and costs. In China, hydrogen is produced mainly from fossil fuels like coal, although there are considerable industrial by-product hydrogen resources available that could also be used.

2.3.2 Midstream Hydrogen Storage and Transport

Hydrogen storage and transport connects upstream hydrogen production with downstream applications. The main methods for storing and transporting hydrogen are as a gas, liquid or solid. Storage and transport are the biggest barriers restricting the development of hydrogen in China at present. Due to constraints such as technological immaturity and inadequate standards, hydrogen storage and transport costs are high, which have nullified the economies of scale and price advantages of upstream hydrogen production. As the advantages cannot be effectively transferred to the end user, it is impossible to promote large-scale hydrogen use downstream.

High-pressure Gaseous Hydrogen Storage

In this method, hydrogen is compressed to increase its density and is then stored in high-

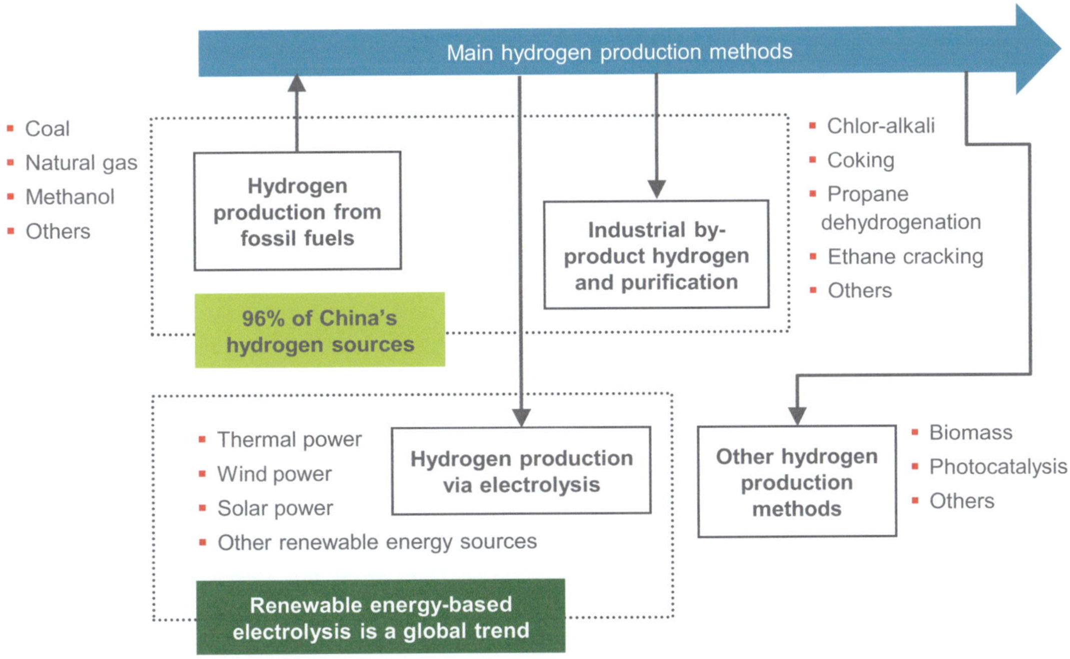

Fig. 4 Main hydrogen production methods. *Source* Produced by the authors

pressure tanks. It enables large volumes of hydrogen to be stored in a small space. Other advantages are its low level of equipment complexity, low energy consumption, fast charging and discharging, and ease of storage and transport. It is the most mature and widely used hydrogen storage technology. However, gaseous hydrogen storage has two disadvantages: it has a low safety level and a small volumetric specific capacity. The hydrogen is stored in tanks or cylinders at high pressure, which is a safety risk. The theoretical hydrogen storage capacity of a cylinder is 70 grams per litre (g/L), but the actual maximum capacity is only 25 g/L. In addition, the carbon fibre hydrogen storage cylinder commonly used across the world is expensive.

High-pressure gaseous hydrogen is usually transported in a tank by truck. However, due to constraints such as a small tank volume and high transport safety risks, it is only suitable for small-volume and short-distance transport, which does not match the needs of hydrogen applications in China. Pipelines can transport large volumes of hydrogen at low cost and little use of energy. They are, therefore, an important means for large-scale, long-distance, point-to-point hydrogen transport, although low-pressure pipeline hydrogen transport is still in the initial stage of development, both in China and abroad. In 2016, the length of hydrogen pipelines was 2,608 kilometres in the USA, 1,598 kilometres in Europe and less than 200 kilometres in China. The total length of hydrogen pipelines worldwide is less than 5,000 kilometres, compared with 2 million kilometres of oil and gas pipelines. As a result, most pipeline hydrogen transport projects blend hydrogen with natural gas.

Liquid Hydrogen Storage

In this process hydrogen is cooled to a temperature below minus 253 degrees Celsius and converted to a liquid state for storage. Compared with high-pressure gaseous hydrogen storage, liquid hydrogen has a high storage density per unit volume (up to 70 kg/m^3). Its disadvantages are: liquefaction uses large amounts of energy (about 20 kWh per kilogram of hydrogen liquefied), it has a short storage life and hydrogen exhaust can leak during long-term storage. The cost of this technology is relatively competitive. Chinese oil and gas companies have rich experience of storing and transporting liquefied natural gas and liquefied petroleum gas, and they own large fleets of transport vehicles. These companies are expected to shape competitiveness in the coming years if costs decline. More than one-third of overseas hydrogen refuelling stations use liquid hydrogen.

Solid Hydrogen Storage

Solid hydrogen is relatively safe to store and transport. A physical or chemical reaction takes place between the hydrogen and the storage material, forming dissolved solids called hydrides. Compared with high-pressure gaseous and liquid hydrogen storage, solid-state hydrogen storage has a larger storage density and is easier to handle and transport. It is also less costly and safer. This makes it a perfect match for fuel cell vehicles and other applications that have strict requirements for hydrogen storage volumes. Solid hydrogen storage is the most promising technology at present.

2.3.3 Downstream Applications

In China, hydrogen is used mainly as a feedstock for making chemicals, such as synthetic ammonia and methanol, and for hydrogenation in refining and chemicals. It is also used as a reducing gas, shielding gas and reaction gas in industries such as iron and steel, electronics, building materials and fine chemicals. Its applications will extend into transport and buildings as hydrogen becomes increasingly used as an energy source.

Hydrogen refuelling stations are key to the uptake of hydrogen in transport. According to China Galaxy Securities, the number of hydrogen refuelling stations has been growing worldwide since 2014, totalling 553 by the end of 2020. Data from the Research Institute of the China Hydrogen Alliance show that there were 128 hydrogen refuelling stations in China at the end of 2020.

3 Outlook for Hydrogen Supply and Demand

This section analyses likely future trends in the hydrogen industry and the barriers to large-scale hydrogen development (policies, standards, research and development and cost effectiveness). It also forecasts the future changes in hydrogen supply and demand in both international and domestic markets.

3.1 Challenges for Large-scale Hydrogen Development

The global energy transition offers many opportunities for China's hydrogen industry. The national government has introduced numerous policies to encourage the development of the hydrogen industry, and local governments are also actively issuing policies to guide and support the development of hydrogen and fuel cells. Thanks to these policies and strong government support, China has become the world's number-one hydrogen producer, with an annual hydrogen output of about 33 million tonnes and more than 100 hydrogen refuelling stations. Although the domestic hydrogen industry has made positive progress, it still faces huge challenges.

First, the industrial policy system for hydrogen is incomplete and actionable implementation rules are absent throughout the value chain, which means a policy system for promoting hydrogen development has not yet been formed. China should, therefore, clarify the responsibilities of the authorities in a timely manner, optimise the approval procedures for hydrogen refuelling stations and hydrogen production companies, and regulate the industry. The country should increase its financial support; encourage investment in hydrogen production, storage, transport and use infrastructure; and facilitate the sound and sustainable development of the entire hydrogen value chain.

Second, the standards system does not meet the requirements of a well-regulated industry. Society questions the safety of hydrogen, but research and experience worldwide show that hydrogen is safe and reliable if used in accordance with science-based standards. As of 2019, China had adopted nearly 90 national standards related to hydrogen, but these are still insufficient to meet the needs of a developing hydrogen industry.

Third, hydrogen technology research and development lags behind, and there is insufficient support to develop it at industrial scale. As China started late in the creation of a hydrogen value chain, the critical core technologies are in the hands of foreign companies, and an advantageous support system for independently developed key technologies is not yet in place. China still lacks the technologies and experience in renewable power generation, hydrogen production through electrolysis, hydrogen storage and transport, hydrogen-blended natural gas transport and the core equipment needed by hydrogen refuelling stations. In addition to developing the materials, equipment and technologies for fuel cell stacks and support systems and avoiding the technological strangleholds of others, China should intensify its R&D efforts in the technologies and equipment needed for large-scale hydrogen production, storage and transport, and downstream hydrogen applications. In this way it can develop a hydrogen value chain in a coordinated manner.

Fourth, China's hydrogen industry is in the initial stage, and its industrial development has yet to become cost effective. The petrochemical and chemical industries represent 96% of total hydrogen production capacity in China, but hydrogen made from fossil fuels and industrial by-product hydrogen dominate at a cost of high carbon emissions. Compared with electrolysis, hydrogen made from fossil fuel offers poor social and environmental benefits, although it is cost efficient. At present, renewable energy-based hydrogen production is immature, and constraints such as low efficiency, instability and high cost make it difficult to commercialise and put into large-scale operation. R&D in high-pressure compressed hydrogen storage, cryogenic liquid hydrogen storage, and hydrogen storage in liquid organic hydrogen carriers have yielded promising results, but their safety, hydrogen storage density and high cost remain a challenge. Large-scale commercial deployment of these technologies still has some way to go. The proportion of locally developed key components required by hydrogen fuel cells and hydrogen refuelling stations is low, but the costs are high. The commercial supply of renewable hydrogen—constrained by technological immaturity, high production costs and low supply capacity—does not yet meet the requirements of large-scale industrial applications.

3.2 Hydrogen Supply and Demand

Hydrogen—when used as a clean, efficient and renewable secondary energy carrier—has high calorific value and zero pollution. It is also an excellent energy storage medium that is safe, can be produced from multiple sources and has many applications. As such, hydrogen is important for the energy transition and China's realisation of its dual-carbon goal. Thanks to hydrogen's growing technology readiness levels and the gradual commercialisation of hydrogen technologies, alongside rising global concern about climate change and natural disasters, more and more countries consider hydrogen an important strategic option for the energy transition. Using research data, this section forecasts the future changes in hydrogen supply and demand in international and domestic markets.

Globally, the hydrogen industry is in a period of rapid growth. Some developed countries and regions have attached great importance to the hydrogen industry. Europe, the USA, Japan and South Korea have included hydrogen in their national strategic energy plans and accelerated commercialisation of the hydrogen industry. According to the Hydrogen Council, there were 131 large-scale hydrogen development projects and 359 hydrogen projects around the world in February 2021. By 2030, global investment in the hydrogen industry is expected to reach \$500 billion. As the hydrogen market improves, 30 million jobs will have been created by 2050; 6 billion tonnes of carbon dioxide emissions will have been avoided; the value of the hydrogen market will be about \$2.5 trillion; and hydrogen will supply 18% of global energy demand.

China's dual-carbon goal and hydrogen development programme make hydrogen an important part of the country's clean and efficient energy production and consumption system. A white paper on China's Hydrogen Energy and Fuel Cell Industry forecasts that by 2060 hydrogen will be widely used in transport, energy storage, industry, buildings and other sectors, and that hydrogen demand will increase to 60 million tonnes per year. The output value of the hydrogen industry value chain will exceed RMB 10 trillion. By 2060, the end-user selling price of hydrogen will have fallen to RMB 20 per kg; the number of hydrogen refuelling stations will be 12,000; there will be 20 million hydrogen fuel cell electric vehicles (see Table 1); around 20,000 hydrogen-fuelled power generation units will be built per year; and some 5.5 million fuel cell systems will be produced annually.

Table 1 Hydrogen development targets in China

Item	2025	2030	2040	2050	2060
Hydrogen demand (million tonnes)	30.00	35.00	40.00	50.00	60.00
Industrial output value (RMB trillion)	1	2	5	7	12
Hydrogen refuelling stations	200	500	2,000	8,000	12,000
Number of hydrogen FCEVs (million vehicles)	0.10	0.30	1.00	15.00	20.00

Note FCEVs = fuel cell electric vehicles
Source China's Hydrogen Industry Development Report 2020

4 Hydrogen Supply Roadmap

To achieve its goal of a green, cost-effective and accessible hydrogen supply system, China is working to gradually achieve breakthroughs in hydrogen production, storage, transport and processing. To reduce end-user prices, hydrogen production, storage, transport and refuelling need to be integrated. Efforts will be made to develop greener and more cost-effective hydrogen production processes and safer transport. To meet growing hydrogen demand in the long term, solutions that combine renewable energy-based production and refuelling, cost-effective production, and liquid hydrogen and other storage and transport technologies will be the main focus of development.

In the Medium- and Long-term Plan for the Development of the Hydrogen Industry (2021–35) of 2022, the tasks and targets of hydrogen development in China were clearly defined. The plan states that China will co-ordinate the development of the hydrogen industry nationwide, understand the processes of hydrogen industry development, avoid disorderly competition, advance the construction of hydrogen infrastructure in a planned manner, strengthen the safety management of hydrogen infrastructure, and accelerate the development of a safe, stable and efficient hydrogen supply network. Hydrogen production processes will be selected according to local resource endowment and the presence of local industries, and a national low-carbon, low-cost and diverse hydrogen production system will be gradually built. Industrial clusters of coking, chlor-alkali and propane dehydrogenation will prioritise industrial by-product hydrogen, and local consumption will be encouraged to drive down supply costs. In areas with abundant wind, solar and hydropower resources, demonstration projects of renewable hydrogen production will be developed and gradually scaled up, and efforts will be made to explore seasonal energy storage and grid balancing. The research and development of technologies such as solid oxide electrolysis, water photolysis and making hydrogen from seawater and in nuclear reactors at high temperatures will be prioritised. Efforts will be made to explore the possibility of building hydrogen production bases in areas with large hydrogen demand.

The plan also states that China will actively drive innovation in technologies, materials and processes, and support the exploration and deployment of storage and transport methods, without compromising security and controllability. China will: (i) enhance the efficiency of high-pressure gaseous hydrogen storage and transport, drive down costs at a faster pace, and improve the commercialisation of this method; (ii) promote the industrial deployment of cryogenic liquid hydrogen storage and explore the application of solid-state, cryogenic and high-pressure hydrogen storage, as well as storage in liquid organic hydrogen carriers; (iii) carry out pilot and demonstration projects of hydrogen-blended natural gas pipeline transport and pure hydrogen pipeline transport; and (iv) gradually build a high-density, lightweight, low-cost and diverse hydrogen storage and transport system.

The plan requires that development be demand-oriented and that China build a hydrogen refuelling network in an orderly manner and co-ordinate the deployment of refuelling stations. The country will put safety first, use land resources efficiently and add hydrogen refuelling facilities to existing service stations (including those selling compressed and liquefied natural gas) in accordance with laws and regulations. New models of hydrogen refuelling station combining on-site hydrogen production, storage and refuelling will be explored.

5 Action Plan for Blue Hydrogen

Although blue hydrogen made from natural gas and green hydrogen made from renewables are widely used in our society, there is no common definition of them. In this section, we use the term blue hydrogen to refer to industrial by-product hydrogen. We analyse the distribution and use of blue hydrogen resources in China and forecast the possible methods for producing industrial by-product hydrogen in the future.

Production is the first link in the hydrogen supply chain and the basis for the safe, orderly and high-quality development of the hydrogen industry. Of the hydrogen production technologies, coal has the lowest cost but can emit large amounts of carbon dioxide. Hydrogen produced from natural gas and methanol are relatively high cost and have low carbon emissions, although supply and prices are uncertain. Electrolysis is expensive, and is impossible to develop at large scale in the short term. In contrast, industrial by-product hydrogen resources are available in large quantities and at low cost; this method is therefore expected to be important for the development of the hydrogen industry in the near and medium terms.

China's industrial by-product hydrogen comes mainly from the petrochemical and chemical industries, principally chlor-alkali, propane dehydrogenation and ethane cracking. Refining and ammonia synthesis also produce a large amount of tail gas that has a low concentration of hydrogen. Coke, caustic soda and other chemical industries have relatively mature production processes and have replaced outdated production facilities in recent years. As a result, the production of coke oven gas and chlor-alkali by-product hydrogen should remain stable. By-product hydrogen made from light hydrocarbons and chlor-alkali (using ion exchange membranes to produce caustic soda) will be the focus of attention in the near and medium terms.

5.1 By-product Hydrogen in the Chlor-alkali Industry

In recent years, the production of chlor-alkali has been growing slowly. The electrolysis of saturated sodium chloride generates caustic soda, chlorine gas and hydrogen, and other chemicals. The main products made include polyvinyl chloride, propylene oxide, epoxy chloropropane, calcium carbide, caustic soda, liquid chlorine and hydrochloric acid, which are widely used in the textile and petrochemical industries. Caustic soda is one of the most important products made in the process and is a common chemical feedstock in industry. As the world's largest producer and consumer of caustic soda, China had a caustic soda production capacity of 43.8 million tonnes per year in 2019, and an output of 34.64 million tonnes, which equals a capacity utilisation rate of about 80%.

Industrial by-product hydrogen has many uses. First, by-product hydrogen is used for the synthesis of hydrochloric acid and polyvinyl chloride, which account for about 60% of output. Second, some chlor-alkali plants use by-product hydrogen to make hydrogen peroxide. Third, some of the residual by-product hydrogen is recycled using hydrogen boilers. Due to its small scale and output, by-product hydrogen from the chlor-alkali industry is more suitable for small-scale applications like hydrogen fuel cells.

A survey of caustic soda companies shows that average by-product hydrogen yield per tonne of caustic soda produced is about 280 cubic metres (m^3). Assuming a hydrogen conversion rate of 80%, the by-product hydrogen production potential of the chlor-alkali industry is around 77.59 million m^3, equivalent to 866,000 tonnes. Shandong province has the highest potential for producing by-product hydrogen, with an annual output of about 240,000 tonnes. Provinces that have an annual by-product hydrogen production of between 50,000 and 100,000 tonnes are Inner Mongolia, Jiangsu, Zhejiang and Xinjiang. There are 15 provinces with an annual by-product hydrogen output of 10,000–50,000 tonnes.

5.2 By-product Hydrogen from Propane Dehydrogenation and Ethane Cracking

Ethylene and propylene are important chemical feedstocks, which play a prominent role in the chemical industry. As industrialisation and urbanisation increase, demand for ethylene and propylene grows. China has become the world's second largest ethylene producer and leading ethylene consumer, as well as the world's biggest propylene producer and consumer.

The propylene production processes include steam cracking naphtha to make ethylene and propylene, methanol-to-olefins, catalytic cracking, heavy olefin cracking and olefin conversion. Propane dehydrogenation—which has several advantages such as low investment threshold, short construction time and low production cost —is one of the most competitive propylene production processes. In 2019, propane dehydrogenation accounted for 1.5 million tonnes per year, or 34% of new propylene production capacity. In 2019, China's total propane dehydrogenation capacity was about 5.88 million tonnes per year, with an additional 3.87 million tonnes per year under construction or in early-stage preparation. According to the Special Committee on Hydrogen of the China Petroleum and Chemical Industry Federation, at the end of 2019, China had produced 21.5 tonnes of by-product hydrogen from propane dehydrogenation. Due to the rapid development of shale gas in North America, the price of light hydrocarbons dropped sharply. As the technologies of ethylene production from propane dehydrogenation and ethane cracking evolve, a global boom in propane dehydrogenation for propylene production will likely result. By 2023, the propane dehydrogenation industry is forecast to produce 445,400 tonnes of by-product hydrogen.

5.3 Reusing "Waste Hydrogen" from Refining and Ammonia Synthesis

Refinery reforming, tail gas hydrogenation, methanol purge gas and purge gas from ammonia synthesis are important sources of by-product hydrogen. Sinopec's first high-purity hydrogen production demonstration plant is designed to produce high-purity hydrogen for fuel cell electric vehicles using low-cost by-product hydrogen from refining units as feedstock. It has achieved a record hydrogen purity of 99.999% treated from refinery by-product hydrogen for the first time in China. It also marked the successful reuse of "waste hydrogen" from refining and chemical production.

Industrial by-product hydrogen production capacity is dispersed across China. With this in mind, the industry should consider the following pathway to future development.

First, map the industrial by-production hydrogen resources available for external supply. These resources should be included in local hydrogen plans to ensure they are used efficiently.

Second, develop a business model for a direct supply channel from producers. At present, intermediary sales companies can easily change by-product hydrogen supplier to maximise their profits, which can destabilise supply and hydrogen quality. Direct supply from by-product hydrogen producers should ensure stable quality and quantities, and help producers develop long-term strategies.

Third, centralise the purification of by-product hydrogen resources (which are currently dispersed) to reduce the purification costs per unit of hydrogen produced. The hydrogen separation process should be continuously optimised to reduce construction costs.

Fourth, connect by-product hydrogen producers with downstream storage and transport and hydrogen refuelling stations to lower costs and stabilise the hydrogen value chain. This is essential for the early-stage development of the hydrogen industry. When selecting sites for hydrogen refuelling stations, the distribution of by-product hydrogen sources in the surrounding area should be taken into account to optimise transport routes and minimise transport costs.

Fifth, improve hydrogen fuel testing standards, set up third-party hydrogen fuel testing centres and improve the hydrogen management system. The purity of industrial by-product hydrogen varies according to the feedstock and production processes used. It is therefore necessary to establish third-party testing centres, improve testing techniques and subject each by-product hydrogen supplier to regular tests to ensure hydrogen quality.

6 Policy Recommendations for Developing the Hydrogen Industry

Looking ahead, the hydrogen industry will develop around China's three strategic goals of "energy transition, deep decarbonisation and high-quality development". First, to increase the scale of hydrogen production and consumption, hydrogen should be fully integrated into the existing energy system as an efficient and clean form of secondary energy, as a smart and flexible energy carrier and as a low-carbon industrial raw material. Industrial end users should reduce their use of fossil fuels and improve energy efficiency. Second, the country should actively promote the development and use of hydrogen produced from renewable energy and industrial by-products, and explore replacing hydrogen made from coal with renewable hydrogen and hydrogen-based chemicals to cut carbon emissions. Third, China should accelerate investment in key technologies, materials and equipment manufacturing to help the hydrogen industry become independent and reduce the costs of the hydrogen supply chain. This section makes policy recommendations for the short, medium and long terms to advance China's hydrogen industry.

6.1 Policy Recommendations for the Short Term

First, adapt the development of blue hydrogen resources to local conditions to drive down hydrogen supply costs. A cost-effective, efficient, low-carbon and diverse hydrogen supply system would lay the foundation for the sustainable development of the hydrogen industry. Hydrogen market demand should be analysed and blue hydrogen resources developed in line with the principles of "supply determined by demand" and "adapting to local conditions". Research and development should be supported to develop technologies and equipment that produce high-purity hydrogen in a low-cost and efficient way to make blue hydrogen competitive. New business models should be created and the entire blue hydrogen supply chain—production, transport and storage—should be opened to enhance efficiency and reduce supply risks.

Second, infrastructure including integrated hydrogen and integrated refuelling stations that offer petrol and compressed and liquefied natural gas fuel should be rolled out in an orderly way to avoid incurring unrecoverable (sunk) costs. Compared with lithium battery electric vehicles, hydrogen fuel cell vehicles have low energy conversion efficiency and a relatively poor economy. In the near and medium terms, the development of hydrogen in the transport sector should focus on basic research and small-scale demonstration projects. Given China's economic scale and first-mover advantages in the manufacture of lithium batteries and related equipment, the penetration of hydrogen in transport is not expected to be as large as the market expects. Therefore, depending on the development of hydrogen vehicles and infrastructure in the regions, the opening of hydrogen and integrated refuelling stations should match demand to avoid opening too many facilities too fast and incurring unrecoverable costs. In the Beijing–Tianjin–Hebei region and the Yangtze and Pearl river deltas, where there is a strong industrial base and demand for fuel cell vehicles, integrated refuelling stations should be rolled out in line with the development plan for hydrogen fuel cell electric vehicles. In western, north-eastern and pilot provinces where battery swapping infrastructure is weak, integrated refuelling stations will require promotion. Above all, the nationwide roll-out of integrated refuelling stations should be based on market prospects.

Third, establish hydrogen quality standards and a testing and certification system. At present, China is behind in developing the technologies and equipment crucial for hydrogen development. The technical standards, testing, certification and regulatory systems are incomplete and unable to give the hydrogen industry the support it needs. China should establish a basic national quality system that integrates standards, testing and certification as soon as possible to enable the growth of the hydrogen industry.

6.2 Policy Recommendations for the Medium and Long Terms

China should research and develop key hydrogen technologies, explore advanced hydrogen pathways such as replacing hydrogen made from coal and hydrogen-based chemicals with renewable hydrogen, extend hydrogen into new applications and promote the sustainable development of the hydrogen industry to support China's dual-carbon goal.

China should develop pilot and demonstration projects for making chemicals from renewable hydrogen and blending hydrogen with natural gas for pipeline transport. Hydrogen-based chemicals is the collective name of processes for which renewable hydrogen is the main feedstock. Coupled with high-carbon media such as carbon dioxide and petroleum coke, it produces olefins, aromatics and other feedstocks. Production processes based on hydrogen are expected to solve the challenge of reducing emissions in difficult-to-decarbonise industries. In recent years, pilot hydrogen projects such as renewable hydrogen chemicals, carbon-sequestration chemicals, and hydrogen blending with natural gas have emerged in China and across the world. Hydrogen, as a low-carbon industrial feedstock and an efficient and clean secondary energy source, will help decarbonise industry. To meet the requirements of the dual-carbon goal, petrochemical and chemical companies should gradually replace hydrogen made with coal with renewable hydrogen in processes such as synthetic ammonia, methanol and oil refining to reduce the carbon intensity of their products. In addition, they should carry out pilot and demonstration projects of producing chemicals from renewable hydrogen and blending hydrogen with natural gas for pipeline transport.

China should research and develop the key technologies and manufacturing equipment for the hydrogen value chain. The technologies for producing hydrogen from renewable energy, for storage and transport and for making hydrogen fuel cells are immature. Proton exchange membranes, high-pressure hydrogen storage tanks, air compressors and other key products still need to be imported. Breakthroughs in technologies crucial to hydrogen production and fuel cell development are needed urgently. China should: first, accelerate the research and development of hydrogen production technologies and equipment, and focus on large-scale, energy-efficient hydrogen electrolysis; research and develop system integration and the main raw materials and components needed for low-cost, long-life and large-capacity alkaline water electrolysis and proton exchange membrane electrolysis, and address the challenges facing the use of intermittent renewable energy with electrolysers. Second, China should develop long-distance and large-capacity hydrogen storage and transport demonstration projects in cryogenic liquid hydrogen storage and pipeline hydrogen transport; carry out commercial demonstrations of long-distance transport of high-pressure gas hydrogen and cryogenic liquid hydrogen over a distance of more than 300 km; and strive to remove the bottlenecks in large-scale, long-distance, high-efficiency and low-cost hydrogen storage and transport in separate pipelines and in blending hydrogen with natural gas in existing pipelines. Third, the country should aim to make breakthroughs in "stranglehold" technologies and materials; increase R&D in catalysts, proton exchange membranes and carbon fibre, and speed up demonstration and commercialisation projects to replace imported products as soon as possible, thus improving the independence of its hydrogen industry.

Chapter 11: The Future of Energy Infrastructure

Yi Zhou, Qian Lu, Xi Zhou, Xiuqi Yang, Xiaohang Gong, Jiale Geng, Taoyuan Zhou, Ye Ning and Peter Webb

1 Introduction

New energy infrastructure is an important component when building a renewable, low-carbon, secure and efficient energy system. It is an unavoidable requirement of the energy transition and a precondition for China to achieve its dual-carbon goal.

Because renewable energy resources are located far from China's load centres, a power grid system in which large power grids and microgrids co-ordinate with and complement each other is needed. Centralised and distributed power grids should interact and support one another, bringing out the complementary benefits of multiple energy forms in a flexible, intelligent, secure and stable power system.

Y. Zhou (✉) · Q. Lu
Department of Industry Economic Research, Development Research Centre of the State Council, Beijing, China

X. Zhou · X. Q. Yang · X. H. Gong · J. L. Geng · T. Y. Zhou
School of Economics and Management, North China Electric Power University, Beijing, China

Y. Ning
Global Cooperation for the Development of the Internet of Energy, Beijing, China

P. Webb
Shell International B.V., The Hague, The Netherlands

By 2030, energy storage will gain momentum, and China's energy storage capacity will rise to 200 gigawatts (GW). Between 2030 and 2050, the energy storage sector will undergo change, and the country's energy storage capacity will climb to 540 GW. From 2050 to 2060, energy storage will mature, with capacity in China increasing to 646 GW, with new energy storage technologies predominating.

The efficiency of hydrogen storage systems will improve and costs will fall significantly. A network of hydrogen refuelling stations will grow, with the number of stations rising from 1,000 in 2030 to 10,000 in 2050 and 15,000 in 2060.

China currently invests about RMB 1.37 trillion in new energy infrastructure annually. In the present decade, investment is forecast to grow to RMB 2.33 trillion per year, reaching RMB 2.46 trillion per year in 2030. Power systems will attract about 60% of this investment. Between 2020 and 2060, investment in the low-carbon power sector is expected to total RMB 54.78 trillion.

Integration of multiple types of energy will be the main feature of the new energy system. Centred on renewable energy generation, the system will integrate hydrogen, nuclear power, oil, natural gas, bioenergy and other types of generation. Smart grids will be the platform for integrating information, communication and control systems with advanced energy technologies. They will support the transition to renewable

Shell International B.V. and Development Research Centre of the State Council (DRC) of People's Republic of China, *Embracing the Future, Powering Growth: An Energy System Renewed for China*, Advances in Oil and Gas Exploration & Production,
https://doi.org/10.1007/978-3-031-73075-7_12

and low-carbon electricity and the efficient and easy connection of multiple actors. This will help improve energy efficiency, reduce energy supply costs and contribute to energy security and the realisation of the dual-carbon goal.

2 New Energy Infrastructure—Developments and Challenges

New energy infrastructure has a very broad scope. It includes power grids, energy storage, hydrogen, bioenergy, carbon capture, utilisation and storage and other new energy infrastructure. Even though China is the world's fastest-growing renewable energy country, there is a discrepancy between its current level of energy infrastructure and that required by the new energy system. This limits the pace of change in the energy transition. Accelerating the development of new energy infrastructure is not only important to achieve the dual-carbon goal and build a new electric power system, it is also a strong driver for developing the new energy industry and increasing the pace of the energy transition.

2.1 Developments

2.1.1 Power Grids

Power grid infrastructure includes transmission and distribution grids and auxiliary systems. China is a global leader in power grid infrastructure and related technologies. From 2019, the power grid business model started to shift to the IT-enabled Energy Internet of Things (EIoT). With growing demand for new energy connections, greater efficiency and new business models, power grid infrastructure will increasingly integrate information and intelligent technologies in the years to come.

China is building ultrahigh voltage (UHV) transmission links and power distribution networks[1] at a faster pace. Key projects include the Qinghai–Henan 800 kV UHV direct current transmission link, the Zhangbei–Xiongan 1,000 kV UHV alternating current power transmission link, and the Zhangbei flexible direct current power grid. During the 14th Five-year Plan (2021–25), the State Grid Corporation of China (SGCC) plans to add 56 GW of transmission capacity and increase the development of smart grids. SGCC will invest more than RMB 450 billion on average annually over the coming five years to upgrade power grids to the Energy Internet and advance the transition to renewable and low-carbon energy. China Southern Power Grid will invest RMB 100 billion in Guangxi from 2021 to 2025 and about RMB 27 billion in Hainan to develop smart grids for the Hainan Free Trade Port.

Microgrids[2] are an important component of the smart grid. They can integrate different types of distributed power generation (such as wind and solar) and different types of energy storage (batteries, thermal and mechanical), which gives them flexibility. They can connect large amounts of new energy to the power grid and improve power quality and grid services for users, while enhancing power supply reliability and security.

2.1.2 Energy Storage

As the installed capacity of wind and solar power increases, the proportion of renewable energy in the power system grows, which creates a need for flexibility. Energy storage can provide regulating capacity for the power system to maintain a balance between supply and demand.

There are five main types of energy storage, which use different technologies and have different cost characteristics: mechanical, electrochemical, electromagnetic, chemical and thermal. Pumped storage, which is a type of mechanical

[1] A distribution network comprises overhead lines or cables, towers, distribution transformers, disconnectors, reactive power compensators and some ancillary facilities. There are high-voltage distribution networks (35–110 kV), medium-voltage distribution networks (6–10 kV, with 20 kV distribution networks in Suzhou), and low-voltage distribution networks (220–380 V).

[2] A microgrid is a small power generation and distribution system comprising distributed power supply, energy storage unit, energy converter, loads and monitoring and protection devices. It usually has a voltage class of 10 kV and 400 V.

energy storage, has the largest deployment scale at present, followed by electrochemical energy storage. By the end of 2019, the installed capacity of pumped storage was about 30 GW, accounting for 93.4% of China's total installed capacity of energy storage. The Fengning pumped storage plant in China is the world's largest, with a total installed capacity of 3.6 GW. Electrochemical energy storage (which includes batteries, fuel cells and electrochemical capacitors) is growing the fastest. In 2019, the new installed capacity of electrochemical energy storage facilities in power grids was 114.2 megawatts. The total installed capacity of electrochemical energy storage that year was about 1.71 GW, including lithium-ion batteries of about 1.38 GW. Lithium-ion batteries provide energy storage cycles of between 4,000 and 5,000 charges and discharges and an energy density of up to 200 watt-hours per kg.

Energy storage is an important part of China's integrated energy demonstration projects. In China, the number of Energy Internet, new energy microgrids and multi-energy complementary demonstration projects that include energy storage systems is considerable. The more renewable energy in the power system, the greater the need for energy storage.

2.1.3 Hydrogen

As of 2022, China is the world's largest hydrogen producer, with an annual output of about 33 million tonnes, of which about 12 million tonnes is of industrial quality. There are more than 270 hydrogen refuelling stations in the country. Hydrogen is mainly transported in tube trailers, pipelines or in cryogenic-insulated tanks, although a compressor station and transfer station are needed at the start and end points, which are expensive. The construction cost of a hydrogen storage and transport centre is about four times that of a natural gas storage and transport centre, and the cost of a hydrogen pipeline is 1.5–2 times that of a natural gas pipeline of the same diameter. A hydrogen transfer station with a daily transfer capacity of 800 kg costs about RMB 20 million, and the construction cost of a hydrogen refuelling station that provides the same refuelling capacity is four times that of a natural gas refuelling station. For a Class II hydrogen refuelling station (under China's national standards) that has a daily capacity of 800 kg—which is sufficient to refuel six buses (45 kg of hydrogen per bus) and 20 small vehicles (5 kg of hydrogen per vehicle)—the construction cost is more than RMB 24 million. China generally uses tube trailers for hydrogen transport, rarely long-distance pipelines. In 2019, there was only one 25-kilometre cross-city hydrogen pipeline in operation,[3] although several long-distance hydrogen pipelines are under construction.

China has not yet built a complete hydrogen value chain from production, storage and transport to use. On the upstream side of the value chain, coal-to-hydrogen technology is relatively mature and low cost, but it emits high levels of carbon dioxide and many impurities, which require purification. Coal-to-hydrogen could be replaced by hydrogen made using renewable energy, but the latter has a relatively weak technological foundation and low efficiency. In the midstream market, gaseous hydrogen storage is the main option. China's vehicle-mounted high-pressure hydrogen storage and transport technology is advancing slowly and the infrastructure needed is inadequate. In addition, the number of hydrogen refuelling stations is insufficient and their storage capacity is small. In the downstream market, hydrogen has a narrow range of small-scale applications, mainly fuel cells for transport. However, technological bottlenecks lead to increasing costs, making the commercialisation of fuel-cell electric vehicles difficult.

2.2 Challenges

2.2.1 Renewable Energy

Renewable energy generation is random and intermittent. The more renewable energy there is in the power system, the greater the risk of instability. In 2019, the daily fluctuation of new

[3] China Hydrogen and Fuel Cell Industry Handbook, 2020.

energy within State Grid Corporation of China's power system was more than 100 GW. Renewable energy resources are located far frow demand centres in China. In recent years, the renewable energy industry has grown rapidly. Hydro, wind and solar energy sources are mostly located in western China, where the local economy is less developed and demand for electricity is small. The availability of installed renewable energy plants is declining; in some areas, the power grid cannot integrate all the renewable energy generated. As a result, solar, wind and hydro curtailments are increasingly common. In 2019, these curtailments were 30 billion kilowatt-hours (kWh), 16.9 billion kWh and 4.6 billion kWh respectively, amounting to more than 50 billion kWh. Curtailments are the main barrier to a sound development of China's renewable energy industry. The development of renewable energy must therefore be co-ordinated with the overall plan for power system development and electricity demand, including scheduling, regional planning, power generation and power grid construction.

2.2.2 Power Grid Stability and Security

Power grids are responsible for transmitting energy from power plants and power generation hubs, for connecting intermittent renewable energy plants stably and reliably to the grid, and for optimising the integration of multiple energy sources. In recent years, the power system has undergone profound changes due to the integration of ultrahigh-voltage AC and DC power links, the connection of large-capacity new energy sources and a growing number of distributed power networks. As a result, many different factors and complexities impact the power system, making it significantly more challenging to ensure the secure and stable operation of power grids.

China's new energy facilities have a low average daily output, which poses difficulties for power supply security. As a result, synergies between local power generation facilities, power grids and flexibility resources to ensure power supply security are needed. When new energy output is low, the power system needs to achieve balance through conventional power resources. Even with high-capacity UHV DC links, the power grid voltage at the receiving end can be too low, making the power system vulnerable. Addressing the issue of voltage stability is therefore urgent.

2.2.3 Technological Bottlenecks and Economic Constraints

China's energy transition is mainly policy-driven; the full potential of innovation has not yet been harnessed. In key new energy areas such as hydrogen, energy storage and industrial electrification, technological barriers and poor cost effectiveness are evident. Energy storage faces barriers in safety, stability, efficiency and cost effectiveness. Key hydrogen components and materials are imported. The design of hydrogen refuelling stations holds back the development of a large-scale hydrogen economy. The energy transition requires innovation and the integration of technologies from multiple fields. Collaboration between different disciplines needs to improve, especially in basic research.

2.2.4 Investment Is Needed, but Financing Is Difficult to Attract

New energy infrastructure requires large amounts of investment, but has difficulty in attracting financing. As a result of the 14th Five-year Plan (2021–25), local governments are actively increasing their investments in new energy infrastructure, especially solar photovoltaic power generation, energy storage and hydrogen. However, China's current financial system for renewable energy development needs to be improved, and the quality of financial services is low. As renewable energy projects generally require a large investment sum over a long period, it is difficult for domestic financing to meet project requirements. Financial incentive policies are inadequate and the loan interest rates for small, micro and private renewable energy companies are usually high. Only through a reduction in financing costs will companies be motivated to accelerate their pace of investment in new energy infrastructure and support the development of new energy.

3 Development Roadmap for Power Grid Infrastructure

Power grids are key to the energy supply system. They determine whether supply matches demand over time and geographically. They enable renewable energy to be used efficiently and a secure and reliable energy supply system to result. At present, China's plans for the generation of new energy and the extension and expansion of power grids are not co-ordinated. The construction of transmission corridors lags behind the development of new energy, which means there is scope for growth in inter-regional transmission capacity. As demand for renewable energy, smart applications and power for electric vehicles increases, the gap between demand and supply becomes bigger. Efforts should therefore be made to connect more large-scale new energy to the grid, speed up the construction of smart distribution networks and microgrids, synchronise power generation with demand, improve the reliability and quality of supply, and build a power system based on renewable and low-carbon energy, one that is secure and reliable, open and interconnected.

3.1 Challenges and Trends

3.1.1 Insufficient Inter-provincial and Inter-regional Grid Connections

China's power system is fragmented and lacks unified planning. There are six regional power grids that are based on administrative divisions and have weak inter-grid connections: North China, North-east China, North-west China, East China, Central China and South China. A national electricity market has not yet been established and power supply and demand are balanced mainly at the provincial level. The combination of insufficient inter-provincial and inter-regional power exchange capacity, barriers to renewable energy use, and weak links between regional power grids leads to grid instability and even outages, which is a supply security risk. This also affects the development and use of renewable energy in the north-western provinces, causing severe wind and solar curtailments (see Fig. 1).

3.1.2 Grid Construction Lags Behind Renewable Energy Development

China's development plans for new energy generation and grid construction are not co-ordinated. There are restrictions on regional power grids, and the construction of transmission corridors lags behind the rapid development of new energy like wind and solar. In addition, inadequate power system flexibility and insufficient transfer capacity among other things restrict the use of renewable electricity.

China's large solar and wind power clusters are located far from demand centres, and the renewable energy generated can only be used through balancing flexibility and ultrahigh-voltage transmission links. Meanwhile, the intermittency of new energy output requires a large number of controllable power sources to perform rapid and extensive balancing to support electricity use. If the power system is not strengthened, its regulating capacity will be exhausted and the interweaving of power shortages and curtailments will occur repeatedly. Going forward, as the proportion of new energy connected to the grid increases, balancing will become ever-more difficult.

3.1.3 The Challenge of Balancing Supply with Demand

China's dual-carbon goal has a tight schedule and many targets, including ensuring power supply security and sound economic and social development. On the one hand, China's electricity demand will continue to grow, increasing by about 400 billion kWh annually during the 14th Five-year Plan (2021–25). On the other hand, China's power system features a high proportion of renewable energy and power electronics, which pose mounting pressure on power supply security.

China's power distribution grid is not strong enough to provide the carrying capacity for new functions and a diverse range of new loads. That

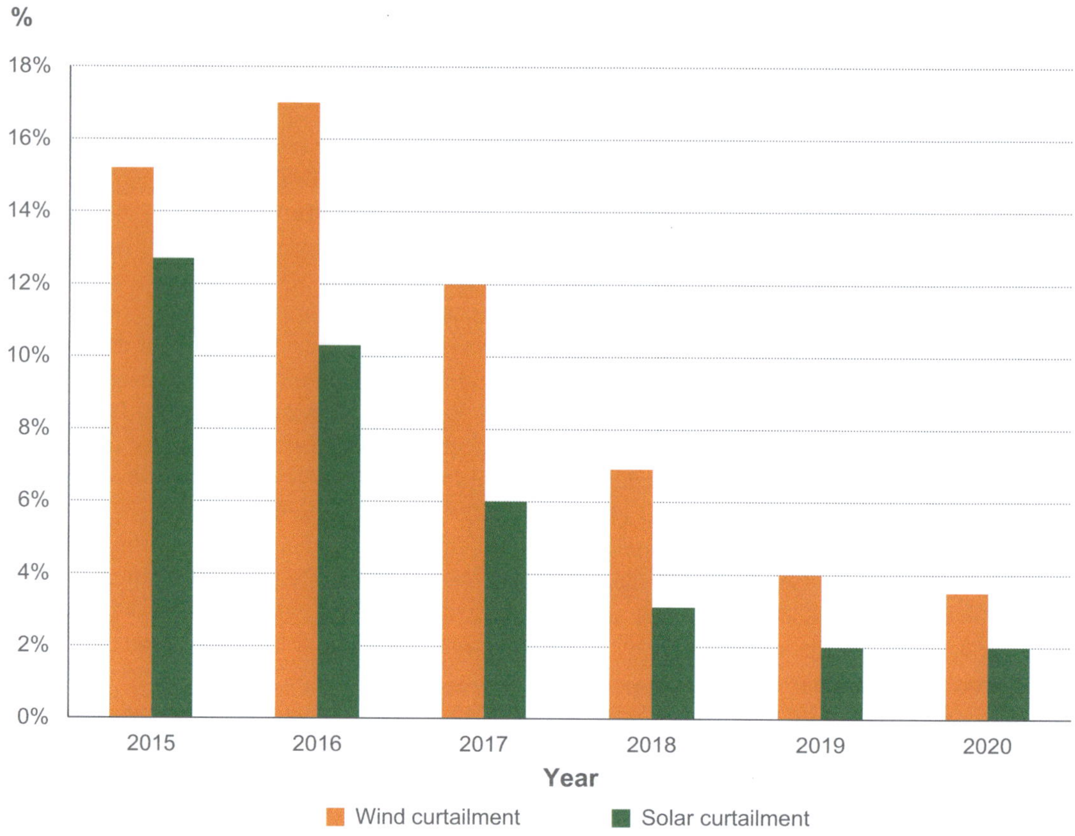

Fig. 1 Wind and solar curtailments in China. *Source* National Energy Administration; CITICS Research; China Renewable Energy Monitoring Centre

makes it difficult to acquire real-time information and perform grid monitoring. More effort is needed to support the Energy Internet. Some cities experience heavy demand, while others are gripped in low-voltage and low-tech strangleholds. All these indicate weak power supply capacity and low reliability. If distribution networks cannot adapt to the new requirements quickly, they will not be able to meet the need for flexible interaction in the future, which could cause a greater imbalance between supply and demand.

3.2 Analysis of Power Supply and Demand

3.2.1 Rapidly Growing Power Demand

As China's economy and advanced manufacturing industries progress, demand for electricity will continue to grow. Looking ahead, electricity will outperform coal, oil and natural gas to dominate the final energy demand mix. Growing power demand requires greater investment in and faster development of power grid infrastructure to meet the future needs of the economy and society.

3.2.2 Significantly More Renewable Power Generation

For China to realise the dual-carbon goal, it needs to build an energy system based on the generation and efficient use of renewables. New energy generation has a low penetration, which is expected to increase steadily in the medium term, becoming the predominant type of power generation over time. An energy system based on renewables poses challenges for the power grid, such as inadequate transmission links between large renewable energy clusters and load centres and insufficient use of renewable energy. Power grid infrastructure needs to be constructed at a faster pace and its flexibility enhanced to ensure intermittent renewable energy can become stable, reliable electricity.

3.2.3 Increasing the Power Flows Between Provinces and Regions

Inter-provincial and inter-regional power transmission is an important way to address the challenge of China's energy resources being located far from demand centres. China has abundant clean energy resources, with 80% of its hydro, wind and solar power resources concentrated in north-western provinces. However, about 70% of China's power demand comes from central and eastern provinces. As a result, China must adapt its path of energy transition to the conditions of its resource endowment and continue to increase the number of inter-provincial and inter-regional power transmission links.

According to China's renewable energy development plan, large-scale renewable energy clusters will be widely deployed in the western and northern provinces where renewable energy resources are abundant, especially in desert areas like the Gobi. More long-distance inter-provincial and inter-regional power links and ultrahigh-voltage projects will therefore be needed (see Fig. 2), as will gird optimisation to integrate them. This will create a strong foundation for the large-scale use of renewable energy and market-oriented electricity trading.

3.2.4 Connecting New Energy and Intelligent Loads to the Grid

By the end of 2022, China topped the world in terms of installed capacity of new energy

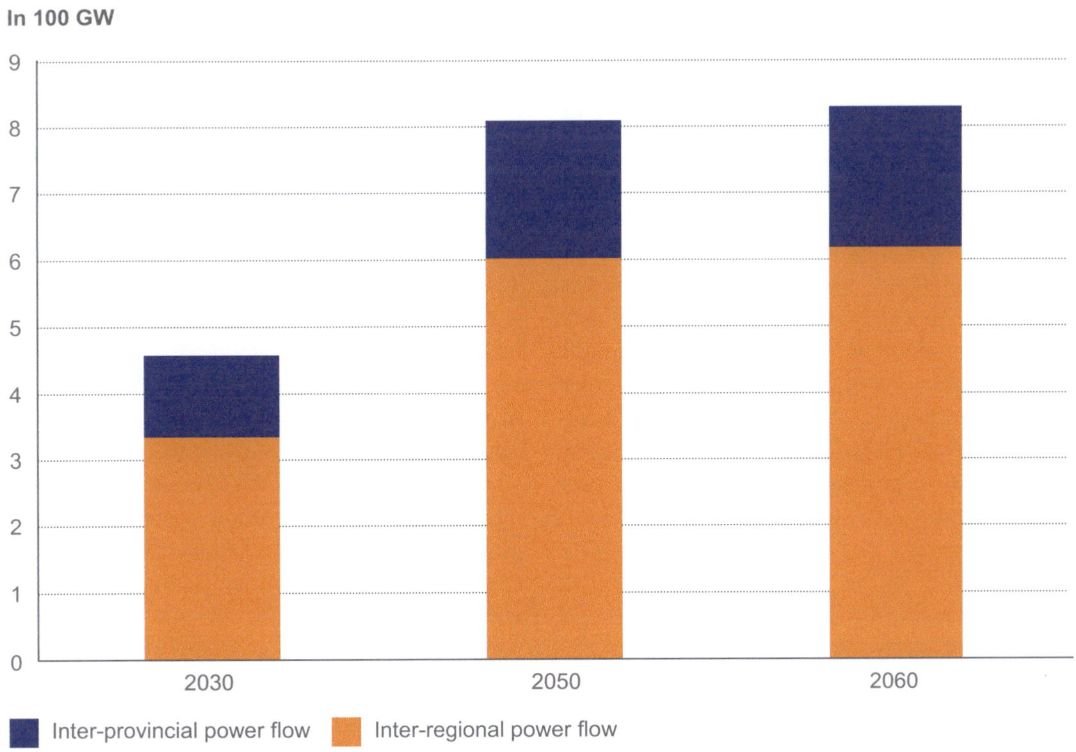

Fig. 2 Inter-provincial and inter-regional power flows in China. Source: 2021 Global Energy Interconnection Development and Cooperation Organization (GEIDCO); The Road to China's Carbon Neutrality, 2021

connected to the grid. During the 14th Five-year Plan (2021–25), more clean energy will be connected to the grid. By 2030, China's total installed capacity of wind and solar power is expected to be more than 1,200 GW.

The combination of the energy transition and advances in power electronics and information networks has made the move towards an intelligent and IT-enabled power grid irreversible. At the beginning of 2019, there were more than 500 million smart energy meters and 800,000 grid-connected electric vehicle charge points in China. By 2060, all energy-efficient household appliances will be IT-enabled. As the number and type of devices connected to the grid increases—from electric vehicles to energy storage units, smart meters and smart homes—the power distribution network will undergo significant change and face new challenges, such as bidirectional power flow and voltage fluctuations. As a result, distribution networks and microgrids need to meet higher requirements for safety, reliability, cost effectiveness and adaptability.

3.2.5 Distributed Power Generation

The growing demand for energy to power China's economic and social development means that distributed renewable power generation will play an important role in the power system thanks to its geographical closeness to users and ability to integrate with other energy sources. Distributed power systems are energy efficient and reduce emissions. They can integrate multiple renewable energy sources such as solar, wind, biomass and wave power with locally available fossil fuels (mainly natural gas). This allows them to perform grid services like peak shaving and guide consumers to use more renewable energy.

3.3 Development Roadmap

For China to achieve its dual-carbon goal, the amount of renewable energy in the electricity mix will have to increase significantly and electricity will have to become the main carrier in final energy consumption. Considering the challenges facing power grid infrastructure, and the geographical divide separating renewable power generation from load centres, China will have to shape its energy transition around these conditions and build a power system based on large grids and microgrids that are co-ordinated and complementary. Grid-to-grid power transfer should optimise energy resource allocation and the flexible regulating capacity of microgrids should meet local demand for distributed power. A bidirectional relationship between transmission and distribution networks should be formed—one that is interactive and complementary, co-ordinated and synergic—to improve the consumption of new energy in the entire power system.

Moving forward, China should build two synchronous power grids that have ultrahigh-voltage transmission links as their backbone in East and West China, forming a grid pattern of west-to-east and north-to-south power transmission. The country should rapidly develop intelligent distribution networks and microgrids, and allow centralised and distributed power grids to interact and support each other, thus harnessing the complementary benefits of multiple energy forms and achieving flexible, intelligent, secure and stable power transmission systems. The formula of digitally controlled "large power grids + small and medium-sized regional power grids + smart distribution networks and microgrids" makes power grids more flexible and controllable, allowing the connection of large amounts of new energy generation adapted to local conditions and resource endowments.

Large power grids will continue to serve as the foundation of the power system. They will play an essential role in transferring power and ensuring electricity security. They will integrate with microgrids, distributed power generation networks, various types of energy storage and electric vehicles. The interconnection of large power grids will enhance capacity for power transfer and sharing, and improve the grids' self-healing ability and resistance to outages.

However, as the amount of grid-connected renewable energy capacity increases, the stability of regional power grids will weaken. Ultrahigh-voltage transmission corridors will be the principal means for regional power transfer, enabling zero-carbon energy to be consumed far from its point of generation.

AC–DC hybrid flexible grids, microgrids and other types of grid will be developed in a collaborative way, and play an important role in optimising energy use. The Energy Internet integrates generation with power grids, loads and storage. It brings multiple energy sources together in a complementary way and enables diverse actors across the power system to interact. Looking ahead, intelligent digital technologies will be further integrated with power grids to optimise end-user energy consumption and enhance supply reliability. Microgrids are an important solution for connecting distributed power. They are self-sufficient, which can counteract the vulnerabilities of large power grids, and they achieve a high penetration of distributed new energy. Their hallmarks are decentralisation and flexibility. In the future, they will be widely deployed in urban centres, remote rural areas in central and western China, economically developed industrial zones, military camps, airports and others.

Ultrahigh-voltage DC transmission corridors will connect renewable energy clusters with load centres and ultrahigh-voltage AC transmission links will enable inter-regional and inter-provincial power transfer, strengthening the power system's resistance to outages. Microgrids will be combined with internet technology to implement the Energy Internet across regions. The market size of China's microgrid industry will expand rapidly in the coming decade. Most microgrids will be built with distributed renewable power generation and energy storage, and will gradually become competitive as the cost of renewable energy generation and energy storage decreases. Smart grids will become the core energy allocation platform. They will feature demand–response, thus enabling interaction between the grid and users and creating an intelligent and interactive user service system.

3.4 Policy Recommendations

3.4.1 Reform the Electricity System

China should deepen its reform of the electricity system, recognise the technical and economic benefits and constraints of real-time, intermittent supply and demand; ensure that electricity generation, transmission and use are balanced dynamically; support the secure and stable operation of the power system; and improve the safety and reliability of electricity supply. When designing the power system, China needs to be aware that electricity is not only an energy carrier, but a commodity and an issue of national energy security as well. The country should steadily reform the electricity retail sector, deregulate it in an orderly manner and open it up to private capital, provide grid connection to different parties fairly, establish a new mechanism for developing distributed power networks, and improve overall planning and supervision. China should reform the power trading system, make the trading mechanism market-oriented, set up independent electricity trading organisations and create a fair and regulated market trading platform.

3.4.2 Enable Inter-provincial and Inter-regional Trading

China should promote the market-oriented reform of the electricity sector in all provinces, establish a unified electricity market and facilitate the development of integrated inter-provincial and intra-regional markets. The country should design a multi-level market system comprising medium- and long-term markets and day-ahead and intra-day markets; it should develop a real-time balancing market and auxiliary services market, and promote renewable energy consumption. In addition, it should create mechanisms for free market participation and inter-provincial and inter-regional sharing of flexibility resources to ensure system balance and stability.

3.4.3 Explore a New Energy Pricing Mechanism

To keep electricity tariffs as stable as possible, China should improve its time-of-use electricity price mechanism and explore electricity pricing mechanisms suitable for connecting large-scale new energy power with the grid. This will support the transition to a new, stable and secure electric power system dominated by new energy. As part of this approach, China should improve the difference between peak and off-peak tariffs and introduce a peak pricing mechanism that reflects actual conditions. The improvement should take into account current differences in peak and off-peak tariffs in the regions, the amount of new energy in the power system, regulating capacity and other factors.

3.4.4 Build More Distributed Power Networks and Microgrids

Efforts should be made to build bi-directional renewable electricity services across multiple channels. This would include bi-directional smart energy meters and innovative interactive services that reflect real-time distributed power supply and demand; improvement of the bi-directional capability of electric vehicles with the distributed power network, and increased numbers of battery charging and battery-swapping facilities; and a green, intelligent and interactive electricity service system centred on new energy and flexible interaction between networks and consumers. China should optimise its market-oriented trading mechanism for distributed power and create systems and mechanisms to enable trading between renewable energy microgrids, small power grids, distribution networks and large power grids. The country should speed up the construction of local distribution networks for industrial parks, build smart renewable energy microgrids, encourage the broad deployment of distributed power generation and microgrid technologies in distribution networks, and enable power transfer between grid and network neighbours.

4 Development Roadmap for Energy Storage

4.1 Challenges and Trends

There are five types of energy storage: mechanical, electrochemical, electromagnetic, chemical and thermal. Pumped storage is a type of mechanical energy storage, and is the most widely used in China. Other types of energy storage used for new energy include electrochemical, compressed air, flywheel and molten salt. In China, mechanical and electrochemical energy storage predominate. Mechanical energy storage includes pumped storage, compressed air and flywheel. Electrochemical energy storage uses batteries to achieve mutual conversion between electrical and chemical energy. The main types of battery are lithium-ion batteries, lead-acid batteries, flow cells and sodium-sulphur cells.

In China, pumped storage has the largest deployment, followed by electrochemical energy storage. By the end of 2019, the installed capacity of pumped storage in the country was about 30 GW, representing 93.4% of China's total installed capacity of energy storage. China's Fengning pumped storage power plant is the world's largest, with an installed capacity of 3.6 GW.

Electrochemical energy storage is growing the fastest. By the end of 2019, the installed capacity of electrochemical energy storage in China was about 1.71 GW, with an average annual growth rate of 80%. The installed capacity of lithium-ion batteries was about 1.38 GW. At present, a lithium-ion battery has energy storage cycles of 4,000–5,000 charges and discharges, and an energy density of 200 watt-hours per kg. According to the China Energy Storage Alliance (CNESA), by the end of 2020, electrochemical energy storage was experiencing explosive growth, with the installed capacity of new facilities at 1,559.6 megawatts, up 145% year-on-year.[4]

Molten salt is the preferred material for high-temperature thermal storage with the lowest power cost. As technologies progress, the service life of equipment is expected to improve. The cost of storing energy in supercapacitors, lithium-ion batteries and pumped storage is currently $7–10 per kilowatt (kW), $300–400/kWh

[4] White Paper on Energy Storage Industry Research, 2021.

and \$700–\$900 per kW respectively.[5] The cost of hydrogen storage and transport depends mainly on the storage method, type of transport and distance transported, which is high.

4.1.1 Constraints Need to Be Removed Urgently

At present, the fastest growing type of electrochemical energy storage, lithium-ion batteries, faces technological challenges, such as their relatively small number of cycles, high cost and short battery life. The challenge in system integration is how to effectively control the operation of a large number of energy storage devices with different parameters in large-scale deployment.

Key technologies, such as low-cost phase–change thermal storage and long-life hydrogen storage will provide crucial support for the interconnection of different types of energy network such as power grids, district heating and hydrogen–natural gas pipelines in the future. Breakthroughs in these technologies could make such interconnections and their joint operation and control possible. In batteries, there need to be breakthroughs in technology development if the range requirements of electric vehicles are to be improved and vehicle-to-grid services widely used.

4.1.2 The Cost Effectiveness of Large-scale Commercial Deployment Is Inadequate

Currently, the large-scale commercial deployment of energy storage is insufficiently cost effective. In 2020, the energy storage cost was about RMB 1.2 per watt-hour. Batteries account for around half of the total, which is more than the combined cost of the battery management, power conversion and energy management systems, including their integration. Even though the battery is the most important part of the energy storage value chain, the energy storage equipment as a whole is not included in the cost of electricity. This makes it difficult for the financial cost to be shared with others, which leads to insufficient cost effectiveness in large-scale commercial deployment. In the future, battery costs will likely decrease and the cost mix of the energy storage system in China will be reshaped. Energy management systems and system integration are expected to overcome difficult and complex application challenges, which will improve the cost effectiveness of large-scale commercial deployment.

4.1.3 Effective Business Models Are Needed

China's energy storage industry has not yet found a suitable market mechanism and business model. Of the energy storage projects that started operation in China in 2020, the installed capacity of those for new energy generation was more than 580 MW, a surge of 438% year-on-year. Although energy storage support facilities for wind and solar power are mandatory, no business models and price transmission mechanisms have yet been introduced. Although these facilities drove rapid growth in the energy storage industry in a short time, there are inadequate drivers for sustained development.

4.2 Analysis of Supply and Demand

Increasing numbers of new energy generation plants and 5G base stations, as well as breakthroughs in technology, will drive up demand for energy storage. On the supply side, pumped storage dominates, although electrochemical storage is leading the way forward in new energy storage growth.[6] According to the China Energy Storage Alliance (CNESA), in 2020, China's total installed capacity of energy storage was 35.6 GW, of which pumped storage was 31.8 GW and lithium-ion batteries (the fastest-growing type of electrochemical storage) was 2.9 GW. These two technologies accounted for 97.47% of the country's total installed energy storage capacity.

[5] GEIDCO, Development Roadmap for Large-scale Energy Storage Technology, September 2020.

[6] China Energy Storage Alliance.

4.2.1 Stabilising the Future Power System

Energy storage is necessary to ensure flexibility in the power system. When demand for flexibility is relatively low, pumped storage has sufficient capacity to meet requirements. With more and more new energy in the power system, the daily fluctuation range of wind and solar power output can be 80% and 100% respectively. In this setting, energy storage is needed across the power system—in power generation, power grids and consumption—to smooth out the randomness and volatility of wind and solar power (see Fig. 3). Given the disadvantages of pumped storage, such as its geographical limitation to mountainous areas and relatively long response times (typically around 2.5 minutes from standby to full load), electrochemical storage, which has a shorter response time, is required to ensure flexibility in the power system.

4.2.2 The Huge Development Potential of Energy Storage Technologies

Energy storage technologies continue to mature, especially those related to electrochemical storage such as lithium-ion batteries, creating new scope for demand. Further technological breakthroughs are expected, mainly in electrochemical and hydrogen storage. Electricity-to-hydrogen, hydrogen storage and hydrogen-to-electricity are in focus. In addition, breakthroughs in battery management systems, for example, improve integration and control. Coupled with the long life, low cost and safety of lithium-ion batteries, demand for new energy storage systems is expected to grow.

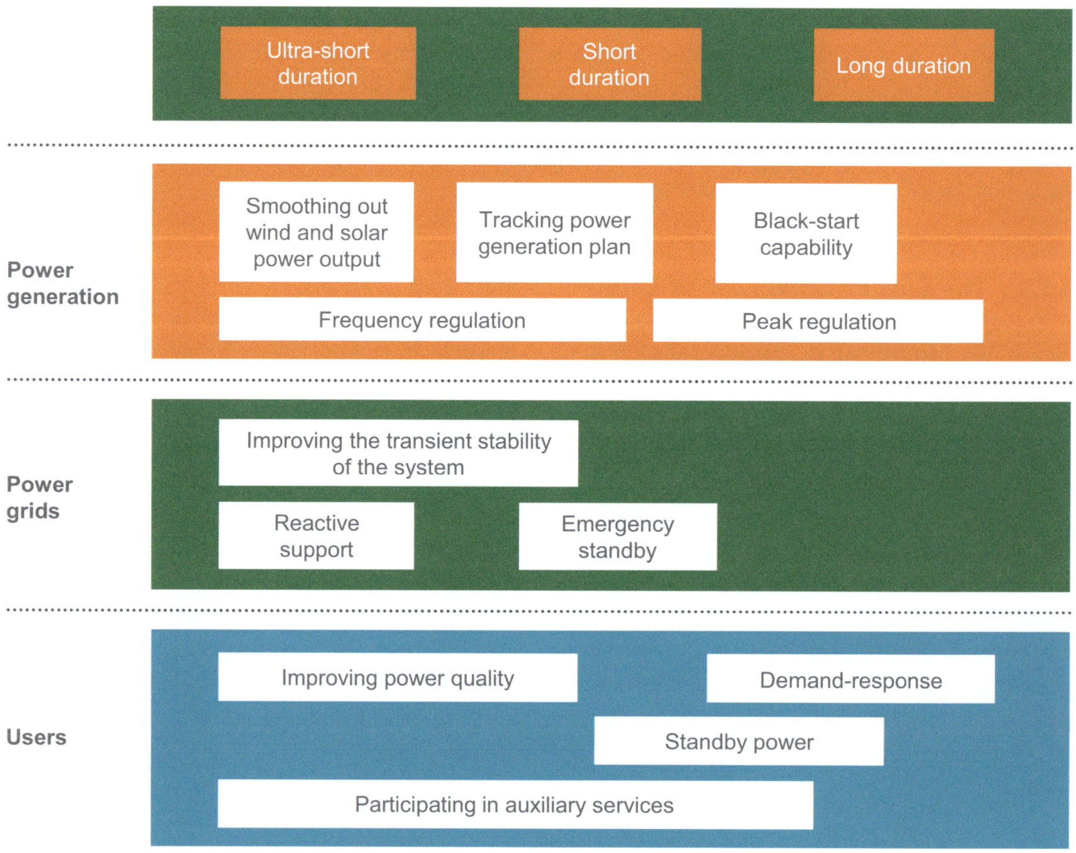

Fig. 3 Energy storage applications in the power system. *Source* Produced by the authors

The pace of maturity is getting faster. First, electrochemical energy storage technologies like lithium iron phosphate batteries and lead-carbon batteries have evolved rapidly and are ready for commercialisation. Second, batteries with a long service life, low cost, high conversion efficiency and a high level of safety are essential for large-scale deployment, requirements that electrochemical energy storage meet. Third, pumped storage has the highest technology readiness level and is already an important part of power grid operations. Fourth, flow batteries are at an early stage of commercialisation and are easy to scale up, which makes them suitable for large-scale energy storage. Fifth, the integration, operation and control technologies for power plants with megawatt-level and 10-megawatt-level battery energy storage capacity have been mastered and meet power system requirements (storage capacity at the 100-megawatt level needs to be improved).

Large-scale, long-duration energy storage supports the emergency supply capacity of the current regulating and energy storage system, power networks and conventional energy reserves. Long-duration energy storage equipment is the focus of energy storage development. In 2020, large-scale, long-duration energy storage capacity was mainly provided by pumped storage combined with lithium-ion batteries. By 2025, large-scale long-duration energy storage is expected to be possible by using a combination of pumped storage, hydrogen storage, compressed-air energy storage, heat pump energy storage, and thermochemical energy storage technologies (see Table 1).

4.2.3 New Energy Penetration Will Increase

As the installed capacity of new energy grows, demand for energy storage will increase. Before 2019, the energy storage system was based on pumped storage, which was sufficient to meet the requirements of a power system with a low penetration of new energy. Guided by the dual-carbon goal, the proportion of new energy will greatly increase by 2050, and energy storage technologies will be widely deployed in different applications in power generation, grids and among users. The need to counter the randomness and volatility of wind and solar power will make energy storage essential.

4.3 Development Roadmap

In a future energy system with a high proportion of clean energy, multiple energy storage units of different sizes, in different locations, with different storage capacities and using different technologies will form a complete energy storage complex that meets the regulating requirements of the power system. By 2035, more short-duration energy storage units will need to be deployed in power generation to smooth out the randomness and volatility of new energy generation; these would include hybrid wind and solar projects and concentrated solar power. On the user side, virtual short-duration energy storage, such as vehicle-to-grid, will play an increasingly important role in the energy system. By 2050, larger-scale energy storage will be needed to serve as flexibility resources, and the energy storage capacity of the

Table 1 Energy storage durations of various technologies

Storage duration	Energy storage types available
Milliseconds	Supercapacitors, superconducting magnetic energy storage, flywheel energy storage
Minutes-to-hours	Lithium-ion batteries (electrochemical energy storage)
Large-scale, long-duration energy storage	Pumped storage, hydrogen storage, compressed-air energy storage, heat pump energy storage, thermochemical energy storage

Source Produced by the authors

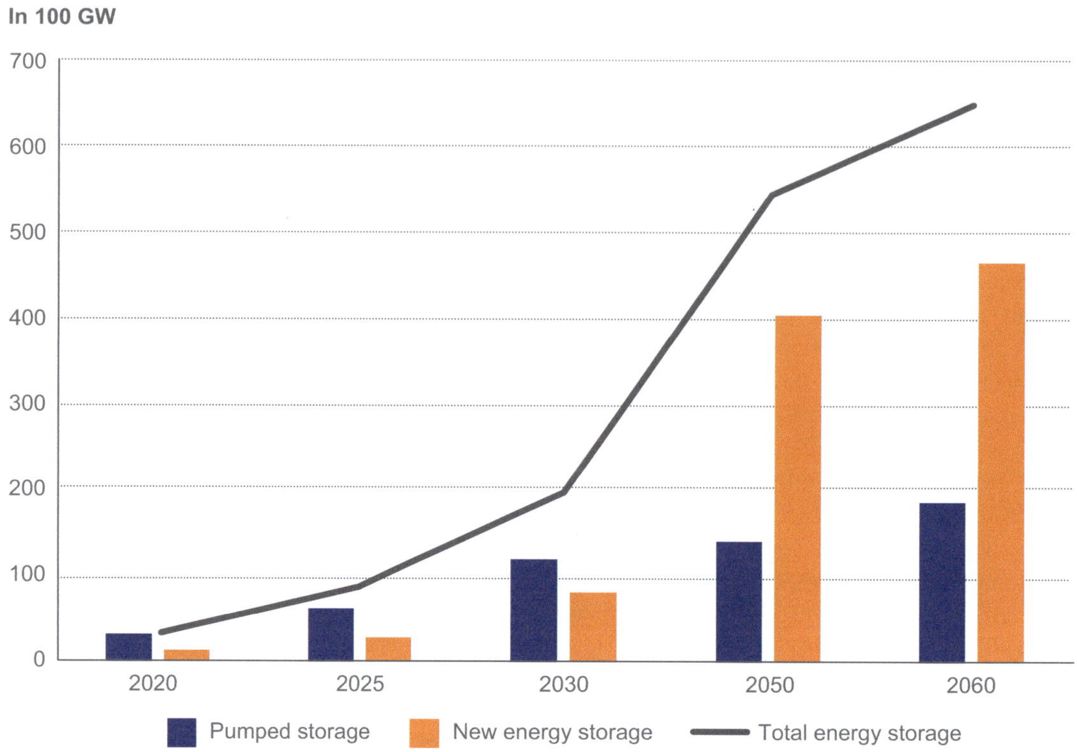

Fig. 4 Changes in the energy storage mix. *Source* Produced by the authors

whole system will reach 30%–40% of the maximum load. Moreover, long-duration energy storage will play an increasingly prominent role in seasonal energy regulation, and the electricity stored will reach 0.5%–2% of annual electricity demand across the system. To switch to a 100% renewable energy system, chemical energy storage technologies will be needed, including electricity-to-fuel, to integrate and optimise storage capacity scattered across different systems.

Between 2020 and 2030, the installed capacity of energy storage in China will grow at a rate close to that of the installed capacity of new energy. From 2031 to 2060, breakthroughs in energy storage technologies, rapidly decreasing energy storage costs and diverse energy storage business models will shift the role of energy storage to one of strategic support in the new power system. Although the power auxiliary services market will face greater challenges in 2060, energy storage will be able to deal with the volatility of the new power system. This, despite increasing volumes of wind and solar power and rising demand for energy storage from users, power grids, 5G base stations and data centres. Fig. 4 shows the predicted change in energy storage mix to 2060, with pumped storage representing conventional energy storage and electrochemical energy storage representing the new.

Between 2020 and 2025, China's energy storage capacity is expected to increase to 98 GW, and the storage-to-new energy ratio,[7] which previously decreased, is projected to increase to 200 GW by 2030. From 2031 to 2050, China's energy storage capacity is expected to climb to 540 GW and by 2060 to 646 GW, as the technologies mature. At that time, energy storage systems will play a strategic role in supporting the dual-carbon target, and will reliably respond to the volatility of the power auxiliary services market.

[7] Storage-to-new energy ratio is the ratio of installed energy storage capacity to installed new energy capacity.

From 2020 to 2030, energy storage is projected to grow rapidly and new energy storage technologies to be commercialised. By 2025, the installed capacity of new energy storage and pumped storage will reach 30 GW and 62 GW respectively. By 2030, the installed capacity of pumped storage will rise to 120 GW, while that of new energy storage will be 80 GW (accounting for 40% of the country's total), which is sufficient to meet the needs of the new energy power system. After 2050, new energy storage is expected to account for 71.2% of the total installed capacity of energy storage.

The cost of short-duration lithium-ion battery storage and long-duration electricity-to-hydrogen-to-electricity storage is estimated to be RMB 953/kWh and RMB 50/kWh respectively by 2030; RMB 635/kWh and RMB 31/kWh by 2050; and RMB 476/kWh and RMB 19/kWh by 2060.

By 2030, energy storage is expected to evolve into an independent market. As the issues related to energy storage technology standards, market positioning and electricity price uncertainty are solved at the policy level, coupled with the development of independent (non-imported) and controllable core technologies, China will be among the leaders in terms of technological innovation and deployment of energy storage. The standards system, market mechanisms and business models of energy storage are expected to be mature and deeply integrated with all parts of the power system.

By 2040, energy storage battery costs are expected to drop by 70%. By 2050, lithium iron phosphate battery technology will have achieved breakthroughs in, for example, extending battery service life; diverse business models for energy storage systems will be possible; lithium-ion batteries will be greatly improved in equipment safety, service life and charge and discharge cycles; pumped energy storage on the other hand will not change significantly, although it will maintain stable growth despite the constraints of geography, efficiency and other factors. By 2060, the value chains of upstream electronic components, midstream system integration and downstream power grids will be integrated. Although the power auxiliary services market will face risks posed by the variability of new energy, energy storage will be able to deal with the volatility of the new power system. Lithium-ion batteries are projected to achieve 140% higher energy density, 150% more cycles and 75% lower costs. Supercapacitors and other power storage technologies will be put into commercial use. The technology maturity of the energy storage battery management system will be greatly improved, as will the energy storage converter industry.

4.4 Policy Recommendations

4.4.1 Deploy Different Energy Storage Technologies and Expand Energy Storage Applications for Users

Using different energy storage technologies can address the limitations of deploying a single energy storage technology. Energy storage applications for users should be expanded by focusing on industrial parks, using 5G base stations for auxiliary power supply, and securing power supply reliability at data centres. On the demand side, an "electricity-heating-cooling-gas hybrid energy storage system" should be deployed, and equal focus should be placed on the development of new vehicle-to-grid technology and the use of decommissioned electric vehicle batteries. Lithium-ion batteries and supercapacitors should be combined for rapid electric vehicle charging and the integrated development of energy storage and electric vehicles. Energy storage applications for big data and microgrids should be explored to strengthen the integration of value chains and find new ways to commercialise energy storage technologies.

4.4.2 Innovate New Business Models for Energy Storage Applications Based on the Electricity and Energy Markets

System scheduling, compensation for power auxiliary services and other steps should be taken to ensure the cost of storing new energy is reasonable. At the same time, the difference between peak and off-peak prices should be widened, and tax incentives introduced to encourage innovative new business models for energy storage. China should refine the time-of-use pricing mechanism to allow new energy companies investing in energy storage to make a profit; it should encourage new business models with energy storage batteries at the core of the industrial, upstream and downstream value chains; it should move beyond the single arbitrage model of peak and off-peak electricity prices; and it should explore diverse business models like household energy storage to unlock the potential of the energy storage industry.

4.4.3 Integrate Distributed Power and Energy Storage Systems

China should draw on the energy storage policies of Germany to improve its standards for integrating distributed and energy storage systems. The country should advance its standards for sodium-ion accumulators and other energy storage technologies, improve regulations and monitoring, and standardise the main parameters of different types of distributed energy storage systems.

4.4.4 Develop the Energy Storage Industry at Scale

For users, the difference between peak and off-peak electricity tariffs should be increased to make up for the construction costs of energy storage systems and transfer the cost of energy storage from power grid utilities to users. With regard to the auxiliary services market and carbon tax and trading, energy storage should be clearly defined as an independent market actor. The cost of alternative energy storage facilities for power grids should be included in transmission tariffs and distribution fees, and a capacity pricing mechanism for power plants with energy storage should be established. The government should formulate separate transaction policies for the electricity purchase price, storage discharge price, transmission and distribution fees and settlement method for energy storage. It should also subsidise the research and development and commercialisation of energy storage technologies to help the industry transition from early commercialisation to large-scale development.

5 Roadmap for Hydrogen Infrastructure

Hydrogen is a low-carbon, efficient, safe and sustainable secondary energy source. It is an important part of China's future energy system and an important carrier of renewable and low-carbon energy to end-use sectors and strategic emerging industries. Hydrogen has the potential to become an energy carrier that integrates different types of infrastructure, thereby improving efficiency, reliability and flexibility. Hydrogen can also provide large-scale, long-duration energy storage and auxiliary grid services for the power sector, such as emergency response, load tracking and regulating reserves. Overall planning and construction of hydrogen infrastructure not only meets the requirements of deep decarbonisation, but also supports the dual-carbon goal.

5.1 Challenges and Trends

5.1.1 Hydrogen Storage and Transport Efficiency Need to be Improved

Unlike conventional petroleum fuels that are easy to transport and can be stored at scale, hydrogen storage and transport technologies face challenges in energy efficiency and safety. Gaseous hydrogen in China is usually stored at high pressure of up to 20 megapascals (MPa) and transported in tube trailers. If daily demand at a hydrogen refuelling station is less than 500 kg, tube trailer transport of gaseous hydrogen saves liquefaction costs and upfront investment in pipeline construction; it also has a higher cost effectiveness within a certain transport distance. If hydrogen demand or the transport distance increases, cost effectiveness can be met only by increasing the gaseous hydrogen transport pressure or by using liquid hydrogen tanker trucks or gaseous hydrogen pipelines.

Tanker trucks are a competitive means of transporting liquid hydrogen over long distances and at large scale. They have large storage and transport capacity, which means fewer tankers and workers are required. In terms of the full cost of liquefaction and transport over long distances and at large scale, liquid hydrogen is significantly more cost competitive than tube trailer transport of high-pressure gaseous hydrogen at 20 MPa. However, liquefaction uses lots of energy and requires heavy fixed investment, which account for more than 90% of the total cost of liquid hydrogen storage and transport.

5.1.2 The Small Number of Hydrogen Refuelling Stations Hinders the Development of the Hydrogen Industry

China has only a small number of hydrogen refuelling stations. In 2020, China had 181 hydrogen refuelling stations built or under construction, including 124 in operation. Fifty-five hydrogen refuelling stations were completed in 2020 alone. Equipment accounts for about 70% of the total investment cost of a hydrogen refuelling station, which is more than RMB 10 million, significantly higher than the investment cost of a conventional service station. Equipment operation and maintenance, labour costs and other factors push up the cost of hydrogen refuelling, which is RMB 13–18 per kg, hindering the large-scale uptake of hydrogen fuel-cell electric vehicles.

5.2 Analysis of Supply and Demand

5.2.1 Hydrogen Enables Deep Decarbonisation

Hydrogen is of great strategic value for China to enhance the security of its energy system and achieve carbon neutrality; it will also play an important role in the decarbonisation of heavy industry and heavy-duty transport and in providing flexibility for the power system.

Hydrogen will help heavy industries on their journey to zero carbon. It can be a direct heat source for many industrial sectors such as iron and steel and metals. Hydrogen can also be used as a reducing agent to produce zero-carbon steel using direct reduced iron technology. Hydrogen can be used for gas welding, providing benefits such as high efficiency and low cost. Zero-carbon hydrogen can meet the rapidly growing demand for synthetic ammonia, including new demand for ammonia as a zero-carbon shipping fuel (a total of 43 million tonnes of ammonia are produced annually in China). In methanol production, catalysts for carbon dioxide hydrogenation have been commercially produced, and some pilot plants have started operating in different parts of the world.

In the transport sector, battery electric vehicles are likely to dominate in the future, but hydrogen fuel cell electric vehicles (FCEVs) will probably attract a small number of users that require a long-range vehicle. Hydrogen FCEVs, which have the advantages of long range and short refuelling time, can play an important role in heavy-duty long-haul road freight. China already has plans to build hydrogen refuelling stations at scale. In shipping, battery-powered and hydrogen fuel cell electric ships will account for an important proportion of short-distance inland and coastal shipping, and the cost effectiveness of these ships will gradually improve. In zero-carbon long-distance shipping, ammonia is likely to serve as a key fuel. Ammonia is produced using hydrogen as the main feedstock, creating annual hydrogen demand of 7.6 million tonnes. In aviation, hydrogen-fuelled aircraft could be a decarbonisation option for short- and medium-distance flights.[8] At present, several hydrogen aircraft models are under development around the world, and some experts believe that hydrogen or batteries can be used to power aircraft with fewer than 100 seats and a flying range of between 300 and 500 km.[9]

[8] International Renewable Energy Agency, Hydrogen from Renewable Power: Technology Outlook for the Energy Transition, 2018.

[9] International Civil Aviation Organization, Secretariat, 2019.

With regard to power system flexibility, hydrogen produced from surplus electricity is expected to be an effective energy storage method. Hydrogen can improve the flexibility of the energy system when combined with natural gas or fuel cells. In our Zero-carbon Scenario, annual hydrogen production of 81 million tonnes is needed in China to meet demand from transport and industry by 2050. Iron and steel, chemical feedstocks and heavy-duty transport will account for the largest demand for hydrogen.

5.2.2 Demand for Large-scale Long-duration Energy Storage Is Growing

In 2021, the U.S. Department of Energy published a report supporting long-duration energy storage, which is defined as energy storage that provides at least 10 hours of continuous operation (discharge) and a service life of 15–20 years. During windless or rainy days, long-duration energy storage is needed to make up for the shortfall in electricity that the diminishing number of thermal power plants once provided.

Hydrogen energy storage features high energy density and low operating and maintenance costs. It can support long-duration storage and an emission-free process. It is a rare technology that can store electricity of more than 100 GWh and also be used to supply power for very short or very long durations. Hydrogen is seen as a highly promising new type of large-scale ultralong-duration energy storage carrier. Hydrogen is an excellent energy storage medium for renewable and sustainable energy systems. It has a wide range of applications and can be used as a fuel gas or chemical feedstock for power generation in fuel cells. As an energy carrier, hydrogen has outstanding advantages: compressed hydrogen, for instance, has a high energy density and can be used to support power grids.

The development of hydrogen energy storage faces barriers, such as the high cost of hydrogen production and storage, the difficulty of ensuring safety, the low efficiency of fuel cells and incomplete policies, regulations and services related to the electricity market. In addition, the overall efficiency of hydrogen energy storage systems is low (about 40%, but it can reach 60% if used with combined heat and power), and the cost of investing in fuel cells is high. Hydrogen energy storage has not yet been put into large-scale commercial operation anywhere.

5.3 Development Roadmap

By 2030, the efficiency of hydrogen storage and transport will have risen to 35%; the hydrogen storage density will have increased to 15–20 moles per litre (mol/l); the continuous discharge duration will be more than 100 hours; and the system cost will have dropped to RMB 9,000–10,000 per kilowatt (kW). By 2050, the efficiency of hydrogen storage and transport will climb to 60%; the hydrogen storage density will exceed 30 mol/l; the continuous discharge duration will reach more than two weeks; and the system cost will be lowered to RMB 8,500 per kW. By 2060, multiple technologies in storage and transport such as liquid hydrogen, high-pressure gaseous hydrogen, pipeline hydrogen, and hydrogen storage in liquid organic hydrogen carriers will develop in parallel; the efficiency of hydrogen storage and transport will be increased to 65% and the system cost will be reduced to RMB 6,000 per kW.

A network of diverse hydrogen refuelling stations and infrastructure will be built, and integrated refuelling stations offering hydrogen, compressed and liquefied natural gas, electric vehicle charge points and petrol will continue to emerge. By 2030, 1,000 hydrogen refuelling stations will be built. By 2050, the number of stations will have increased to 10,000 and by 2060 to15,000 (Fig. 5).

5.4 Policy Recommendations

5.4.1 Produce a National Hydrogen Development Plan

China should incorporate hydrogen into its national strategic energy plans. It should include hydrogen in the development of the renewable energy system, accelerate a national development

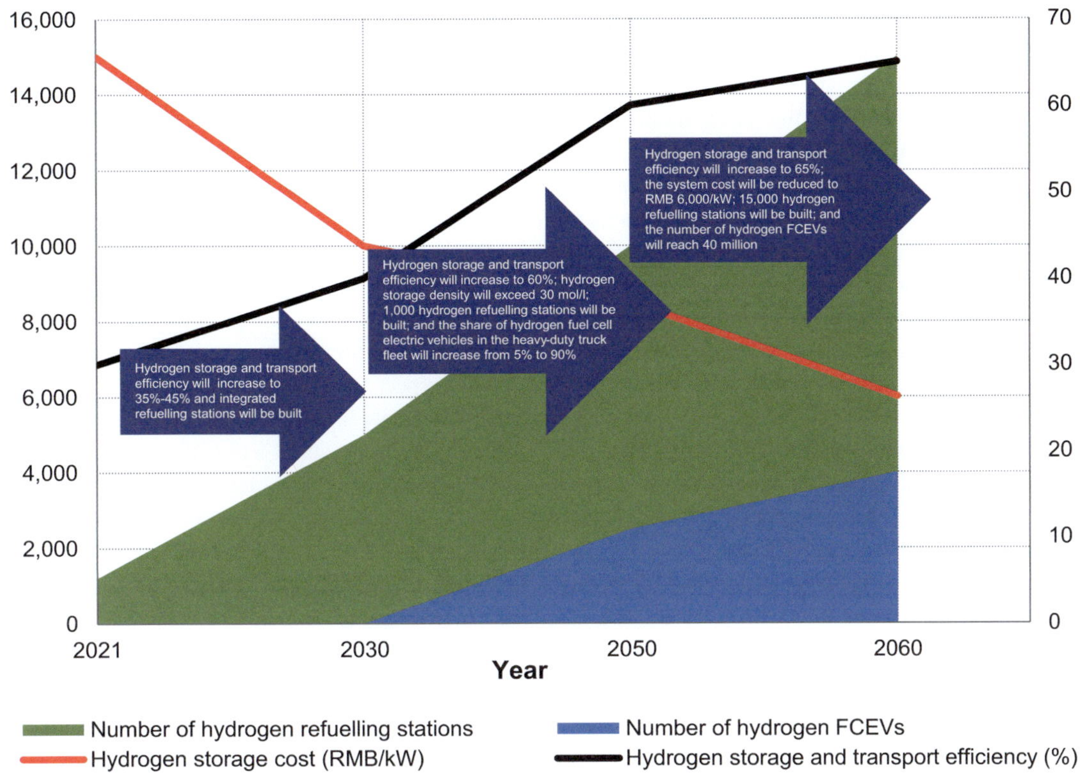

Fig. 5 Development roadmap for hydrogen. *Note* FCEVs = fuel cell electric vehicles

plan for the hydrogen industry and set targets for the uptake of hydrogen in different industrial sectors. China should include hydrogen in the national energy fund and investment management system, establish appropriate financing and guarantee mechanisms, redirect private capital towards hydrogen-related projects and focus on investment in hydrogen infrastructure.

5.4.2 Improve the Research System and Make Breakthroughs in Cryogenic Liquid Hydrogen Storage and Transport and Other Core Technologies

The government, industry associations, scientific research institutions and business should work closely together to establish a joint R&D approach and focus on achieving breakthroughs in core technologies like fuel cell stacks, fuel cell systems, fuel cell vehicles and the hydrogen supply system.

Breakthroughs in the core technologies of hydrogen production, storage and refuelling are needed. Key areas of research are the separation and purification of industrial by-product hydrogen (coke oven gas and chlor-alkali chemicals) for hydrogen production, the production of high-efficiency and low-cost hydrogen from renewables, and the integration of hydrogen production, storage and refuelling. Efforts should be made to research and develop high-pressure gaseous hydrogen storage systems, liquid hydrogen storage and transport systems and other new types of hydrogen storage. Attempts should be made to develop high-pressure refuelling technology of 70 megapascals and key equipment for hydrogen refuelling stations. And research should be made on forward-looking technologies such as cryogenic liquid hydrogen storage, hydrogen storage in liquid organic hydrogen carriers, hydrogen storage in solid materials, and liquid hydrogen refuelling stations.

5.4.3 Build a Low-carbon Hydrogen Supply System

China should clarify its priorities and devise workable action plans for the short, medium and long terms. In the short to medium terms—depending on the current economic, technological and environmental conditions—China should focus on developing the key technologies of high-pressure gaseous and liquid hydrogen storage and transport, promote the development of different types of hydrogen refuelling station (including integrated refuelling stations) and provide policy support.

5.4.4 Introduce Long-term Support and Incentive Policies for the Development of the Hydrogen Industry

China should encourage large enterprises, including central and state-owned enterprises, to lead the co-ordination of the hydrogen value chain. The country should concentrate support on central and state-owned enterprises in the fields of energy and intelligent manufacturing. This will enable those enterprises to use their strengths in resource channels, capital accumulation and expertise. The objective is to enhance competitiveness by integrating upstream and downstream entities, thereby helping the hydrogen value chain to develop rapidly.

China should explore a variety of products and fields suitable for fuel cells and hydrogen, such as large-scale energy storage, standby power supply and combined heat and power. It should encourage the development of pilot projects and guide, for example, those areas with abundant renewable energy resources to build hydrogen and fuel cell industrial development zones. China should also develop new business models to spur the development of the hydrogen and fuel cell industry.

6 New Energy Infrastructure Investment

The power sector will play a pivotal role in the transition of China's energy system towards net-zero emissions. It is imperative, therefore, to construct new smart power transmission and distribution infrastructure, connect renewable energy clusters with major load centres and accelerate the development of microgrids. Large-scale demand–response infrastructure (such as energy storage) that provides power system flexibility is needed to address the periodic and intermittent characteristics of wind and solar. China should also speed up the development of the infrastructure needed for hydrogen production, storage, transport and refuelling, biomass refining facilities and carbon capture, utilisation and storage (CCUS), including carbon dioxide transmission pipelines.

We expect investment in new energy infrastructure to grow rapidly. From 2020 to 2060, China will need to invest about $12 trillion in new energy infrastructure, of which more than half should be invested in the coming two decades. Because electrification is at the core of the net-zero transition, decarbonisation of the power sector will account for the largest share of investment, followed by hydrogen and CCUS. The electricity and financial markets will play a key role in this. They need to provide clear investment and financing models for new energy companies, provide an attractive risk–return balance of projects to stimulate investment by the private sector, issue green bonds and loans linked to sustainable development, and build a broader green financial system.

The investment demand forecasts in this section are based on the investment model in the China's Net-zero Emissions by 2060 Scenario and related international investment and financing case studies. Calculations of investment in new energy infrastructure in the power, biomass, hydrogen and CCUS sectors are based on new production capacity under China's dual-carbon goal and the cost of capital. The data for these calculations were obtained from publicly available sources including the International Energy Agency, the International Renewable Energy Agency and the Hydrogen Council.

6.1 New Energy Infrastructure Investment

To meet its zero-emissions target, China needs to significantly increase investment in new energy infrastructure in the coming two decades. At present, China's new energy infrastructure investment is about RMB 1.37 trillion per year. In the coming decade, the need for investment is estimated to climb to RMB 2.33 trillion per year, peaking at RMB 2.46 trillion in 2030 before declining to about RMB 1.88 trillion per year.

Due to the large-scale deployment of renewable energy, energy storage and network infrastructure, power systems will account for about 60% of new energy infrastructure investment between 2020 and 2060. Investments in hydrogen and CCUS, which will play an important role in decarbonising industries, will also increase significantly.

6.2 Low-carbon Electricity Investment

By 2030, annual capital expenditure in China's low-carbon power sector is expected to grow by about 150% compared with today. Between 2020 and 2060, total investment in the sector is estimated to reach RMB 54.78 trillion. Of this, RMB 19.86 trillion will be used to upgrade existing transmission and distribution infrastructure, which will connect renewable energy with demand centres; RMB 3.15 trillion will be invested in new nuclear power plants to provide low-carbon baseload electricity; RMB 3.42 trillion will be used to expand battery storage capacity, thereby improving power system flexibility; and RMB 23.28 trillion will be invested in wind and solar photovoltaic power generation.[10] Investment demand in the power sector will peak at about RMB 1.71 trillion in 2030, then slow down, with the compound annual growth rate declining from 5% in 2030 to 3% in 2050.

As technology matures and electrification drives China's renewable energy transition, funding for investment is likely to shift to debt financing. Under the combined effect of improved technology and reduced costs and risk, the share of debt financing for renewable energy projects increased from 23% in 2013 to 65% in 2018. Redirecting existing fossil fuel financing towards renewables and other new energy sources can raise the scale of financing.

Due to the relative immaturity of China's electricity market and uncertainties around new energy technologies such as energy storage, it is difficult for private financing to expand its share of funding in the coming decade. Rather, financing will require substantial public investments from state-owned enterprises. In the coming two decades, low-carbon electricity demand is estimated to be 16 GW annually, which will require battery storage capacity of 320 GW to integrate increasing volumes of renewable energy.

6.3 Hydrogen Investment

Because the decarbonisation of industry and transport will create growing demand for hydrogen and bioenergy, annual investment in the low-carbon production sector will soar by 290%. Hydrogen is used mainly in oil refining, chemicals and iron and steel, and is produced largely from fossil fuels (natural gas or coal). With the growing number of hydrogen applications, investment in hydrogen is projected to increase from RMB 5 billion in 2020 to a peak of RMB 171.2 billion in 2030. The money will be used to build network infrastructure and renewable hydrogen production facilities. After 2030, in line with falling technology costs, investment will drop to RMB 82.2 billion in 2050.

6.4 CCUS Investment

In China, CCUS has not yet been deployed at commercial scale, and CCUS investment is at a relatively low level. From 2020 to 2060, average annual investment in CCUS is expected to grow substantially by 180%, driven by investments in power, industry, transport and storage infrastructure. The power and industry sectors require investments of RMB 8.9 trillion and RMB 3.08 trillion respectively to deploy CCUS. Additional investment of RMB 4.39 trillion is required

[10] Only the major investments related to this chapter are listed here.

to improve CCUS transport and storage infrastructure.

Energy storage and demand–response can improve flexibility in the power system. This in turn can significantly reduce the investment required to deploy CCUS in the power sector. Our modelling shows that enhancing the flexibility of renewable energy and the power system provides the most cost-effective path to decarbonising the power sector, which significantly reduces the need for CCUS.

6.5 Financing Channels

Internationally, there are five main ways to solve the financing challenges for new energy infrastructure:

(1) Demonstrate a national commitment to shift to new energy systems by setting clear national targets, such as for carbon pricing.

(2) Increase potential revenue streams for new energy services by developing support policies to complete markets (in, for example, power flexibility, hydrogen supply, CO_2 transport and storage). Completing markets lays the basis for financing new energy infrastructure by establishing clear funding models for private investors. In Texas, this was done through the efficient design of the state electricity market, enabling utility-scale battery storage plants to be financed entirely through private investment.

(3) Ensure an attractive risk–return balance for new energy projects to enable private financing. In the UK, this has been done through policies targeted towards addressing technology-specific risks, including contracts for difference for offshore wind and government debt guarantees for nuclear plants. The UK's Green Investment Bank has also been used to offer more general de-risking tools such as co-investment and concessional finance.

(4) Green the financial system through the development of new financial instruments, mandatory requirements and strategic long-term finance from the public sector. Once the enabling conditions are in place, it is then critical to ensure that climate risks are adequately priced into the financial system. In Europe, this is being done through monetary regulations such as mandatory climate disclosures.

(5) Attract diverse investors by developing secondary markets and encouraging local ownership. Beyond incentivising existing investors to rebalance their portfolios to green investments, additional private finance can be unlocked by attracting a more diverse set of investors. In Germany, the financing of new energy projects through local energy co-operatives is attracting a more diverse set of investors with different priorities. Different technologies are at different levels of progression, which means that the focus of policymakers should vary based on the maturity of each technology. For relatively mature technologies, further finance can be unlocked through the development of new instruments and secondary markets. For less mature technologies, support policy needs to be developed to complete markets and enable finance.

6.5.1 Innovate New Business Models

The overall need for finance can be reduced through a higher utilisation rate of existing infrastructure. This can be achieved with innovative business models for the energy system. In new energy power generation, overall planning and aggregation of distributed new energy infrastructure can be made possible by developing integrated energy services (see Fig. 6). Aggregating distributed energy resources under a single entity allows for the optimal utilisation of existing energy infrastructure. By matching, co-ordinating and planning intermittent renewable energy infrastructure, overall energy utilisation efficiency can be enhanced and idle resources and waste reduced to drive down the total investment requirements. The market for aggregated energy companies is growing and is expected to reach a total value of $4.6 billion by 2023. Other options for reducing the financing need include demand–response and sector coupling.

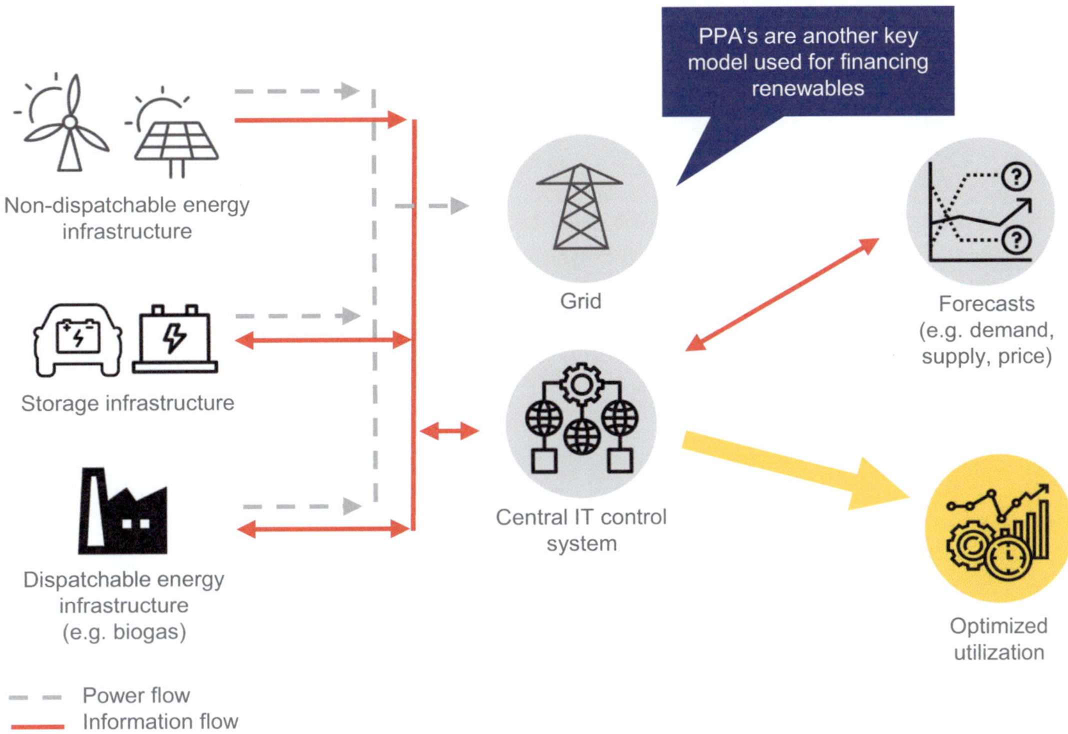

Fig. 6 Innovative business models. *Source* International Renewable Energy Agency, Aggregators Innovation Landscape Brief, 2019.

Case Study 1

In the UK, investment in flexibility can provide net savings for the energy system. Modelling of the UK energy system shows that implementing storage, sector coupling and demand–response technologies could increase the utilisation of the country's existing energy infrastructure. Integrating these technologies into the UK energy system would require £5 billion ($7 billion in average 2021 exchange rates) in capital expenditure for the required infrastructure. However, by increasing the utilisation of existing infrastructure, fewer additional infrastructure projects are needed, providing savings in the form of avoided capital expenditure. Other flexibility options not modelled, such as grid interconnections, could provide further savings.

6.5.2 Complete Markets to Increase Returns

Completing markets ensures that clear funding models for new energy projects are available by creating revenue streams for elements of the low-carbon transition. Carbon pricing creates revenue streams for large-scale hydrogen production and carbon capture by putting a monetary value on the carbon captured. The International Monetary Fund estimates that achieving the 2 °C warming target requires a carbon price of $75 per tonne of CO_2 equivalent by 2030, a price currently reached by only one carbon pricing scheme. Markets can be completed by either creating new markets (such as an energy trading system or a market for increased capacity) or by redesigning existing markets (for instance, by increasing spatial resolution in the electricity wholesale market).

Fig. 7 The higher the financing risks, the higher the financing costs, and vice versa. *Source* OECD, 2017. *Note* CfDs = contracts for difference; MDBs = multilateral development banks.

6.5.3 Ensure an Attractive Risk–Return Balance

Large infrastructure projects involve different types of risk, which occur at different stages of the project. As new energy technologies are more novel, project risks related to development, technology and policy are usually higher for new energy projects than for projects with mature technologies. Investors require higher rates of return for new energy projects with higher technology risks.

Policymakers can cut financing costs substantially by reducing, transferring or compensating for risk (see Fig. 7). De-risking tools include loan syndication, debt subordination, guarantees, insurance, hedging, public co-investment and technical assistance. Ultimately, a combination of de-risking measures is often needed to make projects investable for the private sector. State investment banks provide a useful vehicle for offering and implementing such measures. There are also examples of subnational governments developing their own green investment banks to de-risk investment in local energy infrastructure, such as the Montgomery County Green Bank in the USA.[11]

For mature technologies, a number of standard de-risking tools exist, such as contracts for difference and debt guarantees. However, for less mature technologies, new tools are being developed internationally to deal with the cross-chain risk that is a particular challenge for large infrastructure projects with multiple links, meaning that higher returns are required to secure financing for these projects. Beyond the size of risk, private investors are also less willing to take on risks over which they have limited control, such as those associated with development. Internationally, governments have reduced financing costs through de-risking tools such as subsidies and concessional finance.

[11] Montgomery County Green Bank is a green investment bank with a portfolio of $24 million which is used to mobilise investment in renewable energy projects in Montgomery county, Maryland.

6.5.4 Green the Financial System

Greening the financial system means reshaping it in order to support investment in new energy infrastructure and rapidly reduce activities that increase carbon emissions through comprehensive environment and climate risk management. It is key to unlocking additional finance. Internationally, there are two main trends:

(1) The development of new instruments and markets for green finance, which helps expand the pool of available green capital by attracting a more diverse set of investors with different preferences and appetites for risk.
(2) An increase in mandatory reporting requirements, which helps shift capital towards green investments by helping investors to better understand the financial risks associated with the transition to new energy systems. Understanding these risks is especially important given that, at present, such risks are not adequately priced into the financial system.

Case Study 2

In Germany, energy co-operatives are driving community investment in new energy; they make up a significant proportion of national energy investment. Energy co-operatives have become a popular method for financing small-scale, distributed renewable energy projects in Germany, which has become a world leader in community-based energy projects. While the average investment is small (€5,065), collectively Germany's energy co-operatives have mobilised €2.7 billion in capital for renewable energy infrastructure. In the North Rhine–Westphalia region, co-operatives have installed 2.9 GW of wind energy.

Energy co-operatives are attracting a more diverse range of investors, who are motivated by financial and non-financial factors. Renewable energy projects financed through co-operatives bring several indirect benefits to their home communities, including improving citizen participation and social innovation, contributing towards achieving local climate change goals, increasing social cohesion and promoting local job creation. Investors in German energy co-operatives have indicated that in addition to financial factors they are motivated by non-financial factors such as environmental protection and energy transition support. As such, energy co-operatives have attracted a more diverse set of investors (with different sets of priorities) to green investment, thereby increasing the pool of green capital.

7 Research on an Integrated New Energy System

An integrated new energy system, centred on renewable power generation, is an important part of the future energy system. The integrated new energy system will combine multiple types of energy: hydrogen, nuclear, oil, natural gas, bioenergy and others. Its platform will be the smart grid, a form of Energy Internet that deeply integrates advanced information, communication and control with energy technologies to support the transition to renewable and low-carbon energy. Flexibility and ease of access for multiple actors will be its hallmark. The new energy system will incorporate elements such as energy storage and electrification of industry, transport and buildings, and combine technologies such as big data, cloud computing and the internet of things. It will supply multiple types of energy, improve the efficiency of the energy system, reduce energy production and consumption costs and ensure energy security and economic and social development.

7.1 The Future Energy Mix

7.1.1 The Main Characteristics of the Future Energy System

To achieve the dual-carbon goal, China's energy mix will shift towards the use of large-scale renewable energy and the electrification of end-use demand. Between 2030 and 2050, the share of renewable energy will move beyond that of conventional fossil energy and, by 2060, will dominate the energy mix. At the same time, electrification will advance steadily, supporting the goal of carbon neutrality.

New energy will dominate the energy system. The installed capacity of zero-carbon renewable energy in China is expected to increase greatly, as shown in Table 2. The energy mix will shift from transportable and storable fossil energy to non-storage or non-transportable solar and wind energy that are closely related to weather conditions.

The electrification rate of end-use sectors will increase, driven by electrification itself and rapid growth in electricity consumption. As shown in Fig. 8, the proportion of electricity in final energy use will continue to rise.

The power system will be gradually integrated with modern communication technology to create an open and secure internet of things that features efficient information processing, digital data security, convenient and flexible use and a vibrant digital power ecosystem.

Table 2 Electricity mix, 2020–60

	2020	2030	2060
Total installed capacity (GW)	2,200	3,600–4,100	7,800–8,200
Proportion of installed coal power capacity (%)	49.1	31–36	4
Proportion of installed conventional energy capacity (%)	76	59	23
Proportion of installed non-fossil energy capacity (%)	44.8	52–59	88–89
Proportion of non-fossil energy output (%)	33.9	39–45	86–87

Source Development Report of China's Power Sector, 2020

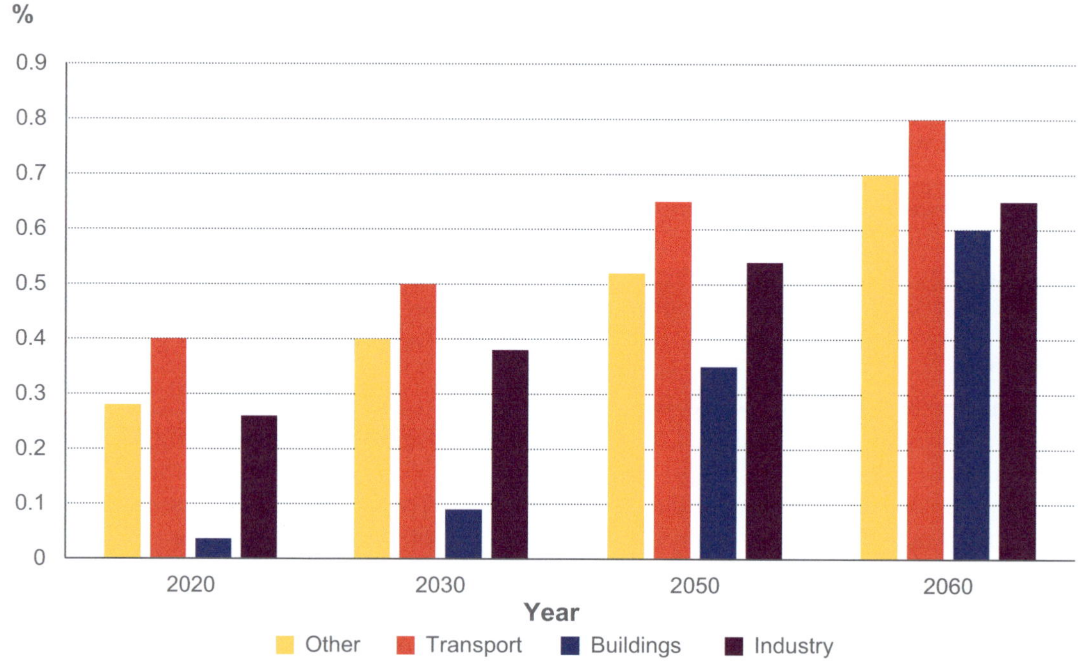

Fig. 8 Electrification rate of end-use sectors. *Source* China Statistical Yearbook 2020; State Grid Corporation of China

7.1.2 Co-ordinated Development of Electric Vehicles and the Power Grid

As forecast by the International Energy Agency in "Net Zero by 2050: A Roadmap for the Global Energy Sector" of 2021, between 2020 and 2030, the number of electric vehicles in the world will grow by a factor of 18, with annual sales reaching 55 million by 2030. In 2020, the number of new energy vehicles in China increased by 29.1% compared with 2019, 80% of which were battery-electric vehicles. As shown in Fig. 9, in our Business as Usual Scenario, the number of battery-electric vehicles is expected to reach 94 million in 2035, 230 million in 2050 and 430 million in 2060. In our Radical Scenario (assuming there is a ban on the sale of internal combustion engines in the future), the number of battery-electric vehicles is estimated to reach 101 million in 2035, close to 350 million in 2050 and 560 million in 2060.

Electric vehicles are a double-edged sword for the power grid. When the number of electric vehicles continues to rise, disorderly charging will increase the load on the power grid, which may cause a short voltage drop below the limit permitted, as shown in Fig. 10. Electric vehicles usually charge at relatively concentrated times, which can have a large impact on the power grid for a short time. Chargers generate power harmonics when they convert AC from the grid and DC from the battery.

Vehicle-to-grid applications can reduce the extreme loads caused by the simultaneous

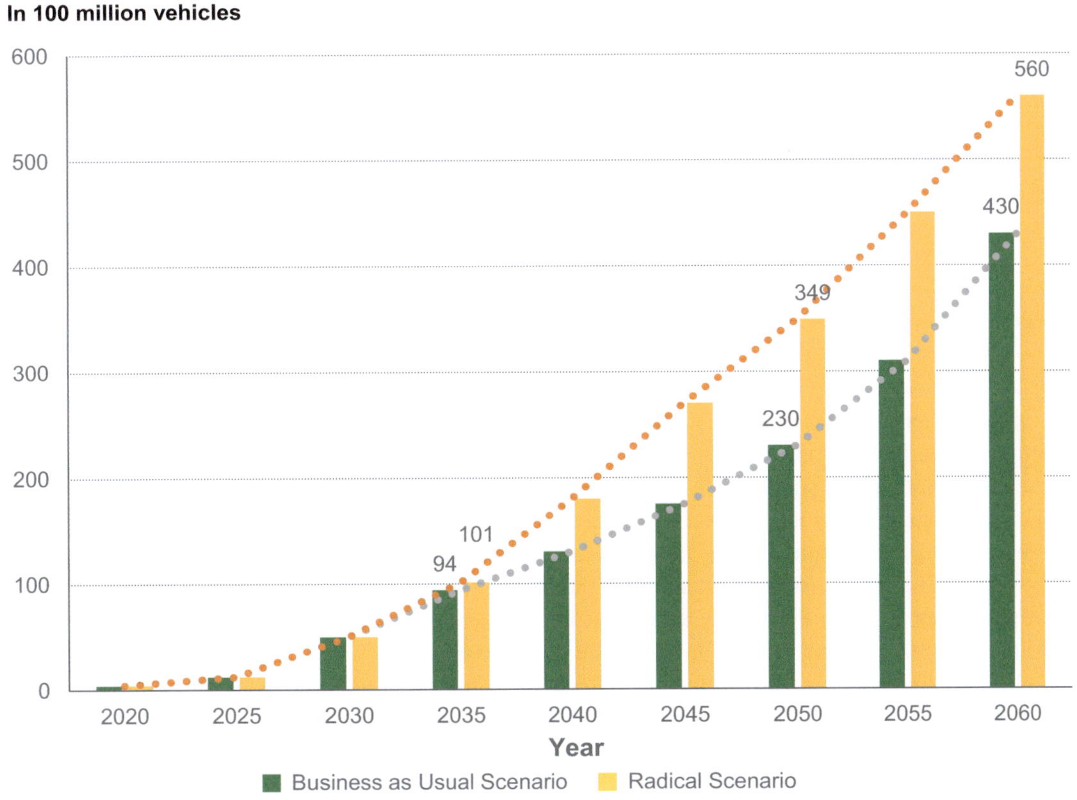

Fig. 9 Forecasts of the number of electric vehicles in China. *Source* World Resources Institute, Analysis of the Impacts of China's Large-scale Uptake of New Energy Vehicles on the Power Grid, 2020

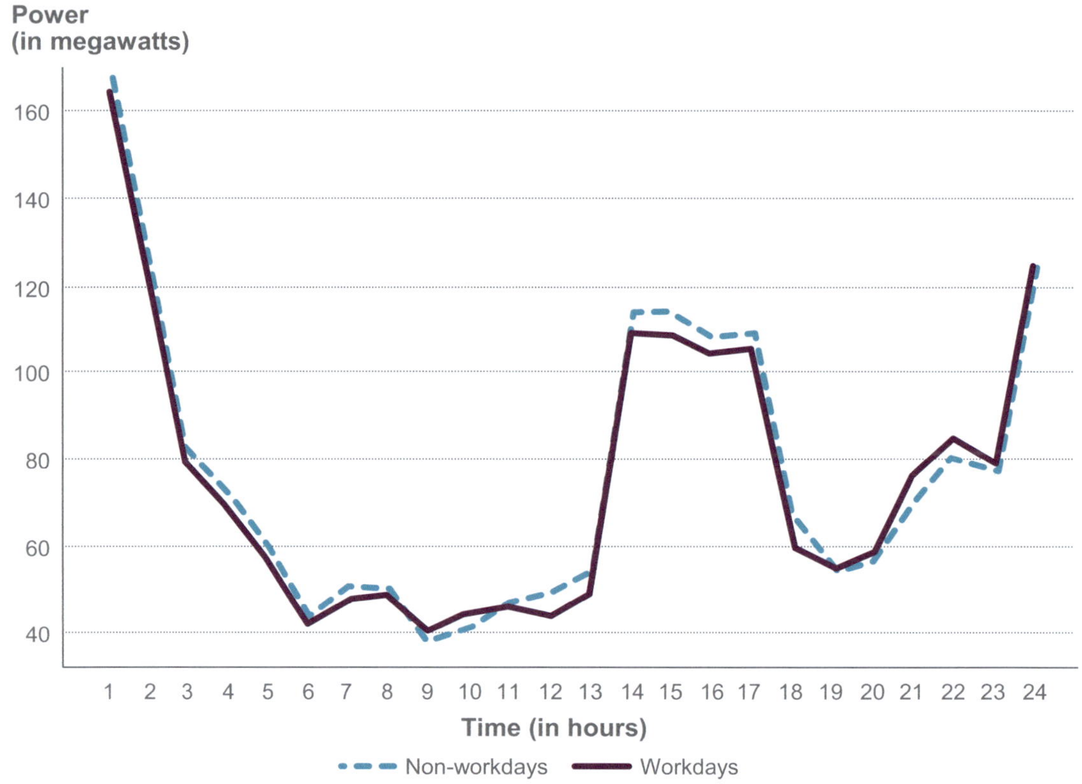

Fig. 10 Electric vehicle charging patterns. *Source* Su Xiaolin, Zhang Yanjuan, Wuzhong et al., Forecasts of Charging Load in the Case of Large-scale Electric Vehicle Uptake and Its Impact on the Power Grid, Modern Electric Power, 2018

charging of vehicles especially at peak times. When demand is low during off-peak hours, vehicles can charge their batteries when tariffs are lower; and when demand is high, they can feed electricity from their batteries to the power grid, helping to maintain grid stability (see Fig. 11),

Vehicle-to-grid technology taps the considerable energy storage potential of electric vehicles. Two types of vehicle-to-grid technology are expected to dominate in the future, as shown in Fig. 12. The first is traditional unidirectional charging, where electricity flows one way from the grid to the car. The second is bidirectional charging and discharging, where the electricity can flow two ways, from the grid to the car and from the car to the grid, as described in the paragraph above. Bidirectional charging unlocks the potential of electric vehicles to act as energy storage devices.

7.1.3 The Role of Synthetic Fuels in Ensuring Energy Supply Stability

As described in the Working Group I Contribution to the Sixth Assessment Report of the IPCC, which was published in 2021, there is no doubt that human influence has warmed the atmosphere, oceans and land. The scale of change across the climate system as a whole—and the present state of many aspects of the climate system—are unprecedented. In the future, extreme weather events will bring new challenges to energy supply: cold waves and heatwaves will intensify electricity and gas demand; and droughts and reduced wind energy will lead to a decline in hydropower and wind power output. For example, in early 2021, the power and water supply systems of Texas were paralysed in the wake of an extreme snowstorm.

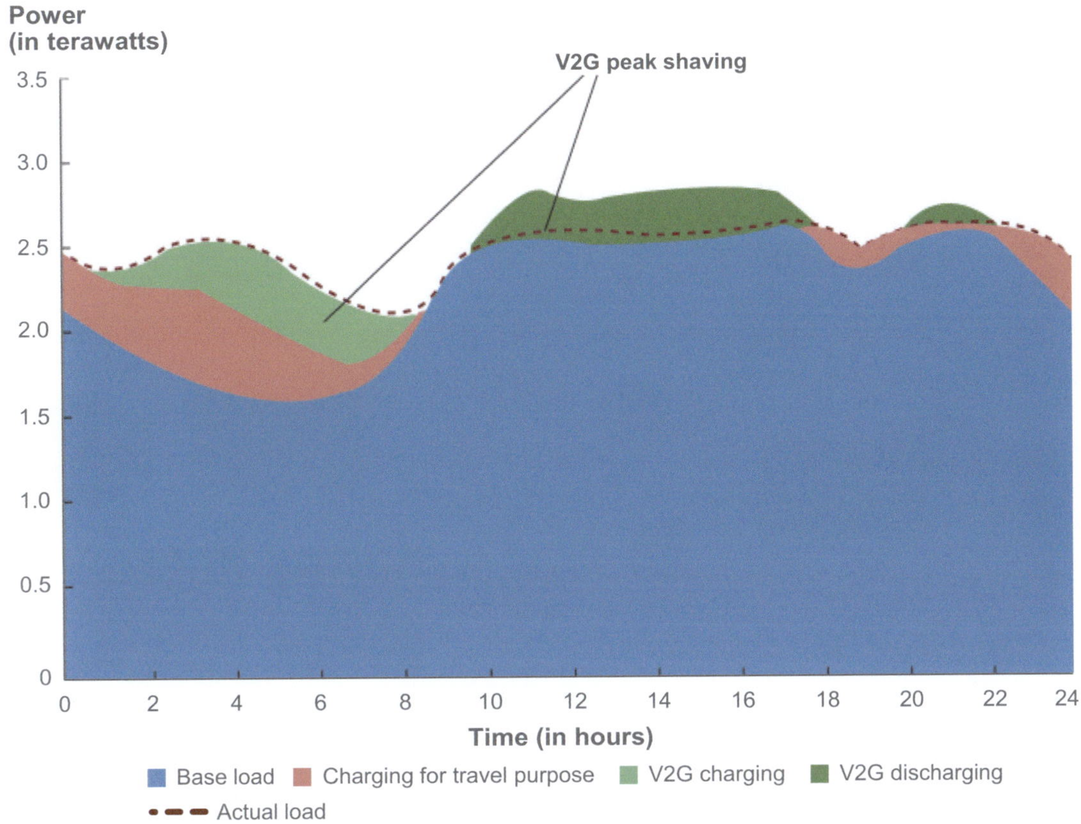

Fig. 11 Electrical load smoothing through vehicle-to-grid (V2G). *Source* Global Energy Interconnection Development and Cooperation Organization, The Road to China's Carbon Neutrality, 2021

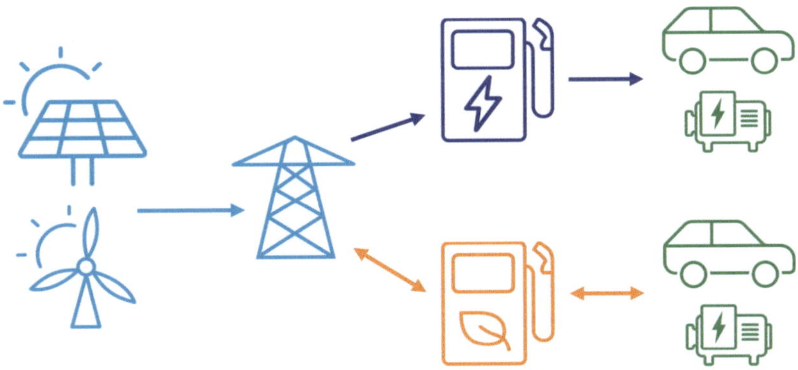

Fig. 12 How vehicle-to-grid works. *Source* Huang Junhui, Zhou Hao, Hanjun et al., Assessment of Distribution Network Reliability Considering V2G, Electric Power Construction, 2017

As shown in Fig. 13, synthetic fuels made with electricity include hydrogen produced through electrolysis, which when mixed with carbon dioxide produces gases such as methane and liquid fuels like methanol. In the future, this technology is expected to be an important energy storage option for the new energy system to combat extreme climate change. Currently, there are three main technologies for producing synthetic fuels from electricity: electrolysis-based hydrogen production with carbon dioxide hydrogenation, electrochemical reduction of

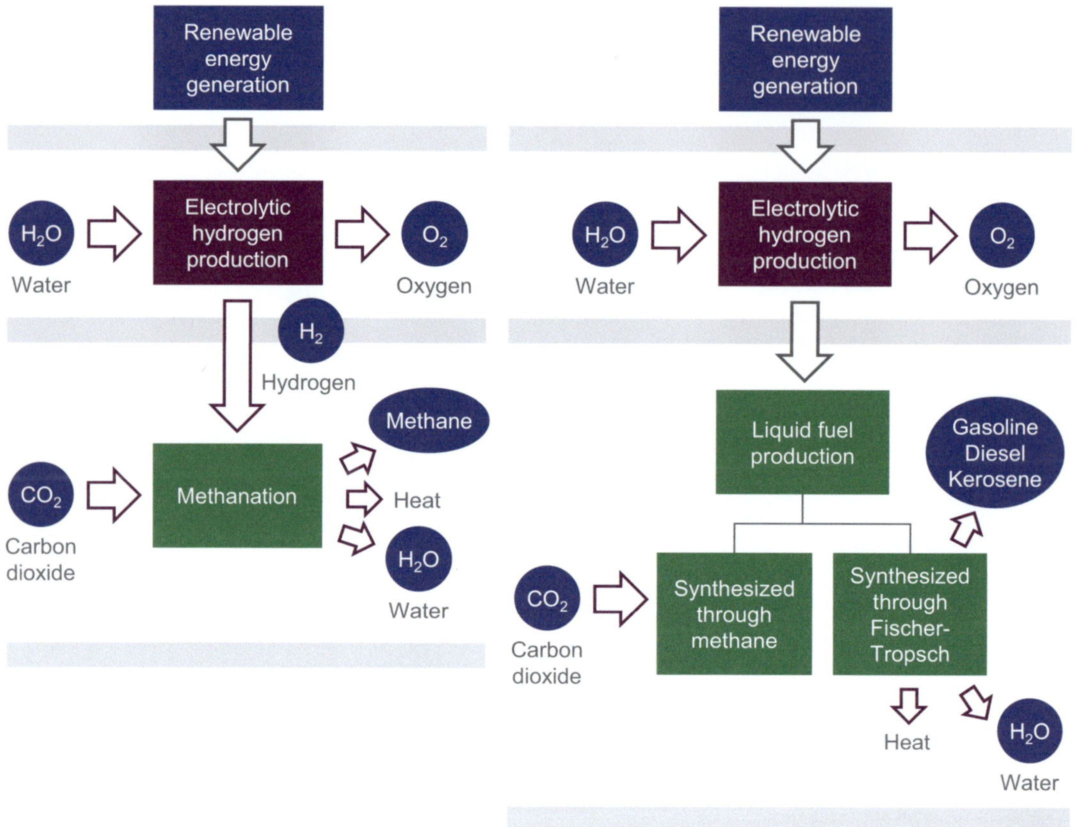

Fig. 13 Making synthetic fuels from electricity. *Source* Song Pengfei, Hou Jianguo, Yao Huichao et al., Electricity-to-Gas Technology Provides an Idea of Large-scale Energy Storage for Power Grids, Modern Chemical Industry, 2016

carbon dioxide, and high-temperature co-electrolysis of carbon dioxide. The first technology—which uses electricity to produce methane, methanol and other synthetic fuels—is relatively mature, and several demonstration plants have been built and are in operation around the world.

Using electricity to make synthetic fuels has advantages. First, converting renewable electricity to methane, methanol and other easy-to-store carbon-based fuels solves the problem of volatile and difficult-to-store renewable energy and lays a foundation for the smooth transition from old to new energy sources. Second, using renewable energy to convert carbon dioxide into useful fuels enables carbon in the environment to be recycled. Third, more methane made from electricity can be stored for longer than with any other energy storage technology. Synthetic fuels made using electricity can serve as a standby energy source to counter the disadvantages of wind and solar power generation in the power system.

7.2 Integrated New Energy System

7.2.1 Characteristics

Compared with the existing energy system, the integrated new energy system is highly digital, interactive, renewable and intelligent. It combines cutting-edge information and communication technologies and is smart and bidirectional. Its purpose is to co-ordinate generation, conversion, transmission, use and storage, increase the consumption of renewable energy, reduce the use of fossil fuels and provide intelligent, low-carbon and secure energy supply. Smart grids provide low-carbon electricity for all kinds of users, equipment and systems. They promote a new energy use system with electricity at the core, one in which electricity, cooling, heat, gas, power and other energy forms complement and integrate with each other efficiently. In this way, interaction and integration between power generation, power grids, demand and storage are made possible, and flexible and stable system regulating is provided.

7.2.2 Operation

Operation of the integrated new energy system centres mainly on analysing the fundamentals of the power system. With large-scale renewables at the core, the integrated new energy system uses ultrahigh-voltage transmission as its principal framework and co-ordinates generation, transmission and distribution, loads and energy storage. This enables the system to shift from the conventional mode of adapting generation to meet demand to bidirectional interaction between generation, power grids, load and storage, which improves system planning and allows flexibility resources to be optimised throughout the system.

Analysing the fundamentals refers to evaluating the basic conditions that affect system operation. These include economic developments and future trends, climate and geography, total demand for energy as well as for the various energy sources, and the fluctuating characteristics that define demand.

System planning includes planning for energy storage, hydrogen and methane, power transmission, as well as construction, within the framework of a new power system. It should solve the technical strangleholds that face new power transmission in the short term and enhance grid flexibility in the medium and long terms. Planning should also be co-ordinated with electric vehicle chargers and hydrogen refuelling stations.

System operation includes control systems, the co-ordination of power generation, grid data sharing and the optimisation of demand-side resources to balance renewable energy supply with demand.

The results of analysis are fed back into system planning and operation, driving the energy system to further optimise planning and operation.

7.2.3 Development Pathways for an Integrated New Energy System

By 2030, renewable energy will play an important role in the energy system. Additional energy demand will be met mainly by renewables like wind and solar and breakthroughs in synthetic fuel technologies will be made at a faster pace. New energy storage technologies like electrochemicals will play a role in achieving the dual-carbon goal. By 2030, a modern power grid that is secure, reliable, renewable, efficient and intelligent will be in operation, and various demand–response projects will enable interaction between smart grids and users. Multiple actors will participate in demand–response, and energy storage, distributed energy and electric vehicles will jointly contribute to peak shaving.

By 2050, renewable energy generation will replace fossil fuels to become the main source of energy supply. It will be part of a multi-energy system of complementary types of power generation in which wind and solar play a predominant role, with nuclear, hydrogen, bioenergy and other flexible carriers providing balancing capacity. The construction of synthetic fuel infrastructure, especially for hydrogen, will be improved. By 2050, strong and reliable synchronous power grids will be fully operational in eastern and western China. Micrograms and smart grids will provide secure and optimised transmission and distribution, improve use efficiency, and balance supply and demand dynamically and flexibly. Energy storage will play a strategic support role in the new electric power system.

By 2060, a renewable energy supply system in which wind and solar power play a dominant role, supplemented by synthetic fuels, will be operational. The energy storage systems will be mature and will stably address volatility in the power auxiliary services market. The eastern and western synchronous power grids will take shape at a faster pace, with greatly improved resource allocation capacity and increased inter-provincial and inter-regional power flows. Modern intelligent technologies in transmission, control, interconnections and new energy storage will be integrated to allow the use of various clean energy sources. They will provide reliable and cost-effective renewable power for all kinds of users, equipment and systems in smart grids that are the biggest and most technologically advanced in the world.

Chapter 12: Carbon Capture, Utilisation and Storage

Xian Zhang, Jingli Fan, Kai Li and Georgios Bonias

1 Introduction

On China's journey to becoming a carbon-neutral country, carbon capture, utilisation and storage (CCUS) will play a major role in the development and use of low-carbon, zero-carbon and negative-emission technologies. Given its substantial mitigation potential, CCUS should be considered a key part of China's technology portfolio for achieving its carbon neutrality goal.

In recent years, China has made impressive progress in technologies like pre-combustion carbon capture, chemical carbon dioxide utilisation and CO_2-enhanced water recovery. Overall, China's CCUS technologies are at a similar level of development as those of other countries, although a large gap exists in some technologies between China and the most advanced nations.

China's geological sequestration potential amounts to trillions of tonnes of CO_2. There are clusters of carbon emission sources near Bohai Bay and several other basins, which could be developed into CCUS hubs. However, barriers to large-scale commercial deployment of CCUS in the near and medium terms remain. These include a lack of large-scale CCUS demonstration projects, CO_2 pipelines and other infrastructure, and the absence of incentive policies and regulatory and technology standards.

Looking ahead, China will intensify its efforts to deploy and commercialise CCUS in energy and industry and develop low-cost, energy-efficient, secure and reliable CCUS technologies and build large-scale full-process CCUS hubs in industrial areas. The country will also use CCUS to achieve near-zero emissions of fossil fuels and promote the deployment of negative-emission technologies, thus helping to build a low-carbon industrial system and achieve the carbon neutrality goal.

CCUS refers to the technology that captures and separates carbon dioxide from emission sources like power generation and industrial production. It transports the captured CO_2 to appropriate sites for use or storage, thereby reducing CO_2 emissions. CCUS is gradually becoming an important strategic option for China to reduce CO_2 emissions, ensure energy security and make its economy and society low carbon. China attaches great importance to the development of CCUS. The Outline of the 14th Five-year Plan (2021–25) for National Economic and Social Development and Vision 2035, sets China the aim of developing major CCUS

X. Zhang (✉)
Administrative Centre for China's Agenda 21, Beijing, China

J. L. Fan · K. Li
China University of Mining & Technology - Beijing, Beijing, China

G. Bonias
Shell International Limited, The Hague, The Netherlands

Shell International B.V. and Development Research Centre of the State Council (DRC) of People's Republic of China, *Embracing the Future, Powering Growth: An Energy System Renewed for China*, Advances in Oil and Gas Exploration & Production,
https://doi.org/10.1007/978-3-031-73075-7_13

demonstration projects. In the Working Guidance for Carbon Dioxide Peaking and Carbon Neutrality in Full and Faithful Implementation of the New Development Philosophy, CCUS is listed as an important means to achieving the dual-carbon goal. The guidance also states that "China will advance the research, demonstration and industrial application of carbon capture, utilisation and storage technologies on a large scale". Looking ahead, CCUS will help meet the goals of carbon neutrality and the development of an ecological civilisation in China.

2 Supply and Demand Analysis

2.1 The Contribution of CCUS to Global Carbon Abatement

CCUS contributes significantly to reducing emissions in most low-carbon development scenarios and plays an important role in tackling climate change. The Fifth Assessment Report of the Intergovernmental Panel on Climate Change (IPCC), published in 2014, found that the vast majority of models that do not consider the use of CCUS find it impossible to achieve the CO_2 equivalent concentration target of 450 parts per million by 2100. Three of the four scenarios in the IPCC's Global Warming of 1.5 °C, published in 2018, require CCUS to achieve the target of keeping global average temperature rise below 1.5 °C (note that for the period 2020–2100, cumulative carbon dioxide emissions need to decrease by between 550 and 1,017 billion tonnes). Climate Change 2022: Mitigation of Climate Change, Working Group III's contribution to the Sixth Assessment Report, published by the IPCC in 2022, reports that most global scenarios that limit warming to 1.5 °C (including those with no or limited overshoot) or 2 °C consider fossil fuels with CCUS and CO_2 removal technologies will play an important role in reaching the target. Climate Change 2022 underlines the importance of CO_2 removal technologies. It estimates that between 2020 and 2100, CO_2 removal technologies will achieve a cumulative reduction of 193–895 billion tonnes and 192–959 billion tonnes respectively under the 2 °C and 1.5 °C scenarios without overshoot.

In CCUS in Clean Energy Transitions, a report published by the International Energy Agency in 2020, the Sustainable Development Scenario—in which global CO_2 emissions from the energy sector decline to net zero by 2070—is a possible pathway forward. The report also states that between 2020 and 2070, CCUS will reduce cumulative emissions by about 240 billion tonnes, which is 15% of the global total. CCUS can help abatement in most carbon-intensive sectors. In the Sustainable Development Scenario, CCUS reduces sector CO_2 emissions in the period 2020–70 by 15% in power generation, 25% in iron and steel, 61% in cement, 28% in chemicals and 90% in fuel processing such as refining, biofuels, hydrogen and ammonia.

The International Energy Agency's Net Zero by 2050: A Roadmap for the Global Energy Sector, published in 2021, proposes a climate target scenario in which global CO_2 emissions from the energy sector are reduced to net zero by 2050 (the Net-Zero Emissions by 2050 Scenario). In this scenario, 7.6 billion tonnes of CO_2 are captured by 2050, almost half of which are from fossil-fuel combustion, 20% are from industrial processes and around 30% are from bioenergy use with CCS (BECCS) and direct air capture. The importance of CCUS in decarbonising coal-fired power plants in emerging economies and gas-fired power plants in developed countries is also underscored in the report. By 2050, around half of the coal power fleet and 7% of the gas power fleet will be retrofitted with CCUS globally.

2.2 The Role of CCUS in China's Journey to Carbon Neutrality

As China and other countries around the world progress in their efforts to combat climate change, the strategic positioning of CCUS as an indispensable emission reduction technology is becoming more and more important. In the two versions of the Development Roadmap for Carbon Capture, Utilisation and Storage Technology in China, issued by the Ministry of Science and Technology in 2011 and 2019 respectively, CCUS is positioned as a "strategic technology option for realising large-scale low-carbon utilisation of fossil fuels". To achieve net-zero emissions, China needs to incorporate large-scale negative-emission technologies in addition to restructuring its energy mix and decarbonising its end-use sectors. In this new setting, CCUS is positioned in accordance with the following five aspects.

First, CCUS is the only technology that can reduce emissions from fossil fuels to net zero. China has a large energy system and diverse energy needs. To achieve the carbon neutrality goal without compromising energy security, China should build a secure and efficient low-carbon energy system in which renewable energy dominates and other energy sources such as nuclear power and fossil fuels play a complementary role. At present, fossil fuels still dominate China's energy demand mix. In 2021, coal, oil and natural gas accounted for 83.5% of China's total primary energy demand. According to the working guidance, up to 20% of China's total energy consumption will continue to come from fossil fuels on the path to carbon neutrality. The carbon emissions from these fossil fuels can only be reduced by CCUS technologies.

Second, CCUS enables thermal power to be part of zero-carbon peak regulation. The abatement policies and plans of developed economies suggest that the power sector needs to be completely decarbonised at least 10 years before the rest of the economy to reach carbon neutrality. China's power system needs to achieve net-zero emissions between 2045 and 2050, and attain negative emissions by 2060, to achieve its carbon neutrality goal. In 2021, China's electricity consumption was 8.3 trillion kilowatt-hours. By 2060, total electricity demand is expected to exceed 17 trillion kilowatt-hours. More renewable energy in the power system creates challenges such as seasonal imbalance and low system inertia, which make stable electricity supply substantially more uncertain. CCUS facilitates thermal power's participation in zero-carbon power balancing; for instance, thermal power can prevent power shortages caused by seasonal or extreme events, while maintaining near-zero emissions and a stable supply of low-carbon electricity.

Third, CCUS is a viable solution for the deep decarbonisation of hard-to-abate sectors such as iron and steel and cement. The long and complex production processes in these sectors emit huge amounts of carbon dioxide, which are difficult to decarbonise through conventional abatement measures such as technological improvements, efficiency gains or the use of alternative fuels and raw materials. For example, about 60% of the carbon emissions in the cement sector come from the thermal decomposition of limestone in the kiln, but it is extremely difficult to replace limestone with other raw materials. Even after conventional emission reduction measures have been adopted, such as improving energy efficiency or reducing energy use, and decarbonisation technologies like hydrogen-based direct reduced iron have been deployed, some 200–300 million tonnes of CO_2 emissions per year will not be fully abated after 2050.

Fourth, CCUS will be the main way to obtain carbon from non-fossil raw materials for chemical production in the future. Carbon, hydrogen and oxygen are the three basic elements of chemical production. Technologies for producing renewable and low-carbon hydrogen and oxygen are slowly maturing. For example, hydrogen produced through electrolysis using renewable electricity can be stored to counter the intermittency of renewable energy, improve the flexibility of power grids, eliminate curtailments and obtain renewable hydrogen for industry. However, fossil fuels remain the main source of carbon in the chemical industry. CCUS captures the CO_2 emissions from the production process and then purifies and recycles the CO_2 to provide high concentrations of carbon for chemical production, while reducing the use of fossil fuels and associated emissions.

Fifth, negative-emission technologies such as bioenergy with CCS and direct air capture serve as the backup solution to support the aim of carbon neutrality. By 2060, it will be difficult for China's hard-to-abate sectors such as iron and steel, cement and long-distance road freight transport to achieve net-zero emissions through conventional means. There will be hundreds of millions of tonnes of non-CO_2 greenhouse gas emissions that are difficult to abate and that need to be offset by negative-emission technologies like bioenergy with CCS and direct air capture. China has relatively abundant biomass resources. The huge negative-emissions potential of these bioresources, in conjunction with CCS, can be unlocked if agricultural and forestry residues and energy crops are used for power generation and the production of iron and steel, cement and biofuels. In addition, direct air capture can be paired with renewable energy and located almost anywhere. The considerable negative-emissions potential of bioenergy with CCS and direct air capture could make a strong contribution to China achieving its carbon neutrality goal.

To summarise, CCUS can reduce emissions from fossil fuels, ensure a flexible and stable supply of electricity from the power system, decarbonise hard-to-abate sectors like cement and iron and steel, and offset non-CO_2 greenhouse gas emissions. CCUS should therefore be clearly positioned as “a key component of China’s technology portfolio for achieving its carbon neutrality goal”.

2.3 Progress in CCUS

Since the 11th Five-year Plan of 2006–10, government funding initiatives such as the National Natural Science Foundation of China, the 973 and 863 programmes and the National Key Research and Development Program have, together with corporate investment and financing, supported research and development on CCUS technologies. As a result, China has made impressive achievements throughout the CCUS value chain, especially in technologies like pre-combustion carbon capture, tanker truck transport, chemical CO_2 use and CO_2-enhanced water recovery. On the whole, China’s CCUS technologies are comparable with those of advanced countries internationally, but there are still gaps in some key technologies such as carbon capture, transport, storage and full-chain optimisation.

Carbon capture refers to the process of separating and concentrating carbon dioxide from different emission sources by absorption, adsorption, membrane separation, low-temperature fractionation or oxygen-enriched combustion. Research on first-generation capture technologies in China has made remarkable progress, and most of these technologies have progressed to demonstration. Some capture technologies are ready for commercialisation, although there is still a lack of engineering experience in multi-million tonne projects. Once the second-generation technologies mature, their energy consumption and costs will be more than 30% lower than those of the first generation. However, second-generation technologies are just starting. New types of membrane separation, adsorption, pressurised oxygen-enriched combustion, chemical chain combustion and others are still at the R&D or laboratory testing stage.

Among the various capture technologies, pre-combustion carbon capture is at the industrial demonstration stage in China, which is comparable with other countries. Integrated gasification combined cycle power generation is a good example of where pre-combustion carbon capture can succeed, as demonstrated by the Tianjin Integrated Gasification Combined Cycle Power Plant project and the Lianyungang Clean Energy Power System Research Facility project. Post-combustion carbon capture technology, on the other hand, which is in the pilot or industrial demonstration stage, lags behind international levels, especially the more widely deployed post-combustion chemical absorption process. Oxygen-enriched combustion technology is one of the most promising large-scale carbon capture options for coal-fired power plants, from which the carbon dioxide generated has a high concentration of 90%–95% and is easy to capture. However, this technology is only in the pilot stage in China and across the world, and is progressing slowly. The State Key Laboratory of Coal Combustion at Huazhong University of Science and Technology has built a pilot-scale test bench and full-process system for oxygen-enriched combustion.

Carbon dioxide transport refers to the process of transporting—mainly via tanker trucks, ships or pipelines—the carbon dioxide captured to sites where the CO_2 can be used or stored. Tanker trucks are typically used for small-scale and short-distance transport. Pipelines offer significant economies of scale, and should be given priority for long-distance large-scale transport and for connecting industrial sites to CCUS hubs. China's tanker truck and ship transport technologies are in commercial use and comparable with those in the international market. Pipeline transport technology is, however, still at the pilot and demonstration stages, and large-scale commercial deployment still has a way to go. Jilin oil field and Qilu Petrochemical are already using onshore pipelines for transport, but are behind international levels.

Biological and chemical utilisation refers to the process of using the different physical and chemical properties of carbon dioxide to produce commercially valuable products and to reduce emissions. The technology is at the demonstration stage both in China and across the world. In the past decade, progress has been made in various biological and chemical utilisation technologies in China, especially in chemicals. Technologies for using carbon dioxide synthesis to produce chemicals are under demonstration. They include the synthesis of organic carbonates, degradable polymers and cyanates, polyurethanes and the production of polycarbonate and polyester materials. Breakthroughs have been made in microalgae-based carbon dioxide fixation and gas fertiliser technologies.

Geological utilisation and storage is a technology that injects the captured carbon dioxide into deep saline formations, depleted oil and gas reservoirs and other geological reservoirs. This technology has been developed to various degrees in both China and across the world. For example, in other countries, enhanced oil recovery and leaching technologies in mining are in commercial operation and CO_2-enhanced water recovery is in industrial-scale demonstration. China has made excellent progress in these technologies in the past 10 years, especially in CO_2-enhanced water recovery which has advanced from proof-of-concept to industrial-scale demonstration quickly in eastern, northern, north-western and western regions and offshore areas of China. It is still noticeably behind the commercial deployment level internationally.

2.4 The Abatement Potential of CCUS in China

2.4.1 The Geological Sequestration Potential of CCUS

Carbon dioxide storage sites include deep saline formations, oil and gas reservoirs and unexploitable coal beds. Globally, the theoretical storage capacity onshore and undersea is 6–42 trillion tonnes and 2–13 trillion tonnes of carbon dioxide respectively. Deep saline formations account for 90% of the potential storage capacity. Moreover, their wide distribution make them relatively ideal locations for carbon dioxide storage. Thanks to good geological exploration earlier and revenue streams from oil and gas displacement, oil and gas reservoirs are also suitable for carbon dioxide storage in the early stage of CCUS deployment. Overall, the global geological storage potential for carbon dioxide is much higher than the cumulative storage needed of nearly 100 billion tonnes under the 2 °C scenario.

China holds geological sequestration potential of between 1.2 and 4.1 trillion tonnes of carbon dioxide. Oil fields are mainly concentrated in Songliao Basin, Bohai Bay Basin, Ordos Basin and Junggar Basin. Deep saline formations are found in 16 large basins in China, which have a storage capacity much higher than that of oil and gas fields. Songliao Basin, Tarim Basin and Bohai Bay Basin have an enormous storage potential, accounting for about half of the total.

2.4.2 The Abatement Potential of CCUS by Source-sink Matching

Matching the emission source with the storage site will decide whether the emissions reduction potential of CCUS technology can be realised. It is therefore better to evaluate the storage potential of CCUS by using the source-sink matching process. The matching process should include screening those emission sources that meet the CCUS retrofit standards, choosing storage sites with suitable geological conditions, assessing the geographical distribution of emission sources and carbon sinks and their environmental constraints, and providing the feasible CCUS solution at scale.

The CCUS abatement potential of China's coal-fired power plants has already been assessed. It takes into consideration plant operating life, CO_2 transport distance, time to deployment, storage capacity and injection capacity. The findings show that transport distance and deployment time have significant impacts on CCUS abatement potential. If the maximum transport distance is 100 km and large-scale CCUS technology deployment does not occur until 2035, China's active coal power plants can achieve a cumulative emissions reduction of 5.5 billion tonnes of CO_2 by 2060. If the maximum transport distance is raised to 800 km and CCUS technology is deployed at large scale in 2025, a much higher cumulative emissions reduction of 38.5 billion tonnes of CO_2 could be achieved.

The CCUS abatement potential of China's coal-fired power plants has also been assessed without taking the operating life of the plants into account. If large-scale CCUS deployment is carried out in 2025 and the maximum transport distance is 100 km, China's active coal-fired power plants could achieve a cumulative emissions reduction of 27 billion tonnes of CO_2 by 2060. If the maximum transport distance is increased to 800 km, a cumulative emissions reduction of 67.7 billion tonnes of CO_2 is possible.

In Bohai Bay Basin, Southern North China Basin, Northern Jiangsu Basin, Songliao Basin and Junggar Basin, the emission sources and storage sites have a high spatial matching, which provides huge opportunities for the development of CCUS hubs. The distance between emission source and carbon sink in Jiangsu, Henan, Anhui, Hebei, Xinjiang, Heilongjiang and Tianjin is relatively short, at between 30 and 140 km on average. This means that these provinces should be prioritised for large-scale CCUS deployment. The south-east coastal areas are the main consumers of energy and industrial raw materials, although Fujian, Guangdong and Guangxi lack good geological sequestration conditions. Due to spatial mismatch between emission source and carbon sink in these regions, offshore sequestration in nearby sedimentary basins is an important alternative for carbon storage.

3 Current Developments and Challenges

3.1 Overview of CCUS Projects

As of 2021, there are 135 commercial CCUS facilities worldwide: 27 in operation, 4 under construction, 58 undergoing testing, 44 at an early stage of development, and 2 that have shut down. The top three countries in terms of the number of CCUS projects are the USA with 71 projects, the UK with 15 and Canada with 8, followed by China and the Netherlands in fourth place. China has six commercial CCUS facilities, including three in operation, two under construction and one at an early stage of development. Global commercial CCUS projects are widely deployed in power generation, natural gas processing and hydrogen production, with a total capture capacity of 149 million tonnes of CO_2 per year.

According to incomplete data, as of 2021 there are 49 CCUS demonstration projects in China, with a capture capacity of about 4.3 million tonnes per year. Among them, the 38 demonstration projects in operation have a combined capture capacity of about 2.96 million tonnes per year. About 2 million tonnes of CO_2 were stored by these projects from 2007 to 2019. The industries that use these demonstration projects are mainly coal chemicals, chemical fertilisers, power generation and cement. Tanker truck transport and deep saline formation storage are the technologies used. Geological technologies such as methane displacement in coalbeds and in-situ leach uranium mining, chemical technologies like degradable polymers, and biological processes such as microalgae-based carbon dioxide fixation are also undergoing demonstration.

3.2 The Cost of CCUS Technologies

3.2.1 A Cost Comparison of Different Carbon Emission Sources

The net emissions reduction cost of full-process CCUS technologies for the main carbon emission sources—coal- and gas-fired power plants, coal chemical plants, natural gas processing plants, steelworks and cement plants—in China is in the range of RMB 150–700 per tonne.

The carbon dioxide avoidance cost of conventional coal-fired power plants and integrated coal gasification combined cycle power plants in China is about RMB 410 per tonne and RMB 560 per tonne respectively, which is low compared with the global average of RMB 410–830 per tonne and RMB 560–1,020 per tonne respectively.

The carbon dioxide avoidance cost of steel and chemical fertiliser production in China is about RMB 510 per tonne and RMB 190 per tonne, which is close to the lowest global average of RMB 460–820 per tonne and RMB 160–230 per tonne.

The carbon dioxide avoidance cost of natural gas combined cycle power plants and cement plants in China is about RMB 680 per tonne and RMB 890 per tonne, which is at a medium-to-low level compared with the global average of RMB 550–1,100 per tonne and RMB 720–1,340 per tonne. The CO_2 avoidance cost of China's natural gas processing sector is about RMB 170 per tonne, which is medium relative to the world average of RMB 140–190 per tonne.

3.2.2 A Cost Comparison of Different Technologies

Carbon dioxide capture (including compression) accounts for between 60% and 80% of the total cost, depending on the emission source and capture technology. The cost of capturing high-concentration and low-concentration carbon dioxide emissions in China is RMB 120–180 per tonne and RMB 220–480 per tonne respectively. The cost of first-generation capture technologies for post-combustion, pre-combustion and oxy-combustion is RMB 300–450 per tonne, RMB 250–430 per tonne and RMB 300–400 per tonne respectively.

The cost of transport technologies varies. Tanker truck technology is mature. It provides transport capacity of less than 100,000 tonnes per year at a cost of about RMB 1.1 per tonne and km. Pipeline technology has the greatest deployment potential and cost-effectiveness, with a transport cost[1] of less than RMB 1 per tonne and km. Some CCUS projects in China are already using pipeline transport technology. They include the CCS-enhanced oil recovery project at China National Petroleum Corporation's Jilin oil field.

The cost of geological utilisation and storage technologies differs. Oil displacement technology costs RMB 70–80 per tonne, in-situ leach uranium mining mainly comprises operating and labour costs, and in-situ mineralisation storage is RMB 300–550 per tonne. The cost of storage in depleted onshore oil and gas fields, onshore saline formations and marine saline formations is about RMB 50 per tonne, RMB 60 per tonne and RMB 300 per tonne respectively.

3.3 Challenges Facing CCUS

3.3.1 Bottlenecks in Key Technologies

The development of CCUS in China lags behind other countries in some key technologies. Pre-combustion physical absorption, tanker truck transport, ship transport, leach mining and degradable polymer synthesis technologies are equal to those of other countries. In the case of post-combustion chemical absorption, pipeline transport, enhanced oil recovery, methane displacement in coalbed gas and enhanced natural gas extraction, commercial demonstration projects have been developed by some countries, whereas China is at the demonstration or pilot stage. In

[1] This cost is based on a gas-phase CO_2 transmission pipeline that has a total length of 70 km and a capacity of 500,000 tonnes per year.

addition, the key technologies for integrating and optimising multiple units of large-scale full-process CCUS projects have reached a bottleneck.

3.3.2 There Are Few Large-scale Full-process Commercial Demonstration Projects

Many new 10-megatonne CCUS industrial hubs are in operation or under construction around the world, the largest of which is the CCS Innovation Zone along the Houston Ship Channel in the USA. It is designed to use multiple CCUS industrial carbon emission sources and sequester 1 million tonnes of CO_2 per year in offshore formations in the Gulf of Mexico. In China, the megaton-class Qilu Petrochemical–Shengli oil field CCUS project has started operations, and the Xinjiang CCUS Industrial Centre plans to provide a capacity of between 200,000 and 3 million tonnes per year. On the whole, China has insufficient commercial deployment of CCUS technology, especially of full-process demonstration projects. In addition, cross-sectoral co-operation models have not yet taken shape, and the source and sequestration of greenhouse gases for demonstration projects remain unclear.

3.3.3 CCUS Infrastructure Is Developing Slowly in China

The USA has built around 8,000 km of carbon dioxide pipelines and other developed economies, including Europe, have set up several carbon sequestration industrial hubs. In contrast, China has not yet developed transport pipelines and storage sites at scale and has not invested sufficiently in transport and storage infrastructure. In addition, infrastructure co-operation and sharing has yet to be introduced.

3.3.4 Policies, Regulations and Standards Need to Be Improved

Compared with other countries, China lacks strong financial and tax incentive mechanisms for CCUS technologies. In 2008, the United States Congress passed the 45Q tax credit, which provides monetary credits for carbon dioxide that is stored geologically, either permanently or through enhanced oil recovery or other methods. The credits were increased and amended in 2018 and again in 2022 as part of the Inflation Reduction Act. A progressive approach was adopted for subsidies: the subsidy for geological storage is up to $50 per tonne and that for non-geological storage (mainly oil displacement) is up to $35 per tonne. In contrast, the absence of financial and tax incentives in China dissuades enterprises from deploying CCUS. In addition, the regulatory system for CCUS is incomplete and a science-based system of standards has not yet been established. In addition, there is no access to preferential credit and concessional loans from financial institutions.

4 Development Pathways

After 2030, as the key technologies gradually mature, CCUS is expected to play an ever-increasing role in a diverse energy supply system in which fossil fuels and renewable energy work in a collaborative and complementary way. From 2050 onwards, the energy consumption and cost of CCUS will have improved, and the widespread deployment of CCUS in multiple sectors will enable low-carbon fossil fuels to be used at scale, while also helping to achieve negative emissions. CCUS will be a key technology for China to achieve carbon neutrality.

4.1 Development Pathways by 2030

By 2030, existing CCUS technologies will be commercially deployed. The cost and energy consumption of first- and second-generation capture technologies will be 20% and 30% lower respectively than their current levels. Breakthroughs will be made in large-scale carbon dioxide pressurisation (equipment) and long-distance pipelines with a transport capacity of millions of tonnes will be built. The existing utilisation technologies will be ready for industrial development and new technology applications will be explored.

4.2 Development Pathways by 2050

By 2050, CCUS will be widely deployed in energy and industry, and many CCUS industrial hubs will be in operation, reducing emissions by 100 million tonnes annually and generating new regional business models. The cost and energy consumption of second-generation capture technologies will be more than 50% lower. Pipeline transport capacity will increase significantly to 100 million tonnes. Utilisation and storage technologies will serve a broad range of commercial applications, and system integration and risk management technologies will mature.

4.3 Development Pathways by 2060

By 2060, negative-emission technologies such as bioenergy with CCS and direct air capture will be fully rolled out, reducing annual emissions by more than 1 billion tonnes. A low-cost, energy-efficient, safe and reliable CCUS system for industrial hubs will be built, enabling the use of fossil fuels with zero-carbon and negative-carbon characteristics and helping to create a renewable and low-carbon industrial system. Technologies such as offshore CO_2 storage, the replacement of methane with carbon dioxide in natural gas hydrates, and CO_2-enhanced shale oil recovery will be widely deployed.

5 Policy Recommendations

5.1 Strengthen the R&D of Next-generation CCUS Technologies

First, clarify the strategic positioning of CCUS under the goal of carbon neutrality. China should increase the role of CCUS in the national "1+N" policy for carbon neutrality and include CCUS in the technology research and development plan for carbon neutrality. China should also carry out multiple CCUS projects to support applied research and the development of technologies, processes and equipment.

Second, intensify the R&D of next-generation low-cost and energy-efficient CCUS technologies. At present, the cost and energy consumption of CCUS technologies is prohibitively high, which discourages many enterprises. To drive down the cost and energy consumption of second- and third-generation technologies significantly, China should make breakthroughs in full-process core technologies such as chemical chain combustion, new membrane separation, pressurised oxygen-enriched combustion and large-scale pipeline transport.

5.2 Develop Large-scale CCUS Demonstration and Industrial Hubs

First, develop full-process, integrated, large-scale CCUS demonstration projects. At present, most existing demonstration projects in China provide a capture capacity of 100,000 tonnes, which is not enough to meet the carbon neutrality goal and is not conducive to driving breakthroughs in core technologies. China should strive to build 3–5 megaton-class CCUS full-process demonstration projects during the 14th Five-year Plan (2021–25), with a focus on oil displacement, gas displacement, solid waste mineralisation and chemical utilisation.

Second, accelerate the planning and deployment of CCUS pipelines and the construction of hub infrastructure. China should build 10-megaton-class CCUS industrial hubs in areas where there are emission sources and carbon sinks. This will harness benefits such as infrastructure sharing, system integration, energy and resource interaction and business model innovation.

Third, explore the feasibility and development potential of CCUS integration. China should integrate CCUS with low-carbon technologies such as renewable energy, energy storage, bioenergy with CCS and other negative-emission technologies.

5.3 Enhance Industrial Policy Support

First, innovate investment and financing mechanisms to speed up the commercialisation of CCUS. At present, many enterprises come up against barriers to commercialisation, such as lack of investment and financing channels and poor connection between the upstream and downstream parts of the value chain. China should use government to guide investment, stimulate market players to invest in renewable and low-carbon technologies, strive to open up CCUS investment and financing channels aligned with the goal of carbon neutrality, and leverage the important role that emission trading systems play in low-carbon resource allocation, risk management and market pricing.

Second, improve incentive policies. Drawing on the 45Q tax credit for carbon sequestration in the USA, China should introduce CCUS tax breaks, subsidies and other incentives to help reach carbon neutrality. Incentives should be focused on the next generation of low-cost and energy-efficient CCUS technologies to form a virtuous circle of increasing investment and finance and reducing costs.

Third, improve policies, regulations and standards. China should develop policies and regulations for storage permitting, safety monitoring, risk assessment and emission-reduction calculations, as well as science-based standards for the construction, operation, supervision and eventual decommissioning of CCUS projects.

Chapter 13: Critical New Energy Minerals

Jiabin Chen, Chao Liu, Binhan Nie,
Dandan Feng, Weiming Li
and Georgios Bonias

1 Introduction

As the response to climate change intensifies and the transition to low-carbon energy accelerates, global demand for critical minerals, which are essential materials for new energy industries, will surge. As a result, competition for the minerals will escalate globally. China has an insufficient resource endowment of critical minerals such as copper, nickel, manganese, lithium and cobalt, most of which are supplied by foreign sources. The stability of China's resource supply chains are, therefore, prone to price volatility, trade restrictions imposed by producing countries, and a complex and changing international climate. This undoubtedly creates new challenges for China's energy and dual-carbon governance system. While moving towards a carbon-neutral future, the USA, Japan, the European Union and other developed economies have re-examined their access to critical minerals, including those essential for new energy development. These economies have published lists of minerals they consider of crucial national strategic significance. At the same time, they have strengthened their supply chains through policy support and joint initiatives with major resource-producing countries to ensure the supply security of these minerals.

China's demand for critical minerals is growing rapidly thanks to the success of its new energy industries. To meet demand for critical minerals and reduce dependence on imported resources, China needs to develop and use its domestic mineral resources and promote international mining co-operation. This two-pronged approach will enable it to develop a diversified and stable supply of mineral resources and help the country advance towards its dual-carbon goal. First, the business environment should be optimised to support and encourage the exploration and development of critical new energy minerals such as lithium ores in Sichuan and lepidolite in Jiangxi. Second, innovations in prospecting technologies for scarce resources such as nickel and cobalt and in recycling technologies for spent batteries should be made to convert potential resources into tangible supply capacity and diversify resource supply channels. Third, international co-operation on resources should be developed in a spirit of mutual political and economic trust with major resource-producing countries. Fourth, a reserve system comprising physical reserves, mineral site reserves and technology reserves should be developed to ensure supply security. Fifth, the

J. B. Chen (✉) · C. Liu · B. H. Nie · D. D. Feng
China Academy of Natural Resources Economics,
Beijing, China

W. Li
DRC Institute of Resources and Environment,
Beijing, China

G. Bonias
Shell International Limited, London, UK

Shell International B.V. and Development Research Centre of the State Council (DRC) of People's Republic of China, *Embracing the Future, Powering Growth: An Energy System Renewed for China*, Advances in Oil and Gas Exploration & Production,
https://doi.org/10.1007/978-3-031-73075-7_14

governance of critical new energy minerals should be incorporated into China's dual-carbon agenda as soon as possible, and China should actively participate in the global resource governance system.

2 Critical New Energy Minerals Are in Short Supply

2.1 China Has Unfavourable Reserves of Critical Minerals

According to the United States Geological Survey, 70% of most mineral reserves are concentrated in five countries. China has an alarmingly poor resource endowment of critical new energy minerals. Even for heavy rare-earth elements, in which China is reserve-rich, the environmental constraints hindering their development result in insufficient domestic supply; consequently, more than half of the country's demand is imported. Except for heavy rare-earth elements, China's reserves of all other mineral resources are at an unfavourable level. China's copper and nickel reserves account for only 3% of the global total; manganese for 4%; lithium for 7%; and China's reserves of cobalt, chromium, the platinum group metals and other critical minerals are not even included in world ranking lists.

2.2 China is Heavily Dependent on Imports and Its Supply Chains Are Vulnerable

According to the United States Geological Survey, the global production of critical new energy minerals—including copper, manganese, nickel, lithium and cobalt—is concentrated in five countries. In 2020, 60% of global copper production was in Chile, Peru, China, the Democratic Republic of the Congo and the USA; 63% of global nickel production was in Indonesia, the Philippines, Russia, New Caledonia and Australia; 68% of cobalt production was concentrated in the Democratic Republic of the Congo; 94% of lithium output was in Australia, Chile, China and Argentina (see Fig. 1); and 68% of manganese production was in South Africa, Australia, Gabon and Brazil. In the major producing countries, large and super-large mines predominate. Taking nickel as an example, despite the relatively scattered nature of nickel mining across the world, the producers that contribute more than 5% of total global production include Norilsk Nickel of Russia, Vale of Brazil and Jinchuan Group of China; other large producers include Glencore, BHP and Sumitomo Corporation. These six companies produce about one third-of global nickel output. Cobalt is produced mainly by Glencore, CMOC Group, Savannah Resources, Jinchuan Group, Sherritt International and Vale. According to these companies' annual reports, in 2019, their total cobalt output amounted to more than 60% of the global total.

The supply–demand picture of critical new energy minerals varies greatly in China. Moreover, most of these minerals are imported from a few countries: the Philippines, Indonesia, Australia, Brazil and the Democratic Republic of the Congo. This exposes China's supply chains to price volatility, trade restrictions imposed by producing countries, complex and changing international situations, and other risks.

2.3 Challenges to China's Supply Security of Critical New Energy Minerals

China's energy governance is based on an energy system dominated by fossil fuels and supplemented by new energy. To achieve the dual-carbon goal, new energy will have to gradually replace conventional energy, which means the supply pressure on critical new energy minerals will increase. This poses new challenges for China's traditional energy governance system. Currently, developed countries have integrated the supply of critical minerals into their national and global resource strategies. The U.S. Department of Energy established a Minerals Sustainability Division to co-ordinate demand and supply of new energy and minerals. China needs to improve the mechanisms and policy system

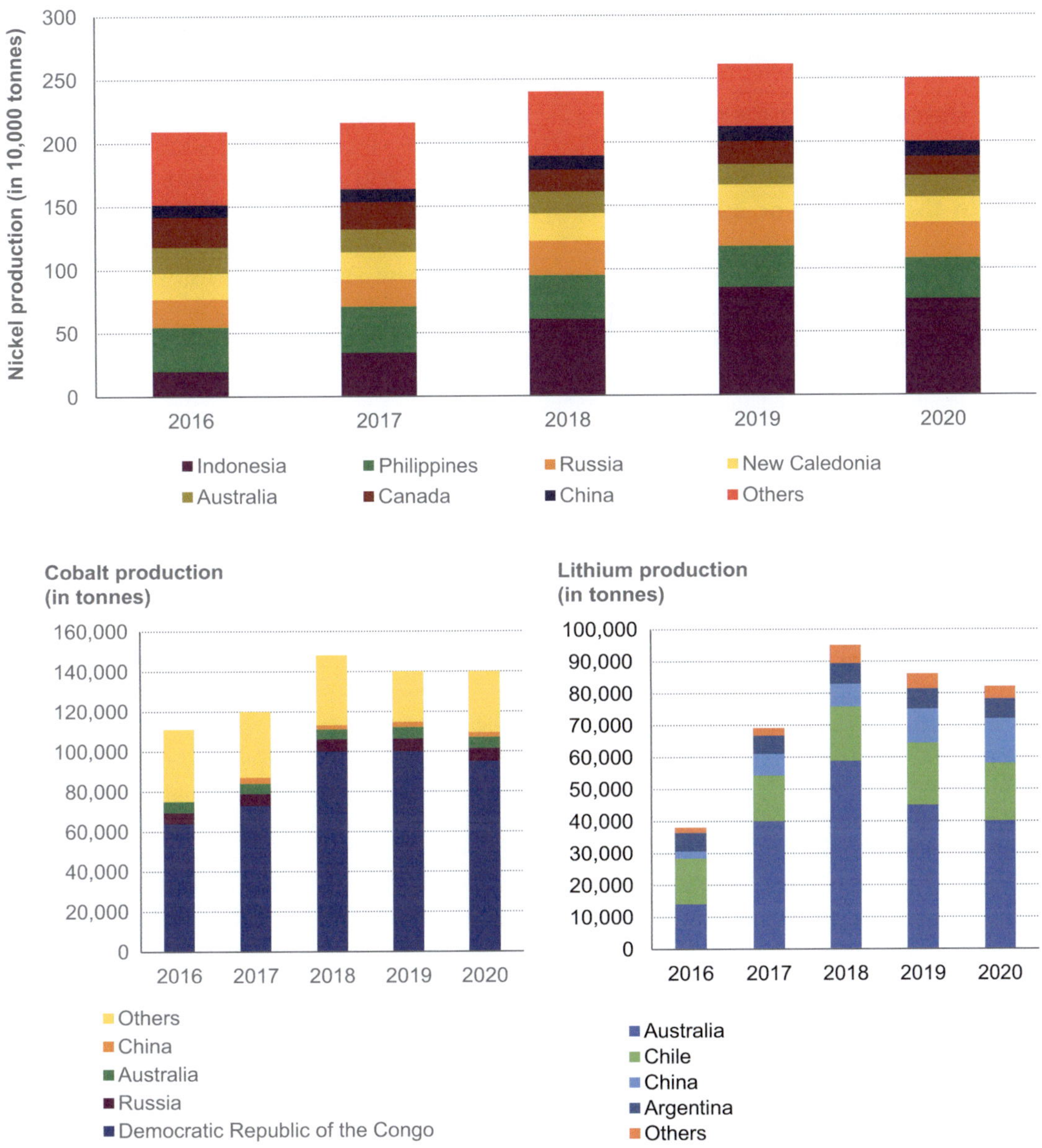

Fig. 1 Global nickel, cobalt and lithium production, 2016–20. *Source* United States Geological Survey, Mineral Commodity Summaries for 2017–21

for planning and implementation, monitoring and evaluation, exploration and development, recycling, the research and development of alternative technologies, emergency reserves and international co-operation governance. These issues need to be planned in an overall programme once the supply of critical new energy minerals is included in the energy and dual-carbon governance system.

3 Demand for Critical New Energy Minerals Will Grow Continuously

3.1 Global Demand for Critical New Energy Minerals Will Increase at Least Fivefold in the Next 20 or 30 Years

In the global transition to renewable energy, demand for the following minerals will rise significantly: nickel, manganese and lithium for new energy batteries; rare-earth elements for permanent magnets used in wind turbines and electric vehicle motors; and copper and aluminium for power grids and electricity-related technologies. According to The Role of Critical Minerals in Clean Energy Transitions, published by the International Energy Agency in 2021, low-carbon energy systems will require far more minerals than fossil-fuel energy systems. By 2040, global demand for lithium, graphite, cobalt and nickel will have grown exponentially. Demand for lithium used in electric vehicles will soar by a factor of 42, and demand for cobalt and graphite will increase more than 20-fold (see Fig. 2).

According to the World Bank report on Minerals for Climate Action: The Mineral Intensity of the Clean Energy Transition, published in 2020, project demand for critical new energy minerals for wind, solar and geothermal energy facilities will rise more than fivefold by 2050 and cumulative demand will be more than 3 billion tonnes by the end of the 21st century. As global efforts to tackle climate change and switch to low-carbon energy increase, the competition between countries for critical new energy minerals is likely to intensify.

3.2 China's Demand for Critical New Energy Minerals Will Grow Fast Under the Dual-carbon Strategy

In the Carbon-neutral Scenario, demand for non-fossil energy will increase to about 25% and 80% of total primary energy demand in 2030 and 2060 respectively. Wind power, solar power and new energy vehicles will drive demand for these minerals. According to conservative forecasts, demand for lithium, cobalt, nickel and copper in China in 2030 will be between 1.1 and 3.5 times more than in 2020. By 2060, under the No Alternative Materials Scenario, demand for most critical new energy minerals will increase by between 20% and 50% on the 2030 level, and demand for some minerals may even double.

3.3 China's Supply Capacity of Critical New Energy Minerals Is Limited

First, China has a poor lithium resource endowment and unfavourable development conditions. For example, the development of lithium ores in Sichuan and Xinjiang is hampered by fragile ecosystems and unfavourable locations. Second, the process of extracting lithium through one-step separation from salt-lake brine or from solid lithium ore requires improvement. Third, the technologies and policy support needed to recycle spent batteries are incomplete, which makes standardisation difficult and limits the supply of recycled minerals. Fourth, geopolitical changes in the world increase the uncertainty of securing overseas resources. For example, more than 80% of China's lithium imports are from Australia and the Lithium Triangle of Chile, Argentina and Bolivia, and more than 90% of China's nickel imports come from the Philippines and Indonesia. Some of these countries restrict the export of raw ore; some wish to increase their own development and use of these resources; and some raise mining taxes and fees. Fifth, mining standards and regulations are becoming increasingly strict. Mining activities are required to be socially and environmentally responsible, renewable and low carbon and to contribute to the development of local communities. This poses higher demands on the enterprises that develop overseas resources.

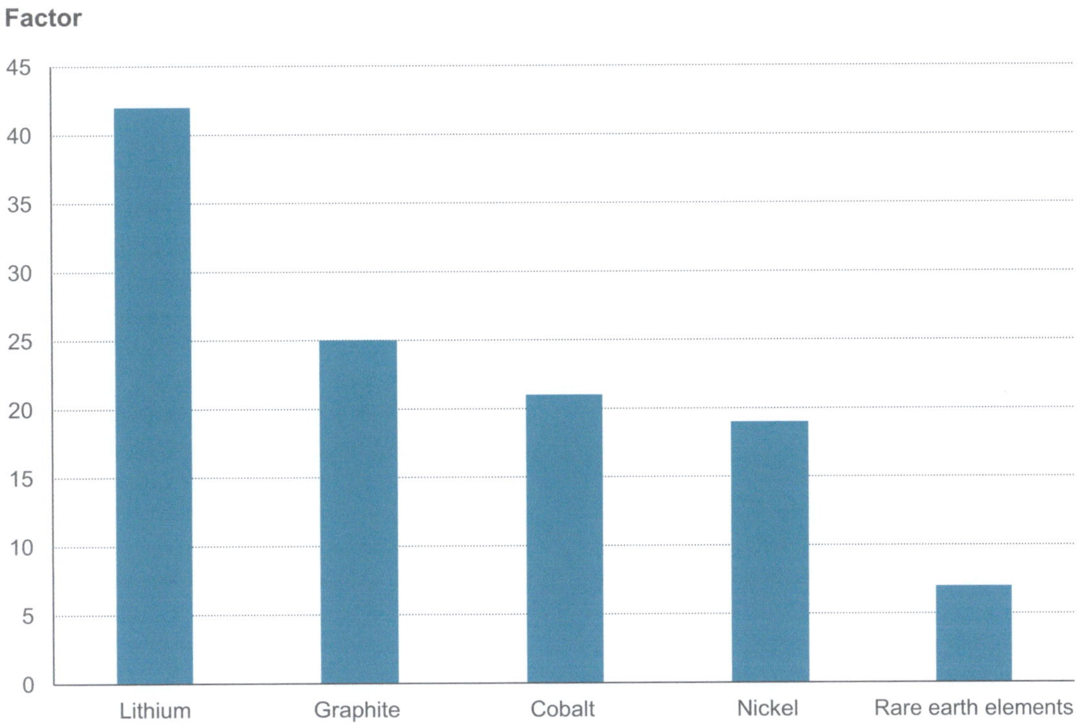

Fig. 2 Demand growth for selected minerals in 2040 under the Sustainable Development Scenario (relative to 2020, 2020 =1). *Source* International Energy Agency

4 International Experience

In their pathway to a carbon-neutral future, major economies lay great stress on the development of new energy industries. For example, the European Commission has adopted the European Green Deal; Japan has a Green Growth Strategy, which includes renewable energy development, and which is part of national energy security policy; and the USA has rejoined the Paris Agreement and initiated a renewable energy programme. These economies are implementing action plans to ensure the supply security of critical minerals for their new energy industries.

4.1 Publish a List of Critical Minerals and Make Supply Security an Issue of National Strategy

The list of critical minerals released by the US government in 2022 covers 50 minerals; the EU's 2020 list of critical raw materials contains 30 minerals; Japan's strategy to secure rare raw materials covers 31 minerals; and Canada's critical minerals list identifies 31 minerals. All these lists include gallium, indium, lithium and other minerals essential to the development of solar power and new energy batteries. These economies also propose to increase exploration and recycling to reduce their reliance on imports of critical minerals.

4.2 Countries Are Co-operating to Improve the Supply of Mineral Resources

First, the USA led the creation of the Energy Resource Governance Initiative (ERGI) in 2019 to build a producer alliance with nine countries including Australia, Brazil and the Democratic Republic of the Congo. In early 2021, the USA and Canada announced a Joint Action Plan for Critical Minerals Collaboration. Second, the

USA, Japan, India and Australia teamed up in 2021 to build a rare-earth supply chain, develop rare-earth refining technologies that are low cost and emit low levels of radioactive waste, and encourage government-owned financial institutions to provide loans to companies mining and refining rare-earth elements. Third, to ensure stable supply chains of critical metals needed by the battery, wind and solar power industries, the USA expects to further expand ERGI and share mining experience with its allies, thereby helping them to explore for and produce minerals such as lithium, copper and cobalt.

4.3 Open New Supply Channels Through Policy Support and Other Measures

In 2020, US President Donald J. Trump signed the Executive Order on Addressing the Threat to the Domestic Supply Chain from Reliance on Critical Minerals from Foreign Adversaries and Supporting the Domestic Mining and Processing Industries. The order requires the country to make a determined effort to prevent the risk of supply disruptions to critical minerals caused by foreign governments and to build autonomous critical mineral supply chains. Moreover, the U.S. Department of Energy formed a Minerals Sustainability Division, which is responsible for the exploitation and use of the critical minerals urgently needed by the USA, including those essential to next-generation nuclear energy, renewable energy, low-carbon fossil energy, energy storage and electric vehicles. In addition, the U.S. government proposed a bill to expand its domestic mineral mining and processing industries; the bill requires federal agencies to speed up the mining permit approval process and instructs the U.S. Geological Survey to reevaluate mineral deposits on federal lands. The country also announced its Strategy to Restore American Nuclear Energy Leadership in 2020, which requires action to be taken immediately to strengthen uranium mining and refining.

To reduce dependence on foreign supplies of critical raw materials, the EU aims to strengthen resource recycling, raw material procurement and processing within the EU, diversify supplies from third countries and promote rules-based open trade of raw materials between countries to eliminate international trade distortions.

The Australian government published Resources Technology and Critical Minerals Processing: National Manufacturing Priority road map in 2021 to co-ordinate the critical mineral strategies, activities and initiatives of all resource jurisdictions of the Council of Australian Governments and to introduce national policies to advance critical minerals trade and investment activities.

To address the challenges of high exploration risks and competition for international rights and interests, Japan has issued policies to enhance strategic reciprocity with resource-rich countries. Japan is also making efforts to develop recycling technologies and policies to facilitate the recycling of critical minerals; establish a mechanism to link enterprises with research institutes and government to develop alternatives to critical minerals; and set up a reserve system and a tracking mechanism to ensure the country has reserve capacity of metals in case of supply disruptions.

5 Policy Recommendations

To satisfy its demand for minerals that are heavily dependent on foreign supplies and crucial to the development of its new energy industries—such as lithium, cobalt, nickel and copper—China should set up overseas resource bases through international co-operation and promote the development and use of domestic resources.

5.1 Improve the Business Environment and Support the Exploration and Development of Critical New Energy Minerals

China should increase its domestic policy support and prioritise the exploration of critical new energy minerals. First, China should focus on, for example, key lithium metallogenic belts in Sichuan and Jiangxi and simplify the approval procedures for exploration and development, mining rights and land use to shorten the period from prospecting to mining and improve approval efficiency. Second, more attention should be paid to the exploration of critical new energy minerals, the supply of which other countries control. Prospecting blocks in China should be divided efficiently and funds provided to help exploration companies find new domestic mineral deposits. Third, the Ministry of Industry and Information Technology, the National Development and Reform Commission, the Ministry of Finance and other central government departments should co-ordinate their incentive policies on tax, finance and technology to encourage the exploration and development of resources in areas with poor natural conditions. Fourth, the policy of shutting down an entire mining area due to pollution in one production site should be stopped. Exploration and development conditions should be optimised to support the mining of hard rock lithium ores in the cold highlands of Sichuan. Fifth, the efficient and sustainable development and use of lepidolite ores should be encouraged and the mining of rocks containing mica should be incentivised. Sixth, extracting lithium from lithium-bearing clay could be a future growth area. The search for lithium-bearing clay should accelerate to identify the location and volume of reserves as soon as possible.

5.2 Innovate New Prospecting and Recycling Technologies, and Diversify Resource Supply Channels

New energy enterprises, mining companies and research institutions should co-ordinate their efforts to develop cutting-edge technologies for the exploration, extraction and smelting of critical new energy minerals. Technological innovation should be incentivised to reduce raw material consumption and increase the use of alternative resources. The research and development of key alternative technologies like cobalt-free batteries should be stepped up to alleviate the tight supply of critical new energy minerals. The management of end-of-life solar panels, wind turbines, electric vehicle batteries and other products and components should be improved, and a recycling system for spent products should be introduced to help relieve the pressure on raw ore supply.

First, exploration for the main types of critical new energy minerals should be carried out in China and deep-mine prospecting technology should be developed. Three-dimensional intelligent prospecting systems should be researched and developed to enable the speedy investigation and evaluation of metallogenic zones that could substantially increase China's mineral reserves.

Second, China should form industrial technology innovation alliances comprising research institutes, universities and domestic enterprises. The country should learn from the advanced technologies used by other countries, adopt the model of "introduction–digestion–assimilation–re-innovation", and develop the technologies needed to commercialise new materials. For example, China should intensify research on efficient and low-cost lithium extraction from salt lakes with a high magnesium–lithium ratio, strengthen the development of cobalt-free or low-cobalt battery cathode materials, encourage technological innovation in the use of low-grade nickel ores, and tackle technological barriers to the sustainable and efficient recycling of lithium, cobalt and nickel in power batteries, thus reducing reliance on these and other scarce resources.

Third, China should improve its policy system for recycling spent batteries, diversify associated supply channels and convert recycled spent power batteries into resource supply capacity. Tax incentives should be introduced to encourage power battery recycling, and the tax and fee burdens on power battery recycling companies should be reduced. Power battery recycling standards should be improved to regulate the market; national power battery recycling demonstration enterprises should be formed to avoid disorderly recycling; battery producers should be required to bear some of the costs and responsibilities of recycling, and encouraged—

along with recycling enterprises—to jointly establish a recycling system; and the import control policies on spent batteries should be relaxed and a list of approved recycling companies made to improve the quality of spent materials imported.

5.3 Develop a Mineral Reserve System with Chinese Characteristics

First, depending on their cost, risk profile, benefits and affordability, priority should be given to cobalt, nickel and other resources in short supply, as well as to strategic metals critical to new energy such as heavy rare-earth elements, indium, germanium and gallium. Second, China should establish a supply early warning system and establish a reserve mechanism that combines government reserves and private sector reserves to ensure supply security. Priority should be given to physical reserves of scarce minerals such as nickel and cobalt.

5.4 Refine the Domestic and Global Governance Systems

First, a national dual-carbon steering group should be formed. The public authorities for industry, natural resources and finance should jointly formulate a national roadmap for the dual-carbon goal to send clear market signals to society on issues such as the pace of energy transition, the growth trajectory for new energy, climate targets and on how targets will be translated into actions. Stable policies play a crucial role in bringing investor confidence to new projects, guiding investment in the exploration and development of critical new energy minerals, accelerating technological innovation and recycling used resources.

Second, China should improve the monitoring and evaluation of critical new energy minerals. China can draw on the experience of the USA and instruct the government departments for natural resources, industry, finance and commerce to work jointly on the exploration and development of critical new energy minerals at home and abroad in a targeted manner. In addition, the value chain for critical new energy minerals should be monitored and assessed regularly and scenarios should be designed to test market supply pressure. To address market risks and supply pressure, strategic reserves of critical new energy minerals should be in place to enhance supply chain resilience.

Third, China should follow global resource governance mechanisms like the Extractive Industries Transparency Initiative and participate in the formulation of global governance rules for mineral resources. China should establish a framework for overseas investment in critical new energy minerals, co-ordinate its policies, release reliable and transparent data, regularly assess its supply chains for vulnerabilities and possible response strategies, promote intellectual property transfer and capacity building, and carry out mining activities overseas in a sustainable and responsible way.

Chapter 14: Electricity Pricing Mechanisms

Jinliang Zhang, Yuzhu Wang, Fan Jia, Siya Wang, Xu Xia and Georgios Bonias

1 Introduction

As China is the largest developing country in the world and is still in the primary stage of socialism, electricity pricing reform should take into account that electricity should be affordable and that it brings economic and social benefits. Electricity prices should therefore be controlled within a reasonable range, preventing prices that are too low from hindering the use of new energy and avoiding prices that are too high from negatively impacting public services and the competitiveness of the economy. The market should play the decisive role in resource allocation, and the role of the government should be to ensure electricity supply is secure and stable, help optimise and modernise industry, build a new electric power system and achieve the dual-carbon goal.

A compensation mechanism for the cost of thermal power capacity should be established. Thermal power plants whose utilisation rate decreases due to more new energy in the system should be compensated. Their role will shift from being baseload suppliers of power with a high utilisation rate to capacity providers during peak load or power shortages. For this reduction in utilisation and earnings, they should be compensated with the funds made available by lower new energy prices.

For new energy that has achieved grid parity, a separate guiding price mechanism is needed that decouples the price of the power sold to the grid from the benchmark price for coal power. This will allow the price of new energy sold to the grid to adjust dynamically in response to price signals and will, in turn, guide investment in a scientific way. In the long term, this will help develop the electricity market, support the creation of a new electric power system and better serve the transition to renewable and low-carbon energy.

Electricity spot markets that include new energy participants should be created to incentivise investment in new energy and generate innovative new business models and growth in renewable electricity consumption. At the same time, a mechanism that links the green certificate trading market with the carbon emissions trading system should be established. This would maximise the environmental value of new electricity generation in production, consumption and emissions reduction, thus helping to achieve the dual-carbon goal through synergies between multiple markets.

Ancillary services are an essential part of the new electric power system. They are needed to

J. L. Zhang (✉) · Y. Z. Wang · F. Jia · S. Wang · X. Xia
North China Electric Power University, Beijing, China

G. Bonias
Shell International Limited, London, UK

Shell International B.V. and Development Research Centre of the State Council (DRC) of People's Republic of China, *Embracing the Future, Powering Growth: An Energy System Renewed for China*, Advances in Oil and Gas Exploration & Production,
https://doi.org/10.1007/978-3-031-73075-7_15

support the connection of large amounts of new energy to the power grid. First, China should encourage more participants to join the ancillary services market. Energy storage, electricity distribution and electricity retail companies, micro-grids, virtual power plants and even individual consumers will all be important participants in the market. Second, it is important to appreciate that ancillary services is a market, so participants need to be incentivised with income expectations.

As part of electricity market reform, an energy storage pricing mechanism should be established to encourage energy storage companies to participate in spot markets and ancillary services. Price incentives should be introduced and a cost compensation mechanism and capacity-based electricity pricing mechanism should be created. Before the market mechanisms are introduced, China should set up development funds to support the construction of energy storage infrastructure.

2 China's Electricity Pricing Mechanisms

Electricity is a pillar of the national economy and electricity prices are closely tied to economic and social development. China's new electric power system will be based on a comprehensive reform of the existing power system. There is a cost to this transition, which will inevitably impact electricity prices. A clear understanding of China's current electricity pricing mechanisms is necessary to develop the new electricity market, optimise the allocation of power resources and ensure the supply of renewable, cost-effective and high-quality electricity for the country's economic and social development.

2.1 On-grid Price Formation Mechanisms

The price of electric power delivered to the grid is determined by the type of power generated. As the reform of the electricity system advances, the policy mechanisms that influence grid prices are changing constantly. The history of grid delivery prices in China is shown in Fig. 1. To ensure the power industry develops stably and meets the interests of the general public, China's grid price mechanism will continue to combine government pricing and market-based pricing.

2.1.1 On-grid Pricing for Coal Power

The on-grid benchmark pricing mechanism for coal power started in 2004. Nine pricing adjustments were made between 2004 and January 1, 2020, when the Guidelines of the National Development and Reform Commission on Deepening the Reform of the On-grid Tariff Formation Mechanism for Coal Power came into effect. The 2004 mechanism linked the price of electricity with the price of coal, whereas its

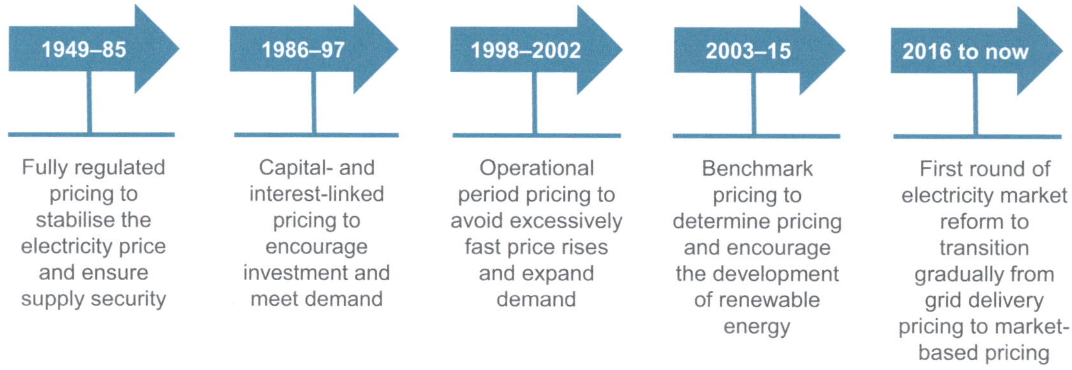

Fig. 1 History of on-grid pricing in China. *Source* Yang Rui, The Evolution and Policy Analysis of Electricity Pricing in China, Electrical Equipment, 22, 2019

replacement is a market-based mechanism of "benchmark price plus market fluctuations". The previous benchmark price included the cost of desulphurisation, denitrification and dedusting, to which the new market-based price added a new variable of ultra-low emissions. Local coal power is now priced in a market-oriented way through participation in electricity market transactions. In 2020, more than 70% of coal power output was sold using on-grid pricing through market-based transactions. To achieve the target of fully deregulating the on-grid tariff for coal power, the National Development and Reform Commission issued the Notice on Further Deepening the Market-oriented Reform of On-grid Tariffs for Coal Power in October 2021. According to this document, all coal power output is to be traded in the electricity market, and on-grid pricing is to be achieved through market transactions in line with the principle of "benchmark price plus fluctuations". The fluctuation rate is not to exceed 20%, although the market transaction price for energy-intensive enterprises is not subject to the 20% limit. With this reform, the remaining 30% of coal power output became part of the market trading scheme, allowing the market-based mechanism to play its role in stabilising the power generation sector.

2.1.2 On-grid Pricing for Wind and Solar Power

The Chinese government has introduced a string of price reduction policies to cut costs and sharpen the market competitiveness of wind and solar power.

China determined the on-grid benchmark tariff mechanisms for onshore wind power in four categories of resource-rich areas in 2009 and the on-grid benchmark price for offshore wind power in 2014. The on-grid benchmark tariff for wind power was lowered four times between 2015 and 2018. In 2019 and 2020, the traditional fixed on-grid tariff mechanism was changed to on-grid pricing determined by competition. In 2021, newly approved onshore wind power projects achieved grid parity without subsidies. The on-grid price of wind power in China since 2009 is shown in Table 1.

Several policies related to the on-grid pricing of solar power have been introduced to match the requirements for building an electricity market while working towards grid parity. The on-grid price of solar power in China is shown in Table 2.

As of 2021, new centralised solar photovoltaic power plants, industrial and commercial distributed solar power projects and new onshore wind power projects no longer qualify for central government subsidies. For the on-grid pricing of new projects in 2021, the local benchmark price for coal power remains, although new projects can voluntarily join on-grid pricing by participating in market-based transactions to better reflect the value of renewable solar and wind power. Local governments are encouraged to formulate support policies for new energy projects and create favourable conditions for the sustainable development of offshore wind, concentrated solar power and other types of new energy. Having done so, local governments will then be in a position to formulate on-grid pricing policies for new energy generation projects, which will spur the development of a new electric power system and help drive the country towards its dual-carbon goal.

The switch in on-grid pricing, from subsidies to parity, is the only way for wind and solar to become the predominant carriers in the future electric power system. This will stabilise industry expectations and speed up investment in, and the development of, a new electric power system in which new energy dominates and the dual-carbon goal is eventually reached.

2.1.3 On-grid Pricing for Nuclear Power

As nuclear power is not suitable technologically for market competition, China set an on-grid price for each nuclear power plant that was in operation in 2012. For nuclear power plants that started operating after January 1, 2013, an on-grid benchmark price policy was implemented, which restricts the price of the power they produce to that of coal power plants. Specifically:

(1) In areas where the national on-grid price for nuclear power is higher than the local on-grid benchmark tariff for coal power (including

Table 1 On-grid prices for wind power (in RMB/kWh)

Policy document	Main requirements	Onshore wind power				Offshore wind power	
		Category 1	Category 2	Category 3	Category 4	Offshore	Intertidal zone
Notice of the National Development and Reform Commission on Improving the On-grid Tariff Policy for Wind Power	2009–14 benchmark price	0.51	0.54	0.58	0.61		
Notice of the National Development and Reform Commission on Adjusting the On-grid Benchmark Tariff for Onshore Wind Power	2015 benchmark price	0.49	0.52	0.56	0.61		
Notice of the National Development and Reform Commission on the On-grid Tariff Policy for Offshore Wind Power	2014–17 benchmark price					0.85	0.76
Notice of the National Development and Reform Commission on Improving the On-grid Benchmark Price Policy for Onshore Wind Power and Solar Power	2016–17 benchmark price	0.47	0.50	0.54	0.60		
Notice of the National Development and Reform Commission on Adjusting the On-grid Benchmark Tariff for Solar Power and Onshore Wind Power	2018 benchmark price	0.40	0.45	0.49	0.57	0.85	0.75
Notice of the National Development and Reform Commission on Improving the On-grid Tariff Policy for Wind Power	2019 guiding price	0.34	0.39	0.43	0.52	0.80	Not higher than the guiding price for onshore wind power
	2020 guiding price	0.29	0.34	0.38	0.47	0.75	

Source Produced by the authors

surcharges for desulphurisation, denitrification and dedusting), the local on-grid benchmark price for coal power applies once the new nuclear power facilities are put into operation.

(2) For nuclear power plants or demonstration projects that introduce new technologies or locally developed key equipment, the on-grid price can be raised as long as it does not exceed the on-grid benchmark price for coal power.

2.1.4 On-grid Tariffs for Hydropower

There are three on-grid pricing mechanisms for hydropower in China:

Table 2 On-grid benchmark price (guiding price) and subsidies for solar power (in RMB/kWh)

Document	On-grid benchmark price for solar power plants (guiding price)			Subsidy for distributed solar power projects	
	Category 1	Category 2	Category 3	Residential	Industrial and commercial
Notice of the National Development and Reform Commission on Improving the On-grid Tariff Policy for Solar Photovoltaic Power	1.15 or 1	1.15 or 1	1.15 or 1		
Notice of the National Development and Reform Commission on Exerting the Price Lever to Promote the Sound Development of the Solar Photovoltaic Industry	0.90	0.95	1	0.42	
Notice of the National Development and Reform Commission on Improving the On-grid Benchmark Price Policy for Onshore Wind Power and Solar Power	0.80	0.88	0.98	0.42	
Notice of the National Development and Reform Commission on Adjusting the Benchmark On-grid Tariff for Solar Power and Onshore Wind Power	0.65	0.75	0.85	0.42	
Notice of the National Development and Reform Commission on the Price Policy for Solar Photovoltaic Power Projects in 2018	0.55	0.65	0.75	0.37	
Notice of the National Development and Reform Commission, the Ministry of Finance, and the National Energy Administration on Matters Related to Solar Photovoltaic Power in 2018	0.50	0.60	0.70	0.32	
Notice of the National Development and Reform Commission on Issues Related to Improvement of the On-grid Tariff Mechanism for Solar Power	0.40	0.45	0.55	0.18	0.10
Notice of the National Development and Reform Commission on Matters Related to the On-grid Tariff Policy for Solar Photovoltaic Power in 2020	0.35	0.40	0.49	0.08	0.05

Source Produced by the authors

(1) The supplier and user determine the inter-provincial and inter-regional transaction price through negotiation. As clearly stated by the government, for hydropower plants that provide inter-provincial and inter-regional power transmission services, the on-grid price of the electricity transferred to other provinces is the settlement price at the receiving province minus the transmission tariff. If no agreement can be reached, the transaction price is determined by the National Development and Reform Commission.

(2) The on-grid price for hydropower follows the benchmark price system. The on-grid benchmark price of hydropower in each province is the same as the average power purchase price paid by provincial power grid companies, although changes in supply and demand and hydropower development costs can be taken into account. For provinces with a relatively high proportion of hydropower, time-of-use prices or graded benchmark prices in wet and dry seasons can be applied.

(3) The on-grid prices for cascade or run-of-the-river hydropower plants in the same drainage basin should be the same. For plants developed by the same investor in the same basin, the same provincial on-grid tariff for each plant applies. For plants developed by

different investors in the same basin, a compensation mechanism for upstream and downstream plants should be applied, after which the same provincial on-grid tariff for each plant can be gradually applied.

Accelerating the development of pumped storage power plants is an important way to improve the capacity, flexibility, economy and security of a power system in which new energy dominates. Currently, the two-part tariff is the principal pricing approach for pumped storage in China. Energy-based pricing is competitive, whereas capacity-based pricing is determined by the generating capacity of the plant. In time, pumped storage power should be included in the market.

The National Development and Reform Commission's report on Further Improving the Price Formation Mechanism of Pumped Storage states that in provinces where an electricity spot market has not yet been established, the price of pumped storage electricity shall be 75% of the benchmark price for coal power; for pumped storage electricity purchased through competitive bidding, the price of the winning bid applies; for the electricity of the winning bid that is not used due to scheduling and other factors, the benchmark price for coal power applies; and the electricity from pumped storage power plants fed into the grid is to be purchased by power grid companies at the benchmark price for coal power.

2.1.5 On-grid Tariffs for Gas Power

There are different on-grid tariff policies for three types of gas power plant in China:

(1) The benchmark price applies to natural gas-fired combined heat and power plants that recently started production. Provincial pricing authorities should take into consideration the cost of generating power from natural gas, its affordability and social and economic benefits when setting the on-grid price for gas-fired power.
(2) The on-grid price for new gas-fired peaking power plants should be determined after comparison with the benchmark on-grid price for combined heat and power plants and the cost difference between the two.
(3) For new natural gas-fired distributed power plants, the on-grid benchmark price refers to the on-grid price of the electricity purchased by power grid companies from distributed power generation facilities. In cases where power users sign purchase agreements with a distributed energy provider, the price and amount of electricity traded is determined by both parties through negotiation.

In addition, there is a mechanism to link the price of natural gas with electricity in China. According to the Guidelines of the National Development and Reform Commission on Deepening the Reform of the On-grid Tariff Formation Mechanism for Coal Power, when the price of natural gas changes significantly, the on-grid price of gas-fired power should also be adjusted. There is a ceiling for that adjustment in that it should not exceed the local on-grid benchmark price for coal power or the average power purchase price that local power grid companies pay of RMB 0.35 per kilowatt-hour.

2.1.6 On-grid Tariffs for Bioenergy

Bioenergy companies are highly sensitive to policy changes. Stable policy and price systems are important for keeping business expectations in balance. Governed by the Renewable Energy Law of the People's Republic of China, the price policies for bioenergy have been adjusted as the industry develops and the external environment changes. The on-grid tariff and subsidy policy for bioenergy has gradually transitioned from the fixed subsidy system of 2006 to the current fixed tariff system. The on-grid tariffs for bioenergy projects set by the government are shown in Table 3.

The Implementation Plan for Improving the Construction and Operation of Bioenergy Projects, jointly issued by the National Development and Reform Commission and other central departments in September 2020, states that for bioenergy projects that fully implement the new

Table 3 On-grid prices set by the government for bioenergy projects (in RMB/kWh)

Document	Agricultural and forestry biomass-fired power generation (direct firing or gasification of agricultural and forestry residues for power generation)	Municipal waste-to-energy power generation	Landfill gas power generation	Biogas power generation
Notice of the National Development and Reform Commission on Issuing the Trial Measures for the Administration of Renewables-generated Power Prices and Cost Allocation	Comprises the on-grid benchmark price for desulfurised coal-fired power in provinces in 2005 plus the subsidised electricity price (RMB 0.25 /kWh)			
Notice of the National Development and Reform Commission on Improving the Price Policy for Agricultural and Forestry Biomass-fired Power	0.75	In accordance with the Notice of the National Development and Reform Commission on Issuing the Trial Measures for the Administration of Renewables-generated Power Prices and Cost Allocation		
Notice of the National Development and Reform Commission on Improving the Price Policy for Municipal Waste-to-Energy Power Generation	0.75	0.65	In accordance with the Notice of the National Development and Reform Commission on Issuing the Trial Measures for the Administration of Renewables-generated Power Prices and Cost Allocation	
Notice on Issuing the Implementation Plan for Improving the Construction and Operation of Bioenergy Projects	The on-grid price is determined by competitive bidding			

Source Produced by the authors

subsidy policy after January 1, 2021, a competitive approach will be adopted when determining their on-grid tariff. This applies to projects that have already been approved but have not yet started construction.

2.2 Transmission and Distribution Tariff Mechanisms

Following the publication of Several Opinions on Deepening the Reform of the Electric Power System in 2015, efforts to reform the power system in China have progressed comprehensively. The reform history of power transmission and distribution tariffs in China is shown in Table 4. By 2020, China's transmission and distribution tariff reform was more or less completed. The separation of power transmission and distribution tariffs is an important step in the reform of electricity pricing. China's transmission and distribution tariffs are divided into inter-provincial and inter-regional transmission tariffs, provincial grid transmission and distribution tariffs, regional grid transmission tariffs, and local grid and distribution grid tariffs.

2.2.1 Inter-provincial and Inter-regional Transmission Tariffs

Inter-provincial and inter-regional special projects are power transmission projects that have relatively clear sending and receiving sides and relatively fixed power flow directions. Special

Table 4 Reform history of power transmission and distribution tariffs in China

2002	Electricity prices are divided into on-grid price, transmission tariff, distribution tariff and selling price. The principles for setting transmission and distribution tariffs are determined by the government
2003	1. Transmission and distribution tariffs are determined by the relevant authorities in accordance with the principles of "reasonable cost, reasonable profit, correct tax payments and fair cost allocation" 2. Transmission and distribution tariffs are divided into common network service price, special service price and ancillary service price 3. The common network service price is determined by the voltage level
2005	1. Transmission and distribution tariffs are determined using the average selling price minus the average purchase price and transmission and distribution losses. A gradual transition to the "cost plus profit" approach takes place 2. The transmission tariff for special projects is a two-part tariff based on the authorised cost approved by the government price authorities
2009	The transmission and distribution tariffs of provincial power grids use a two-part system. The transmission and distribution tariffs are determined by using the average transmission and distribution tariffs of the power grid companies minus the price difference between voltage levels. For the transmission and distribution tariffs for 110 kV (66 kV), a 10% deduction applies; and for 220 kV (330 kV), a 20% deduction applies
2015	Transmission and distribution tariffs are determined separately. The method for determining transmission and distribution tariffs gradually transitions to the principle of "authorised costs and a reasonable profit", with the tariff determined by voltage level. Users or power sellers pay the transmission and distribution tariffs corresponding to the voltage level of the grid to which they are connected. The reform of electricity sales progresses steadily, and access to power distribution is opened to private capital in an orderly manner
2017	1. For distribution network projects whose investors are determined by competitive bidding, the distribution tariff is determined using the bidding-based pricing method 2. For distribution network projects whose investors are not the result of competitive bidding, one or any combination of the three pricing methods—authorised revenues, price caps or yardstick competition—may be used to determine the distribution tariff
2018	1. The two-part transmission tariff applies to the first regulatory cycle of the regional power grid 2. The transmission tariff for the first regulatory cycle of special projects is adjusted and a single-part energy-based price system is implemented 3. The transmission tariff for special inter-provincial and inter-regional projects is adjusted
2020	For inter-provincial and inter-regional transmission tariffs, a single-part energy-based price system applies and the operational period pricing method is used

Source Various

transmission projects are a natural monopoly. It is difficult to determine their transmission tariffs through competitive bidding, rather the tariffs need to be set by government on the basis of cost supervision and assessment.

The Measures for Setting the Transmission Tariffs for Inter-provincial and Inter-regional Special Projects states that the transmission tariff is determined by the pricing authority of the State Council based on a single-part energy-based pricing model. The operational period pricing method is based on covering costs and obtaining a reasonable revenue margin. The transmission tariff is calculated by setting a discount rate for the annual net cash flow of the project, based on the rate of return on capital, to achieve a cash flow balance over the entire operational period. The formula for calculating the transmission tariff is:

$$\text{Transmission tariff (without VAT)} = \frac{\text{Annual revenue}}{\text{Design transmission capacity} \times (1 - \text{Line loss rate for pricing})} \quad (1)$$

The current tariff calculation method for inter-provincial and inter-regional special projects is better and more stringent in terms of workflow and cost supervision and assessment than the previous parameter-based approach. A comparison between the Measures for Setting the

Table 5 Comparison of some pricing parameters for inter-provincial and inter-regional power transmission

Category	Parameters	Current measures	Previous measures
Operation and maintenance cost	Rate level	2%	4%
Operational and depreciation period	Operational and depreciation period	35 years	30 years
Investment	Total investment (temporary tariff)	The approved budget estimate or the budget projected in the design, whichever is lower	Unspecified
	Total investment (official tariff)	The amount identified in cost supervision and assessment	Unspecified
	Capital (temporary tariff)	20% of the investment	See the Notice of the National Development and Reform Commission on Issuing the Measures for Setting Transmission and Distribution Tariffs for Provincial Power Grids (Tentative)
	Capital (official tariff)	Actual amount	Unspecified
Loan interest	Repayment period	25 years	20 years
	Loan interest rate	Actual interest rate vs. market interest rate	See the Notice of the National Development and Reform Commission on Issuing the Measures for Setting Transmission and Distribution Tariffs for Provincial Power Grids (Tentative)
	Estimated capacity utilisation as per the design in hours	No less than 4,500 hours	Government approval document
	Internal rate of return	5% when the project starts operating	See the Notice of the National Development and Reform Commission on Issuing the Measures for Setting Transmission and Distribution Tariffs for Provincial Power Grids (Tentative)
Line loss in the electricity price	Rate level (temporary tariff)	Set line-loss rate	Unspecified
	Rate level (official tariff)	Set line-loss rate and actual line-loss rate	Unspecified

Source Measures for Setting Transmission Tariffs for Inter-provincial and Inter-regional Special Projects; Measures for Setting Transmission and Distribution Tariffs for Provincial Power Grids (Tentative)

Transmission Tariffs for Inter-provincial and Inter-regional Special Projects and the earlier Measures for Setting the Transmission and Distribution Tariffs for Provincial Power Grids (Tentative) is shown in Table 5. By effectively guaranteeing a good return on investment and providing reasonable prices for inter-provincial and inter-regional power transmission—as well as grid-to-grid power transfer, supply security support and other services—the reformed tariff system uses the capacity of existing transmission facilities better than previously and helps guide investment in power generation and power grids. However, there is still room for improvement in the power system's ability to tackle extreme weather in the regions and lay the foundation for inter-regional power transfer transactions.

In terms of transmission revenue allocation and price adjustment mechanisms for the projects participating in inter-provincial and inter-regional incremental spot trading of renewable energy, optimal pricing applies. If a special project has surplus transmission capacity and the power grid company temporarily increases power transmission to improve utilisation of the project, the company may sell the additional power transmission capacity at a price that is not higher than the transmission tariff set for the project. These reform aspects will help increase the proportion of new energy in inter-provincial and inter-regional transmission. A transmission tariff incentive for new energy can also motivate companies to invest in new energy power generation and reduce carbon emissions in the sector.

2.2.2 Transmission and Distribution Tariffs for Provincial Power Grids

According to the definition in the Measures for Setting Transmission and Distribution Tariffs for Provincial Power Grids, the power transmission and distribution tariff for provincial grids is the price of the transmission and distribution services provided by the provincial grid company to users within its scope of business. The tariffs are calculated using the "cost supervision and assessment + incremental plan forecast" method and follows the principle of "authorised costs and a reasonable revenue margin". To determine the transmission and distribution tariffs for a provincial grid, a reasonable revenue for the power grid company is first determined, after which the transmission and distribution tariff for different voltages and user types is set on top of the reasonable revenue margin.

Transmission and Distribution Tariffs by Voltage Level

There are six voltage levels in China's transmission and distribution systems: 500, 220, 110, 35, 10, and 1 kV and below. The transmission and distribution tariff for each voltage level is the ratio of the reasonable revenue determined for that voltage level to the amount of electricity transmitted and distributed at that voltage level.

Transmission and Distribution Tariffs by User Type

There are five user types: residential, agricultural, large industrial, industrial and commercial, and others. The transmission and distribution tariffs for user types are determined by taking user load characteristics and other factors into consideration on the basis of the transmission and distribution tariffs for the relevant voltage level.

2.2.3 Transmission Tariffs for Regional Power Grids

The transmission tariff for regional power grids is the price of transmitting electricity and providing supply security and reliability services by the regional power grid operator. The transmission tariff is determined as follows: revenue is set in accordance with the principle of "authorised costs and a reasonable revenue margin". Typically, the transmission tariff of regional power grids is a two-part tariff in which: the energy-based price reflects the cost of transmission services provided by the regional power grid; and the capacity-based price reflects the cost to the regional power grid of providing reliable power supply, emergency reserves and other supply security services to provincial power grids. The reasonable revenue is set between the energy-based price and the capacity-based price. The energy-based price is calculated using the average load transmitted in the regional power grid and the transmission capacity of the security services it provides, taking into account factors such as the length of the transmission line, power trading and transmission tariff policy. The formula for calculating the ratio of capacity-based price to energy-based price is:

$$\frac{\text{Capacity-based price}}{\text{Kilowatt-hour-based-price}} = \frac{\textit{Depreciation cost} + \textit{labour cost}}{\textit{Operations and maintenance cost}(\textit{excluding labour cost})} \tag{2}$$

2.2.4 Local Grid and Incremental Distribution Grid Tariffs

According to the Guidelines for Determining the Distribution Tariffs of Local Grids and Incremental Distribution Networks, issued by the National Development and Reform Commission, the electricity price for users within a distribution network comprises the on-grid price or market transaction price, the transmission tariff and distribution tariff, as well as the cost of government funding and any surcharges. The price is not be higher than the tariffs for the provincial transmission and distribution voltage levels to which users are connected.

For distribution network projects whose investors are selected by competitive bidding, the distribution tariff should be determined by the bidding process. For distribution network projects whose investors are not determined by competitive bidding, one or any combination of the three pricing methods—authorised revenues, price caps or yardstick competition—may be selected to determine the distribution tariff. The settlement price between the distribution network and the provincial power grid is determined by the transmission and distribution tariffs of the provincial power grid at the relevant voltage levels.

Distribution network companies can also choose to settle the price of power services with the provincial grid company.

2.3 Selling Price Mechanisms

Selling price refers to the final electricity price for users. It consists of the purchase price of the power, transmission and distribution losses, transmission and distribution tariffs, and government funds and surcharges, as shown in Fig. 2. China has two electricity selling price mechanisms: the classified price and the time-of-use price. There are three electricity user types by tariff: residential; agricultural; and industrial, commercial and others. The time-of-use tariff mechanism applies to all users.

2.3.1 Residential Tariff

The residential tariff applies to electricity use by urban and rural households. China's residential tariff uses the single-part price system, a tier-system which is set by the government.

Most provinces in China started to implement the tiered tariff mechanism for residential electricity use from around 2021. The pricing system defines three tiers of monthly electricity use by urban and rural residents: basic electricity use, normal and reasonable electricity use, and electricity demand for a high-quality life. These are set out in the Notice of the National Development and Reform Commission (NRDC) on Issuing the Guidelines for Trial Implementation of the Tiered Tariff Mechanism for Residential Electricity Uses and the Notice of the NRDC on Improving the Residential Tiered Tariff System.

According to these notices, the Tier 1 tariff will continue to be subsidised and affordable to ensure that the basic electricity demand of 80% of households is covered. The Tier 2 tariff reflects the cost of the electricity supplied. It will be gradually adjusted to a level that includes the costs incurred by, and a reasonable profit margin for, the power companies. This should ensure that 95% of residential electricity demand is covered at a tariff that is RMB 0.05/kWh higher than Tier 1. The Tier 3 tariff, which provides electricity for a high-quality life, includes the cost of subsidies for low-income households and the cost of compensating for any environmental damage caused. It also includes the costs incurred by, and a reasonable profit margin for, the power companies. The Tier 3 tariff is RMB 0.3/kWh higher than Tier 1. For low-income households; those who qualify for the five guarantees of food, clothing, housing, medical care and burial; and other households in difficulty, 10–15 kWh of free electricity will be provided to each household per month, or more for multiple-member households. The tiered tariff mechanism is designed to conserve energy; the tariff increases if electricity consumption rises.

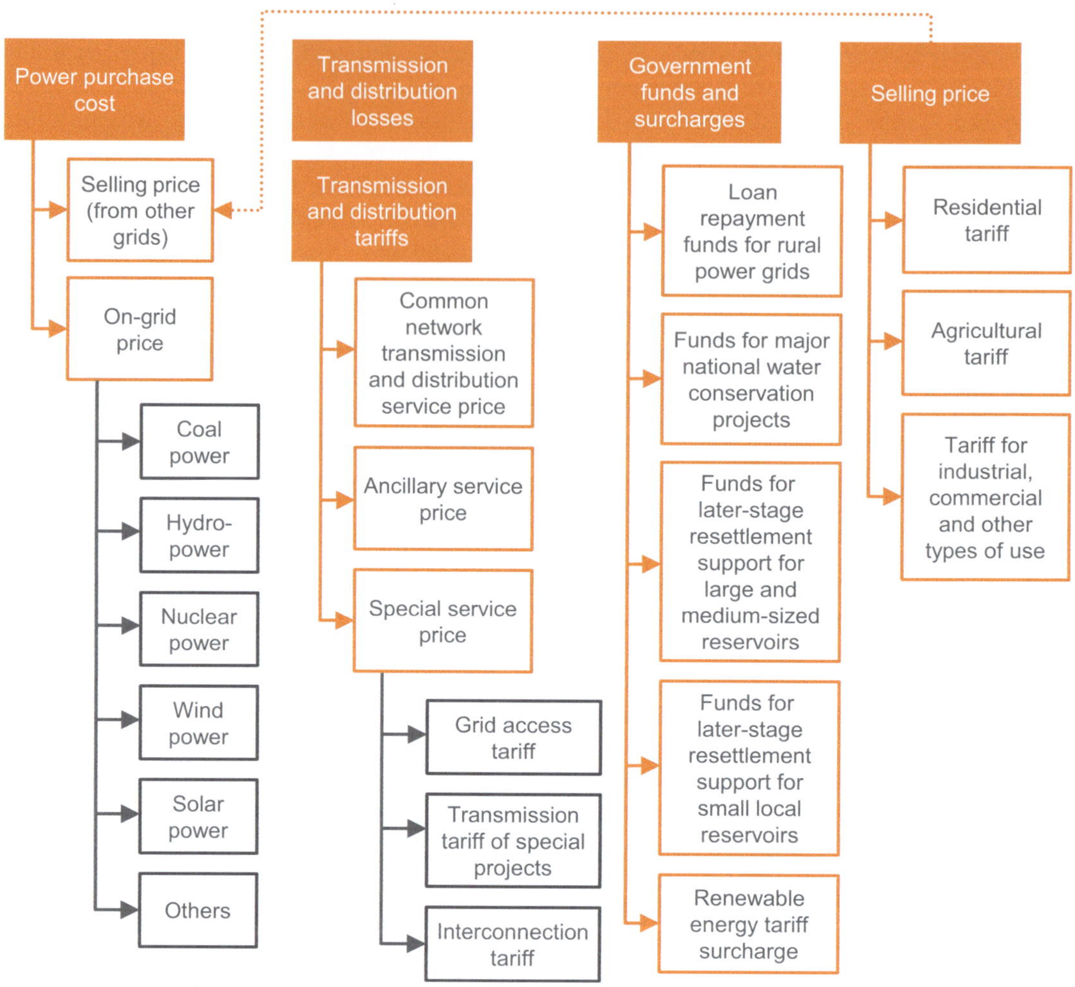

Fig. 2 Composition of the electricity selling price in China. *Source* Ping An Securities

2.3.2 Agricultural Tariff

The agricultural tariff is a concessionary price for the electricity used for irrigation and drainage, tillage and harvesting in rural areas. It is a single-part price based solely on the electricity used, regardless of the capacity of the equipment. The agricultural tariff is based on the average electricity price of all voltage levels and takes affordability for users into account. All provinces should give priority to residential and agricultural electricity demand in their allocations of low-price electricity.

2.3.3 Tariffs for Industrial, Commercial and Other Uses

As stated in the Notice of the National Development and Reform Commission on Deepening the Market-oriented Reform of the On-grid Price for Coal Power, local governments should include all industrial and commercial users in the electricity market, requiring them to purchase electricity at the market price. All users of the 10 kV voltage level and above should enter the market, and other users should also participate in the market as soon as possible. For those companies that have not yet purchased electricity directly from the market, the power grid company should act as the supplier and the power purchase price should be determined through centralised or competitive bidding. When electricity is sold to a user for the first time, the user should be notified at least one month in advance. For users who have participated in market transactions but switched to a supplier other than the power grid company, the price should be 1.5 times the price for those that select the power grid company as supplier. Local governments are encouraged to implement phased preferential policies for electricity demand from small and micro businesses and privately or individually owned businesses. For energy-intensive companies, any rise in the price of electricity price is not limited to 20%. This should encourage them to increase their efforts to save energy, make efficiency gains and modernise their industrial processes and equipment.

For large industrial users in metals, cement, plastics, textiles and paper, interruptible pricing applies. For energy-intensive sectors such as electrolytic aluminium, ferroalloy, calcium carbide, caustic soda, cement, iron and steel, yellow phosphorus and zinc smelting, differential tariffs apply. There are four categories of differential tariffs: eliminated, restricted, permitted and encouraged. For enterprises in the eliminated and restricted categories, the electricity price should be RMB 0.2/kWh and RMB 0.15/kWh higher respectively than the current electricity price. Differential tariffs are a price lever the state uses to eliminate outdated industries. They therefore play a positive role in curbing the disorderly development of energy-intensive sectors. They improve energy efficiency, adjust the country's industrial structure and co-ordinate development of the economy with protection of the environment and sound use of resources.

2.3.4 Time-of-Use Electricity Prices

In the time-of-use price mechanism, the 24 hours of a day are divided into either four periods (critical peak, peak, flat and off-peak) or three periods (peak, flat and off-peak). Different price levels apply to each period. Time of use aims to guide and encourage users to reduce electricity consumption during peak hours of the day, and use electricity during off-peak hours whenever possible. In this way, the difference in power grid load between peak and off-peak hours can be reduced and power grid efficiency can be improved. At present, time-of-use prices apply to all industrial users that have a receiving transformer capacity of 100 kVA and above. Commercial electricity prices apply to residential electricity demand and star-rated hotels in the tourism sector; time-of-use prices apply to agricultural irrigation and drainage, with the price to remain unchanged for first 12 months. The critical peak tariff does not apply to residential households.

3 Challenges for Electricity Pricing Mechanisms

3.1 Supply-side Power Generation

3.1.1 On-grid Tariffs for Power Generation

Inadequate Cost Mechanisms for Power Generation Companies

The on-grid tariff, also known as the wholesale price, is designed to compensate power generation companies for the cost of producing power. However, the costs incurred by all types of power generation cannot be met by the electric power system. The average availability of thermal power generating units operated by the State Grid Corporation of China declined to 4,177 hours in 2020 from 4,263 hours in 2019 and 4,694 hours in 2018. The decrease in operating hours and the increasing costs of thermal power plant investment have created an urgent need for thermal power generating companies to transfer their costs to other parts of the power system. Although the government introduced mechanisms to link on-grid tariffs to the price of coal, the mechanisms were not implemented in a timely manner and the price links did not work as expected. As a result, thermal power plants did not make the revenues they expected and were not able to transfer the cost of generating thermal power to midstream and downstream markets. Natural gas-fired power plants also operate in difficult conditions and show little inclination to invest, which is the result of high fuel costs and low tariffs.

The on-grid tariff mechanisms for hydropower, wind, solar and other types of renewable energy are also inadequate. The pricing only takes into account investment costs; it ignores the social benefits the companies provide, such as peak and frequency regulating. Renewable energy companies have been impacted in recent years by declining tariffs, reduced subsidies and the lack of a unified electricity trading system. The existing price mechanisms do not reflect the social value of renewable energy, which hinders power generation companies from developing sustainably.

Pricing Challenges for New Energy Generation

First, new energy generating plants have almost zero marginal costs. Their biggest challenge is how to recover their substantial upfront investment costs in the tariff. The fact that they have almost zero marginal costs makes it difficult for them to benefit from time-of-use pricing that would encourage consumers to store energy and change their electricity use behaviour, which is one of the biggest potential benefits of new energy. Second, it is difficult to implement a capacity-based pricing mechanism for new energy because of its dependence on the weather and the resulting discrepancy between supply and user demand. Third, there are insufficient price signals to encourage new energy generators to improve supply reliability by reducing their exposure to intermittency. Only price incentives can create efficient and accurate interaction between power generation, power grids, consumer demand and energy storage.

New energy generating units also face challenges from subsidies. To develop renewable energy and reduce the use of fossil fuels, China introduced a compensation mechanism for renewable energy generation. However, due to the extensive deployment of renewable energy plants, it is difficult for the subsidy mechanism to operate as intended. Although a policy document jointly issued by the National Development and Reform Commission, the Ministry of Finance, the People's Bank of China, the former China Banking and Insurance Regulatory Commission, and the National Energy Administration proposed relief measures for the financial constraints facing renewable energy enterprises caused by the monies owed to them in renewable energy subsidies, the problem remains.

There are areas of conflict between the tariff and market-based pricing mechanisms in China. A more comprehensive and systematic renewable energy pricing system is needed, one that reflects the negative externalities of environmental factors in energy prices, such as changes in supply and demand and the scarcity of certain resources. This would encourage a more flexible response to market changes and would optimise resource allocation. Moreover, subsidies, taxes and other government interventions can help balance supply and demand in the electricity market. On the other hand, government subsidies interfere with the market-based evolution of electricity prices, making the price itself unable to reflect the real picture of renewable energy supply and demand. Government subsidies may be of great importance in the early development of renewable energy; but as the market grows, they can hold back the transition to a market-oriented system of renewable energy pricing.

3.1.2 The Pricing Mechanism for Supply-side Energy Storage

Recovering the Cost of Energy Storage

In the new electric power system, new energy generation will change from being a marginal to a predominant carrier, as the installed capacity of solar and wind power continues to grow. Compared with controllable conventional coal power that is capable of continuous output, new energy generation is highly uncontrollable, random and volatile, which is a challenge for power system stability, supply continuity and power quality. Because power demand is growing as a result of rapid economic development, new generating plants are needed to meet this increasing demand. Energy storage facilities can help. They can store surplus energy during off-peak hours and release it during peak periods or when supply is insufficient. This can help smooth new energy output, improve power quality, reduce wind and solar curtailments, increase the use of existing equipment, lessen the need for more installed capacity and reduce costs.

However, the current energy storage investment and recovery mechanism is not clear and the pricing and compensation mechanisms are not well designed. These deficiencies are one of the main barriers to the development of electrochemical energy storage on the grid side. Battery energy storage must therefore participate in the electricity trading market in the same way as thermal power and hydropower.

Calculating the Value of Energy Storage Capacity

Energy storage capacity is an important factor in energy storage pricing. But due to the limited capacity of energy storage facilities, it is difficult to calculate accurately the capacity value of energy storage.

Most standardisation organisations set a direct discount for the continuous discharge of power and the duration it was discharged by the energy storage unit. The capacity value of energy storage systems is determined by their effective load carrying capacity (ELCC). This scheme is currently under discussion by the standardisation authority of the USA, and has gradually been implemented in the UK and adopted by the New York Independent System Operator.

The direct discount scheme offers the advantages of clear criteria and simple operation. However, the requirement for continuous charging and discharging is unreasonable. It requires energy storage facilities to maintain a stable discharge of power during peak hours, whereas in actual operation energy storage units have the ability to increase or decrease the power they discharge in response to changes in load. As a result, the capacity value of energy storage is higher than that after direct discount.

What makes the ELCC model advantageous is that it uses simulation and information such as the shape of the load curve and resource characteristics to calculate a more accurate value of energy storage capacity. However, the disadvantage of this approach is that the simulation programme must interact with the capacity market, which increases the difficulty technologically. In addition, the accuracy of the simulation results depends on the precision of input parameters, which may not be sufficiently accurate.

3.2 Power Grid Transmission and Distribution

3.2.1 Transmission and Distribution Tariffs

New Profit Points for Grid Companies Are Needed

Power grid companies are strictly regulated by the state in accordance with the principle of deregulating power generation from transmission and distribution. Power grids are considered to be a low-profit business, which is extremely detrimental to the commercial development of power grid companies. The transmission and distribution tariff reform has changed the profit model of power grid companies. They have switched from charging the difference between the purchase and selling price of the power they deliver to collecting wheeling charges for transmitting power from one grid to another. This is not good for power grid companies and creates an urgent need for them to find new profit points.

Controversial Transmission and Distribution Tariff Mechanisms

The methods used to calculate transmission and distribution tariffs are in dispute in some provinces, and the practice of charging "market price + approved power transmission and distribution tariffs + government funds and surcharges" is not consistently implemented. The policies for inter-provincial and inter-regional transmission and distribution tariffs are also in difficulty. Although the National Development and Reform Commission issued a notice in 2017 setting inter-provincial and inter-regional transmission tariffs, the tariffs still need to be further adjusted.

Transmission and Distribution Tariffs Should Be Set Separately

The incremental power distribution network is an important building block of power system reform. It is expected to improve the efficiency of grid companies and force transmission and distribution tariffs to be set separately. As indicated in the Guidelines on Determining the Distribution Tariffs for Local Grid and Incremental Distribution Networks, issued by the National Development and Reform Commission in 2017, distribution networks can choose to settle with provincial power grid companies using either the "classified settlement" or "comprehensive settlement" method. Both incremental distribution networks and provincial power grids are supposed to share the basic tariff collected from electricity users. However, most of them use the comprehensive settlement approach, which defines incremental distribution networks as power users who must therefore pay the basic tariff they collect. As a result, the distribution networks do not obtain revenues, but make a loss. The separate setting of transmission and distribution tariffs is therefore key to tackling the survival challenge facing distribution networks.

3.2.2 Pricing Challenges for Energy Storage in Power Grids

In the new electric power system, a shift from centralised to decentralised power grids and from local power generation and use scheduling to a nationwide and open power market will take place. This will pose challenges to power grids. On the one hand, the large-scale connection of new energy requires power grids to balance loads and improve power quality, supply security and stability. On the other hand, the power grid is a public good, so the challenges of supplying power to remote locations and during emergencies need to be addressed. Energy storage is an important part of smart grids. The deployment of mobile or fixed energy storage units at key nodes and remote locations can help with peak shaving frequency regulation. The role of energy storage in the new electric power system is one of providing auxiliary services, although China is still at the exploratory stage of market-oriented mechanisms and pricing models for the ancillary services market.

3.3 Demand Side

3.3.1 Peak and Off-peak Time-of-Use Price Mechanisms

Inadequate Price Differences and Low Peak Prices

First, a reasonable mechanism for peak and off-peak pricing is needed in China. Due to the wide uptake of the peak and off-peak pricing mechanism, users are more likely to consume electricity at low-price hours, which tends to lower the revenues of power supply companies. This in turn can erode the uptake of the peak and off-peak pricing model in the electricity market.

Second, peak shaving is not used effectively due to the small difference between peak and off-peak prices. In other countries, the difference between peak and off-peak prices does not differ greatly from that in China. Under the two-part tariff mechanism typically adopted in overseas markets, the fixed cost related to the power used by consumers is recovered through capacity-based pricing, and because the cost per unit of power produced is similar for different times of the day and for different loads, the price difference is relatively small. At present, for China's large industrial users to which capacity-based pricing applies, most fixed costs are recovered through energy-based pricing, so the price difference should be larger.

The Division into Peak and Off-peak Hours

The division into peak and off-peak prices is performed manually by selecting typical load curves based on subjective experience. Instead, it should be made objectively based on the load characteristics of the power grid and consumer flexibility, using scientific methods like clustering to determine the various peak and off-peak periods. Weather conditions have an increasingly significant impact on the load characteristics of the power grid. For example, in South China, air conditioning in summer can determine the peak and critical peak timing of the power load. If the same fixed peak and off-peak electricity prices apply throughout all seasons, they will not match the changing load characteristics of the power grid. In provinces that have adopted the time-of-use tariff, different electricity prices are set for peak, off-peak and flat rates. The critical peak rate can be 20% higher than in other seasons. In Henan province, for instance, the critical peak rate is applied for four months in winter and summer, with the difference between peak and off-peak prices as much as RMB 0.85/kWh or more. In Shandong province, the critical peak price applies for five months in winter and summer, with the difference between peak and off-peak rates up to RMB 0.78/kWh. In Shanghai, the seasonal factors are reflected in different peak rates, but the average summer peak price of each voltage level for large industrial users is only 1.0393 times that of other months. The seasonal critical peak and peak electricity prices are noticeably low in China.

Peak and Off-peak Pricing for Households

Peak and off-peak pricing encourages electricity consumers to schedule their electricity use. It makes the use of electricity resources more efficient and improves power supply security and stability by smoothing peak loads.

Residential power demand comes mainly from household appliances. The peak and off-peak pricing mechanism not only makes the electricity demand of industrial users more efficient, it can also make a considerable difference for residential power users. China has carried out pilot projects for time-of-use pricing among urban residents. For example, Shandong has introduced voluntary peak and off-peak tariffs. At present, tiered tariffs is the mainstream pricing model for residential electricity use, under which the electricity price rises tier by tier as consumption increases. The current barriers to greater adoption of the time-of-use approach include the difficulties of predicting user response and insufficient flexibility.

Discrepancies Between Time-of-Use Tariffs and the Current Market Transaction Mechanism

Time-of-use tariffs have been introduced in some parts of China to alleviate power shortages, regulate the electricity price and encourage users to shift from peak to off-peak hours. However, it is not realistic for time-of-use pricing to erase the difference between power supply and demand and improve grid and equipment utilisation. In general, China has not yet established a scientific electricity price conversion mechanism that can harmonise time-of-use tariffs with the market-based transaction approach.

3.3.2 End-user Electricity Pricing

Cross-subsidisation of Electricity Prices

Because of long-standing issues in China's power system reform and pricing approach, the issue of cross-subsidisation of electricity prices remains. The average international residential electricity price is 1.5 times that of the industrial electricity price, but the residential electricity price in China is 85% of the industrial electricity price. This is due to the high cross-subsidisation of electricity prices. There are four types of cross-subsidised electricity price in China: (i) cross-subsidisation between regions, when users in developed areas subsidise those in less developed areas; (ii) cross-subsidisation between different voltage levels, when low-voltage power users are subsidised by high-voltage users; (iii) cross-subsidisation between various user types—typically, industrial and commercial users subsidise households and agricultural users; and (iv) cross-subsidisation between different load characteristics, such as when high-load users subsidise low-load users (high-load users are usually industrial and commercial users who are more sensitive to the cost of electricity).

The cross-subsidisation of electricity prices causes a deviation between the cost of delivering electricity and the price users pay. This makes it harder to optimise resource allocation; it also reduces efficiency and removes some of the benefits electricity provides for society. It is therefore imperative to solve the problem of cross-subsidisation.

Preferential Policies for End Users

Power retransfer occurs in many industrial parks, properties and commercial complexes. Some providers of retransfer are not under policy constraints and can determine the retransfer price themselves. Price markup is therefore a common phenomenon. As a result of power retransfer, multiple layers of subcontractor can separate end users from the power grid, which pushes up the price of electricity for them.

Transferring the Cost of Ancillary Services to End Users

Ancillary services became a separate market in the power system reform of 2015. Although some provinces have proposed forming an ancillary services market, only a few have progressed to the pilot stage, and it is difficult for market players to gain access. Many policies stress that ancillary services should follow the principle of "whoever benefits pays". However, not all auxiliary services are reasonably priced, and their costs are not transferred to users but are borne mainly by the power generators.

4 Electricity Price Mechanisms in Other Countries

4.1 United Kingdom

In the UK, a negotiation mechanism for power generation and selling prices has taken shape. The transmission and distribution tariffs are determined by the regulator using the price cap control method; and user prices are determined by the electricity market. The pricing system of the UK electricity market has played a positive role in optimising the allocation of power resources and enhancing the operational efficiency of market players. As Northern Ireland is far from the UK mainland and the UK's power grids are small and independent, the transmission and distribution networks mentioned in this section refer to those of England, Scotland and Wales. There are four types of electricity price in the UK: wholesale price, transmission tariff, distribution tariff and retail price.

(1) **Wholesale Price**

In the UK, the on-grid tariff for renewable energy is linked to the market electricity price to encourage renewable energy companies to take part in market competition. Like conventional power generators, renewable energy companies bid competitively for grid connections. The government subsidises generation companies for the electricity they feed into the grid. In this case, the final price of renewable energy is "market price + government subsidy". As the market electricity price fluctuates, the level of subsidy also changes. However, the sum of these two parts and the feed-in tariff remain at a relatively stable level. For example, the feed-in tariff policy for distributed gas power uses a fixed tariff mechanism. The UK sets fixed tariff levels according to the type of renewable power technology used and the installed capacity of the plant.

(2) **Transmission Tariff**

The transmission tariff comprises two parts: the grid connection price and the transmission price. The transmission service provider charges an access price to the power generation company and an extraction price to the power distribution company. The grid connection price includes the operating cost of the grid connection and the transmission cost includes the cost of infrastructure and service.

(3) **Distribution Tariff**

Like the transmission tariff, the distribution tariff consists of a connection price and a distribution price. The connection price refers to the price for users and power plants connected to the distribution system. It includes the incoming line installation fee, meter installation fee and operation and maintenance fee. The distribution tariff uses the time-of-use price mechanism. It gives time signals to users through morning and evening prices, weekday prices and weekend prices to guide consumers in their use of electricity.

(4) **Retail Price**

In general, the electricity price for users comprises six parts. It includes the wholesale price, the selling price, the transmission and distribution tariffs set by the regulator, an environmental tax, value-added tax and other fees. There are two pricing models for retail price packages in the UK: the two-part tariff and the time-of-use tariff. The retail price package system is shown in Fig. 3.

There are two main tariff packages for business users: fixed tariff and variable tariff. The variable tariff has more flexibility. Under this package, electricity sellers provide price information to users 30 days in advance, so that users can adjust their electricity use and make payment in a variety of ways.

The price and contract terms of the fixed tariff package are relatively stable. Users settle their electricity bill according to the contract price, unless the seller is impacted by factors beyond their control. In the cap tariff package, both parties to the contract set a cap price: when the real-time electricity price rises, the user is not required to pay the price increase; but when the real-time electricity price falls, the seller must lower the selling price.

4.2 France

The European Union (EU) has made efforts to reform the electricity market in France, aiming to eliminate market distortions, intensify competition and strengthen the national power system. Due to a lack of competitive operators, the EU required France to establish two mechanisms to promote market competition. The first is that the transmission system operator Réseau de Transport d'Électricité (RTE) should bid for its transmission tariffs in open auctions. The second is that Électricité de France (EDF) should sell "virtual generating capacity", allowing other market participants to bid for the power generated by EDF and resell it in competitive market transactions. In addition, France has liberalised consumer choice and established a market mechanism to moderate government intervention.

France mainly uses bilateral contracts for electricity trading. France has also set up a

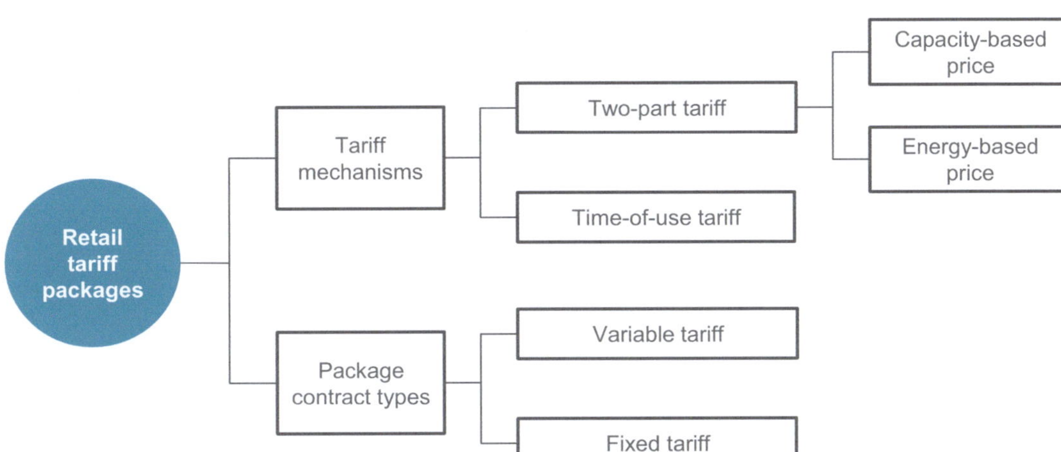

Fig. 3 Retail price packages in the UK. *Source* The authors, based on information provided by the Office of Gas and Electricity Markets (Ofgem)

special power regulatory agency, which is responsible for supervising the implementation of the reforms and ensuring fair market competition.

Feed-in Tariffs

The feed-in tariff is based on a two-track pricing system: a market price and a government-approved regulated price. A balancing fund requires consumers to contribute about 6% of their electricity bill to the fund. This constitutes a revenue stream for EDF to make up for the loss of other revenue streams in the reforms.

Transmission and Distribution Tariffs

The French Energy Regulatory Commission (CRE) regulates the electricity and gas markets in France. It determines the transmission and distribution tariffs using a performance-based regulatory pricing model. It uses postage stamp pricing for voltage level and transmission time, regardless of the transmission distance. The transmission cost is charged to consumers. The tariff has two parts: a capacity-based price and an energy-based price, which are different in winter and summer.

Selling Prices

Electricity prices in France are classed according to user capacity (in kilovolt amps), voltage level and time of use. The time of use and seasonal tariffs are divided into three categories — blue, yellow and green — each of which is further divided into several types of electricity use or special offers. Consumers can choose a price class or offer that meets their electricity use requirements (see Table 6).

4.3 USA

Feed-in Tariff

There are three types of feed-in tariff in the USA. First, in some areas where an electricity market has not yet been formed, the traditional centralised distribution model is used. In these areas, utilities are responsible for delivering electricity to consumers, and the power generation price is based on the cost–revenue ratio (the market-based rate). Second, in those areas where there is a wholesale electricity market and power generation companies participate in electricity auctions, the wholesale price is determined by the market and electricity services are provided by regional utilities. The wholesale price includes transmission and distribution tariffs and public benefit funds. Third, in those areas where there are liberalised wholesale and retail markets, consumers can choose between different electricity sellers and different tariff packages. The price of electricity purchased from electricity retailers is composed of wholesale price, transmission tariff, distribution tariff and public benefit funds.

Transmission and Distribution Tariffs

The transmission and distribution costs of power grid companies are regulated through transmission and distribution tariffs. The tariffs are based on a reasonable level of revenue for the power grid companies and the amount of electricity consumed by the user. The residential electricity price is the highest and the industrial price is the lowest.

Selling Prices

Although electricity pricing is controlled by the government, each power company has a degree of autonomy in structuring its pricing. Generally, consumers are classified as residential users, small users, medium-sized users, large users, and others. Small, medium-sized and large users are industrial and commercial companies. Power capacity and voltage level are also taken into account in the pricing.

4.4 Japan

Japan's power industry has evolved over almost 140 years since Tokyo Electric Lighting, the country's first electric power company, was founded in 1886. Japan's power sector comprises 10 regional integrated generation, transmission and distribution companies. Electricity sales are

Table 6 Classification of electricity selling prices in France

Tariff classification	User capacity	Voltage level	Time period
Blue tariffs			
Small user price	3 kVA	Low	1
Normal price offer	6–36 kVA	Low	1
Off-peak price offer	6–36 kVA	Low	2
Peak-shaving price offer	12–36 kVA	Low	4
Blue–White–Red price offer	18–36 kVA	Low	2
Yellow tariffs			
Normal price offer	36–250 kVA	Low	4
Peak-shaving price offer	36–250 kVA	Low	4
Green tariff A5			
Normal price offer	250–10,000 kVA	Medium–High	5
Peak-shaving price offer	250–10,000 kVA	Medium–High	4
Green tariff A8			
Normal price offer	1–10,000 kVA	Medium–High	8
Peak-shaving price offer	1–10,000 kVA	Medium–High	6
Block price offer	1–10,000 kVA	Medium–High	4
Green tariff B			
Normal price offer	10,000–40,000 kVA	Medium–High	8
Peak-shaving price offer	10,000–40,000 kVA	Medium–High	6
Block price offer	10,000–40,000 kVA	Medium–High	4
Green tariff C			
Normal price offer	Above 40,000 kVA	High–Ultrahigh	8
Peak-shaving price offer	Above 40,000 kVA	High–Ultrahigh	6
Block price offer	Above 40,000 kVA	High–Ultrahigh	4

Source The French Agency for Ecological Transition (ADEME)

entirely market-based. Compared with other countries, Japan's electricity pricing policies are unique.

Feed-in Tariffs

In 2012, Japan launched a feed-in-tariff scheme to stimulate the development of renewable energy. A cost-allocation co-ordination agency, composed of elected associations or companies, is an important part of the scheme. The power companies receive subsidies, which are based on the price of the renewable power they purchase under specific contracts.

Transmission and Distribution Tariffs

Transmission and distribution tariffs are based on incremental cost pricing. In this approach, also known as forward pricing, transmission and distribution prices are based on future changes in incremental costs. The approach takes into account the effects of improvements in technology, management and operations.

Selling Prices

The liberalisation of electricity sales in Japan requires the prices charged by new entrants to the market to be no higher than the regulated price to ensure power supply stability. The pricing mechanisms for electricity sales include a one-part tariff, a two-part tariff, a three-part tariff, a seasonal tariff and a special tariff.

The one-part tariff applies to rated lighting.

The two-part tariff sets a capacity price and an electricity price, and applies to all other uses.

The three-part tariff is a progressive tariff system, which divides household electricity use into three tiers, with the price per kilowatt-hour increasing tier by tier. The first tier is electricity use that is essential to life, which applies to the first 120 kWh consumed per month and is set at the lowest price. The second tier is for electricity use from 121–200 kWh per month, the price of which is the average of the first and third tiers. The third tier applies to electricity use of more than 200 kWh per month, which has the highest price.

Seasonal tariff: To meet seasonal peaks in demand, power companies have to configure their power generation capacity with their power supply carrying capacity and other factors.

They have to invest upfront in power generation facilities, transmission links and other assets to ensure they can meet prolonged spikes in demand. However, when demand decreases, the asset utilisation rate falls, and ultralow-load or even no-load operation can occur, which leads to increased operation and maintenance costs. These additional costs must be reflected in the price of electricity.

Special tariff: This is a progressive electricity price mechanism under which the baseline electricity use for each user category is determined by their electricity use history. For the portion where the contractual electricity use and actual electricity use do not exceed the baseline, a low price applies; for the portion where the contractual electricity use and actual electricity use exceed the baseline, a progressively higher price applies.

4.5 Australia

Australia actively promoted electricity market reform in the 1990s and was one of the first countries to carry out reform. Its retail market reform encourages market competition and incentivises customers to choose a supplier with lower electricity prices and better service quality.

Feed-in Tariffs

Most of Australia's feed-in tariff policies are designed for renewable energy, such as small household solar power installations or large-scale wind power and hydropower projects. These tariffs are mostly paid by the government directly or by electricity retailers. Typically, there are two types of feed-in tariff: a net tariff and a gross tariff. The net feed-in tariff applies to the surplus power sold after the generator has catered for its own needs. The gross feed-in tariff applies to the total electricity produced by the generator.

Transmission and Distribution Tariffs

Transmission services are divided into four categories: entry services, which are transmission network connection services provided to generators; exit services, which are transmission network connection services provided to load customers; transmission use of system services, which are transmission services provided by the transmission network; and common transmission services, which are services that maintain the safety of the power system and provide equivalent benefits to all transmission service providers and which cannot be allocated according to their location.

Table 7 shows the allocation and pricing methods for each category of transmission service.

Selling Prices

The electricity selling price comprises the near-term marginal cost of power companies (which includes generation and transmission costs and the distribution tariff of each state). Retail prices are regulated by the state pricing authorities. Under the time-of-use tariff mechanism, different prices are set for different users for different time periods.

Retail Price Packages

Typical tariff packages in Australia are divided into retail price packages, solar power price packages and green energy price packages (Table 8).

4.6 Implications of Overseas Electricity Pricing Models for China

The UK, France, Japan, Australia and the USA regulate electricity prices differently, based on

Table 7 Pricing methods for each category of transmission service in Australia

Transmission service	Allocation method	Pricing method
Entry services	The entry service cost for each connection point is the sum of the revenue requirements of all assets providing entry services to that connection point. The cost is allocated to the generator connected to that connection point	The entry price is the entry service cost of the node where the generator is located
Exit services	The exit service cost for each connection point is the sum of the revenue requirements of all assets providing exit services to that connection point. The cost is allocated to the load connected to that connection point	The exit price is the exit service cost of the node where the load is located
Transmission use of system services (TUOS)	The revenue requirements for each TUOS asset is allocated to each load node by using the power flow tracking method. The revenue requirements of all TUOS assets allocated to a specific node are combined to arrive at the TUOS charges for the node	The TUOS price can be in one or all of the following forms: load-based price, energy-based price or flat fee
Common transmission services	The cost of common transmission services is allocated to each user node using the postage stamp method, that is, allocation is based on the proportion of annual electricity use at each user node	For each user node, the price of common transmission services that adopt the energy-based price approach is calculated using the energy-based price multiplied by the measured electricity in the settlement period; the price of common transmission services that adopt the capacity-based price approach is calculated using the capacity-based price multiplied by the maximum load capacity under the contract

Source The authors, based on information published by the Australian Energy Market Commission and industry reports

the characteristics of their power industry. Despite their different power sector models, their regulatory measures share many similarities. These similarities reveal a common thread: that the power industry can improve and electricity prices can be reduced through a combination of government regulation and market mechanisms that strike an optimal balance between social and economic benefits. Our analysis of the electricity pricing mechanisms of the five countries is summarised below:

(1) **Greater Importance Is Attached to Efficiency Gains and Renewable Energy Development in the Market-oriented Reform of the Power Sector**

To tackle global climate change, countries have gradually shifted their focus from shaping and improving their power markets to creating market mechanisms that promote the development of renewable energy. Examples include the 10-year plan in Australia, the green energy quota system in Japan and low-carbon energy development in the UK. Other countries are also introducing new market-based products and pricing mechanisms to promote renewable energy development. In China, research on new and renewable energy is inadequate. China should gradually design electricity market mechanisms that are conducive to the development of renewable energy.

(2) **Introduce Competition Mechanisms for Power Generation and Electricity Sales**

Competition stimulates business, accelerates privatisation, regenerates management, improves efficiency and advances electricity pricing reform. The aim of power sector reform is to

Table 8 Retail electricity price packages in Australia

Package type	Specific classifications
Flexible tariff package	Electricity uses are divided into summer and non-seasonal electricity consumption. There are three tiers of electricity use for weekdays and weekends: peak hours, flat hours and off-peak hours
Demand price package	This package comprises two types of electricity use: summer and others. Electrical loads are divided into two time-based categories: 15:00–21:00 on weekdays, and others
Time-of-use tariff package	User loads are divided into two time-based categories: 7:00–23:00 on weekdays, and others
Single-rate tariff package	Electricity use is divided into two parts over a 91-day cycle: the first 1,020 kWh and the rest
Two-rate tariff package	Similar to the single-rate tariff package. The main difference lies in whether users have controllable loads, such as regenerative water heaters or regenerative boilers

Source The authors, based on the Retail Electricity Price Plan of the Australian Energy Market Commission

unlock the potential of enterprises and reduce costs, not to provide government subsidies or increase the price of electricity. The governments of all five countries have moved away from providing subsidies for energy development and have phased out cross-subsidies as their reforms progressed. Encouraging enterprises to adopt a market-based approach, separating power transmission from distribution, improving competition and reducing operating costs are effective ways to ensure that policies and a national power development plan are implemented and electricity pricing reform targets are met. Introducing private equity can make companies even more competitive. Australia's competitive market model, the UK's market-oriented reforms and Japan's green energy quota system are all successful cases from which China can learn.

(3) **A transparent Regulatory System and Well-designed Rules and Regulations Are Needed**

All five countries passed new laws or amended their existing laws to enable power system reform to progress. It is, however, difficult to harmonise the interests of government agencies, power companies, consumers and other actors. Given this, each country formed a regulatory body to oversee power generation and pricing. The pricing authority in each country introduced relatively strict controls on prices. The regulators have legal powers and can impose financial or judicial sanctions if violations of power sector reform or electricity pricing are found. In promoting retail market reform, governments also encourage competition among market participants and incentivise customers to choose their electricity supplier in the hope of obtaining lower electricity prices and better service quality. These practices have provided valuable experience for countries that have carried out electricity reform. Legislation is the foundation of the reform, and rules and regulations help smooth implementation.

(4) **Tariff Packages Broaden Consumer Choice and Respond to Different Needs**

Time-of-use tariffs should be introduced faster and a critical peak tariff mechanism established to guide consumers to use electricity at off-peak hours. Because users consume electricity at different hours and for different purposes, it is necessary to analyse load characteristics and electricity consumption behaviour. Using Australia's experience of price packages and China's reform aims for electricity sales as a foundation, electricity sellers should offer different tariffs for different user types to improve choice for users and attract and retain their loyalty. Market-oriented reform of electricity sales should be gradually deepened to bring about change. Electricity sellers should capture these market changes and design different retail price mechanisms for different stages of deregulation.

(5) **Promote Electricity Markets Comprehensively**

China should deregulate power generation, improve the power market, increase the number of certified power sellers, diversify medium- and long-term trading products, set up electricity spot markets faster, strengthen the ancillary services market and explore market-based capacity compensation mechanisms. The country should create market mechanisms that promote the consumption of new energy, establish a unified national electricity market, optimise the auxiliary services market, and encourage pumped storage, energy storage and interruptible load suppliers to take part in the market. Market-based mechanisms that allow prices to rise and fall for both generators and users should be introduced. New markets such as those for medium- and long-term trading, spot trading, auxiliary services and capacity trading will help speed up reform of the electricity system.

5 Price Mechanisms in the New Electric Power System

5.1 Principles of Pricing Mechanism Design

5.1.1 On-grid Tariffs

On-grid pricing should meet the following principles.

(1) **Electricity Is a Commodity**

Electricity should be treated as a commodity, and the cost of electricity and a reasonable profit for the seller should be reflected in the price.

(2) **Moderate Regulation**

Although electricity is a natural monopoly, any monopoly held by an enterprise can harm the interests of consumers. The regulator should do away with the monopoly and excessive profits to protect the rights and interests of consumers. The principle of moderate regulation is designed to prevent this phenomenon.

(3) **Equal Access**

To realise the principle of equal access to electricity, the correct pricing strategies should be adopted for each part of the power system.

(4) **Ease of Use**

Electricity pricing policies should aim to keep electricity prices steady and avoid large fluctuations, thus providing predictability and stability in the power system. The structure of electricity pricing and the various tariffs should be simple and clear to ensure easy metering and billing. This can help improve the transparency, efficiency and healthy development of the electricity market.

5.1.2 Transmission and Distribution Tariffs

Transmission and distribution networks are a natural monopoly: only one or a few companies in a given region can provide transmission and distribution services. That is why government should regulate transmission and distribution tariffs. The basic principles of setting transmission and distribution prices are as follows:

(1) **Fair, Impartial and Open**

To ensure fair competition and effective operation of the electricity market, all market participants should be treated impartially. To help the electricity market develop soundly and sustainably, measures need to be taken to minimise disruptions to power generation and market participants. This can be achieved by allowing reasonable revenue margins, while ensuring that all participants are able to use the networks efficiently.

(2) **Reasonable Returns**

When determining the grid price, the production cost should be accurately reflected in the price, and grid companies should be allowed a reasonable return on their investment. This will incentivise investors to make future infrastructure investments. The interests of all parties should be well balanced to ensure the sustainable and stable development of power networks.

(3) **Market Supply and Demand Visibility**

Supply and demand and the effects of different power sources and loads on costs should be taken into account when setting power transmission and distribution prices. Power plants and consumers should share the cost of the networks through payment based on their electricity use.

(4) **Ease of Use and Efficiency**

Stable transmission and distribution tariffs help power producers make long-term decisions. The prices should be set in phases and the transition from one to the other should be gradual.

5.1.3 Selling Prices

The principles for determining selling prices are as follows:

(1) **Fair Access and Equal Exchange**

Electricity selling prices should be set in accordance with the principles of fair access, the effective regulation of power demand, and public policy targets. A mechanism that links prices with on-grid tariffs should be established.

(2) **Market-oriented**

Selling prices should be market-oriented to allow reasonable revenue margins for power sellers and ensure electricity and other resources are used sustainably. The market-oriented principle should maintain stable electricity prices for households, agriculture and the public sector.

5.2 On-grid Price Mechanisms

The on-grid price mechanism is affected by many factors, of which power generation is the most important. The cost of ancillary services should be passed on to all parts of the power system through the on-grid price, including consumers.

5.2.1 Connect Electricity Markets with Other Markets

Spot markets should be well designed and well operated to enable an efficient and co-ordinated electricity market. When designing a provincial electricity market system, relations with other power markets, both inside and outside the province, should be co-ordinated and managed. This includes regulating the development of inter-provincial electricity markets and supporting the overall plan of the national electricity regulatory authority. Relations between inter-provincial and provincial power markets should also be co-ordinated. Non-local power generation companies that have signed inter-provincial medium- and long-term priority power generation or trading contracts can become participants in the electricity market by recognising the market clearing price as the settlement basis for transactions. Establishing connections between electricity markets and related equity markets, such as between electricity trading and carbon emissions trading, should also be explored to optimise the value of electricity.

5.2.2 Improve the Commercial Operation of Spot Markets

Efforts should be made to strengthen the flexibility of the market-based price adjustment mechanism, increase the frequency and period of trial operation and improve the transparency and fairness of market pricing. China should accelerate the commercial operation of electricity spot markets, improve market service functions, gradually expand the size of the market and the number of trading products and improve trading methods.

5.2.3 Improve Electricity Pricing Mechanisms

To ensure that electricity prices remain within a reasonable range, it is imperative to improve the mechanisms of market-based and government pricing. Specific measures include strengthening the formulation, implementation and supervision of electricity pricing policies; creating a complete electricity pricing model for calculating, analysing and controlling prices; and improving the efficiency of electricity price monitoring and early warning management. Efforts should also be made to enhance the transparency and fairness of electricity pricing policies and ensure the policies can and will be implemented.

5.2.4 Improve the Pricing Mechanisms for Large-scale Renewable Energy

Improving the on-grid pricing policies for new energy power generation (such as wind and solar) and reforming the hydropower pricing mechanism will support the development of high-quality renewable energy. The model of a "two-part tariff system + a price link between natural gas and electricity" should be adopted to develop on-grid pricing for gas-fired power generation. Gas-fired power generation plays an important role in the energy transition. Its low investment cost and flexible operation allow it to serve as a bridge between fossil fuels and renewable energy, while also strengthening the power system.

The following two documents emphasise the importance of market price signals in helping China to realise its dual-carbon goal: the Working Guidance of the Central Committee of the Chinese Communist Party and the State Council for Carbon Dioxide Peaking and Carbon Neutrality in Full and Faithful Implementation of the New Development Philosophy; and the Notice of the State Council on Issuing the Action Plan for Carbon Dioxide Peaking Before 2030.

As stated in these documents, China will deepen its reform of the energy system; improve the market-based pricing mechanism for electricity and other energy products; continue its market-oriented reform of the coal and oil and gas industries; improve the unified energy market; establish connections between the medium- and long-term markets, spot markets and auxiliary services; increase the size of market-oriented transactions; and fully liberalise electricity pricing.

5.3 Transmission and Distribution Tariff Mechanisms

Transmission and distribution tariffs are the key to electricity market reform. The guiding principle of electricity market reform is "control power transmission and distribution and deregulate power generation and electricity sales". Power transmission and distribution prices will be determined in accordance with the principle of "authorised costs and a reasonable revenue margin"; transmission and distribution costs will be based on voltage level and user category. Although China's pilot projects for transmission and distribution tariffs have yielded good results and valuable experience, challenges remain. As market-oriented power trading increases and different market-based options become available, the transmission and distribution tariffs will need to be further improved.

5.3.1 Optimise the Price Structure and Refine Cost Management

China's power network is huge and complex. The functions and services of some transmission networks are not clearly defined, which may lead to disputes over cost recovery. Prices should be allocated according to functions performed and services provided, and cost recovery should be on the basis of asset ownership and responsibilities.

5.3.2 Optimise the Regulatory Mechanism and Implement Dynamic Pricing

At present, China's transmission and distribution prices follow the principle of "authorised costs and a reasonable revenue margin", which is a relatively traditional government-regulated pricing method. In developed economies, this approach is rarely used, because it is difficult to judge whether the investments made in the grid are reasonable. We therefore recommend that the regulatory approach to monopolistic grid companies should be improved in terms of appraising and approving investments, standardising administration, reducing costs and raising regulatory efficiency. Once these improvements are made, transmission and distribution pricing can become more dynamic.

5.3.3 Introduce Regulatory Measures, Including Incentives and Constraints

When determining the transmission and distribution tariffs for a province, a pricing method suitable for that province should be devised, one that takes into account local expectations for economic development and the condition of the transmission and distribution networks. To ensure that pricing is reasonable, regulatory measures need to be introduced, such as intensifying incentives and constraints and establishing an appropriate and reasonable reward and punishment mechanism. A special grid connection fee should be designed to subsidise the costs incurred when a power plant or user connects to the grid (this may require investment in new transmission lines and equipment to enable the connection and comply with grid codes). These measures will help advance the construction and modernisation of power networks, improve the reliability and security of power supply, sharpen the competitiveness of the electricity market, and promote economic growth and social progress. They will also act as a model for other provinces and help the power sector develop in a comprehensive way.

5.4 Selling Price Mechanisms

5.4.1 Improve the Time-of-Use Tariff Mechanism for Large Industrial Users

To further improve the electricity market, several price mechanisms need to be refined: the peak and off-peak price mechanism, the critical peak price mechanism, the seasonal tariff mechanism, the wet-dry season price mechanism and the time-of-use tariff mechanism. When refining these mechanisms, it is necessary to take into account local power supply and demand, the load characteristics of the network, the proportion of installed new energy capacity and the regulating capacity of the system. Efforts should also be made to clearly define the scope of each mechanism, establish a dynamic adjustment mechanism that adapts to market changes, and improve access for market-oriented power users to enhance competitiveness and efficiency.

5.4.2 Optimise the Two-part Tariff Mechanism for Industrial and Commercial Users

When dividing electricity users into categories, tiered prices can be set by voltage level, user capacity or electricity consumption per unit (hours of electricity use). To strike a balance between maximising the benefits for users and providing full cost recovery for power grid companies in the short term, a two-track pricing optimisation system is needed. Such a system aims to improve social benefits in the short run, minimise the average electricity price for users and ensure the power grid company recovers its transmission and distribution costs and that its new revenue streams are profitable.

5.4.3 Improve Electricity Pricing for Agriculture

(1) **Find Ways to Communicate Pricing to Agricultural Communities**

New pricing policies need to be communicated to agricultural users and village communities in a clear and timely manner.

(2) **Time-of-Use Tariffs for Agriculture**

Agricultural producers can, depending on the characteristics of their electricity consumption, apply for time-of-use tariffs (peak and off-peak pricing) to reduce their electricity bills.

(3) **Reduce the Cross-subsidisation of Agricultural Irrigation and Drainage**

There are two ways to reduce the cross-subsidisation of electricity used for agricultural irrigation and drainage: optimisation and compensation. Optimisation means adjusting and optimising the entire price mechanism; compensation means adjusting only that part that needs to be adjusted.

(4) **Improve the Tiered Tariff Mechanism for Agricultural Production, Irrigation and Drainage**

The tier 1 electricity price, also known as the "lifeline price", covers the average household electricity consumption of the vast majority of irrigation and drainage users. The tier 2 price aims to encourage households with high electricity consumption to save energy. Tier 3 and above compensate for the costs of supplying power and of impacting the environment that are not borne by tier 1 users.

(5) **Develop Pilot Service Funds to Adjust Compensation**

Power service funds are the most efficient way to achieve effective compensation. The service funds should be managed by the government and their operation should be independent of any power company. The collection and subsidy system for such funds should be defined in laws and regulations and be made public.

5.4.4 Refine the Tiered Tariff Mechanism for Residential Users

(1) **Communicate the Tiered Tariff Policy to the Public**

Only when residents have a full understanding of a policy can the objectives of the policy be achieved and the policy benefit the public. Efforts should therefore be made to give residential users a deeper understanding of the tiered tariff system so that its benefits, such as protecting the environment, can be achieved.

(2) **Build a Dynamic Adjustment Mechanism for the Tiered Tariff System**

The existing residential tiered tariff mechanism should be optimised to enable the tier 3 tariff to provide sufficient cross-subsidisation funding sources, thus relieving the pressure on power grid companies to pay subsidies.

(3) **Ensure the Tiered Tariff Policy Is Implemented**

First, accelerate the roll-out of the "one household, one electricity meter" programme. Second, resolve the issue of households with "too many members" through investigation. Third, include the subsidy for the energy-based price in the government's subsidy, as the government is already responsible for subsidising low-income households and those that qualify for the Five Guarantees of food, clothing, housing, medical care and burial. Fourth, roll out the peak and off-peak time-of-use tariff mechanism to properly reflect supply and demand.

6 Policy Recommendations

6.1 Residential Electricity Prices Should Reflect the Cost of Electricity and the Efficiency of Its Use

The tiered tariff policy should be adjusted to better reflect the cost of supplying electricity. It should take into account the electricity demand habits of users, as well as the costs and revenues of power providers. A relatively low price should be set for tier 1 to incentivise users to save electricity and control the costs of providers. The price of tier 2 electricity should allow grid companies to earn reasonable revenues to ensure they remain committed to the system. The tier 3 price should reflect the scarcity of resources to encourage users to reduce electricity consumption and ensure resources are used more efficiently. The peak and off-peak tariff mechanism should be introduced to reflect the relationship between supply and demand. The peak price should be increased to encourage consumers to avoid using electricity during peak hours and to relieve the pressure on electricity suppliers during peak periods. In short, the tiered tariff policy can better distinguish between different types of electricity demand, while controlling grid costs and compensating for the cost of environmental impacts. Finally, the price for residential and industrial and commercial users should be adjusted to reduce cross-subsidisation as much as possible.

6.2 Market-based Electricity Prices Should Be Liberalised and Managed Efficiently

To establish a market-based tariff mechanism that allows prices to rise or fall freely, the government needs to take more regulatory measures. These should include cracking down on market misconduct like price monopoly, fraud and false advertising; enacting strict regulations and mechanisms to ensure the market is fair and transparent and protects the rights and interests of consumers; and supporting power companies to make efficiency gains, cost reductions and service improvements to better meet the requirements of customers and sharpen their market competitiveness. The government should also bring forward the time-of-use tariff policy to encourage consumers to reduce their electricity use during peak hours, thereby alleviating the pressure on the power system. A market-oriented reform of the electricity sector is extremely urgent, and a workable solution is required to address the drawbacks caused by the gap between the market-based coal price and electricity supply.

6.3 Reform the On-grid Tariff Mechanism for New Energy

Although renewable energy is an essential part of the global energy transition, there are barriers to its development, one of which is the high price of renewable electricity. To break down this barrier, technological innovation is needed to improve the efficiency and reduce the cost of renewable energy generation. In terms of policies, more subsidies should be directed to research and development. With regard to technology, the safety, grid-compatibility and related technical requirements of all wind, solar and other new energy projects should be improved to ensure supply stability and reliability. Ultra-large new energy projects should participate in the electricity market as quickly as possible. In addition, plans should be made to phase out subsidies and achieve grid parity to ensure the market-oriented and long-term development of renewable energy continues. The subsidies saved can be used to encourage enterprises to carry out technological improvements and increase their resource utilisation efficiency, especially for wind and solar plants with a generating capacity of 410 GW. This can enhance the efficiency and economic benefits of renewable energy, reduce social and economic costs, and promote the penetration of renewable energy into a wider range of applications.

6.4 Encourage Power Grid Companies to Control Costs

To control costs and improve cost effectiveness, power companies need to take several measures. They need to improve their project assessments to avoid unnecessary waste and losses on high-cost or low-benefit projects. They should also use advanced management accounting such as cost management, budget management and performance assessment to control costs and improve cost effectiveness. By creating a scientific cost management system and a well-designed performance assessment mechanism, companies can better manage expenditure, improve efficiency and achieve sustainable development.

6.5 Optimise the Electricity Price System

The energy transition in China will move forward at a faster pace because new business models and technologies will be introduced on the demand side to improve demand–response and increase electrification.

For this reason, it is crucial to optimise the pricing system to remove the structural issue of selling electricity for less than it costs to produce, allow dynamic peak and off-peak pricing and resolve cross-subsidisation. In view of this, China should improve the industrial and commercial electricity pricing system, reform the cost-based electricity pricing system, establish a dynamic adjustment mechanism for peak and off-peak prices to improve the efficiency of the power system and incentivise users to avoid using electricity during peak hours. If a cross-subsidisation standard can be clearly defined, the attributes of electricity as a commodity can be gradually restored and the cross-subsidisation of electricity prices between provinces can be resolved. Some provinces have already started to tackle the issue of cross-subsidisation.

Chapter 15: Carbon Market

Ke Wang, Mei Lu, Chen Lv,
Wan Yue Xuan and Georgios Bonias

1 Introduction

Achieving peak carbon dioxide emissions by 2030 and carbon neutrality by 2060 are major strategic targets made by the Central Committee of the Communist Party of China (CPC) after careful consideration. The power sector is responsible for more than 50%[1] of China's total carbon dioxide emissions, so effective control and reduction of CO_2 emissions from this sector is essential for China to meet its dual-carbon goal. Thermal power accounts for a large part of China's power generation (69%[2] in 2020), so there is an urgent need to optimise the generation mix. The national emissions trading scheme (ETS) is an important means to optimise the power supply structure, improve power generation efficiency and spur the low-carbon development of the power sector. A well-designed carbon trading mechanism for the power sector can ensure the ETS operates smoothly and emissions from power generation are removed effectively. As the reform of the electricity market progresses, the proportion of renewable energy in the generation mix needs to increase. Greater consideration should therefore be given to enabling the low-carbon development of the power sector in the design of the ETS. China's national ETS initially covered the power sector only, as did the seven regional ETS pilots. The carbon price was relatively low compared with major carbon markets; the trading turnover rate in 2021 was only around 3%,[3] which was considerably lower than the 417% of the European Union Emissions Trading System that same year. There is, therefore, a need to improve the trading activities of the national ETS.

There is a consensus on reducing carbon emissions by market-based means globally. China needs a booming carbon market to make abatement costs visible, resolve potential green trade barriers and enhance the country's say in global climate governance and carbon pricing.

[1] According to the data released by China's Carbon Emissions Accounts and Datasets (CEADs) for 2019, the carbon dioxide emissions from China's power sector were 5,626.1 megatonnes, which equates to 51.7% of the country's total emissions of 10,881.7 megatonnes.

[2] China's thermal power output was 5.33025 trillion kWh and its total power output was 7.77906 trillion kWh in 2020, according to the China Energy Statistical Yearbook.

[3] The turnover rate was calculated by the authors using the ratio of trading volume to emission allowances.

K. Wang (✉) · M. Lu · C. Lv
Beijing Institute of Technology, Beijing, China

W. Y. Xuan
China University of Mining & Technology - Beijing, Beijing, China

G. Bonias
Shell International Limited, The Hague, The Netherlands

Shell International B.V. and Development Research Centre of the State Council (DRC) of People's Republic of China, *Embracing the Future, Powering Growth: An Energy System Renewed for China*, Advances in Oil and Gas Exploration & Production,
https://doi.org/10.1007/978-3-031-73075-7_16

China's carbon markets still have much room for improvement in their efforts to reduce CO_2 emissions and drive down the cost of abatement. Reductions in total carbon emissions and improvements in carbon emissions intensity in the current ETS pilots are credited to traditional command-and-control policies. Optimising the design of China's ETS and strengthening the co-ordination between the ETS and traditional command-and-control policies are necessary to make the cost of carbon emissions reduction visible.

The national ETS should continuously improve its approach to the allocation of emission allowances, the carbon price stability mechanism, the allowance offset mechanism and the supervision and penalty mechanism. Doing so will help form stable incentives for low-carbon investment and accelerate progress towards emission reduction targets. The national ETS should also strengthen its links with carbon capture, utilisation and storage (CCUS) projects to support the development of CCUS technologies.

China should increase the number of sectors covered by the national ETS, improve how enterprises measure their CO_2 emissions, strengthen the links between the green electricity market, the green electricity certificate (GEC) system and the ETS to promote renewable electricity consumption and boost the GEC market. This will also encourage the development of distributed renewable energy, break down the inter-provincial barriers between power systems on the demand side, accelerate the development of the inter-provincial and inter-regional renewable power scheduling mechanism and speed up the formation of a unified national electricity market.

2 The Advantages of the ETS in Reducing Carbon Emissions

2.1 Lowers the Cost of Reducing Emissions

An ETS can clearly define the maximum carbon emissions an economy is prepared to allow by setting a cap on those emissions. Supported by sound and stable regulations, it can achieve the targets set for emission reductions with a high degree of certainty. It does so by setting gradually tighter emission allowances over time. These provide stable and credible signals of long-term abatement policies and targets, thereby shaping market expectations and creating incentives for long-term investment and innovation in low-carbon technologies. The ETS leverages the role of the market in allocating resources to achieve the set emission reduction targets at the lowest possible economic cost. Owners trade scarce emission allowances and a carbon price is generated. Actors participate in the market in response to the carbon price level: the companies with lower emission reduction costs tend to reduce their emissions more than needed and sell the surplus, while companies with higher emission reduction costs reduce their emissions less than needed and buy the surplus allowances of others. As a result, emission reduction efforts are efficiently distributed among participants and the overall abatement cost of the system is at the lowest level. Compared with other policy instruments, an ETS is particularly advantageous when regulators do not have sufficient information on the abatement potential and costs of companies.

2.2 Enhances System Flexibility

An ETS uses various mechanisms to allow participants to decide when and where to invest in reducing their emissions. This flexibility reduces the overall cost of emissions reduction. The allowance storage and borrowing mechanism provides companies with flexibility in the timing of their emission reductions. If the emissions reduction cost is expected to rise, companies can store allowances for future use; if the abatement cost is projected to be lower, companies can borrow future allowances. The creation of offset mechanisms, such as the clean development mechanism, can provide flexibility in where to locate emission reductions. Offset mechanisms allow certified emission reductions to be made outside the scope of the ETS, which provide companies with alternative sources for emission reductions and which drive down the cost of abatement. However, unrestricted offsets will pose a threat to the integrity of the ETS, so carbon markets generally limit the quantity and quality of allowance offsets.

Because an ETS provides dynamic price signals and can adapt to broader economic fluctuations, it can be used as a counter-cyclical regulation instrument. When the economy slows down and economic output is lower than expected, demand for carbon allowances falls and the price of carbon declines, further reducing the burden on the economy. When the economy grows, the price of carbon increases, so that the biggest emission reductions are made when the economy is strong.

2.3 Encourages Technological Innovation

Improving the energy efficiency, productivity and technology of an economy can drive down its carbon emissions intensity and encourage the development of low-carbon technologies. An ETS helps to decouple economic growth from carbon emissions and high levels of carbon intensity. It provides incentives for low-carbon production and the innovation of low-carbon technologies via price signals. In the process chain of technological innovation, an ETS plays a positive role in progressing low-carbon technologies from laboratory-based basic research to commercialisation and market readiness. For technologies that are close to market readiness, a moderate carbon price can be sufficient to trigger its entry into the market. For emerging technologies that are far from market readiness, the ETS can help them progress by encouraging patent applications. Figure 1 shows the effects of ETS on low-carbon innovation, market readiness and patent applications. In terms of the deployment of low-carbon technologies, carbon price signals make emerging low-carbon technologies more competitive, which in turn stimulates investment in well-proven energy-efficient products.

3 China's National ETS

The Decision of the Central Committee of the Communist Party of China on Several Major Issues Concerning Comprehensively Deepening

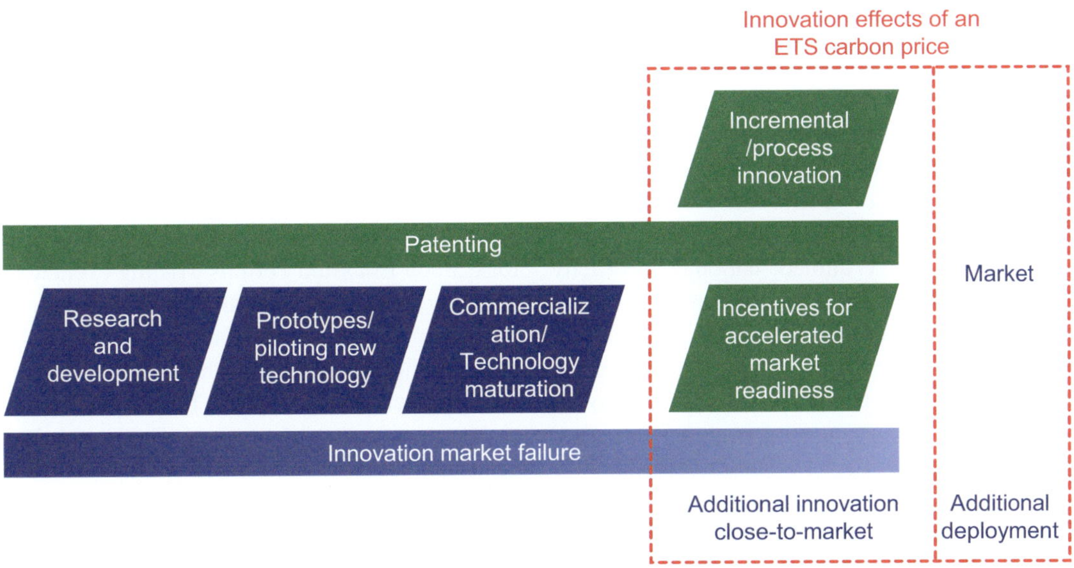

Fig. 1 The benefits of an ETS for innovation. *Source* International Carbon Impact Partnership, Benefits of Emissions Trading: Taking Stock of the Impacts of Emissions Trading Systems Worldwide, 2018

Reform, published in 2013, states that the market should play a decisive role in the allocation of resources. This in turn paves the way for the carbon market to play an important role in curbing carbon emissions. According to the Working Guidance for Carbon Dioxide Peaking and Carbon Neutrality in Full and Faithful Implementation of the New Development Philosophy, published by the Central Committee of the Communist Party of China and the State Council in 2021, for China to reach the dual-carbon goal requires the government and the market to work together and use market-based mechanisms to form effective incentive and constraint mechanisms. This makes clear that the national ETS is an important policy instrument to promote the achievement of the dual-carbon goal using market-based mechanisms.

The power sector was the main contributor to the development of the national ETS. Once formed, the national ETS posed new challenges to power generation companies; it also brought new opportunities to improve the efficiency of coal-fired power generation and progress towards low-carbon and sustainable development. In 2021, the Notice of the National Development and Reform Commission on Further Deepening the Market-oriented Reform of On-grid Tariffs for Coal Power was published, marking another milestone in the market-oriented reform of the power sector. In 2022, the publication of the Guidelines of the National Development and Reform Commission and the National Energy Administration on Accelerating the Construction of a Unified National Electricity Market marked a new stage in the development of a unified national electricity market. As the market-oriented reform of the power sector advances, the ETS mechanisms should be continually optimised to promote the development of China's low-carbon power system.

3.1 Mechanisms of the National ETS

The national ETS initially covered 2,225 power generation companies that emitted 26,000 tonnes[4] of carbon dioxide equivalent and consumed around 10,000 tonnes of coal equivalent annually from 2013 to 2019. As the carbon market is still at an early stage of development, the mechanisms are still undergoing improvement.

3.1.1 Determining and Allocating Emission Allowances

Emission allowances are determined by the Ministry of Ecology and Environment after taking into account in a holistic manner factors such as economic growth, industrial restructuring, energy mix optimisation and the control of air pollutant emissions. Each province's Department of Ecology and Environment allocates allowances to enterprises within its administrative area in line with the allocation plan of the ministry and sector-specific carbon emission benchmarks. China's national ETS uses a performance-based trading design, which means that the allowances allocated to a company change if its output changes. This approach is conducive to reducing carbon emissions intensity. There are four categories of carbon emission benchmark for power generation technologies in the power sector.

3.1.2 Allowance Trading and Offsets

The carbon emission allowances are traded in China's national carbon market, which mainly uses spot trading. The Administrative Measures for Carbon Emission Permits Trading (Tentative) states that the Ministry of Ecology and Environment can add other trading products if and when necessary. Market participants are big emitters. Although the Administrative Measures for Carbon Emission Permits Trading (Tentative) requires trading hubs to take effective measures to prevent excessively speculative trading behaviour, a price stabilising mechanism or allowance storage mechanism has yet to be introduced. Key emitters are allowed to use China Certified Emission Reduction offset credits to meet

[4] See the 2019–20 Implementation Plan for National Carbon Emissions Trading Allowances and Allocations (Power Sector).

compliance obligations for up to 5% of their verified emissions annually.

3.1.3 Carbon Emissions Verification

Key emitters are required to report their greenhouse gas emissions to the Department of Ecology and Environment of the province where each production site is located. This is in accordance with the technical standards for greenhouse gas emission accounting and reporting of the Ministry of Ecology and Environment. Companies have to keep the data used in the report for at least five years. The provincial Department of Ecology and Environment commissions a third party to verify the reports of greenhouse gas emissions by key emitters. That third party is responsible for the authenticity, integrity and accuracy of the verification results submitted.

3.1.4 Supervision, Management and Penalties

The local government department for ecology and environment at or above city level determines if and when to inspect the verification results of the greenhouse gas emission reports and trading settlements of key emitters. The department's approach is one of "random sampling, random selection of inspectors and making results public". The outcome of the inspection is submitted to the Ministry of Ecology and Environment. Should key emitters falsely report, refuse to report or conceal their emissions, the local government department should order them to fulfil their obligations and impose a fine of between RMB 10,000 and RMB 30,000. If a key emitter falsely reports or conceals its emissions, an equivalent amount is subtracted from their emissions allowance for the following year. Where key entities fail to settle their allowances in full and on time, a fine of between RMB 20,000 and RMB 30,000 is imposed.

3.2 Operation of the National ETS

The national ETS came into operation in July 2021. It covered the power sector only. By the end of the year, the ETS had traded 179 million tonnes of emissions at a value of RMB 7.661 billion.[5] This was only 3.98% of the 4.5 billion tonnes of allowances issued for that period, far below the amount traded by the EU ETS in that same period.

3.2.1 China's National ETS Carbon Price Was Lower than Those of Major Carbon Markets Globally

The carbon price and trading volumes in China have been low since ETS operations began. However, whenever the compliance period deadline approaches, the carbon price rises and trading volumes increase significantly. On the whole, the national carbon market mirrors the seven regional ETS pilots. The carbon price of China's national ETS is relatively low compared with those of the major carbon markets globally (see Fig. 2). The low carbon price means that the ETS has failed to provide effective incentives for enterprises to reduce their emissions and invest in or innovate low-carbon technologies.

3.2.2 Trading Volumes Surged when the Compliance Period Deadline Approached

The daily trading volumes of the national carbon market were low in the early stages of operation, mostly less than 500,000 tonnes. From October 2021 however, the daily trading volume gradually increased. In November and December 2021, it reached between 5 million and 10 million tonnes, far above the level of other months. The largest daily trading volume of the year was on December 16 at 2.048 million tonnes (see Fig. 3). Overall, the trading volumes across the country were largest before the deadline of the compliance period, which means the market needs to be improved.

[5] Report on the First Compliance Period of the National Carbon Emissions Trading Market.

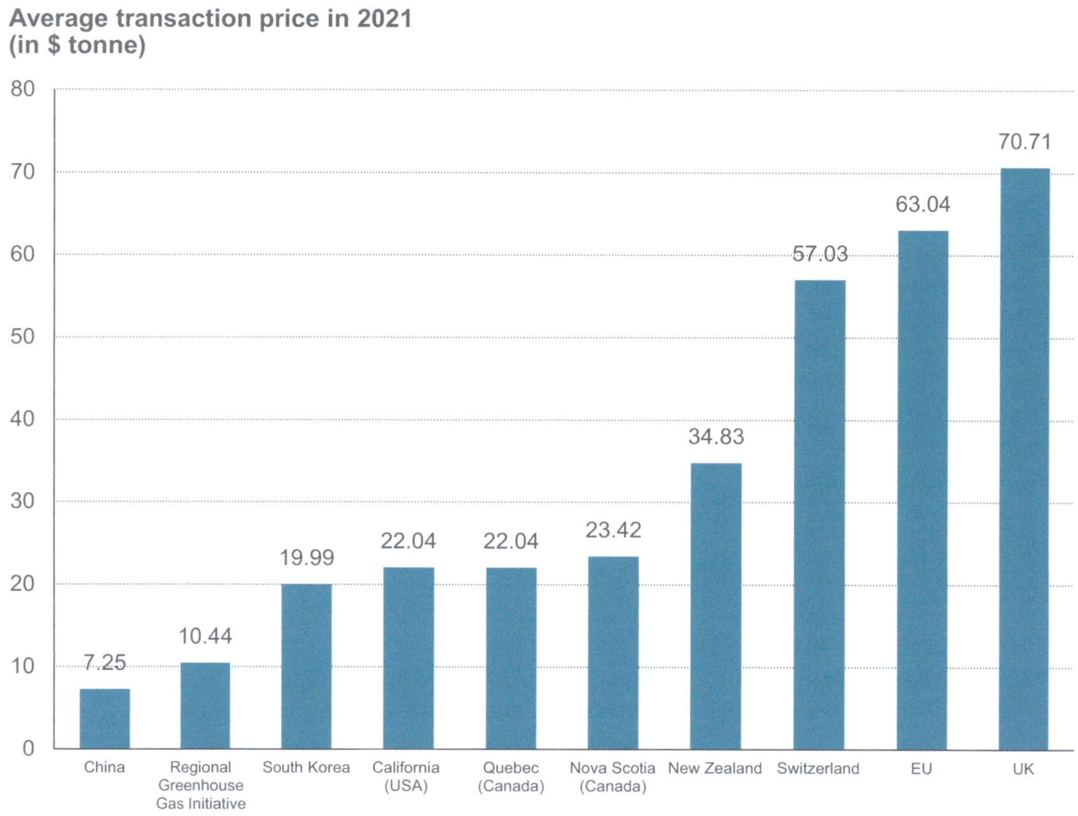

Fig. 2 Carbon price comparison between major global carbon markets in 2021. *Source* International Carbon Action Partnership Allowance Price Explorer

3.2.3 The National ETS Carbon Price Was Relatively Stable in 2021

On July 16, 2021, the first day the national ETS was in operation, the average daily transaction price was RMB 51.2 per tonne, which dipped to RMB 41.9 per tonne on July 28 before rebounding. The price began to decline slowly in mid-August before stabilising at about RMB 40 per tonne from September to early November. As the compliance period deadline approached in mid-December, the carbon price rose abruptly and continued to rise until the end of the year. On the whole, the average daily transaction price was between RMB 40 and RMB 60 per tonne, which was fairly stable (see Fig. 4).

3.2.4 Bulk Trading by Agreement Is the Main Trading Activity

As stated in the Announcement on Matters Related to National Carbon Emissions Trading, carbon emission allowances can be transferred in two ways: listed trading by agreement and bulk trading by agreement. If the transaction volume is less than 100,000 tonnes, listed trading by agreement is used; if the transaction volume is more than 100,000 tonnes, bulk trading by agreement is used. The transaction price of listed trading by agreement is determined in the range of ± 10% of the closing price of the previous trading day, and that of bulk trading by agreement is determined in the range of ± 30% of the closing price of the previous trading day. On the whole, bulk trading by agreement predominates.

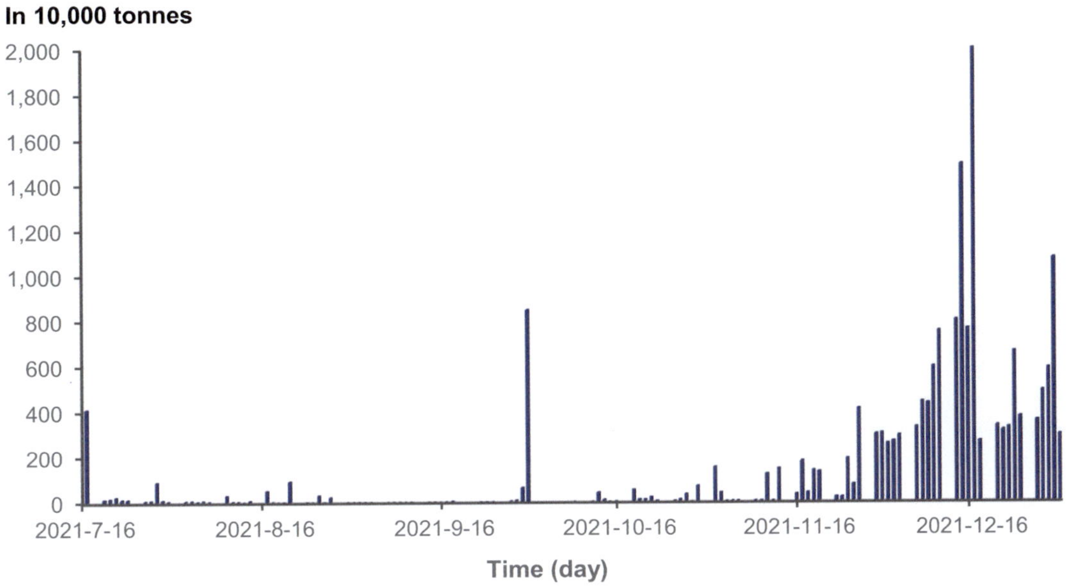

Fig. 3 Distribution of the daily trading volume of China's national carbon market in 2021. *Source* The authors

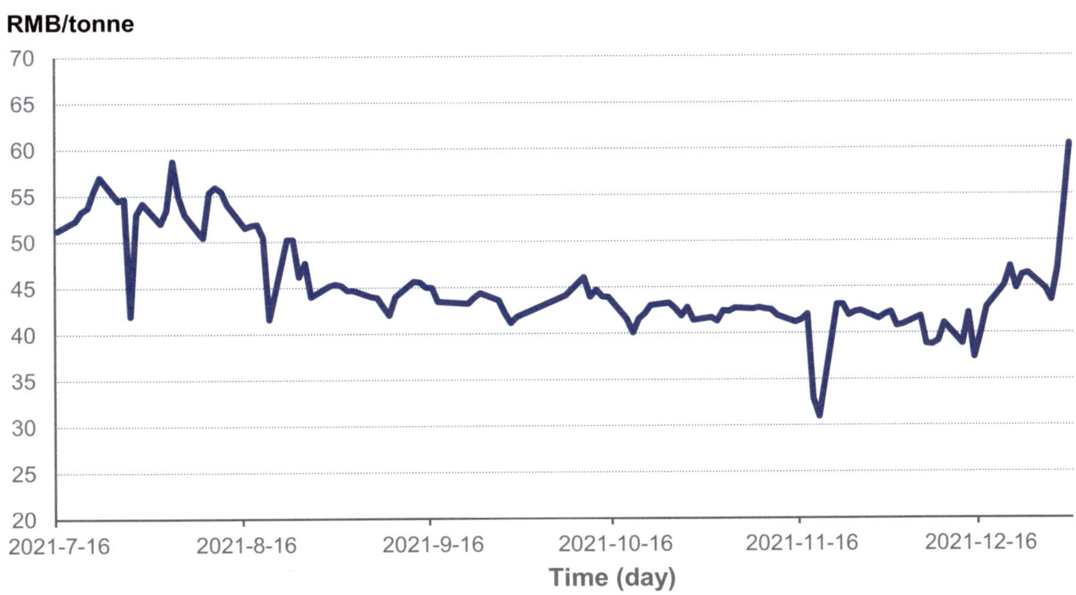

Fig. 4 The average daily transaction price in China's national carbon market, July–December, 2021. *Source* The authors

Table 1 Online trading results of China's seven ETS pilots in 2021

Pilot market	Total trading volume (million tonnes)	Total trading value (RMB million)	Highest transaction price (RMB/tonne)	Lowest transaction price (RMB/tonne)	Average annual transaction price (RMB/tonne)
Beijing	1.8658	135.4426	107.26	24.00	72.59
Tianjin	4.9487	150.5238	34.10	21.00	30.42
Shanghai	1.2743	51.3328	43.66	38.00	40.28
Hubei	3.2903	86.4890	45.47	26.56	35.01
Guangdong	27.5058	1,048.7120	57.70	24.61	38.13
Shenzhen	5.9929	67.6610	36.32	3.12	11.29
Chongqing	1.1506	37.0707	40.00	20.41	32.22

Source Wind Database

As of December 31, 2021, the cumulative volume of bulk trading by agreement accounted for 83%, much higher than that of listed trading by agreement at 17%.[6]

4 China's ETS Pilot Schemes

Compared with the national ETS which opened in 2021, the seven pilot carbon markets in Beijing, Tianjin, Shanghai, Chongqing, Hubei, Guangdong and Shenzhen have been in operation for many years. Each pilot differs in terms of carbon market elements such as the sectors it covers, the methods it uses to set and allocate allowances, and the compliance mechanisms it adopts. In general, however, each ETS pilot faces challenges such as low levels of trading, low liquidity and low carbon prices.

4.1 Operational Profile of the Pilots

By 2021, the five pilot markets in Beijing, Tianjin, Shanghai, Guangdong and Shenzhen had completed eight compliance periods and those in Hubei and Chongqing had completed seven. Table 1 shows the online trading results of each pilot carbon market in 2021. Beijing had the highest average annual transaction price and the most price volatility; Shenzhen registered the lowest average annual transaction price; Guangdong was the most active in terms of both total trading volume and total trading value; and Chongqing was the least active market, with the lowest total trading volume and transaction value.

[6] Shanghai Environment and Energy Exchange.

4.2 Assessment of Emission Reduction Effects

Although the theoretical framework of an ETS is very clear and has been adopted in many countries and regions, its outcome depends on the conditions prevailing in each country or region. As forerunners of China's national ETS, the pilot carbon markets, which have been in operation for many years, provide experience and lessons for the national carbon market, given that they operate against a similar institutional background and in the same national conditions. In addition to the mechanism design of the ETS, the issue of whether the carbon market is well co-ordinated with other energy efficiency and emission reduction policies is also an important factor influencing the functioning of the carbon market.

4.2.1 The Carbon Market Can Do More to Increase Abatements

In theory, the price of emission allowances is determined by supply and demand. For a long time, however, China has adopted command-and-control policies to achieve energy savings and emission reductions. Although this led to a large volume of reductions, it increased the supply of and reduced demand for emission allowances, resulting in a low carbon price. In fact, the performance of the pilot carbon markets indicates that there is much room for improvement in terms of carbon price, transaction concentration and trading volumes. There are two reasons for this. The first is that the pilot carbon markets did not set a more stringent abatement target than that of the command-and-control approach; the second is that with identical abatement targets, the conventional command-and-control approach weakened the functions of the carbon market.

Command-and-control policies generally target sectors with large emissions. These sectors are covered not only by command-and-control policies, but by the pilot carbon markets as well. The effects of command-and-control policies on reducing emissions in those sectors can be determined by comparing them with non-covered sectors that have similar emissions in the same region. Research has shown that the effects of the pilot carbon markets on reducing emissions were not significantly different from those of the command-and-control policies. This implies there is a need to improve the abatement effects of the carbon markets.

The pilot carbon markets were designed from the outset to lower the large social costs of command-and-control policies. The markets provide an alternative for entities with high emission reduction costs, which can fulfil their abatement tasks by purchasing carbon emission allowances. The theory is that under the carbon trading mechanism, production activities are transferred from less efficient companies to those that are more efficient; this drives down the overall abatement cost, compared with the one-size-fits-all approach of command and control. However, research shows that the carbon market plays an insignificant role in reducing emission reduction costs. On the whole, the potential of the carbon market to reduce carbon emissions and abatement costs remains largely unlocked.

4.2.2 Command-and-Control Policies Remain at the Core of the Climate Change Governance System

In the Notice of the State Council on Issuing a Comprehensive Work Plan for Energy Conservation and Emissions Reduction, published in 2007, energy conservation and emission reduction targets were included for the first time in the evaluation of local economic and social development, and the so-called "One Vote Veto" was introduced to assign responsibility for achieving environmental targets to local government officials. The central government set a CO_2 emissions intensity target in the 12th Five-year Plan for 2011–15 and a total energy consumption target in the 13th Five-year Plan for 2016–20. Under this target-oriented responsibility system, the targets and tasks of energy conservation and carbon reduction were broken down and assigned to provinces, prefecture-level cities, counties and key enterprises. This responsibility system has played an important role in achieving China's energy conservation and carbon emission reduction targets. The command-and-control policies, however, impede the control of energy consumption on the demand side. The pressure on local governments to achieve their abatement targets easily results in costly and inefficient one-size-fits-all abatement measures.

For this reason, the Chinese government decided to give a decisive role to the market by creating a trading system for carbon emission allowances. In 2011, the National Development and Reform Commission issued a notice to develop pilot carbon emission trading markets. Subsequently, seven ETS pilot markets were launched successively between the second half of 2013 and the first half of 2014. They are the forerunners of the national ETS, which opened in 2021. Once the pilot carbon markets were in operation, the constraints imposed by the target-oriented responsibility system on energy conservation and carbon emissions reduction were relaxed. However, government at all levels in the pilot markets still have to complete the tasks and targets assigned to them by the higher-level government, and their performance in energy conservation and carbon abatement is still assessed in the same way as it was before the carbon markets opened. Because of this, local government still chooses the command-and-control policies they are familiar with to achieve their targets, especially when they have an inadequate understanding of the carbon market and the energy saving and abatement results achieved by the market are not clear.

4.2.3 The Potential of Carbon Markets Needs to Be Unlocked

Research shows that the sectors that took part in the carbon trading markets significantly reduced their CO_2 emissions and emissions intensity compared with those regions and sectors that did not take part in the pilot markets. The results also show that even if the pilot carbon market had yet to start trading, the sectors in the pilot market reduced their emissions compared with other non-pilot regions and sectors. As companies have no motivation to reduce their carbon emissions voluntarily in the absence of policy incentives, the reason for these emission reductions is likely to be the command-and-control policies under the target-oriented responsibility system. Both approaches—the carbon market and command and control—are working with the same sectors. Because local government continues to assess sector performance and uncertainties remain around the effects of the carbon market on emissions reduction, governments at all levels continue to use the command-and-control approach they are familiar with to achieve their targets.

5 Draw on International Experience to Improve China's National ETS

5.1 The Allocation of Allowances Method

China's national ETS uses mainly the free allocation of emission allowances method, which determines the number of free allowances per tonne of output produced that each site is entitled to (the more emission-efficient installations will produce more output per tonne of CO_2 and therefore will require fewer allowances than less efficient sites). This approach is not conducive to the discovery of the true price of carbon and does not sufficiently spur sites to innovate and deploy emission-reducing technologies. An auction mechanism would make the allocation of allowances fairer. Such a mechanism would not erode the overall competitiveness of the sector, but would spur the transition to renewable and low-carbon power sources. Considering the sector's high emissions, major carbon markets such as the EU ETS, the California Cap-and-Trade Program and the Regional Greenhouse Gas Initiative in the USA have included the power sector in their market coverage. In other major carbon markets, the cost of carbon in the power sector is passed on to downstream customers through electricity prices.

In October 2021, the Notice of the National Development and Reform Commission on Further Deepening the Market-oriented Reform of On-grid Tariffs for Coal Power was published. The notice states that as of October 15, 2021, the on-grid price of all coal power output is deregulated and should be traded on the electricity market. It also states that the market-based transaction price of coal power should not fluctuate more than 20%, but that this limit on price fluctuation does not apply to energy-intensive enterprises. This will potentially increase the market transaction price for energy-intensive enterprises, pushing them to save energy and reduce emissions. The national carbon market is designed not only to ensure that it is cost effective under current conditions, but that it also has the flexibility to respond to the ongoing reforms in the energy and electricity markets. Based on the experience of developed economies, free allowances should be phased out.

5.2 The Price Stability Mechanism

The Administrative Rules for Carbon Emissions Trading (Tentative) states that the Ministry of

Ecology and Environment can create a market regulation and protection mechanism to ensure the sound development of the national carbon market. When abnormal fluctuations in the transaction price trigger the mechanism, the Ministry of Ecology and Environment can take measures such as making market operations public or adjusting the China Certified Emission Reduction scheme to achieve the market regulation needed. However, the national ETS has not specified the upper and lower limits of the carbon price nor clarified the circumstances in which the regulation and protection mechanism can be triggered. Only by incentivising investment in renewable and low-carbon energy sources can emissions intensity in the power sector be significantly reduced. Theoretically, a high carbon price can spur market players to take action to reduce emissions and invest in low-carbon technologies. However, there is a risk that this will weaken industry competitiveness. A key criterion to judge whether the carbon price is reasonable is whether a balance can be struck between the emission reductions achieved and the economic impact on the industry.

The initial design of the EU ETS did not include a price stability mechanism. When the EU carbon market opened, the first two phases included a cocktail of measures to revert the economic downturn, implement renewable energy development plans and policies to promote energy efficiency improvements, all of which resulted in surplus allowances. Due to inadequate preparations for a possible collapse of the carbon market, the carbon price remained low, which inactivated the ability of the carbon market to incentivise investments in low-carbon energy. The only thing an excessively low carbon price will achieve is to increase the operational costs of companies; it will not affect their investment decisions. Major carbon markets such as the EU ETS, the Regional Greenhouse Gas Initiative in the USA, the California Cap-and-Trade Program and the New Zealand Emissions Trading Scheme stabilise carbon prices by monitoring and adjusting the number of allowances in the market, setting upper and lower price limits or setting a reserve price for allowance auctions. China's ETS should learn from these overseas carbon markets by adopting some or all of these measures.

5.3 The Allowance Offset Mechanism

To achieve carbon neutrality, better technologies are required to reduce carbon emissions and to capture and store carbon. The carbon market should play a more important role in the development of carbon capture, utilisation and storage (CCUS) and in spurring technological development. Under the current mechanisms of the national ETS, power generation companies are not allowed to offset the reductions they achieve through CCUS in their carbon emissions accounting. The mechanisms needed to offset CCUS projects are simply not in place. The national ETS needs to do more to promote CCUS.

Europe and the USA are focusing their economic policies on how to sustain prosperity and continue their leadership of the global economy. In the EU, the ETS is the main channel to realise the abatement value of CCUS projects. China should also refine the design of its ETS by including, for example, the reductions achieved by CCUS projects in the carbon emissions data of enterprises or by incorporating CCUS projects in the China Certified Emission Reduction scheme. This would allow such reductions to be part of the allowance offset mechanism. It would enable CCUS projects to obtain carbon credits and encourage technological innovation.

5.4 The Supervision and Penalty Mechanism

Compared with other carbon markets, China's ETS imposes lighter penalties on companies that do not meet their compliance responsibilities. The Administrative Measures for Carbon Emission Permit Trading (Tentative) states that should a key entity fail to settle its allowance in full and on time, a fine of between RMB 20,000 and RMB 30,000 will be imposed. If it still fails to fulfil its responsibilities, the Ministry of Ecology and Environment will reduce the entity's allowance for the coming year by an equivalent amount. To ease the compliance burden on key emitters that do not have sufficient allowances, the Implementation Plan for Setting Allowances and Allocations for National Carbon Emissions Trading in the Power Sector of 2019–20 set a cap on an entity's obligations. It amounts to the entity's free allowances plus 20% of its verified emissions.

In the EU ETS, any entity that fails to make up for its annual excess emissions before the end of the compliance period is fined. Initially, the penalty was EUR 40 per excess tonne of CO_2 emissions, which later rose to EUR 100. In the event of non-compliance of an airline, for example, the host country of the airline may request the European Commission to shut down the airline's operations. Although member states differ in the severity of their penalties, they all impose harsh fines. In Germany, the penalty for excess emissions increases year by year, beginning with EUR 40 per tonne of excess emissions in the first year, rising to EUR 100 in the second year and EUR 200 in the third year. Should an entity fail to surrender its allowances on time, a penalty of up to EUR 500,000 is imposed. In Spain and Ireland, carbon-emitting facilities that start operating without emission allowances are subject to fines of EUR 2 million and EUR 15 million respectively. China's ETS should set higher penalties to create strong constraints on the behaviour of enterprises.

6 Improve the Carbon Market to Promote Renewable Electricity

Changing the power supply mix and increasing the amount of renewable power in the mix is an important way for China to achieve its dual-carbon goal. In the early stage of renewable energy development, China effectively increased the installed capacity of renewable power through subsidy policies. In 2019, the Notice of the National Development and Reform Commission and the National Energy Administration on Establishing and Improving the Renewable Portfolio Standard Scheme was published. The scheme assigned renewable electricity consumption targets to provincial administrations, thus effectively ensuring its use. Power grid companies, independent power sellers and power distribution and sales companies within the provincial administration were the first to be assigned renewable and non-hydro targets in their annual electricity sales. The second group to be assigned renewable targets for their electricity use were the power users who purchase electricity from the thermal power wholesale market and those enterprises with captive power plants. Together, these targets constitute significant policy support for renewable electricity consumption.

The Chinese government has continually made efforts to phase out subsidies for onshore wind power, solar photovoltaic power plants and industrial and commercial distributed solar power and achieve grid parity for these renewable technologies. In 2017, the voluntary trading platform for green electricity certificates was officially launched in Beijing to reward the environmental benefits of renewable energy and reduce renewable energy subsidies. However, due to insufficient incentives for participants to purchase certificates, the voluntary trading market remains inactive. What makes the carbon market advantageous is that it reduces CO_2 emissions by influencing demand-side decisions and enabling energy users to lower their CO_2 emissions by increasing their use of green electricity. The ETS, the green electricity market and the green electricity certificate scheme have all helped optimise the power supply mix in China.

6.1 Green Electricity Certificates

6.1.1 Background

The green electricity certificate (GEC) is an electronic voucher with a unique identification code issued by the state administration for each megawatt-hour of grid-connected non-hydro renewable electricity generated by power producers. It is a unique credential for confirming the generation and use of green electricity and an instrument to discover the environmental and social value of renewable electricity through the market. The certificate is a means for consumers to declare their use of green electricity and demonstrate a reduction in their carbon footprint. The GEC scheme connects the supply and demand sides of green electricity and reduces the cost of its use for consumers.

One reason for the introduction of the GEC trading scheme is that there was a large discrepancy in renewable energy subsidies and that subsidies were not paid on time. Even though renewable energy has grown strongly in China, it has been held back by delayed subsidy payments. Another reason lies in weak incentives for users to consume green electricity and a general lack of awareness on the importance of using green electricity. China has focused its policy measures on the supply side to promote the development of renewable energy and optimise the energy mix. This has resulted in insufficient incentives on the demand side, which is hindering the progress of renewable energy. In the GEC voluntary trading scheme, the purchase price of renewable power is not to be higher than the subsidy for the power output on the certificate. After wind and solar power companies sell their GECs, the electricity output they sell does not qualify for a subsidy and the GECs purchased by users are not to be sold again.

6.1.2 Policies Related to the GEC Scheme

Following the launch of the GEC platform in 2017, the Notice of the National Development and Reform Commission and the National Energy Administration on Actively Promoting the Work Related to Grid Parity of Wind and Solar Power Without Subsidies was published in 2019. The notice enabled projects that have achieved grid parity and have low on-grid prices to earn reasonable revenues through GEC transactions. In 2019, the Notice of the National Development and Reform Commission and the National Energy Administration on Establishing and Improving the Renewable Portfolio Standard Scheme was published, which states that the GEC scheme should be used as a means for market participants to fulfil their renewable energy use targets.

In 2020, the Ministry of Finance, the National Development and Reform Commission, and the National Energy Administration jointly issued Several Opinions on Promoting the Healthy Development of Non-hydro Renewable Energy Generation, which marked the full-scale roll-out of the GEC trading scheme. The GEC scheme started operations on January 1, 2021, although it did not register significant activity. GEC trading-related polices are shown in Table 2.

The Renewable Portfolio Standard scheme aims to ensure that renewable electricity will be used, but it does not reward users sufficiently for the environmental value of their using renewable energy. GECs monetise the environmental value of renewable electricity. Only when electricity users are required to pay for the negative environmental externalities of the electricity they use can the GEC market play its role effectively. In the Implementation Plan for Promoting Green Consumption, jointly issued by central departments including the National Development and Reform Commission in 2022, leading industrial enterprises, large state-owned enterprises and international companies were encouraged to consume green electricity and play an exemplary role to others. The plan also states that efforts will be made to gradually increase the use of green electricity in regions with a large number of export-oriented enterprises and a strong economic base. Persuasive measures to encourage energy-intensive companies to use green electricity will be intensified. Local governments already have the mandate to set a minimum proportion of green electricity use for these enterprises. This has increased demand for green electricity and driven up the price of green electricity and transaction volumes, although the environmental value of a considerable amount of subsidised and non-subsidised green electricity still needs to be paid for. Because the carbon market impacts the demand for energy, a strong link between green electricity, the GEC market

Table 2 GEC trading-related policies

Date	Policy	Main requirements of the policy
February 2016	Guidelines of the National Energy Administration on the Establishment of Renewable Energy Development and Use Target Guiding System	Set clear targets for renewable energy development and use, clarify the responsibilities and obligations of renewable energy development and use, research and improve the system and mechanisms for promoting renewable energy development and use, and issue renewable electricity certificates to operators (including individuals) of non-hydro renewable power
January 2017	Notice of the National Development and Reform Commission, the Ministry of Finance, and the National Energy Administration on the Trial Operation of the Issuance and Voluntary Trading System of Renewable Energy-generated Green Electricity Certificates	Establish a voluntary trading system of renewable energy-generated green electricity certificates and carry out trial issuance of GECs for onshore wind power and solar power producers (excluding distributed solar power facilities). The GEC purchase price shall not be higher than the renewable energy tariff surcharge fund subsidy for the power output corresponding to the certificate. After wind and solar power companies sell their GECs, the corresponding electricity output no longer has access to the national renewable energy tariff surcharge fund subsidy
July 2017	Launch of the voluntary trading platform for GECs	The voluntary trading platform for GECs was officially launched in Beijing
November 2018	Notice of the National Development and Reform Commission and the National Energy Administration on Implementing the Renewable Portfolio Standard Scheme (Draft for Comment)	Solicit opinions on the Renewable Portfolio Standard scheme
May 2019	Notice of the National Development and Reform Commission and the National Energy Administration on Establishing and Improving the Renewable Portfolio Standard Scheme	Set a target for the proportion of renewable electricity in total electricity use. The targeted proportion will be determined by the provinces. Each provincial government will formulate an implementation plan for renewable electricity consumption to determine the consumption target, the mechanisms for achieving the target and how to assess the participants under the Renewable Portfolio Standard scheme. Electricity sellers and users should jointly assume responsibility for reaching the consumption targets. The amount of renewable electricity consumed beyond the target will not be included in the energy consumption assessment of the 13th Five-year Plan period (2016–20)
January 2020	Several Opinions on Promoting Healthy Development of Non-hydro Renewable Energy Generation	As of January 1, 2021, GEC trading under the Renewable Portfolio Standard scheme applies. Efforts will be made to link coal power producers' priority in power generation and enterprises' priority for coal imports with the GEC scheme to expand the GEC market; and multiple market-oriented steps will be taken to promote GEC transactions. Power producers will earn revenues through GEC transactions to replace government subsidies. Quota subsidies will be granted to help households install solar panels to become self-sufficient in power generation and feed surplus electricity into the grid

(continued)

Table 2 (continued)

Date	Policy	Main requirements of the policy
May 2020	Notice of the National Development and Reform Commission and the National Energy Administration on Issuing the 2020 Mandatory Proportion of Renewable Electricity in All Provinces	Each province was assigned a mandatory proportion of renewable energy in its electricity mix. The proportion assigned was based on estimates provided by each province
September 2020	Supplementary Notice on Matters Related to the Several Opinions on Promoting the Healthy Development of Non-hydro Renewable Energy Generation	For those projects on the list of subsidised renewable energy generation installations, their power output that lies within the scope of the life-cycle subsidy will be subsidised as per the following formula: Benchmark on-grid tariff for renewable energy (including the on-grid price determined through bidding and other competitive methods - (minus) the local benchmark on-grid price for coal power / (divided by) 1 + (plus) the applicable value-added tax rate. That portion of their power output that lies outside the scope of the life-cycle subsidy will no longer have access to central subsidy funds, but will participate in GEC trading. Wind and solar power projects that have operated for 20 years from the date of their grid connection, and bioenergy projects that have run for 15 years from the date of their grid connection, will no longer have access to central subsidy funds, whether their power output has reached the electricity output level eligible for a life-cycle subsidy or not, but will participate in GEC trading
January 2021	Trading Rules for Excess Consumption of Renewable Electricity at Beijing Power Exchange Center (Tentative)	The renewable electricity consumption of a market participant will be synchronised—from the power exchange platform to the certificate trading system—and stored in the participant's consumption account. Blockchain technology will be used to generate one excess renewable electricity consumption certificate for every 1 MWh of excess electricity consumed. The excess renewable electricity consumption certificate cannot be included in the following year's mandatory proportion of renewable electricity consumption allocated to the participant. The National Renewable Energy Information Management Centre will synchronise GEC transactions with the certificate trading system of the Beijing Power Exchange Center each month. One GEC equals 1 MWh of non-hydro renewable electricity consumption, which is included in the calculation of the mandatory proportion but which cannot be traded in the excess renewable electricity certificate market. GECs cannot be included in the following year's mandatory proportion of renewable electricity consumption allocated to the participant
May 2022	Implementation Rules for Green Electricity at Beijing Power Exchange Center	This defines the green electricity trading rules. The environmental rights of green electricity producers are transferred to power users, and green electricity certificates are provided to power users who purchase green electricity products

Source Produced by the authors

and the carbon market is needed to help green electricity to grow and relieve the pressure on subsidy funds.

6.1.3 Participants in the GEC Market

There are three main types of participant in the GEC market. The first includes onshore wind power and solar power projects (excluding distributed solar power facilities) that are eligible for national renewable energy tariff surcharge subsidies and for which GECs are issued. Subsidised renewable energy producers can obtain one GEC for each megawatt-hour of non-hydro renewable electricity production they connect to the grid at parity, which they can then sell on the GEC market. The GEC sale price is not to be higher than the renewable energy tariff surcharge fund subsidy for the equivalent power output. Once a power producer has sold a GEC, the power output is no longer eligible for the national renewable energy tariff surcharge fund subsidy.

The second type of participant includes non-subsidised green electricity projects that have achieved grid parity. Power producers can obtain one GEC for each megawatt-hour of power produced.

The third category includes entities to which GECs are automatically issued and transferred in green electricity transactions. The National Renewable Energy Information Management Centre issues and transfers GECs to the relevant market entities, based on green electricity transaction contracts and settlement information.

The participation of wind and solar power output that is not included in the scope of the national renewable energy tariff surcharge fund subsidy policy (new energy without subsidies) has priority in green electricity trading, while wind and solar power output that is included in the scope of the policy (subsidised new energy) can voluntarily participate in green electricity trading. In principle, the price of green electricity should fully reflect the value and environmental benefits of green electricity. As stated in the Rules for Issuance and Voluntary Trading of Green Electricity Certificates (Tentative), the purchaser cannot sell the GEC after purchase, thereby preventing secondary trading of GECs.

6.1.4 Status of the GEC Scheme

On July 1, 2017, China's voluntary trading platform for GECs came into operation in Beijing. As of June 16, 2022, a total of 44.295 million GECs had been issued across China, of which 74.8% were for wind power and 25.2% for solar power. In terms of geographical distribution, the wind power projects were located in 26 provinces, the top three of which were Hebei, Xinjiang and Jilin; the solar power projects were located in 24 provinces, most of which were in Heilongjiang, Shandong and Liaoning. A total of 8.408 million GECs were listed and 2.386 million GECs were purchased, which accounted for 5.39% of the total GECs issued and 28.4% of the total GECs listed. Of the GECs purchased, 78,700 had subsidies (3.3%), 1.8983 million had no subsidy (79.6%) and 408,700 were associated with green electricity trading (17.1%). The number of GECs issued to subsidised wind power projects was 29.0158 million, of which 4.3487 million were listed for trading. 78,700 GECs for subsidised wind power projects were traded, which was 0.27% of the number issued and 1.81% of those listed. The number of GECs issued to subsidised solar power projects was 5.1519 million, of which 364,200 were listed for trading and 182 were traded.[7] The pressure on subsidy funds remained high. As shown in Fig. 5, GECs for wind power projects made up the largest share of both the number of GECs issued and listed, while the GECs for solar power were largest among those traded. The GECs for subsidised green power projects dominated the number of GECs issued and listed, while the GECs traded were mainly those for non-subsidised green power projects. Wind power dominated the number of issued, listed and traded GECs for subsidised green power projects.

[7] China Trading Platform for Green Electricity Certificates.

6.2 Effects of the Carbon Market on Green Electricity

6.2.1 The Influence of China's ETS on Electricity Prices

In a perfectly competitive electricity market, a clear carbon price will theoretically reduce emissions by the following five levers. First, a price on carbon makes low-carbon electricity more cost competitive and encourages generators to shift from fossil fuels to low-carbon alternatives in the power generation mix (supply-side incentive). Second, a carbon price drives up the price of thermal power, prompting consumers to reduce electricity consumption or switch to clean electricity (demand-side incentive). Third, in a well-functioning carbon market, power plants with a low emissions intensity are more profitable, which incentivises investment in low-carbon technologies (investment incentive). Fourth, a carbon price narrows the marginal profit of high-carbon assets and forces their exit (incentive for accelerated exit). Fifth, all these incentives work together to promote investment in new products and technologies (innovation incentive).

The relationship between ETS and electricity price regulation is shown in Fig. 6. The incentive for end consumers to reduce their emissions depends on electricity tariffs and the tariff structure. In the best case, tariffs reflect the marginal cost to generators and an ETS can transmit a price signal to end consumers via the carbon price. Under price regulation, however, where little or no pass-through occurs, the incentive to reduce electricity consumption or switch to less carbon-intensive goods and services does not arise. As stated in the Notice of the National Development and Reform Commission on Further Deepening the Market-oriented Reform of On-grid Tariffs for Coal Power, the on-grid price of all coal power output should be deregulated in an orderly manner, and on-grid pricing should be formed through market transactions in line with the principle of "benchmark price + fluctuation". The current requirements that the market transaction price of coal power should not rise by more than 10% or decline by more than 15% will be increased in both instances to no more than 20%. The market transaction price of energy-intensive enterprises is not subject to the 20% rise limit, nor are electricity spot prices. Residential and agricultural electricity prices should be kept stable. Under the current electricity pricing system, the ETS can affect the relative cost of thermal power and renewable energy through carbon price signals, thus influencing the power source mix. However, as the electricity spot markets are still under development and the carbon price in the national carbon market is low, the ability of the national ETS to transform the power supply structure and encourage clean and efficient dispatch is limited.

In the future, when the spot markets and the medium- and long-term market within the unified national electricity market system are integrated and complementary, and when the carbon price under China's national ETS reaches a high level, the carbon market will play a more prominent role in transforming the power generation mix and clean electricity dispatch. Market-based electricity prices apply to energy-intensive enterprises; it allows power generators to pass the carbon price on smoothly to energy-intensive electricity consumers. For other industrial enterprises, however, the cap on electricity price increases makes it harder for generators to unload the carbon price onto the consumer. It is not possible to pass the carbon price onto agricultural and residential consumers due to the single-part tariff that applies to that category. In general, a low carbon price means the ETS has a very limited ability to transform the power generation mix and change consumer behaviour in the short run. In the long term, increases in the net present value of low-carbon investments will incentivise enterprises to invest in low-carbon alternatives; high-carbon assets will be forced out due to their declining profitability; and greater efforts will be made to introduce new products and technologies. A realistic carbon price is also necessary for the long-term credibility of the national ETS.

Whenever power generators can pass the carbon price on to electricity users, the free allocation of allowances can lead to thermal power plants making a windfall gain. The size of

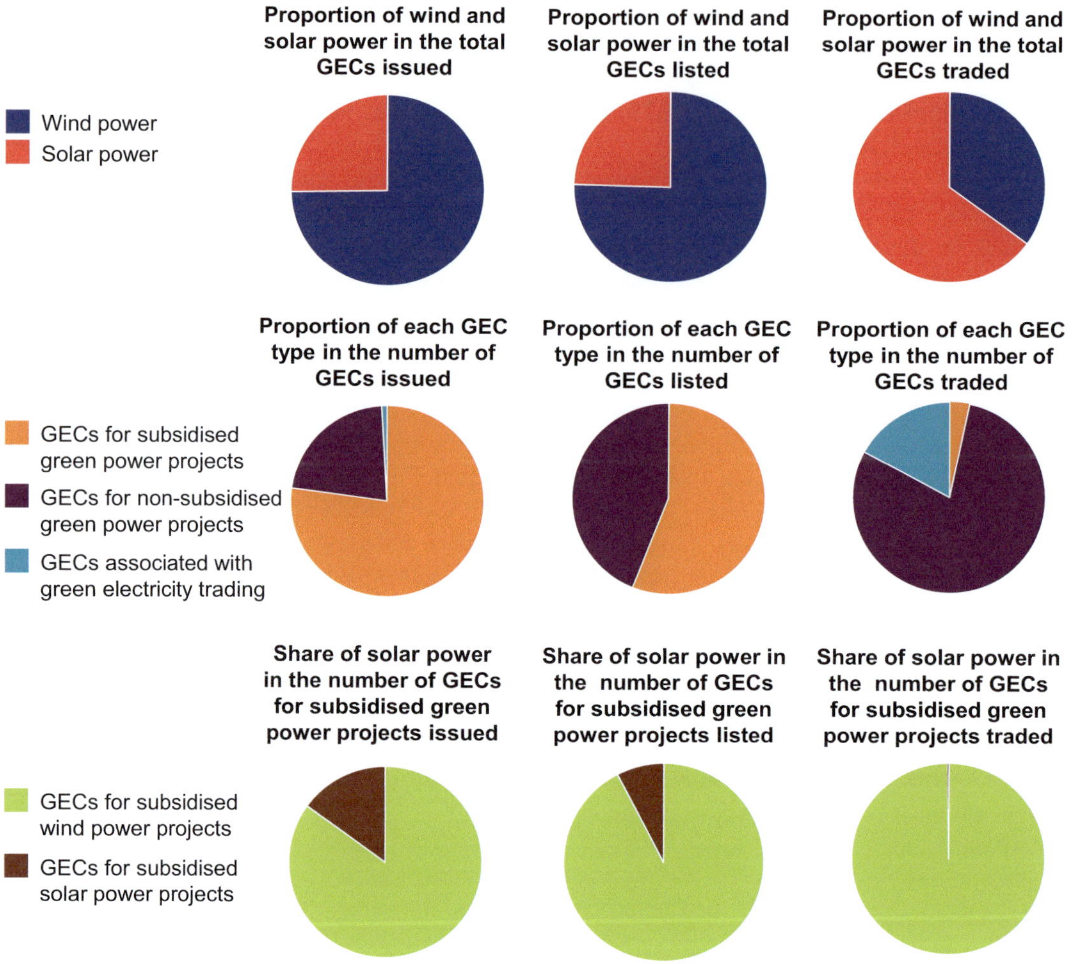

Fig. 5 GECs issued, listed and traded. *Note* The data are cumulative for the number of GECs issued, listed and traded. GECs = green electricity certificates. *Source* Calculated by the authors using data available at China Trading Platform for Green Electricity Certificates

the gain depends on the capability of the plants to transfer the carbon price; enterprises with a strong market position are more likely to pass the cost of carbon onto upstream and downstream users and treat the allowances they are given as a revenue stream. Such windfalls can have an adverse effect on the low-carbon investment and exit decisions of thermal power plants and undermine the efficiency of the ETS. In addition, the free allocation of allowances to thermal power plants will have a second distorting effect. Relative to short-term compliance requirements, thermal power plants that have been allocated a large number of free allowances are less likely to participate in carbon emissions trading. As a result, the carbon market will have low liquidity and companies that need allowances will have no channel to purchase them. This market distortion hampers the discovery of the true price of carbon and weakens the carbon market's ability to provide a reasonable price signal that promotes abatement, investment and innovation. A gradual shift from free allocation to the auctioning of allowances is necessary to enable the carbon market to fulfil its role of promoting low-carbon energy.

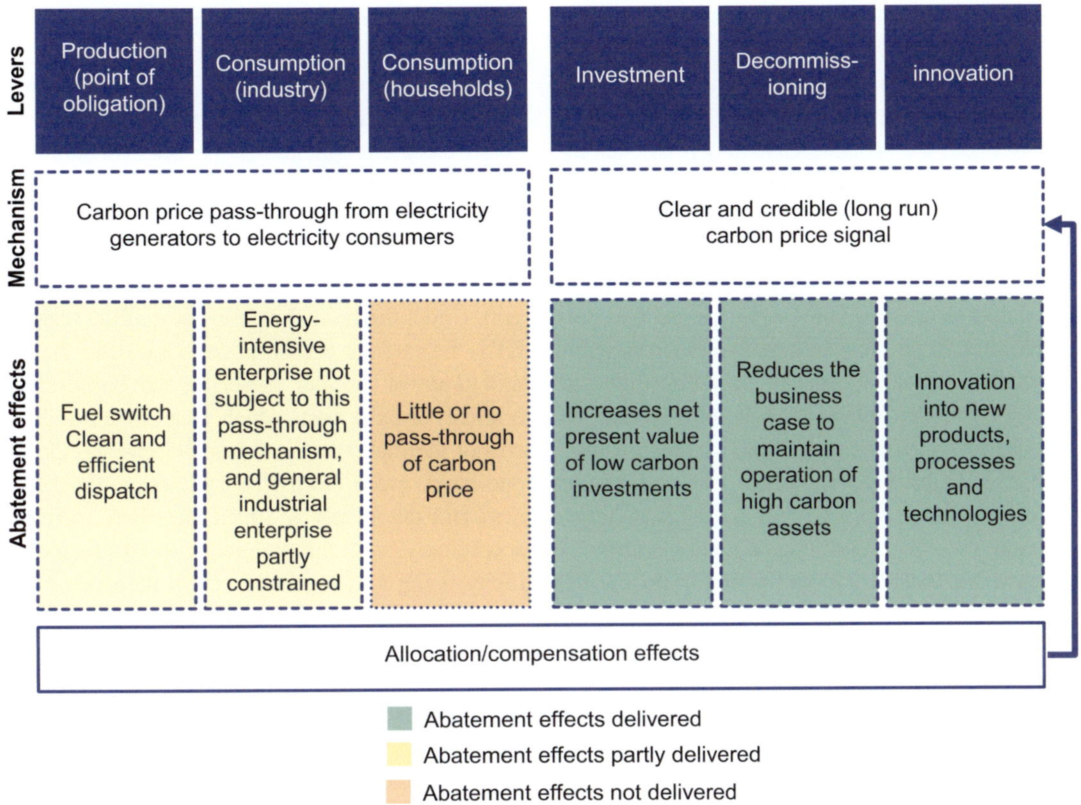

Fig. 6 Relationship between the ETS and electricity price regulation. *Source* International Carbon Action Partnership, Emissions Trading and Electricity Sector Regulation, 2018

6.2.2 The Effects of China's National ETS on Green Electricity

China's intensity-based ETS with free allocation of emission allowances covers thermal power plants only. Thermal power generators with an emissions intensity higher than the benchmarks can purchase surplus allowances only from thermal power generators that have a lower emissions intensity than the benchmarks. Renewable energy plants cannot take part in the ETS except through the very limited route of the China Certified Emission Reduction scheme. In theory, the ETS promotes the development of renewable energy by increasing the cost of thermal power. However, China's ETS has a low carbon price and therefore plays a limited role in driving change in the cost of thermal power and renewable energy. Moreover, the intensity-based free allocation of emission allowances approach of the national ETS sets an emissions intensity benchmark for thermal power plants only; renewables do not participate in the allowances, so enterprises cannot avoid an allowance deficit by increasing their renewable energy generation. Independent renewable power producers do not benefit from the national ETS. The main function of the national ETS is to reduce the emissions intensity of the thermal power plants it covers; it has a very limited effect on transforming the power generation mix. Figure 7 shows the interactions between the national ETS, the GEC market, the electricity market and the Renewable Portfolio Standard scheme.

The Renewable Portfolio Standard is a key instrument to promote the energy transition and increase the use of renewable power. As stated in the Notice of the National Development and Reform Commission and the National Energy Administration on Establishing and Improving the Renewable Portfolio Standard Scheme of 2019,

GECs are a supplementary means to meet the targets for renewable energy use, in addition to actual consumption and the purchase of surplus allowances. In 2021, China's GEC market began trading under the scheme. Essentially, the scheme is designed to increase the use of renewable energy, but it does not reward the environmental and social value of using renewables.

GECs monetise the environmental value of renewable electricity. Only when electricity users are required to pay for the negative environmental externalities of the electricity they consume can the GEC market play its role effectively. As required by the Implementation Plan for Promoting Green Consumption of 2022, the constraints on energy-intensive companies to use green electricity will be intensified and local government will have the authority to set the minimum amount of green electricity use for these enterprises. The mandatory proportion of green electricity consumption set for energy-intensive enterprises is expected to create a boom in the GEC market and relieve the pressure on subsidies. The policy instrument of setting the minimum proportion of green electricity consumption for enterprises provides incentives only with regard to societal responsibility and penalties for non-compliance. The requirement that GECs cannot be traded after purchase limits their financial value and does not give enterprises the incentive of trading and making a profit from certificates. In short, the environmental value of green electricity is not reflected in the price and the incentives for voluntary purchase of GECs are insufficient.

6.3 The Links Between the National ETS and GECs

Theoretically, there should be no difference between the purchase of green electricity direct from the grid and the purchase of GECs. The cost to enterprises should be the same. This section will not distinguish between the two options, but will discuss green electricity by its two attributes: its value as a type of energy and its environmental value. The energy attribute price of green electricity should be determined by supply and demand in the electricity market and the competitiveness of different power sources.

As shown in Fig. 8, in the case of grid parity or competitive bidding, the energy attribute price of green electricity (including the price of auxiliary services) is generally the same as the on-grid price of thermal power. That portion of green power with a marginal cost less than or equal to the on-grid price of thermal power can be fed into the grid. Under the unified carbon price of the national ETS, the implicit carbon price is added to the on-grid price of thermal power on the basis of the original variable cost, which changes the relative cost of thermal power and green electricity. On the one hand, green electricity improves competitiveness and the supply of green electricity increases accordingly, with the marginal cost of green electricity rising to a higher level. On the other hand, the competitiveness of thermal power is weakened and supply decreases accordingly, which means a lower marginal cost when supply and demand are in equilibrium. Different emission coefficients are set for green electricity and the power purchased from the grid to reflect the actual emissions trajectory. Power users can use GECs to offset the CO_2 emissions reduction that corresponds with their electricity use. In the GEC and carbon markets, the linkage mechanism of using GECs to offset CO_2 emissions ensures that the price of green electricity and thermal power are equal at equilibrium, as expressed in the formula: Marginal cost of green electricity + GEC price = Marginal cost of thermal power excluding the carbon price + Implicit carbon price of thermal power.

When a carbon price is in place (as shown in Fig. 9), the GEC price is lower than the implicit carbon price when it is assumed that the green electricity and thermal power markets are cleared and that there is no obvious technological progress to explain a new equilibrium. In fact, when the supply of green electricity exceeds its flow into the grid and new technologies reduce the cost of green electricity, the GEC price may be higher than the implicit carbon price of the same amount of thermal power at a new equilibrium. If a grid emissions factor of 1.12 tonnes of CO_2 per megawatt-hour is used and if the average carbon price of RMB 50 per tonne of CO_2 in the national

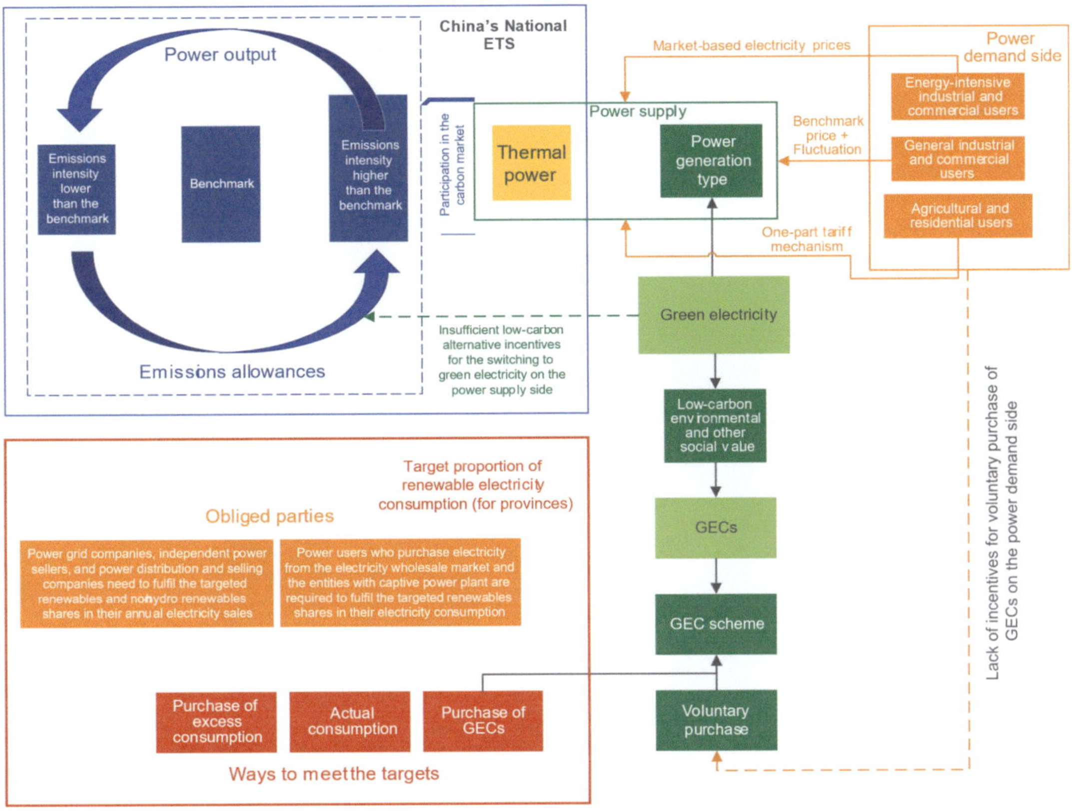

Fig. 7 Interactions between the current ETS, GEC market, electricity market and the Renewable Portfolio Standard scheme. *Source* Produced by the authors

carbon market in 2021 is also used, the implicit carbon price of 1 megawatt-hour of thermal power will be RMB 56. This should be lower than the GEC price because the green electricity market (which is not the same as the green electricity trading market, which refers to the market for all green electricity forms) is far from clearance. Moreover, the national ETS is still in the early stage of improvement. As the national ETS gradually matures, mechanisms like the free allocation of emission allowances will determine both the carbon price and the GEC price.

6.4 Policy Recommendations for Linking China's ETS with Green Electricity

At present, China's national ETS covers only the thermal power sector, so it cannot be linked with the GEC scheme to promote green electricity consumption. The national ETS should cover other sectors as soon as possible to promote the development of the GEC and green electricity markets. GECs reflect the environmental value of green electricity, and the purpose of the GEC market is to discover this environmental value. The environmental value of GEC includes lower CO_2 emissions compared to thermal power. The link between the GEC scheme and the national carbon market will encourage enterprises to make energy use decisions that favour the development of renewables. This will help to break down the inter-provincial barriers to electricity trading, establish a unified national electricity market and encourage renewable energy investment on the demand side.

The link between the national carbon market and the GEC scheme will pose new challenges for the free allocation of emission allowances and

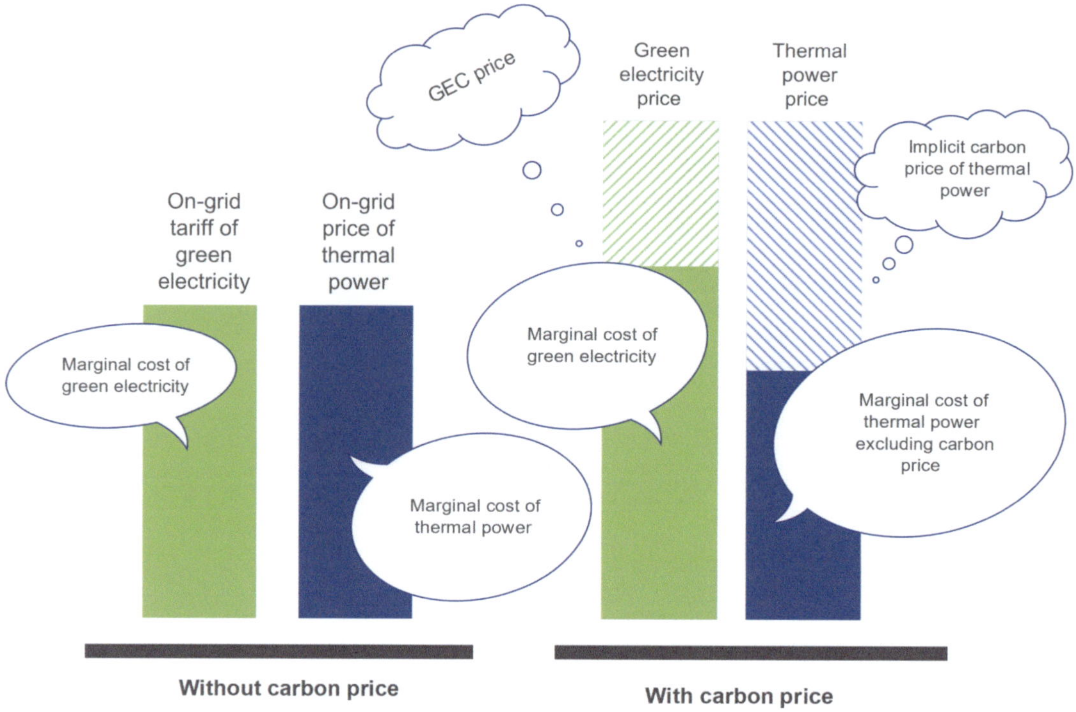

Fig. 8 Analysis of the green electricity and thermal power prices. *Source* Produced by the authors

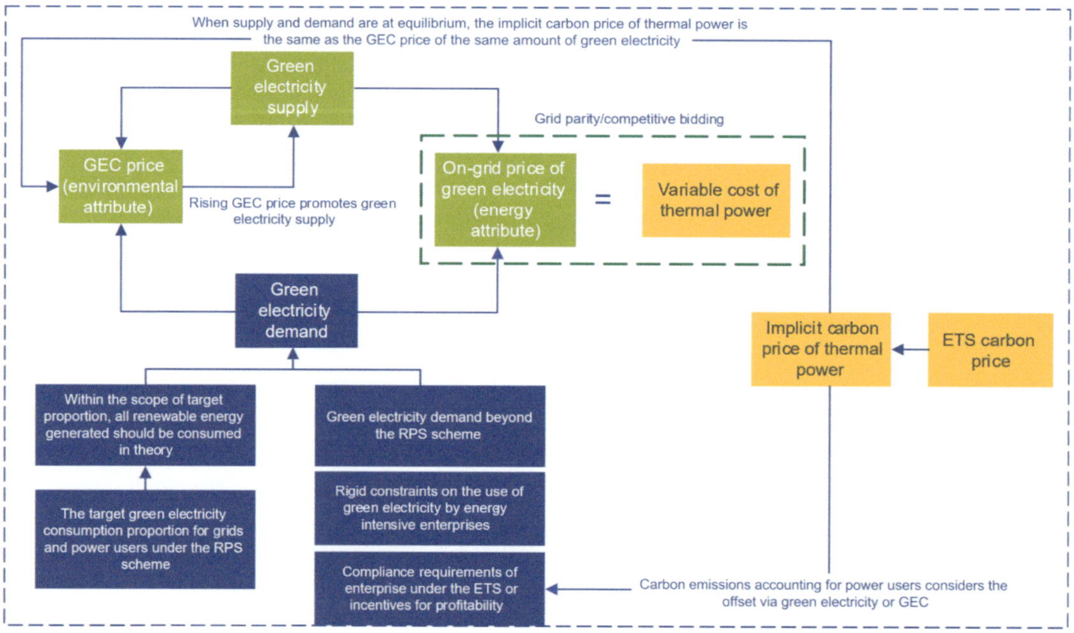

Fig. 9 The supply and demand relationship of green electricity under the mechanism that links the ETS with the GEC scheme. *Source* Produced by the authors

other mechanisms. For enterprises, carbon emissions accounting is where the carbon market interplays with the GEC scheme and the green electricity market to the greatest extent. If an enterprise purchases GECs or green electricity, it means that the enterprise has paid the full cost of green electricity consumption and that the amount of electricity concerned should be excluded from the company's CO_2 emission calculations. China's ETS currently has no provisions to account for CO_2 emissions from enterprises that have purchased green electricity.

As the national ETS gradually expands to cover energy-intensive industries outside the power sector, more enterprises will be subject to the mechanisms of the carbon market. The links between the carbon market and the GEC scheme and green electricity market will therefore play a crucial role in controlling emissions from companies and transforming the power supply mix in China. In the pilot carbon markets, electricity pricing regulations prevent power generators from passing the carbon price onto power consumers. In view of this, enterprises outside the power sector could be incentivised to reduce their power consumption if they were allowed to include both direct and indirect emissions from the electricity and heat they use in their accounting.

As market-based electricity pricing gradually progresses, the cost of carbon paid by power producers will be passed onto power users through market-based tariffs. This means that power users have already paid for the implicit carbon emissions in their electricity consumption, so indirect emissions should not be included in the carbon emission calculations of enterprises outside the power sector. In this case, the carbon emissions from power users with GECs should be offset for the amount that corresponds to those GECs, but the offsets should not exceed the companies' total electricity consumption. Too many GEC offset credits will erode the carbon market's ability to stimulate innovation and the uptake of new technologies across all industries. Enterprises that purchase green electricity should only include direct emissions in their carbon emissions accounting. Similarly, when emission allowances are determined for the various sectors, the CO_2 reductions achieved through higher green electricity consumption under the Renewable Portfolio Standard should also be excluded. Otherwise, there is a risk that excessive emission allowances will disrupt the carbon market and harm the long-term development of green electricity.

7 Improve China's Carbon Market and Response to the CBAM

In recent years, the price of emission allowances in the EU ETS has risen as a result of policies tightening the supply of allowances. In 2021, the European Commission adopted a package of proposals, including extending the coverage of the ETS and establishing the carbon border adjustment mechanism (CBAM). The CBAM is intended to be linked with the EU ETS. EU importers are required to purchase, from the EU member state they import from, CBAM certificates that are equivalent to the carbon content of the goods imported and at a price determined by the auction of allowances under the EU ETS. The CBAM also proposes an exemption policy for those non-EU producers that can prove that a carbon price has already been paid during the production of the goods imported. For enterprises covered by China's ETS, the carbon price they paid during the production of the goods they export to the EU may be deducted from the CBAM tariff.

7.1 China Should Make Abatement Costs Visible

China has long used command-and-control policies to achieve its targets for energy savings and carbon emission reductions. As a result, the cost of abatement is not visible. China should promote the ETS and allow market-based mechanisms to play a greater role in resource allocation to make the abatement costs hidden in traditional policies visible. This is one way to respond to the EU's CBAM and potential carbon tariff initiatives by other countries and regions. China should: (i) co-ordinate the assessment requirements under the Renewable Portfolio

Standard and the development requirements of the ETS; (ii) include the CO_2 emission reductions represented by the purchase and sale of allowances in the calculation of total CO_2 emissions to promote the inter-regional flow of abatements, thus making the abatement costs paid by Chinese participants visible; and (iii) ensure that the various policy objectives reinforce each other and deliver complementary benefits.

7.2 The Effects of Overlapping Policies Should Be Taken into Account when Setting Emission Allowances for China's National ETS

The power sector is a large energy consumer and inevitably is the main target of traditional command-and-control policies. When setting emission allowances for the sector, the abatement effects of other policies should be assessed and a sufficient quantity of allowances issued to allow the ETS to deliver cost benefits to the power market for reducing emissions.

7.3 China Should Formulate Policies to Accommodate the Interconnection of Carbon Markets Globally

At the 26th session of the United Nations Framework Convention on Climate Change (COP26) in 2021, nations concluded the negotiations on Article 6 of the Paris Agreement and reached agreement on reducing carbon emissions by market-based means. Carbon markets across the world have a strong desire to interconnect. The EU and Switzerland have linked their emission trading systems, as have California and Quebec, and four countries and seven subnational governments in the Americas jointly released the Paris Declaration on Carbon Pricing in the Americas to create a platform for carbon trading co-operation in the region. The UK and Mexico have also signalled a willingness to link with other carbon markets. Part of New Zealand's strategy to achieve its national abatement targets is to integrate with the international carbon market, in preparation for which it is adjusting its ETS mechanisms to increase similarities with other carbon markets for ease of interconnection. Pakistan is also actively developing a scheme to link with international carbon markets.

Looking ahead, national carbon trading markets will be further connected to jointly advance global abatement efforts, and more countries and regions will establish emission trading mechanisms to broaden and deepen the global carbon market. China should assess the impact of interconnected international and inter-regional carbon markets on its own ETS, explore a mechanism for connecting its ETS with other carbon markets and identify the support policies needed for such interconnections.

Chapter 16: The Benefits of New Energy Development

Shiji Gao, Jiaofeng Guo, Jifeng Li, Xue Han and Georgios Bonias

1 Introduction

China's development of new energy and optimisation of its power mix have been rapid and comprehensive. At the end of 2020, the country's installed capacity of renewable energy was 935 gigawatts, equivalent to 42.5% of its power generation capacity. That same year, China's output of renewable power reached 2.21 trillion kilowatt-hours, which was 29.1% of the country's total.[1]

All new energy technologies embrace innovation in the search for efficiency gains and cost reductions. By 2050, the system costs per kilowatt of solar, onshore wind and offshore wind power will drop by around 42%, 34% and 45% respectively, compared with 2020. By mid-century, renewable energy will account for 90% of China's power generation capacity.

Renewable energy not only enables China's power sector to transition to lower carbon, it supports the country's economic and social progress. Estimates show that the proportion of added value created by the development of non-fossil energy will gradually increase from 1.3% of GDP in 2020 to 5% in 2060. The number of direct jobs created by the new energy value chain will rise from 3.08 million in 2020 to 6.41 million in 2060. Carbon emissions will also be reduced in the power sector, as will air pollutants and water consumption.

[1] National Bureau of Statistics, National Energy Administration.

S. J. Gao (✉) · J. F. Guo · J. F. Li · X. Han
DRC Institute of Resources and Environment, Beijing, China

G. Bonias
Shell International Limited, London, UK

2 China's New Energy Industry

2.1 The Development and Use of New Energy

In the past decade, the scale of renewable energy development and use in China has continued to expand and the country's cumulative installed capacity of renewable power generation—including hydropower, wind, bioenergy and solar—has topped global rankings for years in a row. Renewable energy is becoming the main supply source for incremental energy demand, providing a solid foundation for China's economic growth. In 2020, all commercial types of renewable energy accounted for 13.8% of the country's primary energy demand, an increase of 6.5 percentage points on the 2010 level (see Fig. 1).

By the end of 2020, the installed capacity of renewable energy was 935 gigawatts (GW), which was 42.5% of China's total installed power generation capacity, an increase of 16.5

Shell International B.V. and Development Research Centre of the State Council (DRC) of People's Republic of China, *Embracing the Future, Powering Growth: An Energy System Renewed for China*, Advances in Oil and Gas Exploration & Production,
https://doi.org/10.1007/978-3-031-73075-7_17

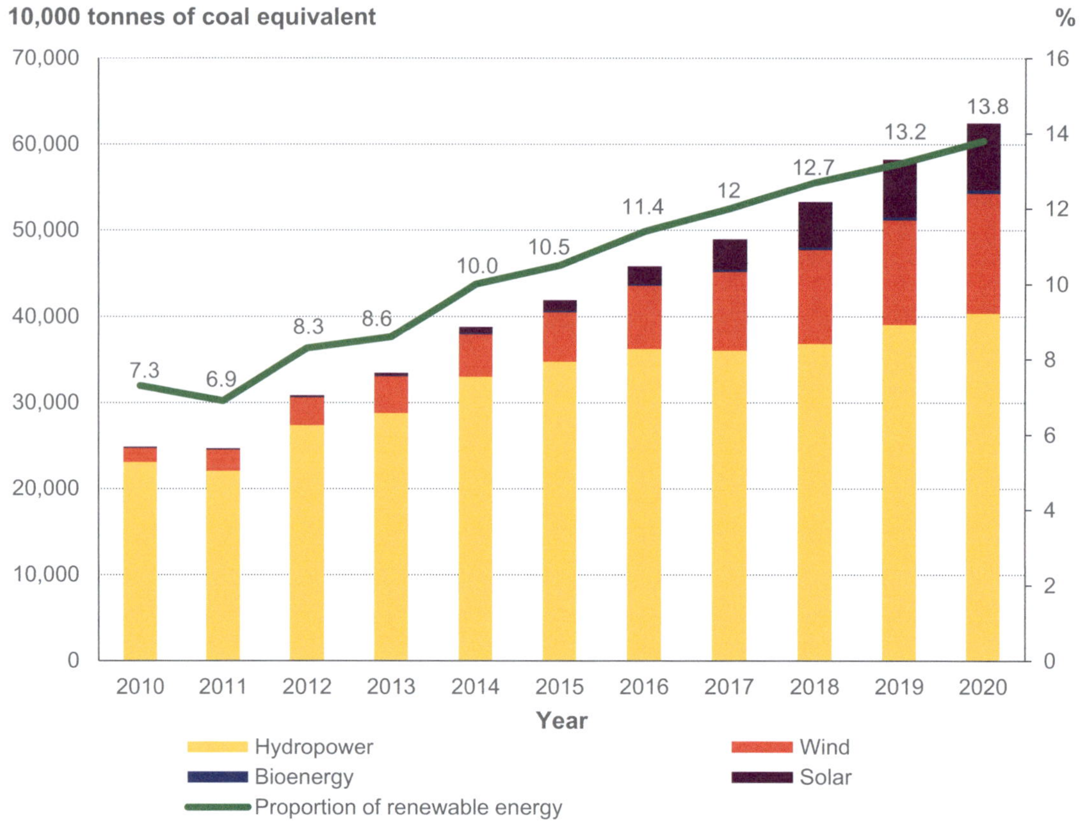

Fig. 1 Consumption of each type of renewable energy, 2010–20. *Source* National Bureau of Statistics, National Energy Administration

percentage points on 2010; and the installed capacity of hydropower, wind, solar and bioenergy was 370 GW (including 31.49 GW of pumped storage), 281 GW, 253 GW and 29.52 GW respectively (see Fig. 2).

Non-fossil electricity output continued to grow. In 2020, renewable power output reached 2.21 trillion kilowatt-hours (kWh), which was 29.1% of the country's total electricity output. Of this, hydropower, wind, solar and bioenergy accounted for 1,355.2 billion kWh, 466.5 billion kWh, 260.5 billion kWh and 132.6 billion kWh respectively (see Fig. 3).[2]

China's wind power market continues to grow rapidly. Centralised, distributed, onshore and offshore wind power projects have been developed in parallel and efficiency improvements in use have been made. Large wind power clusters have been built in an orderly manner, distributed wind energy resources have been harnessed in central and eastern China, and offshore wind power has been developed where appropriate. In 2020, the country's new installed capacity of grid-connected wind power was 71.67 GW, including 68.61 GW of onshore and 3.06 GW of offshore capacity. By the end of 2020, the cumulative installed capacity of wind power in China had risen to 281 GW, 8.5 times larger than that of 2010 and 12.8% of the country's total installed power generation capacity, which was 9.7 percentage points higher than in 2010. Onshore wind power accounted for 271 GW of capacity and offshore wind for about 9 GW. Regionally, the deployment of wind power facilities was continuously optimised. New

[2] National Bureau of Statistics, National Energy Administration.

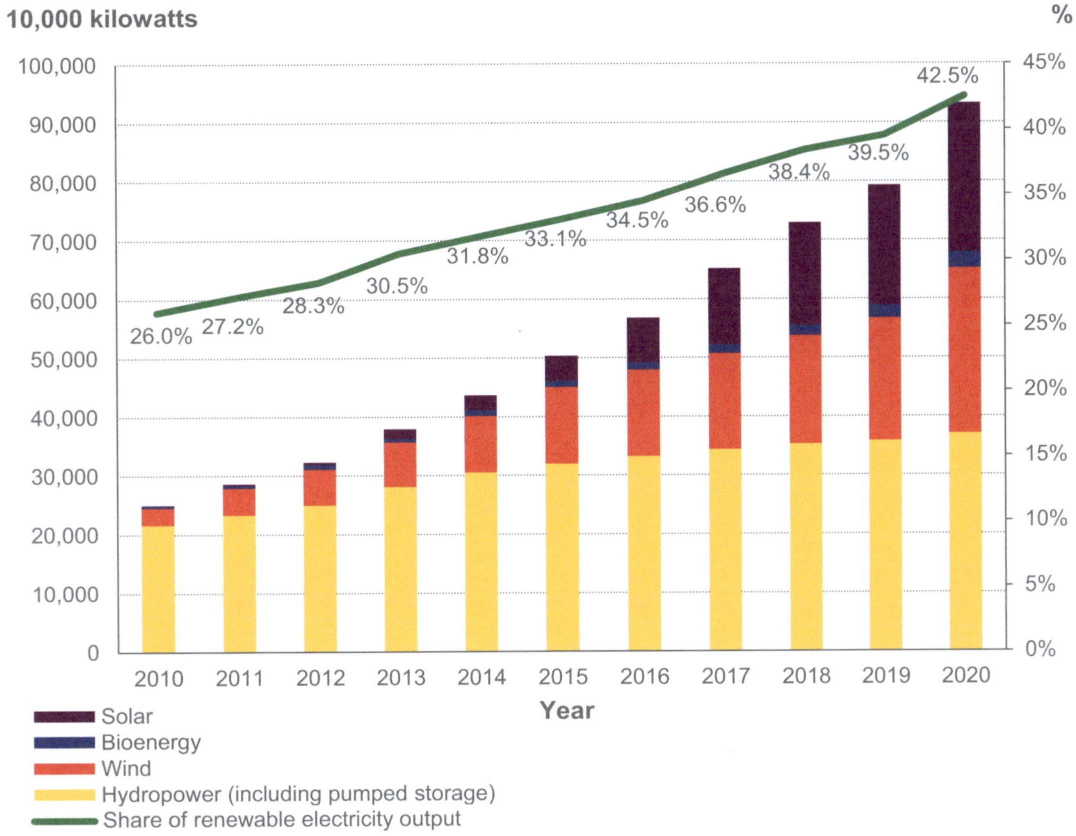

Fig. 2 Changes in the installed capacity of each type of renewable energy, 2010–20. *Source* National Energy Administration

installed capacity of wind power in the central, eastern and southern regions reached 40%, an increase of 3 percentage points on the previous year. In 2020, the average annual hours of wind power availability in China were 2,097 and the total power output was 465.5 billion kWh, which was 6.1% of the country's total electricity output, an increase of 4.9 percentage points on 2010.[3]

The large-scale development and use of wind power has propelled the wind power manufacturing sector, which has improved its innovation capability and international competitiveness. Wind power has reached grid parity; going forward, the priority will be to develop low-cost wind power projects.

The solar photovoltaic (PV) power sector has taken huge steps forward. With the implementation of the "Front Runner" programme to advance the development of solar PV and adopt competitive resource allocation, the sector has achieved technology and efficiency improvements and cost reductions. By developing in parallel centralised and distributed solar power systems and by focusing equally on both development and consumption, the scale of solar PV power capacity and use have greatly increased. China has built up global leadership in solar PV power, which is now low cost and affordable. In 2020, the country's new installed capacity of solar power was 48.2 GW, including 32.68 GW of centralised and 15.52 GW of distributed solar power plants. By the end of 2020, the cumulative installed capacity of solar power had risen to 253 GW, about 290 times larger than in 2010 and

[3] National Energy Administration (NEA), Special Committee of Wind Energy, Chinese Wind Energy Association.

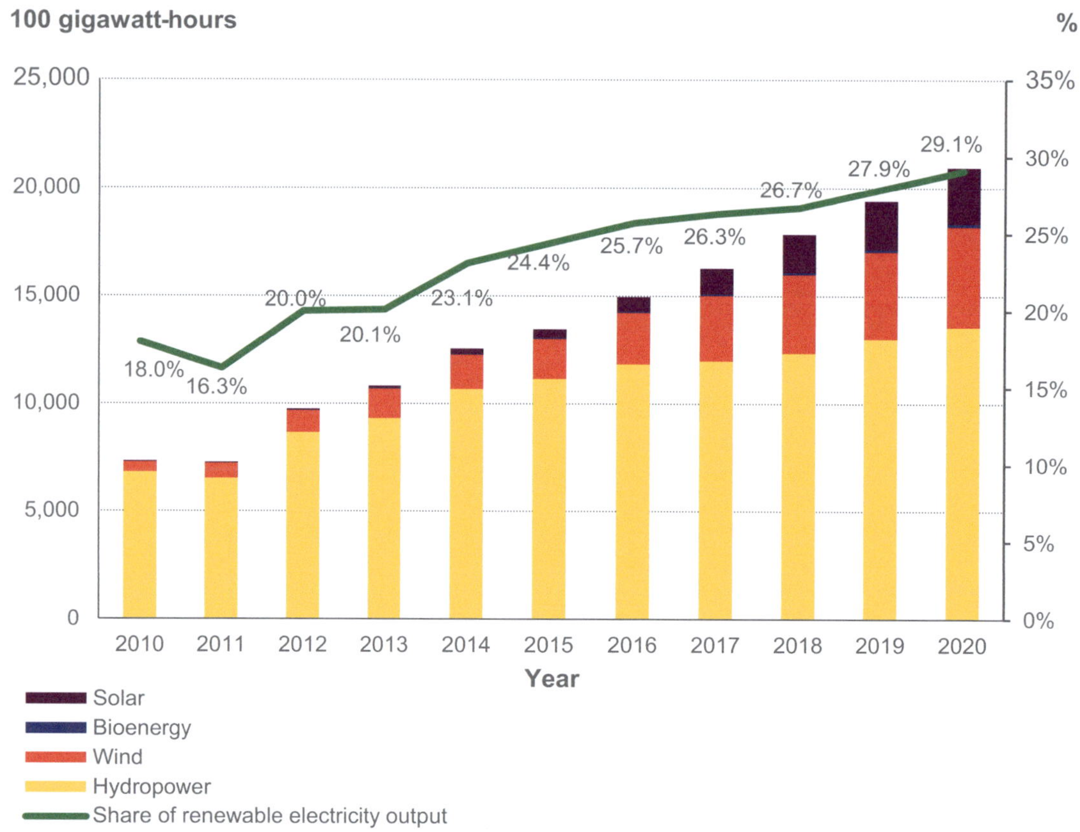

Fig. 3 Changes in the output of each type of renewable energy, 2010–20. *Source* National Energy Administration

11.5% of the country's total installed power supply capacity, an increase of 11.4 percentage points on 2010. China's solar power output was 260.5 billion kWh in 2020, 500 times more than in 2010 and 3.4% of the country's total power output, about 3.4 percentage points higher than in 2010.[4]

Based on local resource endowments, multiple measures were taken to develop biomass, geothermal, ocean energy and other renewable sources. Efforts were made to advance various types of renewable energy generation and non-electricity applications—such as bioenergy, municipal waste-to-power generation, biogas, centralised and distributed geothermal heating and geothermal power—and encourage the large-scale development and use of ocean energy technologies such as tidal energy and wave energy. In 2020, China's total bioenergy consumption was about 60 million tonnes of coal equivalent (tce), including 35 million tce for bioenergy generation, around 10 million tce for solid biomass fuels, about 4.3 million tce for liquid biofuels and about 11 million tce for biogas. Bioenergy remained the pillar of the biomass energy industry. The cumulative installed capacity of bioenergy reached 29.52 GW at the end of 2020, an increase of 5.43 GW (22.5%) year-on-year. Some 132.6 billion kWh of bioenergy were produced, up 19.4% year-on-year.[5]

[4] NEA.

[5] NEA.

2.2 The Development of New Energy Technologies

2.2.1 Technological Development in Solar PV

The PV sector is one of the few fields where China can confidently compete in the global arena. China has held the number-one position globally in terms of new installed capacity since 2013 and cumulative installed capacity since 2015. Its solar PV industry has been the biggest in the world since 2007. Moreover, China leads the way in commercial manufacturing technologies and is more agile and faster in forging a strong presence in new PV technologies. As a result, the main PV equipment is developed and manufactured in China. Of the top 10 manufacturers globally in polysilicon materials, silicon wafers, solar cells and modules, more than five in each segment are Chinese.

The interval between technology upgrades is shortening and battery and component efficiency is continually improving. After more than 10 years of innovation, China has developed world-leading commercial production technologies for PV modules. Trina Solar, Jinko Solar and LONGi have repeatedly broken the world record in commercial PV module production and module conversion efficiency. For example, the mass-produced mono-crystalline solar cell using passivated emitter and rear cell technology achieved an efficiency of 22.8%. In the projects selected for the third batch of Front Runner solar PV bases, the conversion efficiency of photovoltaic cells reached 23.85%.[6] The development of new high-efficiency PV cells and component technologies has progressed rapidly and their commercial manufacturing processes have been continuously improved. Some technologies have achieved mass production status and strong international competitiveness.

Economies of scale and technological progress have driven down costs and boosted efficiency in the PV sector. From 2007 to 2019, cost reductions in PV systems were mainly achieved in the manufacture of modules, the cost of which decreased from 60% to 38.5% of total system costs.[7] As growth in global PV demand and improvements in solar PV generating efficiency slowed, technological advance became the main driver of cost reductions.

China's solar PV industry is the biggest in the world, and its leading companies continue to expand their production capacity. In 2013, China's PV manufacturing sector embarked on large-scale production and rapid growth. A complete photovoltaic product manufacturing chain has taken shape (which China dominates globally) in all four main production segments: silicon materials, silicon wafers, solar cells and modules. In 2020, the production of polysilicon materials, silicon wafers, solar cells and modules in China reached 392,000 tonnes, 161.3 GW, 134.8 GW and 124.6 GW, which represented a year-on-year increase of 14.6%, 19.7%, 22.2% and 26.4% respectively. Each segment contributed more than 70% of global production. Moreover, an array of companies with strong international competitiveness has been fostered. In the 2020 global sales statistics of photovoltaic products, the vast majority of the top 10 manufacturers were Chinese enterprises.

China's PV equipment sector is characterised by sustained and sound development, significantly improved technologies, a shift from low-end to high-end products, gradual improvements in product customisation and continuous enhancement of high-capacity and high-efficiency automation, all of which are fuelling the transition to intelligent manufacturing. The main equipment used in the production of polysilicon materials, silicon wafers, solar cells and modules—including the plasma-enhanced chemical vapor deposition and boron diffusion equipment used in the production of high-efficiency cells—has been developed and manufactured locally, which substantially lowers investment costs. In coming years, continuous improvements in equipment performance, single unit capacity and solar cell efficiency are expected to further reduce investment costs in each part of the PV manufacturing value chain.

[6] China Photovoltaic Industry Association.

[7] China Photovoltaic Industry Association.

2.2.2 Technological Development in Wind Power

China has mastered the core technologies of wind power and is among the global leaders in the design and manufacture of the main components of wind turbines. The country has made impressive breakthroughs in the development of turbines that operate in low wind speeds and harsh environments. According to the Global Wind Energy Council, of the top 15 global wind turbine manufacturers in 2019, eight were Chinese.

China's top three wind turbine manufacturers are among the global top 10 suppliers in new installed capacity. In 2019, 17 wind turbine manufacturers reported new installed capacity of 26.785 GW. Goldwind had new installed capacity of 8.01 GW, which was 30% of the market, followed by Envision Energy, Mingyang Smart Energy, Windey Energy Technology and Dongfang Electric Corporation. These five enterprises had a combined market share of 73.4%. The overall price of wind turbines continues to decline, which is the biggest reason for the reduction in wind power costs. In the past decade, the wind turbine price in China fell from about RMB 5,000 per kilowatt in 2010 to about RMB 3,100 per kilowatt in 2018. However, to ease their project backlogs and meet their grid connection schedules, project owners rushed to install wind turbines in 2019, which caused the price to rise to around RMB 4,000 per kilowatt. The rush cooled in mid-2020, causing the wind turbine price to gradually decrease to about RMB 3,100 per kilowatt by the end of the year. In 2021, fewer new projects started construction, causing the wind turbine price of winning bids to drop below RMB 2,000 per kilowatt.

Wind turbine capacity continues to increase. China's wind power technologies, such as low-speed wind power systems, intelligent wind turbines and smart wind farms advanced, with some reaching international standards. According to the Chinese Wind Energy Association, in 2020, the average wind turbine capacity at new wind power facilities in China was 2,668 kW, an increase of 8.7% year-on-year and 45.2% more than in 2015. In the past five years, the number of wind turbines with a capacity of 3 MW or more has increased significantly. In 2020, 5,978 new wind turbines with a capacity of 3 MW or more were installed, amounting to 20.602 GW, an increase of 184% year-on-year.

Rotor diameter and rotor swept area have also increased. In 2020, the average rotor diameter was 136 meters, which was 58 meters larger than in 2010, a 74% increase. The maximum hub height was 162 meters in 2020, an increase of 15 meters on 2019. The installed capacity of wind turbines with a hub height of 90 meters accounted for the largest proportion in 2020 at 43%, followed by turbines with a hub height of 140 meters at 13.4%. Given that the swept area is proportional to the square of the rotor diameter, thanks to longer blades, the maximum wind energy capture capacity has increased by almost 200% in the past 10 years.

2.3 The Potential of New Energy Sectors and Technologies

2.3.1 Solar PV Technologies and Their Cost Reduction Potential

Looking ahead, China's PV market will continue to make rapid progress. New technologies, new processes and new products will emerge and the efficiency of photovoltaic cells and power generation systems will continue to improve. According to projections by experts and a survey of major Chinese enterprises, the cost of solar power has the potential to drop in the near and medium terms. Turning that potential into actual cost reductions in the solar power value chain will depend mainly on the following three factors.

Improving the Conversion Efficiency of Photovoltaic Cells and Modules. Crystalline silicon photovoltaic cells are expected to achieve an efficiency gain of 2 or 3 percentage points, and the efficiency of dual-junction laminated solar cells is expected to reach 35% in time. The industry believes that thanks to efficiency gains, the non-silicon cost per watt of cell will plunge by about 40% from the current level in the medium to long term, and the capacity of PV modules will increase at a rate of no less than 5 watts per year. Coupled with improvements in the conversion efficiency of cells, the efficiency of each module will improve. The non-silicon cost per watt of module will fall to about 50% of the current level by 2030.

Reducing the Cost of PV Silicon Materials and Increasing the Utilisation Rate of Silicon. The consumption of raw materials per unit weight of polysilicon will decline, which together with the decreasing cost of production lines, improvements in equipment capacity and increasing productivity in manufacturing will gradually reduce the market price of silicon materials. When these cost reductions are combined with improvements in wafer technology, increases in wafer size and higher levels of production efficiency, the processing cost per watt of silicon wafer will decline significantly.

Reducing the Cost of Solar PV Power Plants. Research shows that the cost per unit kilowatt of non-module elements such as brackets, pile foundations, construction and land use will decrease as the cost per unit kilowatt of modules declines. The cost of electrical equipment such as inverters and transformers and the cost of connecting the plant to the power grid will also decline as the efficiency and power capacity of PV modules increase. Advances in electrical equipment manufacturing and system application technologies will also continue to drive down the technology costs of PV power generation.

The investment cost of solar power plants will decrease. Driven by the significantly lower cost of PV modules and improved conversion efficiency, the investment cost of solar power plants is projected to drop by 37% and 53% by 2035 and 2050 respectively. With technological progress and economies of scale, the price of PV modules is expected to be more than 55% and 70% lower in 2035 and 2050 respectively than in 2021. This will reduce the investment cost of solar power plants by more than 60% (see Fig. 4). Improvements in module efficiency will reduce spending on land use, brackets, electrical equipment and construction.

It should be possible to steadily improve system efficiency and increase plant availability. Although mono-facial solar modules (with the cells on one side of the panel) are mainstream, a gradual shift to bifacial solar panels (on which the solar cells are fitted to both sides of the panel) will occur. Bifacial panels are expected to increase the power output of PV modules by between 5% and 25%. Solar trackers can also help to significantly boost power output by about 20%. The efficiency of a solar power system is also affected by such factors as inverter efficiency, collector line losses and booster transformer losses. Improving the performance of each component will therefore increase the efficiency of solar power systems. At present, system efficiency is between 85% and 90%, which is expected to improve by about 5 percentage points.

Based on the above analysis, the reduction in investment cost per unit of installed capacity, along with improvements in system efficiency and reduced light attenuation will increase solar power plant availability. Solar photovoltaic power is expected to become the most cost-effective technology for new power generation capacity by 2025. The average on-grid price for power generated by new centralised solar power plants is projected to fall to about RMB 0.26 per kWh by 2025, a 20% decrease on 2020. The on-grid price in areas with excellent solar energy resources (sunshine duration of 1,800 hours) could even dip to RMB 0.19 per kWh. From 2025 onwards, the cost of solar power is expected to continue its rapid decline, with the average price of electricity generated by new solar power plants set to decrease to RMB 0.19 per kWh by 2035 and RMB 0.13 per kWh by 2060 (see Fig. 5).

2.3.2 Onshore Wind Power Technologies and Their Cost Reduction Potential

This section is based on the results of surveys of companies and interviews with experts. From the technology perspective alone, there is room to lower the cost of onshore wind power. By increasing single unit capacity, unlocking the generating potential of low wind speeds,

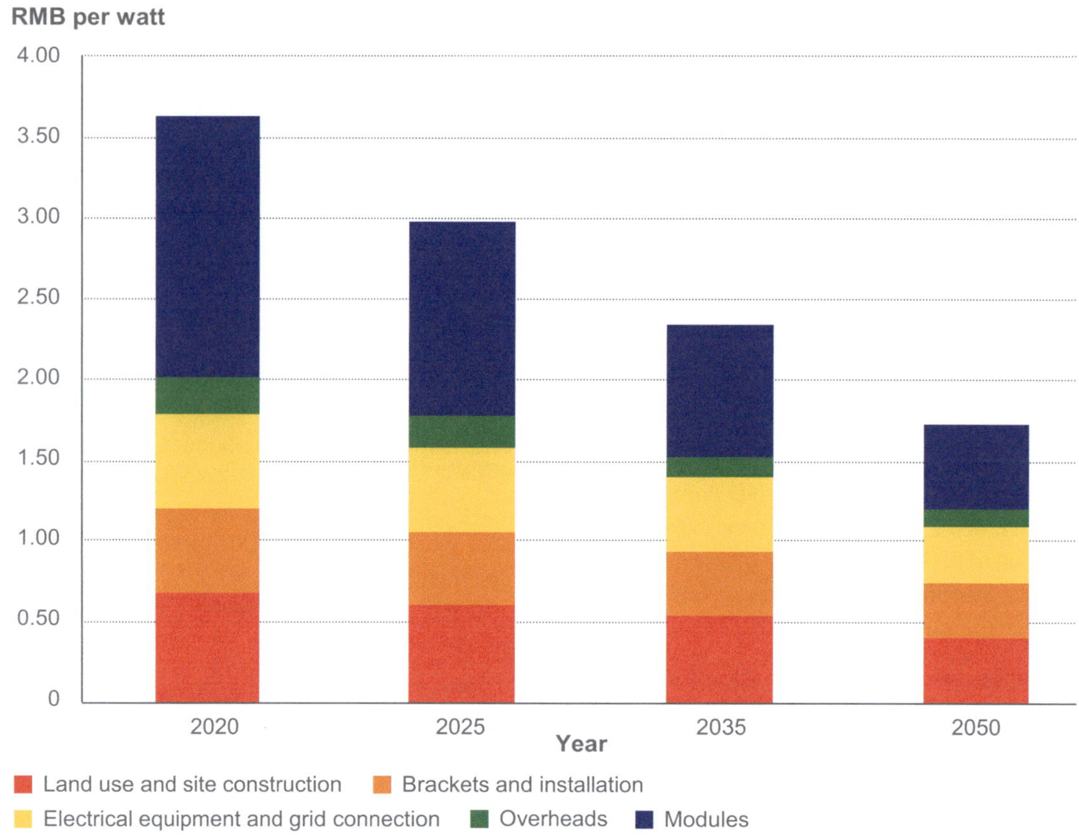

Fig. 4 Solar power plant investment costs. *Source* Survey of companies and interviews with experts

adopting advanced manufacturing technologies and boosting process efficiency, the cost per kilowatt-hour of electricity can be effectively reduced. The deployment of smart wind farms and intelligent operation and maintenance will help improve the operating efficiency and reliability of individual wind turbines and the entire wind farm, thus driving up the overall efficiency of the plant. The economies of scale enabled by the expansion of the wind power industry will also help reduce technology costs.

Due to non-technology factors like the rush to build more wind farms at the end of the 13th Five-year Plan (2016–20), the cost of investing in wind farms has been relatively high. The cost is likely to fall dramatically during the 14th Five-year Plan (2021–25) when it is expected to decrease by about RMB 1,000 per kilowatt compared with 2020. In North, north-east and north-west China and the flat terrains of the eastern–central region, the investment cost can dip to about RMB 5,500 per kilowatt, whereas in the mountainous and hilly areas of the eastern–central region and South China it can fall to around RMB 6,400 per kilowatt. The cost of wind turbines should also decline, from RMB 3,100 per kilowatt at the end of 2020 to about RMB 2,600 per kilowatt in 2025. Tower costs can be reduced by improving management practices, tower quality standards and quality control, and by using the latest lightweight materials. Construction costs can be reduced by using better technologies and deploying large-capacity turbines to lower the installation cost per kilowatt of capacity. In the medium to long term, the initial cost of investment in wind power will gradually decrease by 10% by 2035 and 20% by 2060, compared with 2025.

The operation and maintenance cost per kilowatt of installed capacity will steadily decrease.

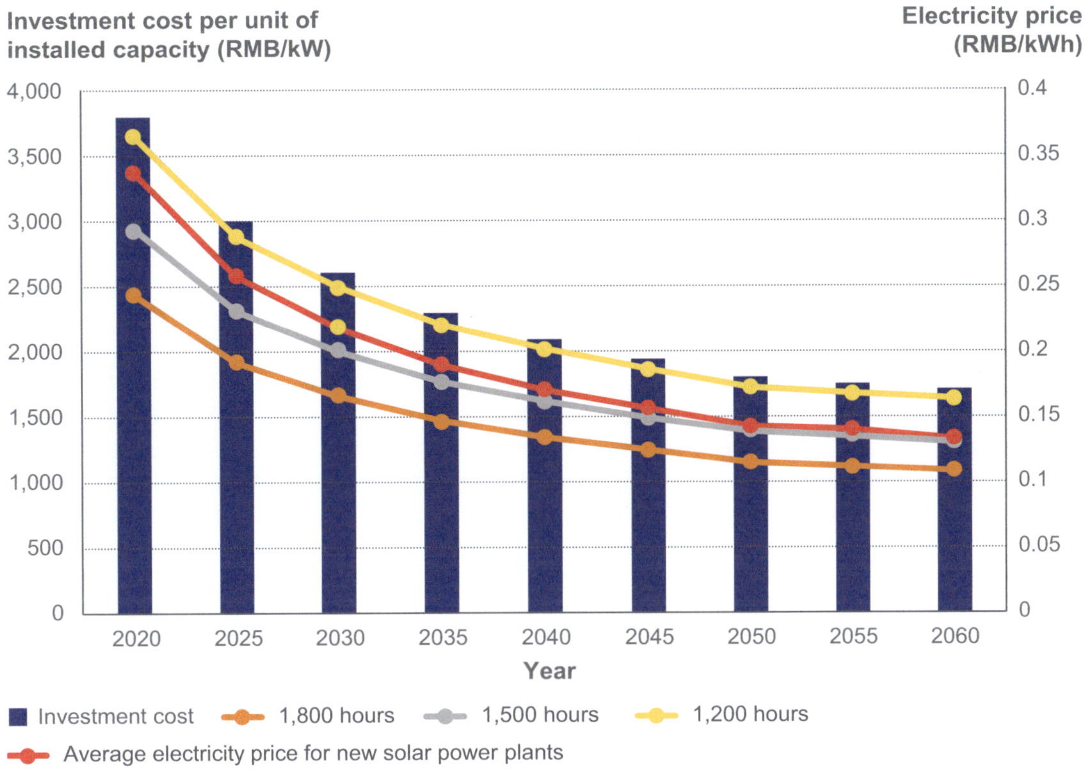

Fig. 5 Forecast unit investment cost and on-grid price for centralised solar power plants. *Note* Estimates are based on an internal rate of return on capital of 8%, an operating life of 20 years, a loan period of 15 years, a loan interest rate of 4.9%, a loan of 80% of the cost of the asset, and a residual value of fixed assets of 5%. *Source* Produced by the authors

The use of intelligent technologies in wind farm design, turbine selection and wind farm and turbine operation and maintenance will improve efficiency and availability and raise power output by between 8% and 10%. Correcting yaw misalignment and increasing the rated power of turbines can further reduce the cost per kilowatt-hour of electricity. The height of the turbine foundation can be increased and the blade design optimised to raise electricity output by 10%–15%. Improvements in wind turbine technologies, wind farm design and asset operation are expected to increase the availability of the onshore wind farm fleet by 150–200 hours in 2025 compared with 2020 and by about 200 hours and 400 hours in 2035 and 2060 respectively compared with 2025.

In addition to the lower unit investment cost and the increase in asset availability, the on-grid price for wind power will continue to decline. By 2025, the average on-grid price for onshore wind power (including tax and reasonable revenue margin) will decrease to RMB 0.31 per kWh, a decline of about 10% from 2020. The price of electricity in areas with flat terrain and high-quality wind resources (with an availability of 3,000 hours) can drop to RMB 0.23 per kWh, while in mountainous and hilly areas with average wind resources it can decrease to RMB 0.39 per kWh. In the medium to long term, the average on-grid price for onshore wind power will fall to RMB 0.25 per kWh by 2035 and RMB 0.21 per kWh by 2060, a decrease of about 20% and 30% respectively compared with 2025. The price in areas with flat terrain and high-quality wind resources will be less than RMB 0.2 per kWh in 2035 and RMB 0.16 per kWh in 2060.

2.3.3 Offshore Wind Power Technologies and Their Cost Reduction Potential

Using the learning curve method, the International Energy Agency predicted a learning rate of 7% for onshore wind power and 9% for offshore wind power by 2050. It also forecast changes in the cost of offshore wind power investment in its 2 °C scenario and Blueprint scenario. Under the 2 °C scenario, offshore wind power investment will decrease by about 25% by 2030 and by about 35% by 2050, compared with 2015 when it was $2,900 per kilowatt. Offshore wind power operating costs will be about 30% and 40% lower by 2030 and 2050 respectively, compared with 2015. The International Energy Agency also predicted there would be three main enablers of technological progress and cost reduction for offshore wind power: first, in wind power technologies, including improvements in the design of wind turbines, advanced components, reliability and testing; second, improved knowledge of wind energy characteristics, including wind energy resource assessment and wind farm site selection and the influence of external conditions on wind turbines; third, in improving the wind power supply chain, including production and installation. The International Renewable Energy Agency reported that by 2025, the levelised cost of electricity (LCOE) of offshore wind power is expected to fall to $0.12 per kWh. The main drivers of this cost reduction include technological advances in construction and installation, process optimisation, growing experience in project development and improved wind farm efficiency. The wide uptake of large wind turbines of 6 MW and more and the development of new wind turbine technologies will become the most important contributors to the reduction in LCOE of offshore wind power.

As with onshore wind power, efficiency gains are a key driver in reducing the cost of offshore wind power in China. In 2019, the new mainstream offshore wind turbine models in China had a capacity of between 4 and 6 MW, with the largest model providing a capacity of 10 MW. In the international market, the largest wind turbines that have been installed and put into operation have a capacity of more than 12 MW. In the early part of the 14th Five-year Plan (2021–25), 6–8 MW wind turbines are expected to become mainstream in new installed capacity, and wind farm capacity is expected to increase by 1–2 percentage points compared with 2020. The operating costs of offshore wind farms will also decrease. Thanks to better wind turbine reliability and advances in offshore construction equipment and technologies, operation and maintenance costs are expected to decline by about 10% between 2021 and 2025.

Considering the combined effects of reduced investment cost and improved generation efficiency, and assuming an operating life of 25 years, the average electricity price is estimated to fall to RMB 0.52 per kWh in 2025, RMB 0.40 per kWh in 2035 and RMB 0.31 per kWh in 2060 (see Table 1).

3 An Economic, Social and Environmental Impact Assessment of New Energy Development

3.1 Economic Effects

The development of new energy not only provides an important industrial foundation for China to achieve its dual-carbon goal, it also represents a

Table 1 The cost reduction potential of offshore wind power and electricity price estimates

	2020	2025	2035	2060
Initial static investment (RMB/kW)	16,500	13,500	11,500	10,000
Annual operating cost (RMB/kW)	320	300	240	200
Annual availability in hours	2,700	2,900	3,300	3,700
Electricity price when the internal rate of return on capital is 8% (RMB/kWh and an operating life of 25 years)	0.71	0.52	0.40	0.31

Source Calculated by the authors from market information and forecasts new driver of economic growth. This study draws the conclusion that the development of non-fossil energy will support China's future economic growth. Considering that the digital and green transformation is considered by many countries to be the main channel for economic and social development, renewable energy is now an indispensable component of economic growth.

In this study, the average annual installed capacity of new renewable energy is expected to be 130 GW between 2021 and 2030 and 160 GW between 2031 and 2060. Annual investment in generator sets, transmission and distribution networks, energy storage and other infrastructure will amount to about RMB 1–1.3 trillion. Annual electricity output will increase from 1.8 trillion kWh in 2020 to 5.8 trillion kWh in 2030 and 16.6 trillion kWh in 2060. Renewable energy generation and its upstream and downstream industries are expected to grow stably for a long time. Energy storage and distributed energy systems will also be developed, which will help to decarbonise the energy system and further drive economic growth. Using projections of the proportion of non-fossil energy in total power output, we have estimated the economic scale of renewable and low-carbon components in the power sector and in electrical equipment manufacturing. Figure 6 shows the preliminary results. The added value of non-fossil energy generation was RMB 1.2 trillion in 2020, which is expected to rise to RMB 3.2 trillion in 2030 and RMB 17.1 trillion in 2060. The share of added value created by non-fossil energy generation will gradually increase from 1.3% of GDP in 2020 to 2.2% in 2030 and 5% in 2060.

3.2 Social Effects (Employment)

In addition to significant beneficial impacts on climate and the environment, the implementation of China's dual-carbon policy will have economic and social implications, especially for industry and employment. On the one hand, more than half of the coal power fleet operating in 2022 will have shut down by 2060, which means some workers in the coal power and coal sectors will need to be re-employed in other industries. On the other hand, the substantial growth of new energy will create new jobs. This section analyses the effects on employment of the transition to a carbon-neutral power system in two aspects: the first is the effect on employment of solar and wind power; the second is to quantify the effect on net employment of the power system transformation.

3.2.1 A Model for Calculating Life-cycle Employment

This study identifies the effect on employment in four stages of the new energy value chain: manufacturing, installation, operation and maintenance, and recycling. The jobs related to various renewable energy technologies are the sum of the jobs in the four stages, as shown in Formula 1.

$$\mathrm{TE}_{t,s} = \sum \mathrm{E}_{t,s,g,p} \tag{1}$$

where: t is year; s is renewable energy technologies; g is the life-cycle stage; p is different products or processes. $\mathrm{E}_{\mathrm{t,s,g,p}}$ is employment opportunities by technology, stage and process; and $\mathrm{TE}_{\mathrm{t,s}}$ is total employment opportunities in a given renewable technology segment in a given year.

The number of jobs in new energy manufacturing is equal to the product of the current manufacturing size in the given year and the employment coefficient. Manufacturing size refers to the sum of new installed capacity and the export volume of the year concerned. This study sets 2020 as the base year for determining the ratio of export volume to new installed capacity and assumes that the ratio will not change as the scale of exports and installed capacity increase over time.

The number of employment opportunities in new energy installation is equal to the product of new installed capacity in the given year and the employment coefficient. The jobs in the operation and maintenance stage are the product of the total installed capacity in the given year and the employment coefficient. And the jobs in the recycling stage are the product of the capacity decommissioned in the given year and the employment coefficient.

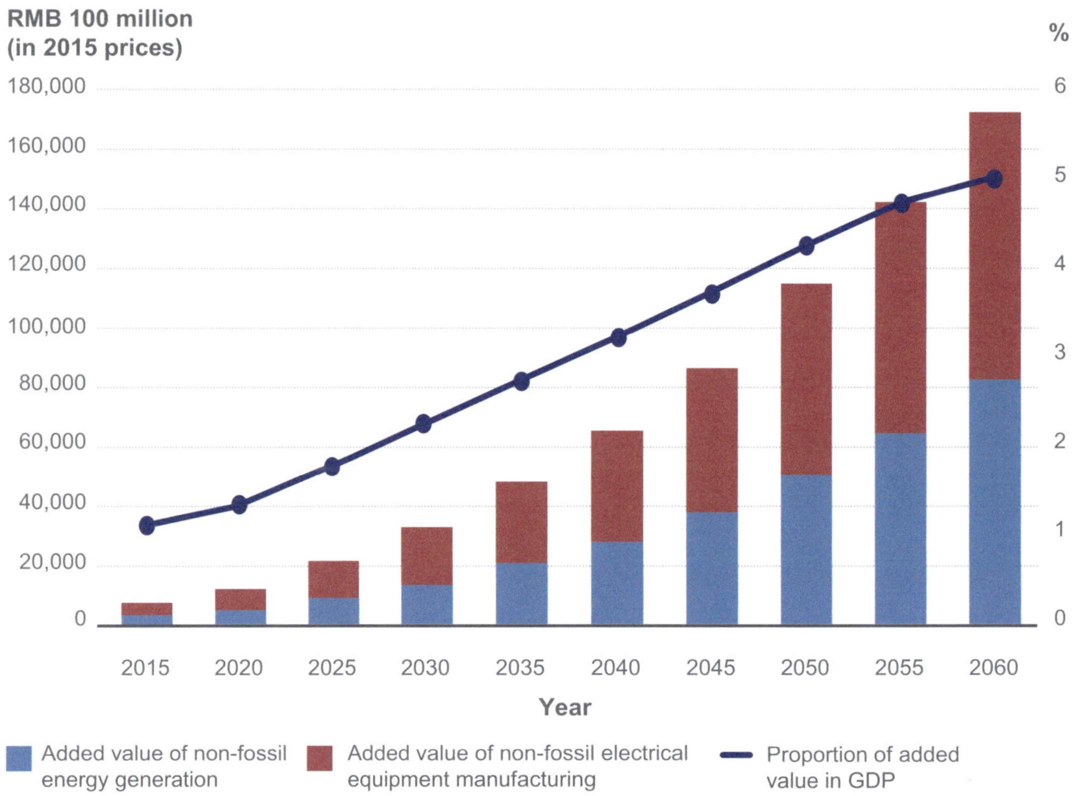

Fig. 6 The direct effects of renewable energy development on China's economy. *Source* Produced by the authors

Wind farms and solar power plants are designed to have an operating life of 20–25 years, so there is a possibility they may be decommissioned before the carbon-neutrality target is achieved, which means there will be a need for recycling. Taking future technological progress into account, this study assumes an operating life of 25 years and calculates the size of manufacturing, new installed capacity and recycling for the given total annual capacity. The operating life of hydropower and nuclear power plants is long, so this study does not consider the recycling stage of those two types of power generation.

To determine the possible range of each employment coefficient in the future, this study used the research results of several papers published by the Energy Research Institute of the National Development and Reform Commission, and the International Renewable Energy Agency. Our study identified the employment effects of each renewable energy technology at each of the four stages and took into account the future decline in employment coefficient brought by technological progress and the effects of scale and the employment coefficient in developed economies.

Based on the above-mentioned methods and data, this study evaluates the employment effects of the power system transformation.

3.2.2 Findings

(1) **The Effects of Wind and Solar Power on Employment**

As the renewable energy sector grows, the number of jobs in wind and solar power will increase, peaking at around 3 million in wind power in 2045 and about 7 million in solar in 2050. Although employment opportunities will decline after the peak, there will still be some 2.3

million jobs in wind power and about 5.2 million jobs in solar power in 2060. The wind and solar power industries will, therefore, create significant social benefits, especially in employment (see Fig. 7).

The rapidly expanding installed capacity of wind and solar power projects will create a growing need for operation and maintenance staff. The number of operation and maintenance jobs in wind power will gradually rise from 106,000 in 2020 to 809,000 in 2060, while the number in solar power will gradually increase from 82,000 in 2020 to 925,000 in 2060 (see Fig. 8).

The number of jobs in manufacturing and installation in wind and solar power grows but then declines. The reason for this is that China's total electricity demand will gradually stabilise around 2050 and the installed capacity of solar and wind power will become saturated, resulting in less new installed capacity. Although the retirement of older installed wind and solar power plants will create a need for replacements, it will be on a smaller scale. As a result, the number of installation and manufacturing jobs in wind and solar power will decrease to 822,000 in manufacturing and 556,000 in installation in wind power in 2060 and 3,131,000 in manufacturing

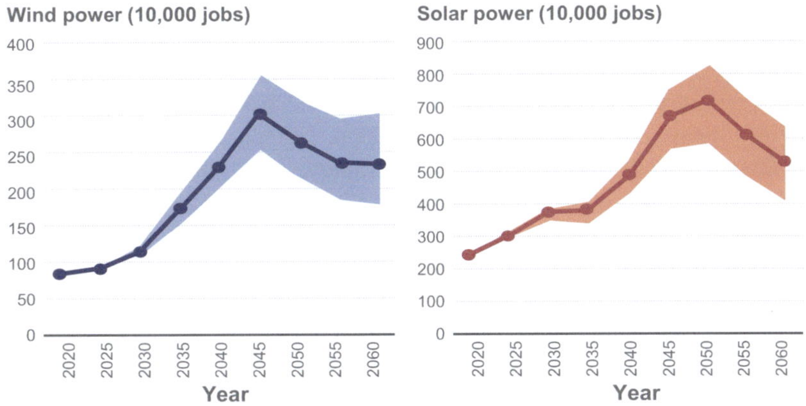

Fig. 7 Employment in the wind and solar power industries. *Source* Produced by the authors

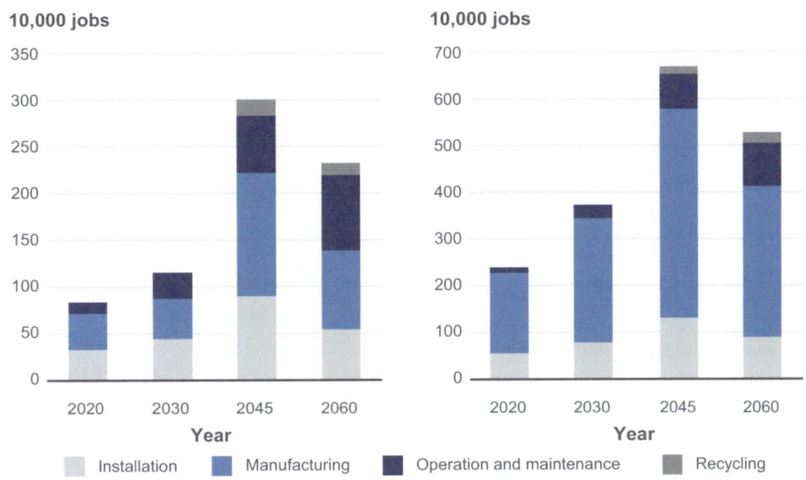

Fig. 8 Employment in the wind (left) and solar (right) power industries by job type. *Source* Produced by the authors

and 948,000 in installation in solar power in 2060. The reasons for the larger number of employment opportunities in solar power are twofold: the first is that the manufacturing process for a solar PV module is more complex (its several stages include the production of crystalline silicon, silicon wafers, solar cells and modules) and therefore requires more people; the second is that China's PV industry has a large production capacity and is a leading exporter. In 2020, China's solar module production was 124.6 GW. However, the new installed capacity of grid-connected solar power in China was only 48.2 GW, which means more than half of the PV modules manufactured in China were exported. Chinese solar PV enterprises are required to be highly competitive in export markets and maintain a high proportion of their sales as exports. Looking ahead, the gradually maturing PV industries of South-east Asia will compete with Chinese companies and possibly slow down China's growth in PV exports. In an extreme scenario, if China's net PV exports were to fall to zero, the quantity of PV manufacturing jobs would decrease to 1,212,000, a level lower than that of 2020. Maintaining the global competitiveness of China's PV industry is, therefore, important for maintaining employment at a sustainable level.

As wind turbines and PV modules are mainly made of recyclable materials such as steel, precious metals and glass, their recycling value is very high. A recycling wave of wind turbine and PV modules is forecast to begin after 2035 when many plants will reach the end of their service life. The need for recycling will increase, creating a total of 126,000 and 203,000 recycling jobs in wind and solar respectively by 2060.

(2) **The Effects of the Power System Transformation on Employment**

There is some uncertainty about the number of jobs needed by the power sector during the power system transformation. In 2060, the number of direct jobs is estimated to be 7,815,000, compared with 8,600,000 in 2020. The main reason for the decrease is that with technological progress and economies of scale, the employment coefficient of each power generation technology declines. In the coal and coal power industries for example, the number of workers has dropped sharply from more than 5 million in 2010 to about 3 million in 2020. This was due to the closure of small coal mines and the adoption of more advanced, labour-efficient and higher-efficiency mining techniques by large coal mines. However, the employment coefficient of the coal industry in developed countries is even lower. In view of this, even without the power system transformation, the number of workers in the coal industry will still decline, as will the number of jobs. Similarly, technological progress in the solar and wind power industries will lead to a dip in the employment coefficient. Although the number of jobs in the power sector will continue to rise to 2050, employment in wind and solar power will decline from 2050 to 2060, which indicates an overall drop in employment in the power sector by 2060.

Despite the uncertainty about the total number of jobs in the power sector, a rapid shift in employment from coal power to wind and solar power will be significant and certain. The total number of jobs in fossil fuel power generation will drop from 4,677,000 in 2020 to 333,000 by 2060. At the same time, the number of jobs in renewable energy generation will climb from 3.08 million in 2020 to a peak of 8.28 million, and then decline to 6.401 million in 2060. This means that between 3 and 5 million new jobs will be created in renewable energy generation. Such significant employment flows bring with them social risk. It is therefore important to train and find new employment for workers in the coal power industry in the short term.

3.3 Environmental Effects (Pollutant Emissions)

China set its dual-carbon goal to mitigate the increasingly severe threat of climate change, thereby initiating a transition from a high-carbon to a low-carbon power system. This transformation of its power system will bring multiple environmental benefits. First, air pollutants and carbon dioxide are often produced simultaneously and with significant homologies, so the implementation of the dual-carbon policy will result in noticeable improvements in air quality. Second, different power technologies have different impacts on water resources. Renewable energy generation, especially wind and solar power, use less water than fossil fuel generation and relieve the stress on water resources.

The power sector reduces its emission of air pollutants mainly in two ways. The first is the deployment of tail-end treatment technologies such as desulphurisation, denitrification and dedusting to reduce the concentration of air pollutants in power generation. The second is by replacing fossil fuel power generation with zero-emission technologies like wind and solar power, thus avoiding the production of pollutants at the "front end" of power generation. In 2015, the former Ministry of Environmental Protection, the National Development and Reform Commission and the National Energy Administration jointly issued a document requiring all eligible coal-fired power plants to achieve ultralow emissions by 2020. Such a transformation would significantly reduce emissions from coal power.[8] However, structural adjustments to the industry will also be needed.

Due to progress in energy conservation and emission reduction technologies and the introduction of policies like the transformation of coal power to ultralow emissions, the emission coefficient of various power generation technologies tends to show a downward trend. To calculate the improvements in air quality of the power system transformation it is critical to know how the dynamic emission coefficient is changing over time. To meet environmental regulatory requirements, the vast majority of power plants in China have installed continuous emission monitoring systems of flue gas. These allow plant emissions to be calculated dynamically and over a long time resolution, as with Formula 2.

$$EF_{p,t,e,y,i} = \overline{C_{p,t,y}} \times V_{t,e} \times M_{i,y} \qquad (2)$$

where: $EF_{p,t,e,y,i}$ is the emission coefficient of pollutant p in thermal power plant i using technology t and fuel e in year y, and the year is determined by the time resolution of this study; $\overline{C_{p,t,y}}$ is the average annual concentration of flue gas pollutant; $V_{t,e}$ is the volume of flue gas produced per unit of fuel consumption when technology t and fuel e are used; and $M_{i,y}$ is the average fuel consumption of the power supply from thermal power plant i in year y.

Tang et al. [1] used monitoring data to carry out a dynamic assessment of thermal power emission coefficients and tested the effectiveness of this method. Li et al. [2] and Zhang et al. [3] also used monitoring data to calculate the emission coefficients, and the results support the findings of Tang et al. These three research papers provide the benchmark parameters used in the analysis of the emission scenarios for the future power system in this study. As shown in Table 2, the average annual emission coefficient of each power generation technology was obtained from a plant emissions assessment.

Based on the emission coefficient calculations, scenarios were used to make reasonable assumptions about how the emission coefficient would trend over time. According to the ultralow emission transformation standard, the concentration of particulate matter (PM), nitrogen oxide (NO_x) and sulphur dioxide (SO_2) in emissions should not be higher than 10 mg per cubic metre (mg/m^3), 35 mg/m^3 and 50 mg/m^3 respectively, and the average amount of coal used to generate power should be less than 310 g per kilowatt-hour (g/kWh). The theoretical emission coefficients of coal power after ultralow emission transformation can be calculated using Formula 2. The results are as low as 0.106 g/kWh, 0.152 g/kWh and 0.030 g/kWh respectively, which is lower than the existing monitoring results. For this reason, this study assumes that by 2020, the PM emission coefficient of coal power will remain unchanged, while the SO_2 and

[8] Former Ministry of Environmental Protection, the National Development and Reform Commission and the National Energy Administration: Notice on Issuing the Work Plan for Full Implementation of the Ultralow Emission and Energy-efficiency Transformation of Coal-fired Power Plants.

Table 2 Summary of the emission coefficient assessment results using monitoring data (g/kWh)

	Coal power	Gas power	Biomass power	Coal power	Coal power (300–600 MW)
SO_2	0.165	0.026	0.435	0.152	0.282
NOx	0.239	0.151	0.881	0.208	0.350
PM	0.028	0.008	0.146	0.024	0.060
Year	2017	2017	2017	2018	2015
References	Tang et al	Tang et al	Tang et al	Li et al	Zhang et al

Source Created by the authors

NO_x emission coefficient will decrease to 0.120 g/kWh and 0.170 g/kWh respectively, because the study considers it impossible for some small plants to perform the ultralow emission retrofit.

To assess the effects of tail-end treatment, this study set two tail-end abatement scenarios: the Established Policy Scenario and the Best Technology Scenario. In the Established Policy Scenario, only the existing ultralow emission retrofit policy is considered, so the emission coefficients of coal power will remain at the 2020 level after 2021. Because the emission coefficients of gas-fired power generation are already very low, they will remain unchanged. Given that the current emission coefficients of bioenergy are very high, this study assumes there will be a moderate decline in emission coefficients and does not take into account ultralow emission retrofits. As a result, the emission coefficient of SO_2, NO_x and PM will reach 0.260 g/kWh, 0.576 g/kWh and 0.116 g/kWh respectively by 2030, and will remain unchanged thereafter. In the Best Technology Scenario, it is assumed that the emission coefficients of coal power, gas power and bioenergy will decrease to match the air pollutant concentration requirements of carbon capture and storage deployment. By 2060, the emission coefficient of SO_2, NO_x and PM will be lower than 0.030 g/kWh, 0.152 g/kWh and 0.032 g/kWh respectively.

To assess the air quality benefits of the power system transformation, we created two power system scenarios: the Reference Scenario and the Carbon-neutral Scenario. These two scenarios assume the same total power output. In the Reference Scenario, it is assumed that the power output of various power technologies will increase proportionally, that is, no structural transformation will occur. In the Carbon-neutral Scenario, a rapid and profound transformation of the power system takes place, including substantial growth in renewable power output and the deployment of carbon capture and storage, making zero or even negative emissions possible over time. The scenario combination settings are shown in Table 3. Since no carbon capture and storage facilities are deployed in the Reference Scenario, it is assumed that tail-end treatment remains in the Established Policy Scenario but not in the Best Technology Scenario.

Even without the power system transformation, the dramatic growth in power demand will bring about a continuous increase in the emission of various air pollutants. However, thanks to ultralow emission retrofits, emissions of conventional air pollutants including SO_2, NO_x and PM will not rise significantly at 1,424,000 tonnes, 2,146,000 tonnes and 307,000 tonnes respectively. Emissions of SO_2 and PM will be lower than in 2015 (1,627,000 tonnes and 316,000 tonnes), while NO_x emissions will be slightly higher than in 2015 (1,875,000 tonnes). This indicates that thanks to the ultralow emission transformation, the baseline emissions of air pollutants from power generation will be controlled, albeit at a low level. In the long run, however, the power system transformation can produce substantial improvements in air quality.

When the power system transformation is taken into consideration, the rapid decline in coal power output will be offset by the power technologies that have zero air pollutant emissions, such as wind, solar, hydro and nuclear. The trend

Table 3 Summary of scenario combination settings

	Reference scenario (REF)	Carbon–neutral scenario (CN)
Established Policy Scenario	REF	CN-Current
Best technology scenario		CN-Max

Source Produced by the authors

in emissions will be one of peaking followed by a rapid fall. SO_2, NO_x and PM emissions will peak at 6.83 billion tonnes, 10.86 billion tonnes and 1.58 billion tonnes respectively, and then stabilise at 1.84 billion tonnes, 3.80 billion tonnes and 0.62 billion tonnes by 2060. This represents a decrease of 12.7% (499,000 tonnes), 17.7% (786,000 tonnes) and 19.4% (96,000 tonnes) respectively. Since bioenergy facilities have not yet achieved ultralow emissions, the remaining emissions will be mainly generated by new bioenergy plants.

In the CN-Current Scenario, due to the high emission coefficients of bioenergy, emissions of SO_2, NO_x and PM will be 990 million tonnes, 2,200 million tonnes and 440 million tonnes, which are 53.7%, 57.7% and 70.9% of total emissions from the power system respectively. If there is no improvement in its tail-end treatment technology, bioenergy, rather than fossil fuels, will be responsible for most power system emissions in the Carbon-neutral Scenario. It is, therefore, crucial to improve the tail-end treatment of bioenergy to transform the power mix. Assuming the best technology is deployed (to carbon capture and storage standards), the SO_2 emission concentration required will be extremely low and total SO_2 emissions will be as low as 400 million tonnes, which is only 21.5% of the level in the CN-Current scenario. Biomass and coal power will partly contribute to this reduction. Because carbon capture and storage requires a relatively low NO_x emission concentration and the PM emission concentration of thermal power already meets carbon capture and storage requirements, the NO_x and PM emissions from coal power and gas power will remain largely unchanged, but the emissions from bioenergy will decline sharply. This shows that the power system transformation needs bioenergy plants to retrofit ultralow emission systems to achieve near-zero air pollutant emissions. In this way, the carbon neutrality and air quality management targets can be achieved in tandem.

From 2015 to 2020, the ultralow emission transformation of coal power plants achieved a significant reduction in air pollutant emissions from existing thermal power plants. However, in the long run, achieving in all China's provinces the $PM_{2.5}$ concentration limit of 10 μg per cubic metre (μg/m^3) set by the World Health Organization will require a comprehensive industry-wide reduction in air pollutant emissions. This study found that when this target is achieved, total SO_2, NO_x and PM emissions will be reduced to 1.428 million tonnes, 4.772 million tonnes and 2.413 million tonnes respectively. Despite the relatively low accompanying benefits related to a PM emission reduction, the reduction of SO_2 and NO_x pollutants will reach 34.9% and 14.2% of the remaining emissions respectively, indicating significant benefits in relative terms.

To achieve the dual-carbon goal requires a transition to renewable energy and a substantial increase in electrification. Renewable energy will replace both coal power generation and fossil fuel consumption by industry, thereby also reducing emissions from fossil energy use in other sectors. Given that most sectors have not yet implemented an ultralow emission transformation, their emission coefficient per unit of fossil energy consumed is much higher than that of the power sector. If this assessment method is followed, the replacement of fossil energy by renewable energy will produce larger air quality benefits. Due to reasons of space and the fact that this book is about the power sector, this chapter will not quantify the results of this assessment.

3.4 Water Resource Benefits

3.4.1 Research Method

The power system inevitably creates demand for water in the power generation process. However, water use varies greatly among the different power technologies, so it is important to quantify the benefits of less water use in the power system transformation.

Generally, two measures are used to evaluate the impact of power generation technologies on water resources: water withdrawal and water consumption. Water withdrawal is when water is removed directly from the water source. Water consumption does not include the water discharged into the water system after waste-water treatment, it only includes the water lost through use or evaporation. Water consumption is therefore often lower than the amount withdrawn.

The volume of water withdrawn and consumed is highly related to the type of power generation and the cooling technology it uses. There are three principal cooling technologies used to generate power in China. First, once-through cooling, in which the water withdrawn is cooled and then discharged directly into the downstream area of the water intake. This method requires a large volume of water, only a very small part of which is used, which makes it suitable for coastal and river locations rich in water resources. In China, this type of cooling is used mainly in the Yangtze River delta. Second, open recirculating cooling, in which the cooling water is returned to the cooling tower for reuse after passing through the condenser. Due to continuous circulation, the volume of water consumed is often low, most of which is through evaporation. This type of cooling is widely used in northern China. Third, air cooling, also known as dry cooling, requires little water. This type of cooling is often used in North and north-west China [4].

The results of a literature review on the water withdrawal and water consumption coefficients of various power generation and cooling technologies are shown in Tables 4 and 5 respectively. For thermal power and coal power, the water consumption of open recirculating cooling is greater than that of once-through and air cooling, but the water withdrawal of once-through cooling is far greater than the other two. Solar and wind power technologies require very little water, most of which is used for cleaning rather than cooling. Hydropower is hugely controversial in the way it uses water, so this study assumes water withdrawal and consumption are zero.

Having found the water withdrawal and water consumption per unit of power produced by the various power generation technologies, we used bottom-up calculation and scenario analysis to calculate water consumption and withdrawal and assess the benefits for water of a power system transformation, as shown in Formula 3, 4, 5 and 6:

$$\mathrm{WC}_T = \sum_S \mathrm{Gen}_T \times \mathrm{Coolshare}_{T,S} \times \mathrm{WCI}_{T,S} \quad (3)$$

$$\mathrm{TWC} = \sum \mathrm{WC}_T \quad (4)$$

$$\mathrm{WW}_T = \sum_S \mathrm{Gen}_T \times \mathrm{Coolshare}_{T,S} \times \mathrm{WWI}_{T,S} \quad (5)$$

$$\mathrm{TWW} = \sum \mathrm{WW}_T \quad (6)$$

where: $\mathrm{WC_T}$ is the water consumption of T technology; $\mathrm{WW_T}$ is the water withdrawal of T technology; TWC is the total water consumption of the power generation system in a given year; and TWW is the total water withdrawal of the power generation system in a given year. $\mathrm{Gen_T}$ is the power generated by T technology; $\mathrm{Coolshare_{T,S}}$ is the application share of each cooling technology; and $\mathrm{WCI_{T,S}}$ and $\mathrm{WWI_{T,S}}$ are respectively the water consumption and water withdrawal intensity of each cooling technology deployed in a given power generation technology.

According to Li et al. [7] in 2018, the output from coal power plants that use once-through cooling, open recirculating cooling and air cooling was 1.13 trillion kWh, 2.18 trillion kWh and 0.91 trillion kWh respectively. On that basis, the application share of these three cooling technologies was 26.8%, 51.6% and 21.6% respectively. Gas power mainly uses air cooling; nuclear power mainly uses once-through cooling; and bioenergy is assumed to use the same proportions as coal power. In the Future Scenario, it is assumed that the proportion of the three

Table 4 Results of a literature review on water withdrawal and consumption in coal power (m^3/MWh)

		Ref [5]	Ref [4]	Ref [6]	This study
Water consumption	Once-through cooling	0.15–1.2	0.34	0.43	0.34
	Open recirculating cooling	1.2–20	2.02	1.87	1.87
	Air cooling	0.2–0.4	0.39	/	0.39
Water withdrawal	Once-through cooling	75.7–259.1		102.5	102.5
	Open recirculating cooling	1.5–7.6		2.4	2.4
	Air cooling	0.3–0.5			0.5

Source The authors, based on relevant literature

cooling technologies will remain unchanged for thermal power. However, considering the rapid increase in the installed capacity of nuclear power, it is assumed that 50% of open recirculating cooling units will be installed in power generation by 2040, which will remain unchanged thereafter.

3.4.2 Findings

Water consumption refers to the direct consumption of water resources by the power system. If the low-carbon transition does not occur, the growing use of thermal and nuclear power generation will increase China's water consumption from 110 million cubic metres in 2020 to 15,500 million cubic metres in 2060, placing greater pressure on water resources. The power system transformation, the purpose of which is to help meet the dual-carbon goal, will reduce water consumption. This is because a large amount of electricity will be generated by wind and solar power technologies that use very low amounts of water, mainly for cleaning PV modules and wind turbines. The water consumption per megawatt-hour of solar and wind power production is only 20 litres and 5 litres respectively. Although total water consumption will rise again after 2050 due to increases in nuclear and bioenergy output, it will eventually reach 5.97 billion cubic metres in 2060, which is lower than in 2020. This indicates that the power system transformation can help brake the trend of rising water consumption. It is also 9.5 billion cubic metres lower than the water consumption level in the Reference Scenario. Total water withdrawal in the Reference Scenario will continue to climb to 405.3 billion cubic metres, most of which will be used to meet the water requirements of once-through cooling. In the Power System Transformation Scenario, total water withdrawal will start to decrease slowly after 2035, falling to 251.4 billion cubic metres in 2060, a decrease of 153.9 billion cubic metres compared to that in the Carbon-neutral Scenario.

The decrease in water consumption will be mainly due to declining coal power output. Even so, water consumption and water withdrawal by the power system will remain at a high level, mainly because of the huge need for cooling water in nuclear power, as shown in Fig. 9. By 2060, water consumption and water withdrawal for coal power generation will be 770 million cubic metres and 18,140 million cubic metres respectively, which is a very low proportion of the total. Nuclear power will be responsible for most of the water consumed and withdrawn. The volume of cooling water required per unit of nuclear power generated will increase, which will put further pressure on the need for the nuclear power industry to reduce water consumption and withdrawal. Improvements in nuclear power cooling technology and reductions in water withdrawal and consumption will be essential to relieve the pressure on water resources.

The development of renewable energy will alleviate the pressure on water resources in the regions. The distribution of water resources in China is uneven, with more water resources in eastern and southern regions. There is a strong correlation between the choice of cooling technology and local water resources. For example, once-through cooling technology for coal power is used mainly in eastern and central China, while open recirculating cooling and air cooling are used mainly in north-west and northern

Table 5 Results of a literature review on water withdrawal and consumption of various power generation technologies (m^3/MWh)

		Gas power		Bioenergy		Nuclear power		Solar power		Wind power	
		Water consumption	Water withdrawal	Water consumption	Water withdrawal	Water consumption	Water withdrawal	Water consumption	Water withdrawal	Water consumption	Water withdrawal
Ref [5]	Once-through cooling			1.4	127.3	0.2–1.5	172.9–178				
Ref [5]	Open recirculating cooling			2.1–2.2	2.5–2.6	2.3–3.1	4.2–7.0				
Ref [5]	Air cooling			0.2	0.3	0.3–0.5	0.4–0.5				
Ref [5]	Average	0.8–1.2	/	1.8		3.0–4.2	/				
Ref [6]	Once-through cooling	0.9	43.1	1.1	132.5	1.0	167.9				
Ref [6]	Open recirculating cooling	1.5	1.9	2.1	3.3	2.5	4.2				
Ref [6]	Air cooling	0.008	0.008	0.1	0.1	/	/				
Ref [7]	Average							0.02	0.02	0.005	0.006

Source The authors, based on relevant literature

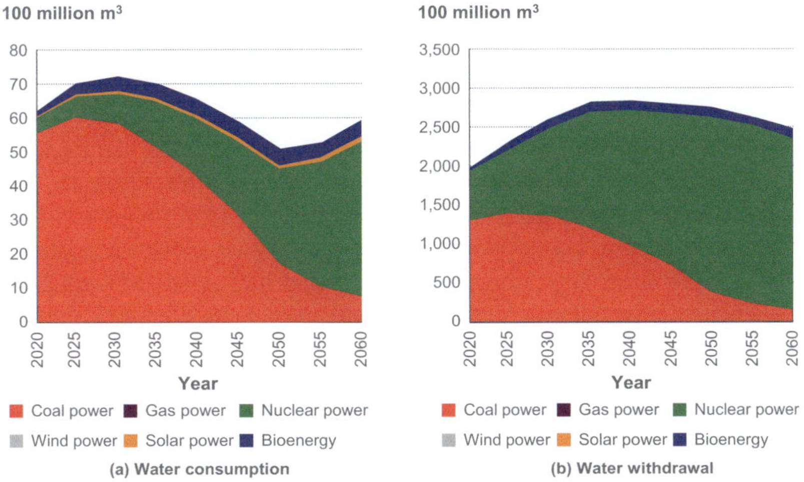

Fig. 9 Sources of water consumption and water withdrawal under the power system transformation. *Source* Analysed and calculated by the authors

China. The power system transformation aims for a rapid exit from coal power and the large-scale deployment of wind and solar power, which will greatly ease the pressure on water supply for power generation in the north-west. Nuclear power plants will be built mainly alongside the coast, rivers and lakes where there are relatively abundant water resources, so no significant water stress should occur. In view of this, the power system transformation will not only reduce total water consumption and withdrawal, but also enable a spatial shift away from areas with scarce water resources like north-west and North China.

4 Policy Recommendations for Promoting New Energy Development

4.1 Build an Electric Power System in Which New Energy Predominates

China should speed up its efforts to build a new electric power system that is characterised by "a high proportion of new energy, the flexible and reliable allocation of resources in a highly resilient power grid, interaction between diverse highly electrified loads, and the integration of digitalised power grid infrastructure".

In power generation, the main tasks include building a peak regulating power supply system to address the high proportion of new energy connected to the grid, use thermal power to provide flexible peak shaving and ensure new energy facilities are intelligent, flexible and easy to connect to the grid. Urban and rural user needs should be integrated with distributed power systems. In power grids, the main tasks include building highly resilient power networks and enabling large-scale bidirectional resource allocation. Specific measures include incorporating both AC and DC long-distance transmission, interconnecting regional grids, integrating microgrids with the main grid and establishing a next-generation scheduling system that features grid-wide co-ordination, one that is data-driven, secure and allows informed decision-making. In power consumption, the main tasks include enabling interaction between diverse highly electrified loads, unlocking user-side potential and improving system efficiency by enabling user interaction and demand–response. More effort should be made to develop and deploy diverse energy storage technologies such as pumped storage, battery storage and heat and hydrogen storage, and increase the amount of new energy consumed.

During the 14th Five-year Plan (2021–25), China should define the operational requirements of a new, flexible electric power system that contains a large amount of new energy. China should do this by researching and developing the required mechanisms for integrating a high proportion of new energy; integrating generation, grids, loads and storage in a power balancing system that allows collaboration between the parts; developing and demonstrating DC distribution networks that contain new energy facilities; revising the technical standards that support power system planning; and by accelerating the transition to a flexible and resilient power system. By 2035, the new electric power system should be well developed and by 2060 it should be fully operational.

4.2 Promote the Development of a New Energy Industry

4.2.1 Create a Priority Development Mechanism for Renewable Energy

China should implement a planning system to prioritise the development and use of renewable energy and to control investment in new fossil energy projects. Efforts should be made to increase renewable energy development in the central and eastern regions and expand the renewable energy clusters in the west. The technical standards for power system planning should be revised to meet the requirements of a power system containing a high proportion pf renewable energy. And plans for generation, grid, load, storage and other flexibility resources should be made and benchmarked against leading international power systems.

China should implement the Renewable Portfolio Standard and announce targets for renewable power consumption for each province. The country should improve its monitoring system for renewable energy consumption and assess all provinces yearly. Efforts should be made to improve the green electricity certificate management system and clarify the market trading rules to create favourable conditions for the large-scale implementation of the certificate scheme.

China should encourage the entire country to prioritise the use of renewable energy. The country should guide enterprises, social institutions and individuals to consume products made from green energy; improve green service policies and regulations and make green products more competitive; and optimise the voluntary trading mechanism for green electricity certificates.

4.2.2 Integrate New Energy Development with Spatial Planning

To address the bottleneck in land use that could prevent billions of kilowatts of wind and solar power from coming online in the future, China should co-ordinate the development of renewable energy with spatial planning. Renewable energy and related infrastructure projects should be included in spatial planning at provincial, city and county levels to ensure that land is made available for new energy projects.

4.2.3 Investigate the Security Challenges to the New Electric Power System and Provide Possible Solutions

China is building a new electric power system in which new energy predominates. Because hundreds of gigawatts of new energy will be installed each year, the peak regulating capacity and flexibility of the power system need to be enhanced. China should strictly control the amount of new coal power built and make its role in the new power system more flexible. Installed coal power capacity does not need to be shut down. The different types of new energy are complementary. Power generation, grids, loads and storage should be integrated and the inter-provincial and inter-regional transmission of new energy made possible to extend its range of use.

China should investigate the reliability of new energy generation and develop improvements for a power system that contains a high proportion of renewable energy and power electronics. China should increase its R&D efforts in basic research, base materials and control and equipment technologies to solve key issues in power supply security such as power grid structure, power system simulation, power stability control during random and large fluctuations, and cyber security.

4.2.4 Improve the New Energy Trading Mechanism

China should form a strong link between the guaranteed purchase price and market-based transactions of new energy-generated electricity. Subsidised projects and grid parity demonstration projects should take part in the guaranteed purchase price system and trade the excess energy they generate on the market once they have fulfilled their subsidy obligations for plant availability. China should encourage new energy generation projects to take part in electricity market trading and sign long-term power purchase agreements under the Renewable Portfolio Standard. In pilot electricity spot markets, local government is encouraged to support the development of new energy generation through contracts for difference.

4.2.5 Accelerate Efforts to Strengthen Supply Chains

China should combine major projects with intensive scientific and technological research to reduce its dependence on imports of key components, raw materials and testing equipment like large-capacity insulated-gate bipolar transistors. The same applies to future technologies that have yet to be commercialised such as advanced photovoltaic cells; large-capacity floating wind turbines and related technologies; wind farm and turbine simulation; low-cost renewable hydrogen production, storage and transport; and large-capacity, long-duration and fast start–stop energy storage. By 2025, China will see a significant improvement in the innovation and competitiveness of its renewable energy technologies, and will maintain its global leadership in solar power, onshore wind power and power batteries and reach a globally advanced level in offshore wind power and renewable hydrogen.

China should shape a national energy science and technology programme and public service platform led by national energy laboratories. To achieve the dual-carbon goal and build a new type of energy system, China should focus on developing forward-looking low-carbon technologies, build national low-carbon innovation capacity in a co-ordinated manner, expand and make good use of national laboratories, plan and invest in strategic energy innovation platforms and infrastructure, and conduct research on key technologies and equipment. Learning from the National Renewable Energy Laboratory in the USA and the Fraunhofer-Gesellschaft in Germany, China should build national energy laboratories and support collaborative innovation and the sharing of knowledge to achieve its innovation objectives.

4.2.6 Reduce the Cost of New Energy Development

China should accelerate the adoption of peak and off-peak pricing and time-of-use tariffs, reduce transmission and distribution prices and cross-subsidies, implement tax and fee reduction policies and increase government funding to support the development and demonstration of original technologies and core equipment. The country should standardise renewable energy resource fees and urban land use taxes, regulate land use fees and adopt differentiated land use policies to reduce the cost of land for renewable energy projects (project owners who use state-owned unused land to develop renewable energy projects should be able to use the land rent-free). China should also reduce the cost of transporting large items of equipment like wind turbines and improve the green finance system by establishing a credit assessment mechanism for renewable energy projects and by reducing the cost of financing such projects.

4.2.7 Streamline Administration and Improve the Regulatory System for New Energy

Government approval procedures should be simplified. China should create a "green channel" for central approval of new energy projects to improve administrative efficiency. For wind power projects, the approval system should be shifted to the filing system; the approval process for connecting wind and solar power plants to the grid should be clarified and simplified; and integrated energy projects such as generation–grid–load–storage integration should be assessed as a single entity.

The regulations for grid connection, network transmission and fair scheduling services of renewable energy (electricity, heat and gas) should be improved. China should explore ways to establish an overall planning and unified supervisory system for urban heating networks, gas pipelines and power grids; implement a disclosure system that requires power grid companies above city level to establish a one-stop information platform for the planning and construction of new energy-generated power systems; and clearly formulate the necessary technical standards for connecting distribution networks to new energy generation. The supervision of transmission and distribution networks should be improved and a performance-based mechanism should be created to allow grid companies that accept a high proportion of new energy and distributed power to obtain a higher return on their investment.

4.3 Suggestions for New Energy Co-operation Between China and the EU

4.3.1 China–EU Co-operation Today

(1) **Mechanisms for China–EU Renewable Energy Co-operation**

China and the EU already have energy co-operation and exchange mechanisms. In 2005, the National Energy Administration of China and the European Commission established an inter-governmental energy exchange and co-operation mechanism under which dialogue between the two parties on energy policies is held annually and alternately in China and the EU. The scope of exchange covers energy policies, energy security, energy infrastructure and renewable energy.

In recent years, China has established bilateral energy co-operation mechanisms with many EU member states, including the China–France Energy Dialogue, the China–Swiss Energy Working Group, the China–Finland Energy Cooperation Working Group, the Sino–German Working Group on Energy, and the Sino–Danish Offshore Wind Cooperation. Both sides are working to continuously expand and deepen co-operation on the energy transition, nuclear power, advanced solar PV technology, energy storage, power system flexibility, clean heating and cooling, and energy technology innovation.

(2) **Low-carbon Energy Trade and Investment Between China and the EU**

China and the EU's co-operation on energy is primarily in nuclear, solar and hydropower. To implement the relevant requirements in the memorandum of understanding of the Sino–German Working Group on Energy, the National Energy Administration of China and the Federal Ministry for Economic Affairs and Energy of Germany jointly launched Phase 2 of their energy transition project, under which both sides will conduct in-depth research on energy system transformation, electricity market reform and renewable energy.

Chinese and EU companies have carried out long-term co-operation on the construction of nuclear power plants and the joint development of nuclear power markets in third countries. China–EU nuclear power investment projects are shown in Table 6.

China has established bilateral co-operation with Denmark and the Netherlands on policy, planning, technology and standards for offshore wind power. Chinese enterprises such as SDIC Power, Shanghai Electric and China Three Gorges Corporation have joined forces with EU

Table 6 China–EU nuclear power projects

Project	Development stage	Parties	Location
Taishan Nuclear Power Plant	Completed in 2019	China General Nuclear Power Group, EDF	Taishan, Guangzhou, China
Hinkley Point C nuclear power plants	Construction is under way	CGN, EDF and the UK government	Hinkley Point C, Bradwell B, and Sizewell C
Central and Eastern Europe nuclear power market development	Feasibility study	China and some EU member states	Bulgaria, Czech Republic, and other Central and Eastern European countries

Source The authors, based on public data

companies on wind power projects. In the first half of 2019, EDF partnered with CHN Energy Investment Group to jointly invest in the Dongtai IV and V offshore wind farms located off the coast of Jiangsu province. They are the first offshore wind power projects developed by EDF in China, and also the first offshore wind farms that have used foreign investment in China.

China's exports of PV modules to the EU continue to grow. Jinko Power, CHINT and other companies have won bids for solar power plant projects in the Netherlands, Spain and other European countries. In addition, China and the EU have also developed rich and mutually beneficial co-operation on geothermal energy, biomethane and hydropower.

4.3.2 Further Opportunities for China–EU Co-operation

The respective advantages of China and the EU in renewable energy are complementary. China has advantages in the production of photovoltaic components. For example, China's monocrystalline silicon and polysilicon production is in a leading position globally, with 36 of the world's top 40 photovoltaic companies in 2019 from China. China accounted for 37% of the world's total installed capacity of onshore wind power in 2019. However, China still has some way to go in terms of the core technologies in energy systems, storage, transport and control, and has yet to develop a relatively strong position in hydrogen and bioenergy. In contrast, the EU has long invested in the critical equipment and technologies of renewable energy, and has developed many advanced technologies. The EU has built up a strong competitive position in precision instruments and technologies for hydrogen and biogas storage and transport. Germany has launched its National Hydrogen Strategy and has made breakthroughs in electrolysis and hydrogen storage and transport. It provides world-class intelligent power grid control systems. In 2018, Germany achieved 100% renewable power supply for a short period, proving the effectiveness of intelligent control systems for connecting distributed energy systems to the grid. Norway, although not part of the EU, has reduced the cost of bioenergy applications and built a complete biomass technology system covering power generation, fermentation and biofuel production. Denmark is a leading nation in offshore wind turbines and power generation control systems.

Both China and the EU have a vast renewable energy market. China still has difficulty controlling carbon emissions and tackling climate change. According to the China 2050 High Renewable Energy Penetration Scenario and Roadmap Study, renewable energy should account for 62% of China's total energy use by 2050. That is a huge gap to bridge considering the current proportion is less than 20%. The EU still needs to step up renewable energy investment and development in order to achieve its 2030 abatement target and 2050 carbon-neutral goal, and prevent a return to fossil energy use.

The post-epidemic green economic recovery and improved climate governance around the globe have created historic opportunities. China and the EU can use their respective advantages in new energy to invest in the partner market and benefit from stronger trade relations in renewable energy technologies and equipment. This will further increase the share of renewable energy in total energy production and consumption. Both parties can collaborate on technology R&D and scientific research and encourage students and researchers to learn from each other's renewable energy technologies and development approaches by studying or working in the partner country or community. The strengths and advantageous technologies of both sides can be combined in renewable energy projects in third countries, thereby promoting the global energy transition.

4.3.3 Policy Recommendations for Deepening China–EU Co-operation

It is China's hope that both sides will stick to the four principles of peaceful co-existence, openness and co-operation, multilateralism, and dialogue and negotiation to take their co-operation on renewable energy to the next level.

(1) **Peaceful Co-existence**

Economic and industrial security are essential to national security. Not all imports of goods are a threat to domestic industries. A case in point is the EU's anti-dumping and anti-subsidy investigations into China's photovoltaic equipment since 2012, which hit trade in photovoltaic equipment between China and the EU heavily. The rapid rebound in trade after the investigations were cancelled in 2018 indicate that trade in photovoltaic equipment is important to the development of new energy in both markets. The weak points of a country in a renewable energy sector do not necessarily mean it has vulnerabilities in national energy security. Trade in renewable technologies and equipment should, therefore, not be excessively restricted by national security concerns.

(2) **Openness and Co-operation**

The post-pandemic economic recovery and the renewable energy transition are not a zero-sum game for China or the EU. Building on their collaboration on large-scale nuclear power and wind power projects, both sides can gradually extend their co-operation into other fields such as bioenergy, hydrogen, renewable energy storage and transport, the energy internet and smart energy, and jointly develop third-party markets in the relatively developed regions of South America, South Asia and South-east Asia.

At the same time, China will continue its efforts to reform its energy industry and improve the phase-in and phase-out mechanisms for renewable energy subsidies, tax incentives and other policies to restore the commodity attributes of energy. For example, China will set a timetable for phasing out subsidies for renewable energy technologies at different stages of development as soon as possible, in order to ensure fair competition in domestic and international markets. China is building new infrastructure to drive demand, which can effectively help renewable energy consumption. China will open up its renewable energy market to the outside world and encourage energy companies and countries, including the EU and the UK, to enter the renewable energy industry in China without trade barriers. For instance, the EU has world-leading technologies and expertise in bioenergy. Given China's huge market potential, China and the EU can collaborate on R&D, technology investment and technology transfer in China's bioenergy projects.

(3) **Multilateralism**

Since 1992, countries have signed international conventions like the United Nations Framework Convention on Climate Change and the Paris Agreement, which have played a critical role in tackling the adverse effects of global warming on the economy and society. In recent years, voices and practices that go against multilateralism and which aim to replace the global governance system with bilateralism or even unilateralism have made an impact. Both China and the EU should make clear that it is impossible to achieve global carbon abatement and global warming mitigation targets through the efforts of one or a few countries alone. Both sides should work together to function as defenders of the global multilateral governance system and ensure that the Paris Agreement and other international conventions on climate change can be fulfilled.

(4) **Dialogue and Negotiation**

China and the EU should resolve misunderstandings through dialogue, address difficulties through development and properly manage differences. Going forward, the EU is expected to discuss, build and share new energy results with China through the Belt and Road Initiative, and jointly lead the world to achieve green economic recovery and the energy transition. More detailed co-operation should be established for different types of renewable energy and more extensive exchange should be carried out by, for example, encouraging the mutual exchange of students and researchers. On issues such as the green transformation of residential communities and energy efficiency improvements to manufacturing equipment, both sides should carry out comprehensive, multi-level exchange in government, industry, universities, research institutions and finance, and explore points of common interest and possible co-operation.

References

1. Tang L, Xue X, Qu J et al (2020) Air pollution emissions from Chinese power plants based on the continuous emission monitoring systems network. Sci Data 7:325
2. Li J et al (2020) Incorporating health co-benefits in decision-making for the decommissioning of coal-fired power plants in China. Environ Sci Technol 54 (21):13935–13943
3. Zhang Y, Bo X, Zhao Y et al (2019) Benefits of current and future policies on emissions of China's coal-fired power sector indicated by continuous emission monitoring. Environ Pollut 251:415–424
4. Zhang X et al (2017) China's coal-fired power plants impose pressure on water resources. J Cleaner Prod. S0959652617307485
5. Wang C, Tian L, Yu M et al (2018) Review of studies on the water-energy relationship in the power sector. China Environ Sci 12
6. Macknick J, Newmark R, Heath G, Hallett KC (2012) Operational water consumption and withdrawal factors for electricity generating technologies: a review of existing literature. Environ Res Lett 7(4):189–190
7. Li H, Cui X, Hui J, He G, Weng Y, Nie Y et al (2021) Catchment-level water stress risk of coal power transition in China under 2°C /1.5°C targets. Appl Energy 294

MIX
Papier aus verantwortungsvollen Quellen
Paper from responsible sources
FSC® C105338

If you have any concerns about our products,
you can contact us on
ProductSafety@springernature.com

In case Publisher is established outside the EU,
the EU authorized representative is:
Springer Nature Customer Service Center GmbH
Europaplatz 3, 69115 Heidelberg, Germany

Printed by Libri Plureos GmbH
in Hamburg, Germany